Developments in Mathematics

VOLUME 27

Series Editors:

For further volumes:
http://www.springer.com/series/5834

Saïd Abbas • Mouffak Benchohra
Gaston M. N'Guérékata

Topics in Fractional Differential Equations

Saïd Abbas
Laboratoire de Mathématiques
Université de Saïda
Saïda, Algeria

Mouffak Benchohra
Laboratoire de Mathématiques
Université de Sidi Bel-Abbès
Sidi Bel-Abbès
Algeria

Gaston M. N'Guérékata
Department of Mathematics
Morgan State University
Baltimore, MD, USA

ISSN 1389-2177
ISBN 978-1-4899-9547-6 ISBN 978-1-4614-4036-9 (eBook)
DOI 10.1007/978-1-4614-4036-9
Springer New York Heidelberg Dordrecht London

Mathematics Subject Classification (2010): 26A33, 34A60, 34A37, 39A12, 34A37, 39A12, 34A60, 34B15, 26A33, 34K05

Printed on acid-free paper

Springer is part of Springer Science+Business Media (www.springer.com)

We dedicate this book to our family members. Inparticular, Saïd Abbas dedicates to the memory of his father, to his mother, his wife Zoubida and his children Mourad, Amina, and Ilyes; Mouffak Benchohra makes his dedication to the memory of his father Yahia Benchohra and Gaston N'Guérékata to the memory of his father Jean N'Guérékata.

Preface

Fractional calculus (FC) generalizes integrals and derivatives to non-integer orders. During the last decade, FC was found to play a fundamental role in the modeling of a considerable number of phenomena, in particular, the modeling of memory-dependent phenomena and complex media such as porous media. FC emerged as an important and efficient tool for the study of dynamical systems where classical methods reveal strong limitations. This book is devoted to the existence and uniqueness of solutions for various classes of Darboux problem for hyperbolic differential equations or inclusions involving the Caputo fractional derivative, the best fractional derivative of the time. Some equations present delay which may be finite, infinite, or state-dependent. Others are subject to impulsive effect. The tools used include classical fixed point theorems as well as sharp (new) ones such as the one by Dhage on ordered Banach algebras and the fixed point theorem for contraction multivalued maps due to Covitz and Nadier, as well as some generalizations of the Gronwall's lemma. Each chapter concludes with a section devoted to notes and bibliographical remarks and all abstract results are illustrated by examples.

The content of this book is new and complements the existing literature in fractional calculus. It is useful for researchers and graduate students for research, seminars, and advanced graduate courses, in pure and applied mathematics, engineering, biology, and all other applied sciences.

We owe a great deal to R.P. Agarwal, L. Górniewicz, J. Henderson, J.J. Nieto, B.A. Slimani, J.J. Trujillo, A.N. Vityuk, and Y. Zhou for their collaboration in research related to the problems considered in this book. We express our appreciation to Professor George Anastassiou who strongly supported our project. Finally, we thank the editorial assistance of Springer, especially Elizabeth Loew and Jacob Gallay.

Saïda, Algeria S. Abbas
Sidi Bel-Abbès, Algeria M. Benchohra
Baltimore, Maryland, USA G.M. N'Guérékata

Contents

Chapter 1
Introduction

Fractional calculus is a generalization of the ordinary differentiation and integration to arbitrary non-integer order. The subject is as old as the differential calculus and goes back to times when Leibniz and Newton invented differential calculus. One owes to Leibniz in a letter to L'Hôspital, dated September 30, 1695 [181], the exact birthday of the fractional calculus and the idea of the fractional derivative. L'Hôspital asked the question as to the meaning of $d^n y/\mathrm{d}x^n$ if $n = \frac{1}{2}$; i.e., what if n is fractional? Leibniz replied that $d^{\frac{1}{2}}x$ will be equal to $x\sqrt{\mathrm{d}x : x}$. In the letters to J. Wallis and J. Bernoulli (in 1697), Leibniz mentioned the possible approach to fractional-order differentiation in that sense that for non-integer values of n the definition could be the following: $\frac{d^n \mathrm{e}^{mx}}{\mathrm{d}x^n} = m^n \mathrm{e}^{mx}$. In 1730, Euler mentioned interpolating between integral orders of a derivative and suggested to use the following relationship: $\frac{d^n x^m}{\mathrm{d}x^n} = \frac{\Gamma(m+1)}{\Gamma(m-n+1)} x^{m-n}$, where $\Gamma(.)$ is the (Euler's) Gamma function defined by $\Gamma(\xi) = \int_0^\infty t^{\xi-1}\mathrm{e}^{-t}\mathrm{d}t,\ \xi > 0$. Also for negative or non-integer (rational) values of n. Taking $m = 1$ and $n = \frac{1}{2}$, Euler obtained: $\frac{d^{\frac{1}{2}}x}{\mathrm{d}x^{\frac{1}{2}}} = \sqrt{\frac{4x}{\pi}} = \frac{2}{\sqrt{\pi}} x^{\frac{1}{2}}$. In 1812, Laplace [1820 vol. 3, 85 and 186] defined a fractional derivative by means of an integral, and in 1819 there appeared the first discussion of a derivative of fractional order in a calculus text written by Lacroix [171]. The first step to generalization of the notion of differentiation for arbitrary functions was done by Fourier (1822) [125]. After introducing his famous formula

$$f(x) = \frac{1}{2\pi} \int_{-\infty}^{+\infty} f(z)\mathrm{d}z \int_{-\infty}^{+\infty} \cos(px - pz)\mathrm{d}p,$$

Fourier made a remark that

$$\frac{d^n f(x)}{\mathrm{d}x^n} = \frac{1}{2\pi} \int_{-\infty}^{+\infty} f(z)\mathrm{d}z \int_{-\infty}^{+\infty} \cos\left(px - pz + n\frac{\pi}{2}\right)\mathrm{d}p,$$

S. Abbas et al., *Topics in Fractional Differential Equations*, Developments in Mathematics 27, DOI 10.1007/978-1-4614-4036-9_1,

and this relationship could serve as a definition of the nth order derivative for non-integer n. In 1823, Abel [38], considered the integral representation

$$\int_0^x \frac{S'(\eta)}{(x-\eta)^\alpha}\mathrm{d}\eta = \psi(x),$$

for arbitrary α and then wrote

$$S(x) = \frac{\sin(\pi\alpha)}{\pi}x^\alpha \int_0^1 \frac{\psi(xt)}{(1-t)^{1-\alpha}}\mathrm{d}t = \frac{1}{\Gamma(1-\alpha)}\frac{d^{-\alpha}\psi(x)}{sx^{-\alpha}}.$$

The first great theory of fractional derivation is due to Liouville (1832) [185].

I. In his first definition, according to exponential representation of a function $f(x) = \sum_{n=0}^{\infty} c_n \mathrm{e}^{a_n x}$, he generalized the formula $\frac{d^\nu f(x)}{\mathrm{d}x^\nu} = \sum_{n=0}^{\infty} c_n a_n^\nu \mathrm{e}^{a_n x}$.

II. Second type of his definition was Fractional Integral

$$\int^\mu \Phi(x)\mathrm{d}x^\mu = \frac{1}{(-1)^\mu \Gamma(\mu)}\int_x^\infty (\tau - x)^{\mu-1}\Phi(\tau)\mathrm{d}\tau,$$

$$\int^\mu \Phi(x)\mathrm{d}x^\mu = \frac{1}{\Gamma(\mu)}\int_{-\infty}^x (x-\tau)^{\mu-1}\Phi(\tau)\mathrm{d}\tau.$$

III. Third definition includes Fractional derivative

$$\frac{d^\mu F(x)}{\mathrm{d}x^\mu} = \frac{(-1)^\mu}{h^\mu}\left(F(x) - \frac{\mu}{1}F(x+h) + \frac{\mu(\mu-1)}{1.2}F(x+2h) - \cdots\right),$$

$$\frac{d^\mu F(x)}{\mathrm{d}x^\mu} = \frac{1}{h^\mu}\left(F(x) - \frac{\mu}{1}F(x-h) + \frac{\mu(\mu-1)}{1.2}F(x-2h) - \cdots\right).$$

But the formula most often used today, called Riemann–Liouville integral, was given by Riemann (1847). His definition of Fractional Integral is

$$D^{-\nu} f(x) = \frac{1}{\Gamma(\nu)}\int_c^x (x-t)^{\nu-1} f(t)\mathrm{d}t + \psi(t).$$

On these historical concepts, one will be able to refer to work of Dugowson [117]. According to Riemann–Liouville the notion of fractional integral of order α, $\alpha > 0$ for a function $f(t)$, is a natural consequence of the well-known formula (Cauchy–Dirichlet), which reduces the calculation of the n–fold primitive of a function $f(t)$ to a single integral of convolution type.

$$I^n_{a+} f(t) = \frac{1}{(n-1)!} \int_a^t (t-\tau)^{n-1} f(\tau)\mathrm{d}\tau, \; n \in \mathbf{N},$$

vanishes at $t = a$ along with its derivatives of order $1, 2, \ldots, n-1$. One requires $f(t)$ and $I^n_{a+} f(t)$ to be causal functions, that is, vanishing for $t < 0$. Then one extends to any positive real value by using the Gamma function, $(n-1)! = \Gamma(n)$. So, the left-sided mixed Riemann–Liouville integral of order α of f is defined by

$$I^\alpha_{a+} f(t) = \frac{1}{\Gamma(\alpha)} \int_a^t (t-\tau)^{\alpha-1} f(\tau)\mathrm{d}\tau.$$

The operator of fractional derivative $D^\alpha f(t)$ can be defined by the Transform of Laplace integrals, the derivative of order $\alpha < 0$ a causal function $f(t)$ is given by the Riemann–Liouville integral:

$$D^\alpha f(t) = \int_0^t \frac{\xi^{-\alpha-1}}{\Gamma(-\alpha)} f(t-\xi)\mathrm{d}\xi. \tag{1.1}$$

If $\alpha > 0$, we can pose

$$D^\alpha f(t) = PF \int_0^t \frac{\xi^{-\alpha-1}}{\Gamma(-\alpha)} f(t-\xi)\mathrm{d}\xi, \; \alpha > 0, \; \alpha \notin \mathbf{N},$$

where PF represents the finite part of the integral (Schwartz). In 1867, Grünwald and Letnikov joined this definition which is sometimes useful

$${}_aD^\alpha_t f(t) = \lim_{h\to 0} h^{-\alpha} \sum_{k=0}^{[\frac{t-a}{h}]} (-1)^k \binom{\alpha}{k} f(t-kh) = \sum_{k=0}^{\infty} (-1)^k \binom{\alpha}{k} f(t). \tag{1.2}$$

This definition of fractional derivative of a function $f(t)$ based on finite differences is obtained from the classical definition of integer order derivative (Grünwald [137]). We can get an idea of the equivalence of definitions (1.1) and (1.2) using the factorial function $\Gamma(\alpha)$ by Gauss: $\Gamma(\alpha) = \lim\limits_{k\to\infty} \dfrac{k!k^\alpha}{\alpha(\alpha+1)\cdots(\alpha+k)}$. A list of mathematicians, who have provided important contributions up to the middle of the last century, includes N.Ya. Sonin (1869), A.V. Letnikov (1872), H. Laurent (1884), P.A. Nekrassov (1888), A. Krug (1890), J. Hadamard (1892) , O. Heaviside (1892–1912), S. Pincherle (1902), G.H. Hardy and J.E. Littlewood (1917–1928), H. Weyl (1917), P. Lévy (1923), A. Marchaud (1927), H.T. Davis (1924–1936), E. L. Post (1930), A. Zygmund (1935- 1945), E.R. Love (1938–1996), A. Erdélyi (1939–1965), H. Kober (1940), D.V. Widder (1941), M. Riesz (1949), W. Feller (1952), and K. Nishimoto (1987-). They considered the Cauchy Integral formula

$$f^{(n)}(z) = \frac{h!}{2\pi i} \int\limits_{c} \frac{f(t)}{(t-z)^{n+1}} \mathrm{d}t,$$

and substituted n by ν to obtain

$$D^{\nu} f(x) = \frac{\Gamma(\nu+1)}{2\pi i} \int\limits_{c}^{x^+} \frac{f(t)}{(t-z)^{\nu+1}} \mathrm{d}t.$$

The Riemann–Liouville definition of fractional calculus is the popular definition, it is this which shows joining of two previous definitions.

$${}_aD_t^{\alpha} f(t) = \frac{1}{\Gamma(n-\alpha)} \left(\frac{d}{\mathrm{d}t}\right)^n \int\limits_{a}^{t} \frac{f(\tau)\mathrm{d}\tau}{(t-\tau)^{\alpha-n+1}}; \quad n-1 \leq \alpha < n.$$

The Riemann–Liouville derivative has certain disadvantages when trying to model real-world phenomena with fractional differential equations. Therefore, we shall introduce a modified fractional differential operator ${}_a^cD_t^{\alpha}$ proposed by Caputo (1967) first in his work on the theory of viscoelasticity [91] and 2 years later in his book [92]. Caputo's definition can be written as

$${}_a^cD_t^{\alpha} = \frac{1}{\Gamma(n-\alpha)} \int\limits_{a}^{t} \frac{f^{(n)}(\tau)\mathrm{d}\tau}{(t-\tau)^{\alpha-n+1}}; \quad n-1 \leq \alpha < n.$$

The Mittag-Leffler function is a generalization of the exponential function that plays an important role in fractional calculus. The function was developed by the Scandinavian mathematician Mittag-Leffler (1846–1927) [195, 196], who was a contemporary of Oliver Heaviside(1850–1925). In 1993, Miller and Ross used differential operator D as $D^{\overline{\alpha}} f(t) = D^{\alpha_1} D^{\alpha_2} \cdots D^{\alpha_n} f(t);\ \overline{\alpha} = (\alpha_1, \alpha_2, \ldots, \alpha_n)$, in which D^{α_i} are Riemann–Liouville or Caputo definitions. The idea of fractional calculus and fractional order differential equations and inclusions has been a subject of interest not only among mathematicians but also among physicists and engineers. Indeed, we can find numerous applications in rheology, porous media, viscoelasticity, electrochemistry, electromagnetism, signal processing, dynamics of earthquakes, optics, geology, viscoelastic materials, biosciences, bioengineering, medicine, economics, probability and statistics, astrophysics, chemical engineering, physics, splines, tomography, fluid mechanics, electromagnetic waves, nonlinear control, control of power electronic, converters, chaotic dynamics, polymer science, proteins, polymer physics, electrochemistry, statistical physics, thermodynamics, neural networks, etc. [115, 133, 134, 151, 187, 189, 191, 208, 209, 220, 229, 231, 250]. The problem of the existence of solutions of Cauchy-type problems for ordinary differential equations of fractional order and without delay in spaces of integrable functions was studied in some works [164, 228]. The similar problem in spaces of continuous functions was studied in [243]. Recently several papers have been devoted to the study of hyperbolic partial integer order differential equations and inclusions with local and nonlocal conditions; see for instance [85–88, 176], the nonlocal conditions of this type can be applied in the theory of elasticity with better effect than the initial or Darboux conditions. For similar results with set-valued right-hand side we refer to [74–76, 89, 158, 213]. During the last 10 years, hyperbolic ordinary and partial differential equations and inclusions of fractional order have been intensely studied by many mathematicians; see for instance [4–6, 15, 244–247].

In recent years, there has been a significant development in fractional calculus techniques in ordinary and partial functional differential equations and inclusions, some recent contributions can be seen in the monographs of Anastassiou [52], Baleanu et al. [61], Diethelm [113], Kaczorek [156], Kilbas et al. [166], Lakshmikantham et al. [175], Miller and Ross [192], Podlubny [214], Samko et al. [225], the papers of Abbas et al. [25, 30, 32, 35, 36], Abbas and Benchohra [5, 6, 9, 10], Agarwal et al. [39, 43, 45, 46], Ahmad and Nieto [47], Ait Dads et al. [49], Almeida and Torres [50, 51], Araya and Lizama [53], Arshad and Lupulescu [54], Balachandran et al. [59, 60], Baleanu and Vacaru [62], Bazhlekova [64], Belarbi et al. [66], Benchohra et al. [67–69, 71, 73], Burton [83], Chang and Nieto [94], Darwish et al. [100], Danca and Diethelm [99], Debbouche [102], Debbouche and Baleanu [103], Delbosco and Rodino [105], Denton and Vatsala [106], Diagana et al. [112], Diethelm [114, 115], Dong et al. [116], El-Borai [118, 119], El-Borai et al. [120, 121], El-Sayed [122–124], Furati and Tatar [131, 132], Henderson and Ouahab [144, 145], Herzallah et al. [149, 150], Ibrahim [154], Kadem and Baleanu [157], Kaufmann and Mboumi [163], Kilbas and Marzan [165], Kirane et al. [167], Kiryakova and Luchko [168], Li et al. [183], Labidi and Tatar [170], Lakshmikanthan [173], Lakshmikantham and Vatsala [178], Li and N'Guérékata [182], Luchko [186], Magin et al. [188], Mainardi [189], Moaddy et al. [197], Mophou [198], Mophou et al. [199–204], Muslih and Agrawal [205], Muslih et al. [206], Nieto [207], Podlubny et al. [216], Ramrez and Vatsala [217], Razminia et al. [218], Rivero et al. [219], Sabatier et al. [221], Salem [222–224], Samko et al. [226], Tarasov [232], Tarasov and Edelman [233], Tenreiro Machado [234–236], Tenreiro Machado et al. [237–239], Trigeassou et al. [240], Vzquez [241], Wang et al. [248], Vityuk [242], Vityuk and Golushkov [244], Yu and Gao [249], Zhang [251], Zhou et al. [253–255], and the references therein.

Applied problems require definitions of fractional derivative allowing the utilization of physically interpretable initial conditions. Caputo's fractional derivative, originally introduced by Caputo [90] and afterwards adopted in the theory of linear viscoelasticity, satisfies this demand. For a consistent bibliography on this topic, historical remarks, and examples, we refer to [41,48,49,77,214,215].

The method of upper and lower solutions has been successfully applied to study the existence of multiple solutions for initial and boundary value problems of the first-and second-order partial differential equations. This method plays an important role in the investigation of solutions for differential and partial differential equations and inclusions. We refer to the monographs by Benchohra et al. [70], the papers of Abbas and Benchohra [7, 8, 12, 14], Heikkila and Lakshmikantham [143], Ladde et al. [172], Lakshmikantham and Pandit [176], Lakshmikantham et al. [177], Pandit [213], and the references cited therein.

The theory of impulsive integer order differential equations and inclusions has become important in some mathematical models of real processes and phenomena studied in physics, chemical technology, population dynamics, biotechnology, and economics. The study of impulsive fractional differential equations and inclusions was initiated in the 1960s by Milman and Myshkis [193, 194]. At present the foundations of the general theory are already laid, and many of them are investigated in detail in the books of Benchohra et al. [70], Lakshmikantham et al. [174], Samoilenko and Peresyuk [227], and the references therein. There was an intensive development of the impulse theory, especially in the area of impulsive differential equations and inclusions with fixed moments. The theory of impulsive differential equations and inclusions with variable time is relatively less developed due to the difficulties created by the state-dependent impulses. Some interesting extensions to impulsive differential equations with variable times have been done by Bajo and Liz [58], Abbas et al. [2, 26, 27], Abbas and Benchohra [13], Belarbi and Benchohra [65], Benchohra et al. [70, 72], Frigon and O'Regan [128–130], Kaul et al. [160], Kaul and Liu [161, 162], and the references cited therein. In the case of non-integer order derivative, impulsive differential equations and inclusions have been initiated in the papers [41, 77]. See also [40, 48, 49].

Functional differential equations with state-dependent delay appear frequently in applications as model of equations and for this reason the study of these types of equations has received great attention in the last year; see for instance [140, 141] and the references therein. The literature related to partial functional differential equations with state-dependent delay is limited; see for instance [11, 37, 148]. The literature related to ordinary and partial functional differential equations with delay for which $\rho(t,.) = t$ or $(\rho_1(x,y,.), \rho_2(x,y,.)) = (x,y)$ is very extensive; see for instance [5, 6, 139] and the references therein.

Implicit differential equations involving the regularized fractional derivative were analyzed by many authors, in the last year; see for instance [16, 17, 33, 34, 246, 247] and the references therein.

Integral equations are one of the useful mathematical tools in both pure and applied analysis. This is particularly true of problems in mechanical vibrations and the related fields of engineering and mathematical physics. There has been a significant development in ordinary and partial fractional integral equations in recent years; see the monographs of Miller and Ross [192], Podlubny [214], Abbas et al. [18–23, 28, 29], Banaś et al. [63], Darwish et al. [100], Dhage [107–111], and the references therein.

In this book we are interested by initial value problems (IVP for short) for partial hyperbolic functional differential equations and inclusions with Caputo's fractional derivative and partial hyperbolic implicit differential equations involving the regularized fractional derivative. Our results may be interpreted as extensions of previous results of Dawidowski and Kubiaczyk [101], Kamont [158], Kamont and Kropielnicka [159] obtained for "classical" hyperbolic differential equations and inclusions with integer order derivative and those of Kilbas and Marzan [165] considered with fractional derivative and without delay. In fact, in the proof of our theorems we essentially use several fixed-point techniques. This book is arranged and organized as follows:

In Chap. 2, we introduce notations, definitions, and some preliminary notions. In Sect. 2.1, we give some notations from the theory of Banach spaces and Banach algebras. Section 2.2 is concerned to recall some basic definitions and facts on partial fractional calculus theory. In Sect. 2.3, we give some properties of set-valued maps. Section 2.4 is devoted to fixed-points theory, here we give the main theorems that will be used in the following chapters. In Sect. 2.5, we give some generalizations of Gronwall's lemmas for two independent variables and singular kernel.

In Chap. 3, we shall be concerned by fractional order partial functional differential equations. In Sect. 3.2, we study initial value problem for a class of partial hyperbolic differential equations. We give two results, one based on Banach fixed-point theorem and the other based on the nonlinear alternative of Leray–Schauder type. We present two similar results to nonlocal problems. An example will be presented in the last illustrating the abstract theory. Section 3.3 is concerned to study a system of perturbed partial hyperbolic differential equations. We give two results, one based on Banach fixed-point theorem and other based on a fixed-point theorem due to Burton and Kirk for the sum of contraction and completely continuous operators. Also, we give similar results to nonlocal problems and we present an illustrative example. Section 3.4 is devoted to study initial value problem for partial neutral functional differential equations. We present some existence results using Krasnoselskii's fixed-point theorem. Also we present an example illustrating the applicability of the imposed conditions. In Sect. 3.5, we shall be concerned by partial hyperbolic differential equations in Banach algebras. We shall prove the existence of solutions, as well as the existence of extremal solutions. Our approach is based, for the existence of solutions, on a fixed-point theorem due to Dhage under Lipschitz and Carathéodory conditions, and for the existence of extremal solutions, on the concept of upper and lower solutions combined with a fixed-point theorem on ordered Banach algebras established by Dhage under certain monotonicity conditions. An example is presented in the last part of this section. In Sect. 3.6, we investigate the existence of solutions for a class of initial value problem for partial hyperbolic differential equations by using the lower and upper solutions method combined with Schauder's fixed-point theorem. In Sect. 3.7, we study a system of partial hyperbolic differential equations with infinite delay. We present two results, one based on Banach fixed-point theorem and the other based on the nonlinear alternative of Leray–Schauder type. Section 3.8 is devoted to study the existence and uniqueness of solutions of some classes of partial functional and neutral functional hyperbolic differential equations with state-dependent delay. Some examples will be presented in the last part of this section. In the last section of this paper, we shall be concerned by global uniqueness results for partial hyperbolic differential equations. We investigate the global existence and uniqueness of solutions of four classes of partial hyperbolic differential equations with finite and infinite delays and we present some illustrative examples.

In Chap. 4, we shall be concerned by functional partial differential inclusions. In Sect. 4.2, we investigate the existence of solutions of a class of partial hyperbolic differential inclusions with finite delay. We shall present two existence results, when the right-hand side is convex as well as nonconvex valued. The first result relies on the nonlinear alternative of Leray–Schauder type. In the second result, we shall use the fixed-point theorem for contraction multivalued maps due to Covitz and Nadler. In Sect. 4.3, we prove a Filippov-type existence result for a class of partial hyperbolic differential inclusions by applying the contraction principle in the space of selections of the multifunction instead of the space of solutions. The second result is about topological structure of the solution set, more exactly, we prove that the solution set is not empty and compact. Section 4.4 is devoted to an existence result of solutions for functional differential inclusions. Our proof relies on the nonlinear alternative of Leray–Schauder combined with lower and upper solutions method. Section 4.5 deals with the existence of solutions for the initial value problems for fractional-order hyperbolic and neutral hyperbolic functional differential inclusions with infinite delay by using the nonlinear alternative of Leray–Schauder type for multivalued operators. In Sect. 4.6, we investigate the existence of solutions for a system of integral inclusions of fractional order. Our approach is based on appropriate fixed-point theorems, namely Bohnenblust–Karlin fixed-point theorem for the convex case and Covitz-Nadler for the nonconvex case.

In Chap. 5, we shall be concerned with functional impulsive partial hyperbolic differential equations. Section 5.2 deals with the existence and uniqueness of solutions of a class of partial hyperbolic differential equations with fixed time impulses. We present two results, the first one is based on Banach's contraction principle and the second one on the nonlinear alternative of Leray–Schauder type. As an extension to nonlocal problems, we present two similar results. Finally we present an illustrative example. In Sect. 5.3, we investigate the existence and uniqueness of solutions of a class of partial hyperbolic differential equations with variable time impulses. We present two results, the first one is based on Schaefer's fixed-point and the second one on Banach's contraction principle. As an extension to nonlocal problems, we present two similar results. An example will be presented in the last illustrating the abstract theory. Section 5.4 deals with the existence of solutions and extremal solutions to partial hyperbolic differential equations of fractional order with impulses in Banach algebras under Lipschitz and Carathéodory conditions and certain monotonicity conditions. Finally we present an illustrative example. Section 5.5 deals with the existence of solutions to partial functional differential equations with impulses at variable times and infinite delay. Our works will be considered by using the nonlinear alternative of Leray–Schauder type and we present an illustrative example. Section 5.6 is devoted to study the existence and uniqueness of solutions of two classes of partial hyperbolic differential equations with fixed time impulses and state-dependent delay. We present two results for each of our problems, the first one is based on Banach's contraction principle and the second one on the nonlinear alternative of Leray–Schauder type. In Sect. 5.7, we investigate the existence and uniqueness of solutions of two classes of partial hyperbolic differential equations with variable time impulses and state-dependent delay, we present existence results for our problems based on Schaefer's fixed-point. In Sect. 5.8, we investigate the existence of solutions for a class of initial value problem for impulsive partial hyperbolic differential equations by using the lower and upper solutions method combined with Schauder's fixed-point theorem.

In Chap. 6, we shall be concerned with impulsive partial hyperbolic functional differential inclusions. Section 6.2 deals with the existence of solutions of a class of partial hyperbolic differential inclusions with fixed time impulses. We shall present existence results when the right-hand side is convex as well as nonconvex valued. We present three results, the first one is based on the nonlinear alternative of Leray–Schauder type. In the second result, we shall use the fixed-point theorem for contraction multivalued maps due to Covitz and Nadler. The third result relies on the nonlinear alternative of Leray–Schauder type for single-valued map combined with a selection theorem due to Bressan and Colombo for lower semicontinuous multivalued operators with closed and decomposable values. In Sect. 6.3, we investigate the existence of solutions of some classes of partial impulsive hyperbolic differential inclusions at variable times by using the nonlinear alternative of Leray–Schauder type. In Sect. 6.4, we use the upper and lower solutions method combined with fixed-point theorem of Bohnnenblust-Karlin for investigating the existence of solutions of a class of partial hyperbolic differential inclusions at fixed moments of impulse.

In Chap. 7, we shall be concerned with implicit partial hyperbolic differential equations.In Section 7.2 we investigate the existence and uniqueness of solutions for implicit partial hyperbolic functional differential equations. We present two results, the first one is based on Banach's contraction principle and the second one on the nonlinear alternative of Leray–Schauder type. Section 7.3 deals with a global uniqueness result for fractional-order implicit differential equations, we make use of the nonlinear alternative of Leray–Schauder type for contraction maps on Fréchet spaces. To illustrate the result an example is provided. In Sect. 7.4, we shall be concerned with implicit partial hyperbolic differential equations with finite delay, infinite delay, and with state-dependent delay. We present two results for each of our problems, the first one is based on Banach's contraction principle and the second one on the nonlinear alternative of Leray–Schauder type. We illustrate our results by some examples. Section 7.5 deals with the existence and uniqueness of solutions of a class of implicit impulsive partial hyperbolic differential equations. We present two results for our problem, the first one is based on Banach's contraction principle and the second one on the nonlinear alternative of Leray–Schauder type. To illustrate the results an example is provided. In Sect. 7.6, we shall be concerned with the existence and uniqueness of solutions of two classes of partial implicit impulsive hyperbolic differential equations with fixed time impulses and state-dependent delay. We present two results for each of our problems, the first one is based on Banach's contraction principle and the second one on the nonlinear alternative of Leray–Schauder type. Also, we present some illustrative examples.

In Chap. 8, we shall be concerned with Riemann–Liouville integral equations of fractional order. In Sect. 8.2 we study the existence and uniqueness of solutions of a certain Fredholm-type Riemann–Liouville integral equation of two variables by using Banach contraction principle. Section 8.3 deals with the existence and uniqueness of solutions for a system of integral equations of fractional order with multiple time delay by using some fixed-point theorems. We illustrate our results with some examples. In Sect. 8.4 we prove an existence result for a nonlinear quadratic Volterra integral equation of fractional order. Our technique is based on a fixed-point theorem due to Dhage [109]. Finally, an example illustrating the main existence result is presented in the last section. Section 8.5 deals with the existence and global asymptotic stability of solutions of a class of fractional order functional integral equations by using the Schauder fixed-point theorem. Also, we obtain some results about the asymptotic stability of solutions of the equation in question. Finally, we present an example illustrating the applicability of the imposed conditions. In Sect. 8.6 we prove the existence and local asymptotic attractivity of solutions for a functional integral equation of Riemann–Liouville fractional order in Banach algebras, by using a fixed-point theorem of Dhage [109]. Also, we present an example illustrating the applicability of the imposed conditions.

Chapter 2
Preliminary Background

In this chapter, we introduce notations, definitions, and preliminary facts that will be used in the remainder of this book.

2.1 Notations and Definitions

Let $J := [0,a] \times [0,b]$; $a,b > 0$ and $\rho > 0$. Denote $L^{\rho}(J,\mathbb{R}^n)$ the space of Lebesgue-integrable functions $u : J \to \mathbb{R}^n$ with the norm

$$\|u\|_{L^{\rho}} = \left(\int_0^a \int_0^b \|u(x,y)\|^{\rho} \mathrm{d}y\mathrm{d}x\right)^{\frac{1}{\rho}},$$

where $\|.\|$ denotes a suitable complete norm on $\mathbb{R}^n$. Also $L^1(J,\mathbb{R}^n)$ is endowed with norm $\|.\|_{L^1}$ defined by

$$\|u\|_{L^1} = \int_0^a \int_0^b \|u(x,y)\|\mathrm{d}y\mathrm{d}x.$$

Let $L^{\infty}(J,\mathbb{R}^n)$ be the Banach space of measurable functions $u : J \to \mathbb{R}^n$ which are bounded, equipped with the norm

$$\|u\|_{L^{\infty}} = \inf\{c > 0 : \|u(x,y)\| \leq c,\ a.e.\ (x,y) \in J\}.$$

As usual, by $AC(J,\mathbb{R}^n)$ we denote the space of absolutely continuous functions from J into $\mathbb{R}^n$, and $C(J,\mathbb{R}^n)$ is the Banach space of all continuous functions from J into $\mathbb{R}^n$ with the norm

$$\|u\|_{\infty} = \sup_{(x,y)\in J} \|u(x,y)\|.$$

Also $C(J,\mathbb{R})$ is endowed with norm $\|.\|_{\infty}$ defined by $\|u\|_{\infty} = \sup_{(x,y)\in J} |u(x,y)|$.

S. Abbas et al., *Topics in Fractional Differential Equations*, Developments in Mathematics 27, DOI 10.1007/978-1-4614-4036-9_2,

Define a multiplication "$\cdot$" by

$$(u \cdot v)(x, y) = u(x, y)v(x, y), \quad \text{for } (x, y) \in J.$$

Then $C(J, \mathbb{R})$ is a Banach algebra with above norm and multiplication.

If $u \in C([-\alpha, a] \times [-\beta, b], \mathbb{R}^n);\ a, b, \alpha, \beta > 0$ then for any $(x, y) \in J$ define $u_{(x,y)}$ by

$$u_{(x,y)}(s, t) = u(x + s, y + t),$$

for $(s, t) \in [-\alpha, 0] \times [-\beta, 0]$. Here $u_{(x,y)}(.,.)$ represents the history of the state from time $(x - \alpha, y - \beta)$ up to the present time (x, y).

2.2 Properties of Partial Fractional Calculus

In this section, we introduce notations, definitions, and preliminary lemmas concerning partial fractional calculus theory.

Definition 2.1 ([216, 225]). The Riemann–Liouville fractional integral of order $\alpha \in (0, \infty)$ of a function $h \in L^1([0, b], \mathbb{R}^n);\ b > 0$ is defined by

$$I_0^\alpha h(t) = \frac{1}{\Gamma(\alpha)} \int_0^t (t - s)^{\alpha - 1} h(s) \mathrm{d}s.$$

Definition 2.2 ([216, 225]). The Riemann–Liouville fractional derivative of order $\alpha \in (0, 1]$ of a function $h \in L^1([0, b], \mathbb{R}^n)$ is defined by

$$\begin{aligned} D_0^\alpha h(t) &= \frac{\mathrm{d}}{\mathrm{d}t} I_0^{1-\alpha} h(t) \\ &= \frac{1}{\Gamma(1-\alpha)} \frac{\mathrm{d}}{\mathrm{d}t} \int_0^t (t - s)^{-\alpha} h(s) \mathrm{d}s; \text{ for almost all } t \in [0, b]. \end{aligned}$$

Definition 2.3 ([216, 225]). The Caputo fractional derivative of order $\alpha \in (0, 1]$ of a function $h \in L^1([0, b], \mathbb{R}^n)$ is defined by

$$\begin{aligned} {}^c D_0^\alpha h(t) &= I_0^{1-\alpha} \frac{\mathrm{d}}{\mathrm{d}t} h(t) \\ &= \frac{1}{\Gamma(1-\alpha)} \int_0^t (t - s)^{-\alpha} \frac{\mathrm{d}}{\mathrm{d}s} h(s) \mathrm{d}s; \text{ for almost all } t \in [0, b]. \end{aligned}$$

Definition 2.4 ([166, 225]). Let $\alpha \in (0,\infty)$ and $u \in L^1(J,\mathbb{R}^n)$. The partial Riemann–Liouville integral of order α of $u(x,y)$ with respect to x is defined by the expression

$$I^{\alpha}_{0,x}u(x,y) = \frac{1}{\Gamma(\alpha)}\int_0^x (x-s)^{\alpha-1}u(s,y)ds; \text{ for almost all } (x,y) \in J.$$

Analogously, we define the integral

$$I^{\alpha}_{0,y}u(x,y) = \frac{1}{\Gamma(\alpha)}\int_0^y (y-s)^{\alpha-1}u(x,s)ds; \text{ for almost all } (x,y) \in J.$$

Definition 2.5 ([166, 225]). Let $\alpha \in (0,1]$ and $u \in L^1(J,\mathbb{R}^n)$. The Riemann–Liouville fractional derivative of order α of $u(x,y)$ with respect to x is defined by

$$(D^{\alpha}_{0,x}u)(x,y) = \frac{\partial}{\partial x}I^{1-\alpha}_{0,x}u(x,y); \text{ for almost all } (x,y) \in J.$$

Analogously, we define the derivative

$$(D^{\alpha}_{0,y}u)(x,y) = \frac{\partial}{\partial y}I^{1-\alpha}_{0,y}u(x,y); \text{ for almost all } (x,y) \in J.$$

Definition 2.6 ([166, 225]). Let $\alpha \in (0,1]$ and $u \in L^1(J,\mathbb{R}^n)$. The Caputo fractional derivative of order α of $u(x,y)$ with respect to x is defined by the expression

$${}^cD^{\alpha}_{0,x}u(x,y) = I^{1-\alpha}_{0,x}\frac{\partial}{\partial x}u(x,y); \text{ for almost all } (x,y) \in J.$$

Analogously, we define the derivative

$${}^cD^{\alpha}_{0,y}u(x,y) = I^{1-\alpha}_{0,y}\frac{\partial}{\partial y}u(x,y); \text{ for almost all } (x,y) \in J.$$

Definition 2.7 ([244]). Let $r = (r_1, r_2) \in (0,\infty)\times(0,\infty)$ and $u \in L^1(J,\mathbb{R}^n)$. The left-sided mixed Riemann–Liouville integral of order r of u is defined by

$$(I^r_0 u)(x,y) = \frac{1}{\Gamma(r_1)\Gamma(r_2)}\int_0^x\int_0^y (x-s)^{r_1-1}(y-t)^{r_2-1}u(s,t)dtds.$$

In particular,

$$(I_0^0 u)(x,y) = u(x,y),\ (I_0^\sigma u)(x,y) = \int\limits_0^x \int\limits_0^y u(s,t)\mathrm{d}t\mathrm{d}s;\ \text{for almost all } (x,y) \in J,$$

where $\sigma = (1,1)$.

For instance, $I_0^r u$ exists for all $r_1, r_2 > 0$, when $u \in L^1(J, \mathbb{R}^n)$. Note also that when $u \in C(J, \mathbb{R}^n)$, then $(I_0^r u) \in C(J, \mathbb{R}^n)$; moreover,

$$(I_0^r u)(x,0) = (I_0^r u)(0,y) = 0;\ x \in [0,a],\ y \in [0,b].$$

Example 2.8. Let $\lambda, \omega \in (-1, \infty)$ and $r = (r_1, r_2) \in (0,\infty) \times (0,\infty)$, then

$$I_0^r x^\lambda y^\omega = \frac{\Gamma(1+\lambda)\Gamma(1+\omega)}{\Gamma(1+\lambda+r_1)\Gamma(1+\omega+r_2)} x^{\lambda+r_1} y^{\omega+r_2};\ \text{for almost all } (x,y) \in J.$$

By $1-r$ we mean $(1-r_1, 1-r_2) \in [0,1) \times [0,1)$. Denote by $D_{xy}^2 := \frac{\partial^2}{\partial x \partial y}$, the mixed second-order partial derivative.

Definition 2.9 ([244]). Let $r \in (0,1] \times (0,1]$ and $u \in L^1(J, \mathbb{R}^n)$. The Caputo fractional-order derivative of order r of u is defined by the expression $({}^c D_0^r u)(x,y) = (I_0^{1-r} D_{xy}^2 u)(x,y)$ and the mixed fractional Riemann–Liouville derivative of order r of u is defined by the expression $(D_0^r u)(x,y) = (D_{xy}^2 I_0^{1-r} u)(x,y)$.

The case $\sigma = (1,1)$ is included and we have

$$({}^c D_0^\sigma u)(x,y) = (D_0^\sigma u)(x,y) = (D_{xy}^2 u)(x,y);\ \text{for almost all } (x,y) \in J.$$

Example 2.10. Let $\lambda, \omega \in (-1, \infty)$ and $r = (r_1, r_2) \in (0,1] \times (0,1]$, then

$$D_0^r x^\lambda y^\omega = \frac{\Gamma(1+\lambda)\Gamma(1+\omega)}{\Gamma(1+\lambda-r_1)\Gamma(1+\omega-r_2)} x^{\lambda-r_1} y^{\omega-r_2};\ \text{for almost all } (x,y) \in J.$$

Definition 2.11 ([247]). For a function $u : J \to \mathbb{R}^n$, we set

$$q(x,y) = u(x,y) - u(x,0) - u(0,y) + u(0,0).$$

By the mixed regularized derivative of order $r = (r_1, r_2) \in (0,1] \times (0,1]$ of a function $u(x,y)$, we name the function

$$\overline{D}_0^r u(x,y) = D_0^r q(x,y).$$

The function

$$\overline{D}_{0,x}^{r_1} u(x,y) = D_{0,x}^{r_1}[u(x,y) - u(0,y)],$$

is called the partial r_1–order regularized derivative of the function $u(x,y): J \to \mathbb{R}^n$ with respect to the variable x. Analogously, we define the derivative

$$\overline{D}_{0,y}^{r_2} u(x,y) = D_{0,y}^{r_2}[u(x,y) - u(x,0)].$$

Let $a_1 \in [0,a]$, $z^+ = (a_1, 0) \in J$, $J_z = [a_1, a] \times [0,b]$. For $w \in L^1(J_z, \mathbb{R}^n)$, the expression

$$(I_{z^+}^r w)(x,y) = \frac{1}{\Gamma(r_1)\Gamma(r_2)} \int_{a_1}^{x} \int_{0}^{y} (x-s)^{r_1-1}(y-t)^{r_2-1} w(s,t) \mathrm{d}t \mathrm{d}s,$$

is called the left-sided mixed Riemann–Liouville integral of order r of w. The Caputo fractional-order derivative of order r of w is defined by $({}^c D_{z^+}^r w)(x,y) = (I_{z^+}^{1-r} D_{xy}^2 w)(x,y)$ and the mixed fractional Riemann–Liouville derivative of order r of w is defined by $(D_{z^+}^r w)(x,y) = (D_{xy}^2 I_{z^+}^{1-r} w)(x,y)$.

Let $f, g \in L^1(J, \mathbb{R}^n)$.

Lemma 2.12 ([5,6]). *A function $u \in AC(J, \mathbb{R}^n)$ such that its mixed derivative D_{xy}^2 exists and is integrable on J is a solution of problems*

$$\begin{cases} ({}^c D_0^r u)(x,y) = f(x,y); \ (x,y) \in J, \\ u(x,0) = \varphi(x); \ x \in [0,a], \ u(0,y) = \psi(y); \ y \in [0,b], \\ \varphi(0) = \psi(0), \end{cases}$$

if and only if $u(x,y)$ satisfies

$$u(x,y) = \mu(x,y) + (I_0^r f)(x,y); \ (x,y) \in J,$$

where

$$\mu(x,y) = \varphi(x) + \psi(y) - \varphi(0).$$

Lemma 2.13 ([35]). *A function $u \in AC(J, \mathbb{R}^n)$ such that the mixed derivative $D_{xy}^2(u-g)$ exists and is integrable on J is a solution of problems*

$$\begin{cases} {}^c D_0^r[u(x,y) - g(x,y)] = f(x,y); \ (x,y) \in J, \\ u(x,0) = \varphi(x); \ x \in [0,a], \ u(0,y) = \psi(y); \ y \in [0,b], \\ \varphi(0) = \psi(0), \end{cases}$$

if and only if $u(x,y)$ satisfies

$$u(x,y) = \mu(x,y) + g(x,y) - g(x,0) - g(0,y) + g(0,0) + I_0^r(f)(x,y); \ (x,y) \in J.$$

Let $h \in C([x_k, x_{k+1}] \times [0, b], \mathbb{R}^n)$, $z_k = (x_k, 0)$, $0 = x_0 < x_1 < \cdots < x_m < x_{m+1} = a$ and

$$\mu_k(x, y) = u(x, 0) + u(x_k^+, y) - u(x_k^+, 0); \;\; k = 0, \ldots, m.$$

Lemma 2.14 ([7,8]). *A function $u \in AC([x_k, x_{k+1}] \times [0, b], \mathbb{R}^n)$; $k = 0, \ldots, m$ whose r-derivative exists on $[x_k, x_{k+1}] \times [0, b]$, $k = 0, \ldots, m$ is a solution of the differential equation*

$$({}^c D^r_{z_k} u)(x, y) = h(x, y); \; (x, y) \in [x_k, x_{k+1}] \times [0, b],$$

if and only if $u(x, y)$ satisfies

$$u(x, y) = \mu_k(x, y) + (I^r_{z_k} h)(x, y); \; (x, y) \in [x_k, x_{k+1}] \times [0, b].$$

Let $J_0 = [0, x_1] \times [0, b]$, $J_k = (x_k, x_{k+1}] \times [0, b]$; $k = 1, \ldots, m$, $I_k : \mathbb{R}^n \to \mathbb{R}^n$; $k = 0, 1, \ldots, m$ and denote $\mu(x, y) := \mu_0(x, y)$; $(x, y) \in J$.

Lemma 2.15 ([7, 8]). *Let $h : J \to \mathbb{R}^n$ be continuous. A function u whose r-derivative exists on J_k; $k = 0, \ldots, m$ is a solution of the fractional integral equation*

$$u(x, y) = \begin{cases} \mu(x, y) + \frac{1}{\Gamma(r_1)\Gamma(r_2)} \int\limits_0^x \int\limits_0^y (x - s)^{r_1 - 1}(y - t)^{r_2 - 1} h(s, t) \mathrm{d}t \mathrm{d}s; \\ \text{if } (x, y) \in [0, x_1] \times [0, b], \\ \mu(x, y) + \sum\limits_{i=1}^k (I_i(u(x_i^-, y)) - I_i(u(x_i^-, 0))) \\ + \frac{1}{\Gamma(r_1)\Gamma(r_2)} \sum\limits_{i=1}^k \int\limits_{x_{i-1}}^{x_i} \int\limits_0^y (x_i - s)^{r_1 - 1}(y - t)^{r_2 - 1} h(s, t) \mathrm{d}t \mathrm{d}s \\ + \frac{1}{\Gamma(r_1)\Gamma(r_2)} \int\limits_{x_k}^x \int\limits_0^y (x - s)^{r_1 - 1}(y - t)^{r_2 - 1} h(s, t) \mathrm{d}t \mathrm{d}s; \\ \text{if } (x, y) \in (x_k, x_{k+1}] \times [0, b], \; k = 1, \ldots, m, \end{cases}$$

if and only if u is a solution of the fractional IVP

$$\begin{cases} {}^c D^r_{x_k} u(x, y) = h(x, y); \quad (x, y) \in J_k, \quad k = 0, \ldots, m. \\ u(x_k^+, y) = u(x_k^-, y) + I_k(u(x_k^-, y)); \; y \in [0, b]; \quad k = 1, \ldots, m. \end{cases}$$

Lemma 2.16 ([2]). *Let $h : J \to \mathbb{R}^n$ be continuous. A function u whose r-derivative exists on J_k; $k = 0, \ldots, m$ is a solution of the fractional integral equation*

$$u(x,y) = \begin{cases} \mu(x,y) + \frac{1}{\Gamma(r_1)\Gamma(r_2)} \int_0^x \int_0^y (x-s)^{r_1-1}(y-t)^{r_2-1} h(s,t) \mathrm{d}t\mathrm{d}s; \\ \text{if } (x,y) \in [0,x_1] \times [0,b], \\ \varphi(x) + I_k(u(x_k,y)) - I_k(u(x_k,0)) \\ + \frac{1}{\Gamma(r_1)\Gamma(r_2)} \int_{x_k}^x \int_0^y (x-s)^{r_1-1}(y-t)^{r_2-1} h(s,t) \mathrm{d}t\mathrm{d}s; \\ \text{if } (x,y) \in (x_k, x_{k+1}] \times [0,b],\ k = 1, \ldots, m, \end{cases}$$

if and only if u is a solution of the fractional IVP

$$\begin{cases} {}^cD^r_{x_k} u(x,y) = h(x,y); \quad (x,y) \in J_k;\ k = 0, \ldots, m, \\ u(x_k^+, y) = I_k(u(x_k,y));\ y \in [0,b], \quad k = 1, \ldots, m. \end{cases}$$

Let $f \in C(J, \mathbb{R}^*)$, $g \in L^1(J, \mathbb{R})$ and $\mu_0(x,y) = \frac{\varphi(x)}{f(x,0)} + \frac{\psi(y)}{f(0,y)} - \frac{\varphi(0)}{f(0,0)}$.

Lemma 2.17 ([32]). *A function $u \in AC(J, \mathbb{R})$ such that the mixed derivative $D^2_{xy}(\frac{u}{f})$ exists and is integrable on J is a solution of problems*

$$\begin{cases} {}^cD^r_0\left(\frac{u(x,y)}{f(x,y)}\right) = g(x,y),\ (x,y) \in J, \\ u(x,0) = \varphi(x);\ x \in [0,a],\ u(0,y) = \psi(y);\ y \in [0,b], \\ \varphi(0) = \psi(0), \end{cases}$$

if and only if $u(x,y)$ satisfies

$$u(x,y) = f(x,y)\Big(\mu_0(x,y) + (I^r_0 g)(x,y)\Big);\ (x,y) \in J.$$

Let $f \in C([x_k, x_{k+1}] \times [0,b], \mathbb{R}^*)$, $g \in L^1([x_k, x_{k+1}] \times [0,b], \mathbb{R})$, $z_k = (x_k, 0)$, and

$$\mu_{0,k}(x,y) = \frac{u(x,0)}{f(x,0)} + \frac{u(x_k^+, y)}{f(x_k^+, y)} - \frac{u(x_k^+, 0)}{f(x_k^+, 0)};\ k = 0, \ldots, m.$$

Lemma 2.18 ([3]). *A function $u \in AC([x_k, x_{k+1}] \times [0,b], \mathbb{R})$, $k = 0, \ldots, m$ such that the mixed derivative $D^2_{xy}(\frac{u}{f})$ exists and is integrable on $[x_k, x_{k+1}] \times [0,b]$, $k = 0, \ldots, m$ is a solution of the differential equation*

$${}^cD^r_{z_k}\left(\frac{u}{f}\right)(x,y) = g(x,y);\ (x,y) \in [x_k, x_{k+1}] \times [0,b],$$

if and only if $u(x, y)$ *satisfies*

$$u(x, y) = f(x, y)\Big(\mu_{0,k}(x, y) + (I^r_{z_k} g)(x, y)\Big);\ (x, y) \in [x_k, x_{k+1}] \times [0, b].$$

Let $\mu' := \mu_{0,0}$.

Lemma 2.19 ([3]). *Let* $f : J \to \mathbb{R}^*$, $g : J \to \mathbb{R}$ *be continuous. A function* u *such that the mixed derivative* $D^2_{xy}(\frac{u}{f})$ *exists and is integrable on* J_k; $k = 0, \ldots, m$ *is a solution of the fractional integral equation*

$$u(x, y) = \begin{cases} f(x, y)[\mu'(x, y) + \dfrac{1}{\Gamma(r_1)\Gamma(r_2)} \displaystyle\int_0^x \int_0^y (x - s)^{r_1 - 1}(y - t)^{r_2 - 1} g(s, t)\mathrm{d}t\,\mathrm{d}s]; \\ \text{if } (x, y) \in [0, x_1] \times [0, b], \\ f(x, y)[\mu'(x, y) + \displaystyle\sum_{i=1}^{k} \left(\dfrac{I_i(u(x_i^-, y))}{f(x_i^+, y)} - \dfrac{I_i(u(x_i^-, 0))}{f(x_i^+, 0)} \right) \\ + \dfrac{1}{\Gamma(r_1)\Gamma(r_2)} \displaystyle\sum_{i=1}^{k} \int_{x_{i-1}}^{x_i} \int_0^y (x_i - s)^{r_1 - 1}(y - t)^{r_2 - 1} g(s, t)\mathrm{d}t\,\mathrm{d}s \\ + \dfrac{1}{\Gamma(r_1)\Gamma(r_2)} \displaystyle\int_{x_k}^{x} \int_0^y (x - s)^{r_1 - 1}(y - t)^{r_2 - 1} g(s, t)\mathrm{d}t\,\mathrm{d}s]; \\ \text{if } (x, y) \in (x_k, x_{k+1}] \times [0, b],\ k = 1, \ldots, m, \end{cases}$$

if and only if u *is a solution of the fractional IVP*

$$\begin{cases} {}^cD^r_{x_k}\left(\frac{u}{f}\right)(x, y) = g(x, y); \quad (x, y) \in J_k;\ k = 0, \ldots, m, \\ u\left(x_k^+, y\right) = u\left(x_k^-, y\right) + I_k\left(u\left(x_k^-, y\right)\right);\ y \in [0, b]; \quad k = 1, \ldots, m. \end{cases}$$

2.3 Properties of Set-Valued Maps

Let $(X, \|\cdot\|)$ be a Banach space. Denote $\mathcal{P}(X) = \{Y \in X : Y \neq \emptyset\}$, $\mathcal{P}_{cl}(X) = \{Y \in \mathcal{P}(X) : Y \text{ closed}\}$, $\mathcal{P}_b(X) = \{Y \in \mathcal{P}(X) : Y \text{ bounded}\}$, $\mathcal{P}_{cp}(X) = \{Y \in \mathcal{P}(X) : Y \text{ compact}\}$ l, and $\mathcal{P}_{cp,cv}(X) = \{Y \in \mathcal{P}(X) : Y \text{ compact and convex}\}$.

Definition 2.20. A multivalued map $T : X \to \mathcal{P}(X)$ is convex (closed) valued if $T(x)$ is convex (closed) for all $x \in X$. T is bounded on bounded sets if $T(B) = \cup_{x\in B} T(x)$ is bounded in X for all $B \in \mathcal{P}_b(X)$ (i.e. $\sup_{x\in B} \sup_{y\in T(x)} \|y\| < \infty$). T is called upper semicontinuous (u.s.c.) on X if for each $x_0 \in X$, the set $T(x_0)$ is a nonempty closed subset of X, and if for each open set N of X containing $T(x_0)$, there exists an open neighborhood N_0 of x_0 such that $T(N_0) \subseteq N$. T is lower semi-continuous (l.s.c.) if the set $\{x \in X : T(x) \cap A \neq \emptyset\}$ is open for any open subset $A \subseteq X$. T is said to be completely continuous if $T(\mathcal{B})$ is relatively compact for every $\mathcal{B} \in P_b(X)$. T has a fixed point if there is $x \in X$ such that $x \in T(x)$. The fixed point set of the multivalued operator T will be denoted by $FixT$. A multivalued map $G : X \to \mathcal{P}_{cl}(\mathbb{R}^n)$ is said to be measurable if for every $v \in \mathbb{R}^n$, the function $x \longmapsto d(v, G(x)) = \inf\{\|v - z\| : z \in G(x)\}$ is measurable.

For more details on multivalued maps see the books of Aubin and Cellina [56], Aubin and Frankowska [57], Deimling [104], Gorniewicz [135], Hu and Papageorgiou [153], and Kisielewicz [169].

Lemma 2.21 ([153]). *Let G be a completely continuous multivalued map with nonempty compact values, then G is u.s.c. if and only if G has a closed graph (i.e., $u_n \to u$, $w_n \to w$, $w_n \in G(u_n)$ imply $w \in G(u)$).*

Definition 2.22. A multivalued map $F : J \times \mathbb{R}^n \to \mathcal{P}(\mathbb{R}^n)$ is said to be Carathéodory if

(i) $(x, y) \longmapsto F(x, y, u)$ is measurable for each $u \in \mathbb{R}^n$.
(ii) $u \longmapsto F(x, y, u)$ is upper semicontinuous for almost all $(x, y) \in J$.

F is said to be L^1-Carathéodory if $(i), (ii)$ and the following condition holds;

(iii) For each $c > 0$, there exists $\sigma_c \in L^1(J, \mathbb{R}_+)$ such that

$$\begin{aligned} \|F(x, y, u)\|_{\mathcal{P}} &= \sup\{\|f\| : f \in F(x, y, u)\} \\ &\leq \sigma_c(x, y) \quad \text{for all } \|u\| \leq c \text{ and for } a.e.\ (x, y) \in J. \end{aligned}$$

For each $u \in C(J, \mathbb{R}^n)$, define the set of selections of F by

$$S_{F,u} = \left\{w \in L^1(J, \mathbb{R}^n) : w(x, y) \in F(x, y, u(x, y)) \ \text{a.e.}\ (x, y) \in J\right\}.$$

Definition 2.23. Let $\mathcal{A}$ be a subset of $J \times \mathbb{R}^n$. $\mathcal{A}$ is $\mathcal{L} \otimes \mathcal{B}$ measurable if $\mathcal{A}$ belongs to the σalgebra generated by all sets of the form $\mathcal{N} \otimes \mathcal{D}$ where $\mathcal{N}$ is Lebesgue measurable in J and $\mathcal{D}$ is Borel measurable in $\mathbb{R}^n$. A subset $\mathcal{I}$ of $L^1(J, \mathbb{R}^n)$ is decomposable if for all $u, v \in \mathcal{I}$ and $N \subset J$ measurable, $u\chi_{\mathcal{I}} + v\chi_{J-\mathcal{N}} \in \mathcal{I}$, where χ_J stands for the characteristic function of J.

Definition 2.24. Let $F : J \times \mathbb{R}^n \to \mathcal{P}(\mathbb{R}^n)$ be a multivalued map with nonempty compact values. Assign to F the multivalued operator $\mathcal{F} : C(J, \mathbb{R}^n) \to \mathcal{P}(L^1(J, \mathbb{R}^n))$ by letting $\mathcal{F}(u) = S_{F,u}$. We say F is l.s.c. if $\mathcal{F}$ is l.s.c. and has nonempty closed and decomposable values.

The above operator $\mathcal{F}$ is called the Niemytzki operator associated to F.

Definition 2.25. Let X be a separable metric space and $N : X \to \mathcal{P}(L^1(J, \mathbb{R}^n))$ be a multivalued operator. We say N has property (BC) if

1. N is lower semicontinuous (l.s.c.)
2. N has nonempty closed and decomposable values

Next we state a selection theorem due to Bressan and Colombo.

Lemma 2.26 (Bressan-Colombo [82]). *Let X be a separable metric space and $N : X \to \mathcal{P}(L^1(J, \mathbb{R}^n))$ be a multivalued operator which has property (BC). Then N has a continuous selection , i.e., there exists a continuous function (single-valued) $g : X \to L^1(J, \mathbb{R}^n)$ such that $g(u) \in N(u)$ for every $u \in X$.*

Let (X, d) be a metric space induced from the normed space $(X, \|\cdot\|)$. Consider $H_d : \mathcal{P}(X) \times \mathcal{P}(X) \longrightarrow \mathbb{R}_+ \cup \{\infty\}$ given by

$$H_d(A, B) = \max\left\{\sup_{a\in A} d(a, B), \sup_{b\in B} d(A, b)\right\},$$

where $d(A, b) = \inf_{a\in A} d(a, b)$, $d(a, B) = \inf_{b\in B} d(a, b)$. Then $(\mathcal{P}_{b,cl}(X), H_d)$ is a metric space and $(\mathcal{P}_{cl}(X), H_d)$ is a generalized metric space (see [169]).

Definition 2.27. A multivalued operator $N : X \to \mathcal{P}_{cl}(X)$ is called

(a) γ-Lipschitz if and only if there exists $\gamma > 0$ such that

$$H_d(N(u), N(v)) \leq \gamma d(u, v) \text{ for each } u, v \in X,$$

(b) A contraction if and only if it is γ-Lipschitz with $\gamma < 1$.

Now, we recall some basic definitions and facts on the theory of Banach algebras. Let X be a Banach algebra with norm $\|\cdot\|$.

Definition 2.28. An operator $T : X \to X$ is called *compact* if $T(X)$ is a relatively compact bounded subset of X. Similarly $T : X \to X$ is called *totally bounded* if T maps a bounded subset of X into the relatively compact subset of X. Finally $T : X \to X$ is called *completely continuous* operator if it is continuous and totally bounded operator on X.

It is clear that every compact operator is totally bounded, but the converse may not be true.

Definition 2.29. A function $f : J \times \mathbb{R} \to \mathbb{R}$ is called *Chandrabhan* if

(i) The function $(x, y) \to f(x, y, z)$ is measurable for each $z \in \mathbb{R}$
(ii) The function $z \to f(x, y, z)$ is nondecreasing for almost each $(x, y) \in J$

Definition 2.30. A nonempty closed set K in a Banach algebra X is called a *cone* if

(i) $K + K \subseteq K$
(ii) $\lambda K \subseteq K$ for $\lambda \in \mathbb{R}$, $\lambda \geq 0$
(iii) $\{-K\} \cap K = 0$, where 0 is the zero element of X

The cone K is called to be *positive* if

(iv) $K \circ K \subseteq K$, where "$\circ$" is a multiplication composition in X

We introduce an order relation $\leq$ in X as follows. Let $u, v \in X$. Then $u \leq v$ if and only if $v - u \in K$. A cone K is said to be *normal* if the norm $\|\cdot\|$ is monotone increasing on K. It is known that if the cone K is normal in X, then every order-bounded set in X is norm-bounded. For any $v, w \in X, v \leq w$, the order interval $[v, w]$ is a set in X given by

$$[v, w] = \{u \in X : v \leq u \leq w\}.$$

Lemma 2.31 (Dhage [108]). *Let K be a positive cone in a real Banach algebra X and let $u_1, u_2, v_1, v_2 \in K$ be such that $u_1 \leq v_1$ and $u_2 \leq v_2$. Then $u_1u_2 \leq v_1v_2$.*

2.4 Fixed Point Theorems

By $\overline{U}$ and ∂U we denote the closure of U and the boundary of U respectively. Let us start by stating a well-known result, the nonlinear alternative.

Theorem 2.32 (Nonlinear alternative of Leray Schauder type [136]). *Let X be a Banach space and C a nonempty convex subset of X. Let U a nonempty open subset of C with $0 \in U$ and $T : \overline{U} \to C$ continuous and compact operator.*
Then either

(a) T has fixed points or
(b) There exist $u \in \partial U$ and $\lambda \in (0, 1)$ with $u = \lambda T(u)$

The multivalued version of nonlinear alternative:

Lemma 2.33 ([136]). *Let X be a Banach space and C a nonempty convex subset of X. Let U a nonempty open subset of C with $0 \in U$ and $T : \overline{U} \to \mathcal{P}(C)$ an upper semicontinuous and compact multivalued operator.*
Then either

(a) T has fixed points or
(b) There exist $u \in \partial U$ and $\lambda \in (0, 1)$ with $u \in \lambda T(u)$

Theorem 2.34 (Schaefer [146]). *Let X be a Banach space and $N : X \to X$ completely continuous operator. If the set*

$$E(N) = \{u \in X : u = \lambda N(u) \text{ for some } \lambda \in (0, 1)\}$$

is bounded, then N has fixed points.

The nonlinear alternative of Schaefer type recently proved by Dhage [107] is embodied in the following theorem.

Theorem 2.35 (Dhage [107]). *Let X be a Banach algebra and let $A, B : X \to X$ be two operators satisfying*

(a) A is Lipschitz with a Lipschitz constant α
(b) B is compact and continuous
(c) $\alpha M < 1$, where $M = \|B(X)\| := \sup\{\|Bz\| : z \in X\}$.

Then either

(i) The equation $\lambda[Au\,Bu] = u$ has a solution for $0 < \lambda < 1$, or
(ii) The set $\mathcal{E} = \{u \in X : \lambda[Au\,Bu] = u,\ 0 < \lambda < 1\}$ is unbounded.

Other single-valued fixed point theorems:

Theorem 2.36 (Burton-Kirk [84]). *Let X be a Banach space, and $A, B : X \to X$ two operators satisfying:*

(i) A is a contraction
(ii) B is completely continuous.

Then either

(a) The operator equation $u = A(u) + B(u)$ has a solution, or
(b) The set $\mathcal{E} = \{u \in X : u = \lambda A(\frac{u}{\lambda}) + \lambda B(u)\}$ is unbounded for $\lambda \in (0, 1)$

Theorem 2.37 (Krasnoselskii [44]). *Let X be a Banach space, let D be a bounded closed convex subset of X and let A, B be maps of D into X such that $Au + Bv \in D$ for every $(u, v) \in D$. If A is a contraction and B is completely continuous, then the equation*
$Aw + Bw = w$ has a solution w on D.

Next we state two multivalued fixed point theorems

Lemma 2.38 (Bohnenblust-Karlin [81]). *Let X be a Banach space and $K \in \mathcal{P}_{cl,cv}(X)$ and suppose that the operator $G : K \to \mathcal{P}_{cl,cv}(K)$ is upper semicontinuous and the set $G(K)$ is relatively compact in X. Then G has a fixed point in K.*

Lemma 2.39 (Covitz–Nadler [96]). *Let (X, d) be a complete metric space. If $N : X \to \mathcal{P}_{cl}(X)$ is a contraction , then N has fixed points.*

We will make use of the following fixed point theorems of Dhage [108] for proving the existence of extremal solutions for our problems in Sects. 3.5 and 5.4 under certain monotonicity conditions.

Theorem 2.40. *Let K be a cone in a Banach algebra X and let $v, w \in X$. Suppose that $A, B : [v, w] \to K$ are two operators such that*

(a) A is completely continuous
(b) B is totally bounded
(c) $Au_1 Bu_2 \in [v, w]$ for all $u_1, u_2 \in [v, w]$
(d) A and B are nondecreasing

Further if the cone K is positive and normal, then the operator equation $Au\,Bu = u$ has a least and a greatest positive solution in $[v, w]$.

Theorem 2.41. *Let K be a cone in a Banach algebra X and let $v, w \in X$. Suppose that $A, B : [v, w] \to K$ are two operators such that*

(a) A is Lipschitz with a Lipschitz constant α
(b) B is totally bounded
(c) $Au_1\,Bu_2 \in [v, w]$ for all $u_1, u_2 \in [v, w]$
(d) A and B are nondecreasing

Further if the cone K is positive and normal, then the operator equation $Au\,Bu = u$ has a least and a greatest positive solution in $[v, w]$, whenever $\alpha M < 1$, where $M = \|B([v, w])\| := \sup\{\|Bu\| : u \in [v, w]\}$.

Remark 2.42. Note that hypothesis (c) of Theorems 2.40 and 2.41 holds if the operators A and B are positive monotone increasing and there exist elements v and w in X such that $v \leq Av\,Bv$ and $Aw\,Bw \leq w$.

2.5 Gronwall Lemmas

In the sequel we will make use of the following generalizations of Gronwall's lemmas for two independent variables and singular kernel.

Lemma 2.43 ([142]). *Let $\upsilon : J \to [0, \infty)$ be a real function and $\omega(.,.)$ be a nonnegative, locally integrable function on J. If there are constants $c > 0$ and $0 < r_1, r_2 < 1$ such that*

$$\upsilon(x, y) \leq \omega(x, y) + c \int_0^x \int_0^y \frac{\upsilon(s, t)}{(x - s)^{r_1}(y - t)^{r_2}} dt ds,$$

then there exists a constant $\delta = \delta(r_1, r_2)$ such that

$$\upsilon(x, y) \leq \omega(x, y) + \delta c \int_0^x \int_0^y \frac{\omega(s, t)}{(x - s)^{r_1}(y - t)^{r_2}} dt ds,$$

for every $(x, y) \in J$.

Lemma 2.44 ([142]). *Let $\upsilon, \omega : J \to [0, \infty)$ be nonnegative, locally integrable functions on J. If there are constants $c > 0$ and $0 < r_1, r_2 < 1$ such that*

$$\upsilon(x, y) \leq \omega(x, y) + c \int_0^x \int_0^y \frac{\upsilon(s, t)}{(x - s)^{1-r_1}(y - t)^{1-r_2}} dt ds,$$

then, for every $(x, y) \in J$,

$$\upsilon(x, y) \leq \omega(x, y) + \int_0^x \int_0^y \sum_{j=1}^{\infty} \frac{\left(c\Gamma(r_1)\Gamma(r_2)\right)^j \omega(s, t)\mathrm{d}t\mathrm{d}s}{\Gamma(jr_1)\Gamma(jr_2)(x-s)^{1-jr_1}(y-t)^{1-jr_2}}.$$

If $\omega(x, y) = \omega$ *constant on* J, *then the inequality (3.8) is reduced to*

$$\upsilon(x, y) \leq \omega E_{(r_1,r_2)}\left(c\Gamma(r_1)\Gamma(r_2)x^{r_1}y^{r_2}\right),$$

where $E_{(r_1,r_2)}$ *is the Mittag-Leffler function [166], defined by*

$$E_{(r_1,r_2)}(z) := \sum_{k=1}^{\infty} \frac{z^k}{\Gamma(kr_1+1)\Gamma(kr_2+1)}; \quad r_j, z \in \mathbb{C},\ \Re e(r_j) > 0;\ j = 1, 2.$$

Chapter 3
Partial Hyperbolic Functional Differential Equations

3.1 Introduction

In this chapter, we shall present existence results for some classes of IVP for partial hyperbolic differential equations with fractional order.

3.2 Partial Hyperbolic Differential Equations

3.2.1 Introduction

This section is concerned with the existence and uniqueness of solutions to fractional order IVP, for the system

$$({}^cD_0^r u)(x,y) = f(x,y,u_{(x,y)}), \text{ if } (x,y) \in J := [0,a] \times [0,b], \tag{3.1}$$

$$u(x,y) = \phi(x,y), \text{ if } (x,y) \in \tilde{J} := [-\alpha,a] \times [-\beta,b] \backslash (0,a] \times (0,b], \tag{3.2}$$

$$u(x,0) = \varphi(x),\ x \in [0,a],\ u(0,y) = \psi(y),\ y \in [0,b], \tag{3.3}$$

where $\alpha, \beta, a, b > 0$, ${}^cD_0^r$ is the standard Caputo's fractional derivative of order $r = (r_1, r_2) \in (0,1] \times (0,1]$, $f : J \times C \to \mathbb{R}^n$ is a given function, $\phi : \tilde{J} \to \mathbb{R}^n$ is a given continuous function, $\varphi : [0,a] \to \mathbb{R}^n, \psi : [0,b] \to \mathbb{R}^n$ are given absolutely continuous functions with $\varphi(x) = \phi(x,0)$, $\psi(y) = \phi(0,y)$ for each $x \in [0,a]$, $y \in [0,b]$ and $C := C([-\alpha,0] \times [-\beta,0], \mathbb{R}^n)$ is the space of continuous functions on $[-\alpha,0] \times [-\beta,0]$.

Next we consider the following nonlocal initial value problem

$$({}^cD_0^r u)(x,y) = f(x,y,u_{(x,y)}), \text{ if } (x,y) \in J, \tag{3.4}$$

S. Abbas et al., *Topics in Fractional Differential Equations*, Developments in Mathematics 27, DOI 10.1007/978-1-4614-4036-9_3,

$$u(x,y) = \phi(x,y), \text{ if } (x,y) \in \tilde{J}, \tag{3.5}$$

$$u(x,0) + Q(u) = \varphi(x), \; x \in [0,a], \; u(0,y) + K(u) = \psi(y), \; y \in [0,b], \tag{3.6}$$

where f, ϕ, φ, ψ are as in problem (3.1)–(3.3) and $Q, K : C(J, \mathbb{R}^n) \to \mathbb{R}^n$ are continuous functions.

3.2.2 *Existence of Solutions*

Let us start by defining what we mean by a solution of the problem (3.1)–(3.3).

Definition 3.1. A function $u \in C_{(a,b)} := C([-\alpha,a] \times [-\beta,b], \mathbb{R}^n)$ whose mixed derivative D^2_{xy} exists and is integrable is said to be a solution of (3.1)–(3.3) if u satisfies (3.1) and (3.3) on J and the condition (3.2) on $\tilde{J}$.

Further, we present conditions for the existence and uniqueness of a solution of problem (3.1)–(3.3).

Theorem 3.2. *Assume that the following hypotheses hold:*

(3.2.1) $f : J \times C \to \mathbb{R}^n$ *is a continuous function.*
(3.2.2) For any $u, v \in C$ *and* $(x,y) \in J$, *there exists* $k > 0$ *such that*

$$\|f(x,y,u) - f(x,y,v)\| \le k\|u - v\|_C.$$

If

$$\frac{k a^{r_1} b^{r_2}}{\Gamma(r_1+1)\Gamma(r_2+1)} < 1, \tag{3.7}$$

then there exists a unique solution for IVP (3.1)–(3.3) on $[-\alpha,a] \times [-\beta,b]$.

Proof. Transform the problem (3.1)–(3.3) into a fixed-point problem. Consider the operator $N : C_{(a,b)} \to C_{(a,b)}$ defined by,

$$N(u)(x,y) = \begin{cases} \phi(x,y), & (x,y) \in \tilde{J}, \\ \mu(x,y) + \frac{1}{\Gamma(r_1)\Gamma(r_2)} \int\limits_0^x \int\limits_0^y (x-s)^{r_1-1}(y-t)^{r_2-1} & \\ \times f(s,t,u_{(s,t)}) \mathrm{d}t\,\mathrm{d}s, & (x,y) \in J. \end{cases} \tag{3.8}$$

Let $v, w \in C_{(a,b)}$. Then, for $(x, y) \in [-\alpha, a] \times [-\beta, b]$,

$$\begin{aligned}
&\|N(v)(x,y) - N(w)(x,y)\| \\
&\leq \frac{1}{\Gamma(r_1)\Gamma(r_2)} \int_0^x \int_0^y \|f(s,t,v_{(s,t)}) - f(s,t,w_{(s,t)})\| \\
&\quad \times |(x-s)^{r_1-1}||(y-t)^{r_2-1}| \mathrm{d}t\mathrm{d}s \\
&\leq \frac{k}{\Gamma(r_1)\Gamma(r_2)} \int_0^x \int_0^y (x-s)^{r_1-1}(y-t)^{r_2-1} \|v_{(s,t)} - w_{(s,t)}\|_C \mathrm{d}t\mathrm{d}s \\
&\leq \frac{k}{\Gamma(r_1)\Gamma(r_2)} \|v - w\|_J \int_0^x \int_0^y (x-s)^{r_1-1}(y-t)^{r_2-1} \mathrm{d}t\mathrm{d}s \\
&\leq \frac{k x^{r_1} y^{r_2}}{\Gamma(r_1+1)\Gamma(r_2+1)} \|v - w\|_J.
\end{aligned}$$

Consequently,

$$\|N(v) - N(w)\|_J \leq \frac{k a^{r_1} b^{r_2}}{\Gamma(r_1+1)\Gamma(r_2+1)} \|v - w\|_J.$$

By (3.7), N is a contraction, and hence N has a unique fixed point by Banach's contraction principle. □

Theorem 3.3. *Assume that (3.2.1) and the following hypothesis hold:*

(3.3.1) There exist $p, q \in C(J, \mathbb{R}_+)$ such that

$$\|f(x,y,u)\| \leq p(x,y) + q(x,y)\|u\|_C,$$

for $(x, y) \in J$ and each $u \in C$. Then the IVP (3.1)–(3.3) have at least one solution on $[-\alpha, a] \times [-\beta, b]$.

Proof. Transform the problem (3.1)–(3.3) into a fixed point problem. Consider the operator N defined in (3.8). We shall show that the operator N is continuous and completely continuous.

Step 1: *N is continuous.* Let $\{u_n\}$ be a sequence such that $u_n \to u$ in $C_{(a,b)}$. Let $\eta > 0$ be such that $\|u_n\| \leq \eta$. Then

$$\|N(u_n)(x,y) - N(u)(x,y)\|$$

$$\leq \frac{1}{\Gamma(r_1)\Gamma(r_2)} \int_0^x \int_0^y |x-s|^{r_1-1}|y-t|^{r_2-1}\|f(s,t,u_{n(s,t)}) - f(s,t,u_{(s,t)})\|dtds$$

$$\leq \frac{1}{\Gamma(r_1)\Gamma(r_2)} \int_0^a \int_0^b |x-s|^{r_1-1}|y-t|^{r_2-1}$$

$$\times \sup_{(s,t)\in J} \|f(s,t,u_{n(s,t)}) - f(s,t,u_{(s,t)})\|dtds$$

$$\leq \frac{a^{r_1}b^{r_2}\|f(.,.,u_{n(.,.)}) - f(.,.,u_{(.,.)})\|_\infty}{r_1 r_2 \Gamma(r_1)\Gamma(r_2)}.$$

Since f is a continuous function, we have

$$\|N(u_n) - N(u)\|_\infty \leq \frac{a^{r_1}b^{r_2}\|f(.,.,u_{n(.,.)}) - f(.,.,u_{(.,.)})\|_\infty}{\Gamma(r_1+1)\Gamma(r_2+1)} \to 0 \text{ as } n \to \infty.$$

Step 2: *N maps bounded sets into bounded sets in $C_{(a,b)}$.* Indeed, it is enough to show that, for any $\eta^* > 0$, there exists a positive constant $\widetilde{\ell}$ such that, for each $u \in B_{\eta^*} = \{u \in C_{(a,b)} : \|u\|_\infty \leq \eta^*\}$, we have $\|N(u)\|_\infty \leq \widetilde{\ell}$. By (H_3) we have for each $(x,y) \in J$,

$$\|N(u)(x,y)\|$$

$$\leq \|\mu(x,y)\| + \frac{1}{\Gamma(r_1)\Gamma(r_2)} \int_0^x \int_0^y (x-s)^{r_1-1}(y-t)^{r_2-1}\|f(s,t,u_{(s,t)})\|dtds$$

$$\leq \|\mu(x,y)\| + \frac{1}{\Gamma(r_1)\Gamma(r_2)} \left\| \int_0^x \int_0^y (x-s)^{r_1-1}(y-t)^{r_2-1}p(s,t)dtds \right\|$$

$$+ \frac{1}{\Gamma(r_1)\Gamma(r_2)} \int_0^x \int_0^y (x-s)^{r_1-1}(y-t)^{r_2-1}q(s,t)\|u_{(s,t)}\|_\infty dtds$$

$$\leq \|\mu(x,y)\| + \frac{\|p\|_\infty}{\Gamma(r_1)\Gamma(r_2)} \int_0^x \int_0^y (x-s)^{r_1-1}(y-t)^{r_2-1}dtds$$

$$+ \frac{\|q\|_\infty \eta^*}{\Gamma(r_1)\Gamma(r_2)} \int_0^x \int_0^y (x-s)^{r_1-1}(y-t)^{r_2-1}dtds.$$

Thus

$$\|N(u)\|_\infty \leq \|\mu\|_\infty + \frac{\|p\|_\infty + \|q\|_\infty \eta^*}{\Gamma(r_1+1)\Gamma(r_2+1)} a^{r_1}b^{r_2} := \widetilde{\ell}.$$

Step 3: *N maps bounded sets into equicontinuous sets in $C_{(a,b)}$.* Let (x_1, y_1), $(x_2, y_2) \in (0,a] \times (0,b]$, $x_1 < x_2$, $y_1 < y_2$, B_{η^*} be a bounded set of $C_{(a,b)}$ as in Step 2, and let $u \in B_{\eta^*}$. Then

$$\|N(u)(x_2, y_2) - N(u)(x_1, y_1)\| = \|\mu(x_1, y_1) - \mu(x_2, y_2)\|$$

$$+\frac{1}{\Gamma(r_1)\Gamma(r_2)}\Bigg\| \int_0^{x_1}\int_0^{y_1} [(x_2 - s)^{r_1-1}(y_2 - t)^{r_2-1} - (x_1 - s)^{r_1-1}(y_1 - t)^{r_2-1}]$$

$$\times f(s,t,u_{(s,t)})\mathrm{d}t\,\mathrm{d}s+\frac{1}{\Gamma(r_1)\Gamma(r_2)}\int_{x_1}^{x_2}\int_{y_1}^{y_2}(x_2-s)^{r_1-1}(y_2-t)^{r_2-1} f(s,t,u_{(s,t)})\mathrm{d}t\,\mathrm{d}s$$

$$+\frac{1}{\Gamma(r_1)\Gamma(r_2)}\int_0^{x_1}\int_{y_1}^{y_2}(x_2 - s)^{r_1-1}(y_2 - t)^{r_2-1} f(s,t,u_{(s,t)})\mathrm{d}t\,\mathrm{d}s$$

$$+\frac{1}{\Gamma(r_1)\Gamma(r_2)}\int_{x_1}^{x_2}\int_0^{y_1}(x_2 - s)^{r_1-1}(y_2 - t)^{r_2-1} f(s,t,u_{(s,t)})\mathrm{d}t\,\mathrm{d}s\Bigg\|$$

$$\leq \|\mu(x_1, y_1) - \mu(x_2, y_2)\| + \frac{\|p\|_\infty + \|q\|_\infty\eta^*}{\Gamma(r_1)\Gamma(r_2)}$$

$$\times\int_0^{x_1}\int_0^{y_1} [(x_1 - s)^{r_1-1}(y_1 - t)^{r_2-1} - (x_2 - s)^{r_1-1}(y_2 - t)^{r_2-1}]\mathrm{d}t\,\mathrm{d}s$$

$$+\frac{\|p\|_\infty + \|q\|_\infty\eta^*}{\Gamma(r_1)\Gamma(r_2)}\int_{x_1}^{x_2}\int_{y_1}^{y_2}(x_2 - s)^{r_1-1}(y_2 - t)^{r_2-1}\mathrm{d}t\,\mathrm{d}s$$

$$+\frac{\|p\|_\infty + \|q\|_\infty\eta^*}{\Gamma(r_1)\Gamma(r_2)}\int_0^{x_1}\int_{y_1}^{y_2}(x_2 - s)^{r_1-1}(y_2 - t)^{r_2-1}\mathrm{d}t\,\mathrm{d}s$$

$$+\frac{\|p\|_\infty + \|q\|_\infty\eta^*}{\Gamma(r_1)\Gamma(r_2)}\int_{x_1}^{x_2}\int_0^{y_1}(x_2 - s)^{r_1-1}(y_2 - t)^{r_2-1}\mathrm{d}t\,\mathrm{d}s$$

$$\leq \|\mu(x_1, y_1) - \mu(x_2, y_2)\|$$

$$+\frac{\|p\|_\infty + \|q\|_\infty\eta^*}{\Gamma(r_1 + 1)\Gamma(r_2 + 1)}[2y_2^{r_2}(x_2 - x_1)^{r_1} + 2x_2^{r_1}(y_2 - y_1)^{r_2}$$

$$+x_1^{r_1}y_1^{r_2} - x_2^{r_1}y_2^{r_2} - 2(x_2 - x_1)^{r_1}(y_2 - y_1)^{r_2}].$$

As $x_1 \to x_2$, $y_1 \to y_2$ the right-hand side of the above inequality tends to zero. The equicontinuity for the cases $x_1 < x_2 < 0$, $y_1 < y_2 < 0$ and $x_1 \le 0 \le x_2$, $y_1 \le 0 \le y_2$ is obvious. As a consequence of Steps 1–3, together with the Arzela–Ascoli theorem, we can conclude that N is continuous and completely continuous.

Step 4: *A priori bounds.* We now show there exists an open set $U \subseteq C_{(a,b)}$ with $u \ne \lambda N(u)$, for $\lambda \in (0,1)$ and $u \in \partial U$. Let $u \in C_{(a,b)}$ and $u = \lambda N(u)$ for some $0 < \lambda < 1$. Thus for each $(x,y) \in J$,

$$u(x,y) = \lambda\mu(x,y) + \frac{\lambda}{\Gamma(r_1)\Gamma(r_2)} \int_0^x \int_0^y (x-s)^{r_1-1}(y-t)^{r_2-1} f(s,t,u_{(s,t)})dtds.$$

This implies by (3.3.1) that, for each $(x,y) \in J$, we have

$$\begin{aligned} \|u(x,y)\| &\le \|\mu(x,y)\| + \frac{1}{\Gamma(r_1)\Gamma(r_2)} \int_0^x \int_0^y (x-s)^{r_1-1}(y-t)^{r_2-1} \\ &\quad \times [p(s,t) + q(s,t)\|u_{(s,t)}\|_C]dtds \\ &\le \|\mu(x,y)\| + \frac{\|p\|_\infty}{\Gamma(r_1+1)\Gamma(r_2+1)} a^{r_1} b^{r_2} \\ &\quad + \frac{1}{\Gamma(r_1)\Gamma(r_2)} \int_0^x \int_0^y (x-s)^{r_1-1}(y-t)^{r_2-1} q(s,t)\|u_{(s,t)}\|_C dtds. \end{aligned}$$

We consider the function τ defined by

$$\tau(x,y) = \sup\{\|u(s,t)\| : -\alpha \le s \le x,\ -\beta \le t \le y,\ 0 \le x \le a,\ 0 \le y \le b\}.$$

Let $(x^*, y^*) \in [-\alpha, x] \times [-\beta, y]$ be such that $\tau(x,y) = \|u(x^*, y^*)\|$. If $(x^*, y^*) \in J$, then by the previous inequality, we have for $(x,y) \in J$,

$$\begin{aligned} \tau(x,y) &\le \|\mu(x,y)\| + \frac{\|p\|_\infty}{\Gamma(r_1+1)\Gamma(r_2+1)} a^{r_1} b^{r_2} \\ &\quad + \frac{1}{\Gamma(r_1)\Gamma(r_2)} \int_0^x \int_0^y (x-s)^{r_1-1}(y-t)^{r_2-1} q(s,t)\tau(s,t)dtds \\ &\le \|\mu(x,y)\| + \frac{\|p\|_\infty}{\Gamma(r_1+1)\Gamma(r_2+1)} a^{r_1} b^{r_2} \\ &\quad + \frac{\|q\|_\infty}{\Gamma(r_1)\Gamma(r_2)} \int_0^x \int_0^y (x-s)^{r_1-1}(y-t)^{r_2-1} \tau(s,t)dtds. \end{aligned}$$

If $(x^*, y^*) \in \tilde{J}$, then $\tau(x,y) = \|\phi\|_C$ and the previous inequality holds.

By Lemma 2.43, there exists a constant $\delta = \delta(r_1, r_2)$ such that

$$\tau(x,y) \le \left[\|\mu\|_\infty + \frac{a^{r_1}b^{r_2}\|p\|_\infty}{\Gamma(r_1+1)\Gamma(r_2+1)}\right] \times \left[1 + \frac{\delta\|q\|_\infty}{\Gamma(r_1)\Gamma(r_2)}\int_0^x\int_0^y (x-s)^{r_1-1}(y-t)^{r_2-1}dtds\right]$$
$$\le \left[\|\mu\|_\infty + \frac{a^{r_1}b^{r_2}\|p\|_\infty}{\Gamma(r_1+1)\Gamma(r_2+1)}\right] \times \left[1 + \frac{\delta a^{r_1}b^{r_2}\|q\|_\infty}{\Gamma(r_1+1)\Gamma(r_2+1)}\right] := M.$$

Since for every $(x, y) \in J$, $\|u_{(x,y)}\|_C \le \tau(x, y)$, we have

$$\|u\|_\infty \le \max(\|\phi\|_C, M) := R.$$

Set

$$U = \{u \in C_{(a,b)} : \|u\|_\infty < R + 1\}.$$

By our choice of U, there is no $u \in \partial U$ such that $u = \lambda N(u)$, for $\lambda \in (0, 1)$. As a consequence of the nonlinear alternative of Leray–Schauder type [136], we deduce that N has a fixed point u in $\overline{U}$ which is a solution to problem (3.1)–(3.3). □

Now we present two similar existence results for the nonlocal problem (3.4)–(3.6).

Definition 3.4. A function $u \in C_{(a,b)}$ is said to be a solution of (3.4)–(3.6) if u satisfies (3.4) and (3.6) on J and the condition (3.5) on $\tilde{J}$.

Theorem 3.5. *Assume that (3.2.1), (3.2.2), and the following conditions,*

(3.5.1) There exists $\tilde{k} > 0$ *such that*

$$\|Q(u) - Q(v)\| \le \tilde{k}\|u - v\|_\infty, \text{ for any } u, v \in C(J, \mathbb{R}^n),$$

(3.5.2) There exists $k^* > 0$ *such that*

$$\|K(u) - K(v)\| \le k^*\|u - v\|_\infty, \text{ for any } u, v \in C(J, \mathbb{R}^n)$$

hold. If

$$\tilde{k} + k^* + \frac{ka^{r_1}b^{r_2}}{\Gamma(r_1+1)\Gamma(r_2+1)} < 1, \tag{3.9}$$

then there exists a unique solution for IVP (3.4)–(3.6) on $[-\alpha, a] \times [-\beta, b]$.

Theorem 3.6. *Assume that (3.2.1), (3.3.1), and the following conditions*

(3.6.1) There exists $\tilde{d} > 0$ *such that*

$$\|Q(u)\| \leq \tilde{d}(1 + \|u\|_\infty), \text{ for any } u \in C(J, \mathbb{R}^n),$$

(3.6.2) There exists $d^* > 0$ *such that*

$$\|K(u)\| \leq d^*(1 + \|u\|_\infty), \text{ for any } u \in C(J, \mathbb{R}^n)$$

hold, then there exists at least one solution for IVP (3.4)–(3.6) on $[-\alpha, a] \times [-\beta, b]$.

3.2.3 An Example

As an application of our results we consider the following partial hyperbolic functional differential equations of the form:

$$({}^c D_0^r u)(x, y) = \frac{1}{(3e^{x+y+2})(1 + |u(x-1, y-2)|)}, \quad \text{if } (x, y) \in [0, 1] \times [0, 1], \tag{3.10}$$

$$u(x, 0) = x, \; u(0, y) = y^2; \; x, y \in [0, 1], \tag{3.11}$$

$$u(x, y) = x + y^2, \; (x, y) \in [-1, 1] \times [-2, 1] \backslash (0, 1] \times (0, 1]. \tag{3.12}$$

Set

$$f(x, y, u_{(x,y)}) = \frac{1}{(3e^{x+y+2})(1 + |u(x-1, y-2)|)}, \quad (x, y) \in [0, 1] \times [0, 1].$$

For each $u, \; \overline{u} \in \mathbb{R}$ and $(x, y) \in [0, 1] \times [0, 1]$ we have

$$|f(x, y, u_{(x,y)}) - f(x, y, \overline{u}_{(x,y)})| \leq \frac{1}{3e^2} \|u - \overline{u}\|_C.$$

Hence condition (3.2.2) is satisfied with $k = \frac{1}{3e^2}$. We shall show that condition (3.7) holds with $a = b = 1$. Indeed

$$\frac{k a^{r_1} b^{r_2}}{\Gamma(r_1 + 1)\Gamma(r_2 + 1)} = \frac{1}{3e^2 \Gamma(r_1 + 1)\Gamma(r_2 + 1)} < 1,$$

which is satisfied for each $(r_1, r_2) \in (0, 1] \times (0, 1]$. Consequently, Theorem 3.2 implies that problem (3.10)–(3.12) has a unique solution defined on $[-1, 1] \times [-2, 1]$.

3.3 Perturbed Partial Differential Equations

3.3.1 Introduction

In this section we discuss the existence of solutions to the Darboux problem for the fractional order IVP, for the system

$$({}^cD_0^r u)(x,y) = f(x,y,u_{(x,y)}) + g(x,y,u_{(x,y)}), \text{ if } (x,y) \in J, \tag{3.13}$$

$$u(x,y) = \phi(x,y), \text{ if } (x,y) \in \tilde{J} := [-\alpha,a]\times[-\beta,b]\backslash(0,a]\times(0,b], \tag{3.14}$$

$$u(x,0) = \varphi(x),\ x \in [0,a],\ u(0,y) = \psi(y),\ y \in [0,b], \tag{3.15}$$

where $J = [0,a]\times[0,b]$, $a,b,\alpha,\beta > 0$, $f,g : J\times C \to \mathbb{R}$ are given functions, $\phi : \tilde{J} \to \mathbb{R}$ is a given continuous function, $\varphi : [0,a] \to \mathbb{R}$, $\psi : [0,b] \to \mathbb{R}$ are given absolutely continuous functions with $\varphi(x) = \phi(x,0)$, $\psi(y) = \phi(0,y)$ for each $x \in [0,a]$, $y \in [0,b]$ and $C := C([-\alpha,0]\times[-\beta,0],\mathbb{R})$. Next we consider the following nonlocal initial value problem:

$$({}^cD_0^r u)(x,y) = f(x,y,u_{(x,y)}) + g(x,y,u_{(x,y)}), \text{ if } (x,y) \in J, \tag{3.16}$$

$$u(x,y) = \phi(x,y), \text{ if } (x,y) \in \tilde{J}, \tag{3.17}$$

$$u(x,0) + Q(u) = \varphi(x),\ x \in [0,a],\ u(0,y) + K(u) = \psi(y),\ y \in [0,b], \tag{3.18}$$

where f,g,ϕ,φ,ψ are as in problem (3.13)–(3.15) and $Q,K : C(J,\mathbb{R}) \to \mathbb{R}$ are continuous functions.

3.3.2 Existence of Solutions

Let us start by defining what we mean by a solution of the problem (3.13)–(3.15).

Definition 3.7. A function $u \in C_{(a,b)} := C([-\alpha,a]\times[-\beta,b],\mathbb{R})$ whose mixed derivative D_{xy}^2 exists and integrable is said to be a solution of (3.13)–(3.15) if u satisfies (3.13) and (3.15) on J and the condition (3.14) on $\tilde{J}$.

Theorem 3.8. *Assume that the following hypotheses*

(3.8.1) The functions $f,\ g : J\times C \to \mathbb{R}$ are continuous,
(3.8.2) There exists $k > 0$ such that for $(x,y) \in J$

$$\|g(x,y,u) - g(x,y,v)\| \le k\|u-v\|_C, \textit{ for any } u,\ v \in C,$$

(3.8.3) There exist $p,q \in C(J,\mathbb{R}_+)$ such that

$$|f(x,y,u)| \le p(x,y) + q(x,y)\|u\|_C, \; for \; (x,y) \in J \; and \; u \in C,$$

hold. If

$$\frac{ka^{r_1}b^{r_2}}{\Gamma(r_1+1)\Gamma(r_2+1)} < 1, \tag{3.19}$$

then the IVP (3.13)–(3.15) have at least one solution on $[-\alpha, a] \times [-\beta, b]$.

Proof. Consider the operators $F, G : C_{(a,b)} \to C_{(a,b)}$ defined by

$$F(u)(x,y) = \begin{cases} \phi(x,y), & (x,y) \in \tilde{J}, \\ \mu(x,y) + \frac{1}{\Gamma(r_1)\Gamma(r_2)} \int_0^x \int_0^y (x-s)^{r_1-1}(y-t)^{r_2-1} & \\ \quad \times f(s,t,u_{(s,t)}) \mathrm{d}t\mathrm{d}s, & (x,y) \in J, \end{cases}$$

and

$$G(u)(x,y) = \begin{cases} 0, & (x,y) \in \tilde{J}, \\ \frac{1}{\Gamma(r_1)\Gamma(r_2)} \int_0^x \int_0^y (x-s)^{r_1-1}(y-t)^{r_2-1} & \\ \quad \times g(s,t,u_{(s,t)}) \mathrm{d}t\mathrm{d}s, & (x,y) \in J. \end{cases}$$

Then the problem of finding the solutions of the IVP (3.13)–(3.15) is reduced to finding the solutions of the operator equation $F(u)(s,t) + G(u)(s,t) = u(s,t)$, $(s,t) \in J$. We shall show that the operators F and G satisfy all the conditions of Theorem 2.36 The proof will be given in several steps.

Step 1: *F is continuous.* Let $\{u_n\}$ be a sequence such that $u_n \to u$ in $C_{(a,b)}$. Let $\eta > 0$ such that $\|u_n\|_\infty \le \eta$ and $\|u\|_\infty \le \eta$. Then

$$\begin{aligned} &|F(u_n)(x,y) - F(u)(x,y)| \\ &\le \frac{1}{\Gamma(r_1)\Gamma(r_2)} \int_0^x \int_0^y |x-s|^{r_1-1}|y-t|^{r_2-1}|f(s,t,u_{n(s,t)}) - f(s,t,u_{(s,t)})| \mathrm{d}t\mathrm{d}s \\ &\le \frac{\|f(\cdot,\cdot,u_{n(\cdot,\cdot)}) - f(\cdot,\cdot,u_{(\cdot,\cdot)})\|_\infty}{\Gamma(r_1)\Gamma(r_2)} \int_0^a \int_0^b |x-s|^{r_1-1}|y-t|^{r_2-1} \mathrm{d}t\mathrm{d}s. \end{aligned}$$

Since f is a continuous function, we have

$$\|N(u_n) - N(u)\|_\infty \le \frac{a^{r_1}b^{r_2}\|f(\cdot,\cdot,u_{n(\cdot,\cdot)}) - f(\cdot,\cdot,u_{(\cdot,\cdot)})\|_\infty}{\Gamma(r_1+1)\Gamma(r_2+1)} \to 0 \text{ as } n \to \infty.$$

Step 2: *F maps bounded sets into bounded sets in $C_{(a,b)}$.* Indeed, it is enough show that, for any $\eta^* > 0$, there exists a positive constant ℓ such that, for each $u \in B_{\eta^*} = \{u \in C_{(a,b)} : \|u\|_\infty \le \eta^*\}$, we have $\|N(u)\|_\infty \le \ell$. By (3.8.3) we have for each $(x, y) \in J$,

$$
\begin{aligned}
|F(u)(x,y)| &\le |\mu(x,y)| + \frac{1}{\Gamma(r_1)\Gamma(r_2)} \int_0^x \int_0^y (x-s)^{r_1-1}(y-t)^{r_2-1} \\
&\quad \times |f(s,t,u_{(s,t)})| \mathrm{d}t\mathrm{d}s \\
&\le |\mu(x,y)| + \frac{\|p\|_\infty + \|q\|_\infty \eta^*}{\Gamma(r_1)\Gamma(r_2)} \int_0^x \int_0^y (x-s)^{r_1-1}(y-t)^{r_2-1} \mathrm{d}t\mathrm{d}s.
\end{aligned}
$$

Thus

$$
\|F(u)\|_\infty \le \|\mu\|_\infty + \frac{\|p\|_\infty + \|q\|_\infty \eta^*}{\Gamma(r_1+1)\Gamma(r_2+1)} a^{r_1} b^{r_2} := \ell.
$$

Step 3: *F maps bounded sets into equicontinuous sets in $C_{(a,b)}$.* Let (x_1, y_1), $(x_2, y_2) \in (0, a] \times (0, b]$, $x_1 < x_2$, $y_1 < y_2$, B_{η^*} be a bounded set of $C_{(a,b)}$ as in Step 2, and let $u \in B_{\eta^*}$. Then

$$
\begin{aligned}
&|F(u)(x_2,y_2) - F(u)(x_1,y_1)| \\
&\le |\mu(x_2,y_2) - \mu(x_1,y_1)| + \frac{1}{\Gamma(r_1)\Gamma(r_2)} \left| \int_0^{x_1} \int_0^{y_1} [(x_2-s)^{r_1-1}(y_2-t)^{r_2-1} \right. \\
&\quad \left. - (x_1-s)^{r_1-1}(y_1-t)^{r_2-1}] f(s,t,u_{(s,t)}) \mathrm{d}t\mathrm{d}s \right| \\
&\quad + \left| \frac{1}{\Gamma(r_1)\Gamma(r_2)} \int_{x_1}^{x_2} \int_{y_1}^{y_2} (x_2-s)^{r_1-1}(y_2-t)^{r_2-1} f(s,t,u_{(s,t)}) \mathrm{d}t\mathrm{d}s \right| \\
&\quad + \left| \frac{1}{\Gamma(r_1)\Gamma(r_2)} \int_0^{x_1} \int_{y_1}^{y_2} (x_2-s)^{r_1-1}(y_2-t)^{r_2-1} f(s,t,u_{(s,t)}) \mathrm{d}t\mathrm{d}s \right| \\
&\quad + \left| \frac{1}{\Gamma(r_1)\Gamma(r_2)} \int_{x_1}^{x_2} \int_0^{y_1} (x_2-s)^{r_1-1}(y_2-t)^{r_2-1} f(s,t,u_{(s,t)}) \mathrm{d}t\mathrm{d}s \right| \\
&\le |\mu(x_2,y_2) - \mu(x_1,y_1)| \\
&\quad + \frac{\|p\|_\infty + \|q\|_\infty \eta^*}{\Gamma(r_1+1)\Gamma(r_2+1)} \left[2y_2^{r_2}(x_2-x_1)^{r_1} + 2x_2^{r_1}(y_2-y_1)^{r_2} \right. \\
&\quad \left. + x_1^{r_1} y_1^{r_2} - x_2^{r_1} y_2^{r_2} - 2(x_2-x_1)^{r_1}(y_2-y_1)^{r_2} \right].
\end{aligned}
$$

As $x_1 \to x_2$, $y_1 \to y_2$ the right-hand side of the above inequality tends to zero. The equicontinuity for the cases $x_1 < x_2 < 0$, $y_1 < y_2 < 0$ and $x_1 \le 0 \le x_2$, $y_1 \le 0 \le y_2$ is obvious. As a consequence of Steps 1–3, together with the Arzela–Ascoli theorem, we can conclude that $F : C_{(a,b)} \to C_{(a,b)}$ is continuous and completely continuous.

Step 4: *G is a contraction.* Let $u, v \in C_{(a,b)}$. Then, for each $(x, y) \in J$ we, have

$$\begin{aligned}\|G(u)(x,y) - G(v)(x,y)\| &\le \frac{1}{\Gamma(r_1)\Gamma(r_2)} \int_0^x \int_0^y (x-s)^{r_1-1}(y-t)^{r_2-1} \\ &\quad \times |g(s,t,u_{(s,t)}) - g(s,t,v_{(s,t)})| dt ds \\ &\le \frac{k\|u-v\|_\infty}{r_1 r_2 \Gamma(r_1)\Gamma(r_2)} \int_0^x \int_0^y (x-s)^{r_1-1}(y-t)^{r_2-1} dt ds.\end{aligned}$$

Thus,

$$\|G(u) - G(v)\|_\infty \le \frac{k a^{r_1} b^{r_2}}{\Gamma(r_1+1)\Gamma(r_2+1)} \|u - v\|_\infty,$$

Since by (3.19), G is a contraction.

Step 5: *A priori bounds.* Now it remains to show that the set

$$\mathcal{E} = \left\{ u \in C(J, \mathbb{R}) : u = \lambda F(u) + \lambda G\left(\frac{u}{\lambda}\right), \quad \text{for some } 0 < \lambda < 1 \right\}$$

is bounded. Let $u \in \mathcal{E}$, then $u = \lambda F(u) + \lambda G(\frac{u}{\lambda})$ for some $0 < \lambda < 1$. Thus, for each $(x, y) \in J$ we have

$$\begin{aligned}u(x,y) &= \lambda \mu(x,y) + \frac{\lambda}{\Gamma(r_1)\Gamma(r_2)} \int_0^x \int_0^y (x-s)^{r_1-1}(y-t)^{r_2-1} f(s,t,u_{(s,t)}) dt ds \\ &\quad + \frac{\lambda}{\Gamma(r_1)\Gamma(r_2)} \int_0^x \int_0^y (x-s)^{r_1-1}(y-t)^{r_2-1} g\left(s,t,\frac{u_{(s,t)}}{\lambda}\right) dt ds.\end{aligned}$$

This implies by (3.8.2) and (3.8.3) that, for each $(x, y) \in J$, we have

$$
\begin{aligned}
|u(x,y)| &\le |\mu(x,y)| + \frac{\|p\|_\infty}{\Gamma(r_1)\Gamma(r_2)} \int_0^x \int_0^y (x-s)^{r_1-1}(y-t)^{r_2-1} dt ds \\
&+ \frac{\|q\|_\infty}{\Gamma(r_1)\Gamma(r_2)} \int_0^x \int_0^y (x-s)^{r_1-1}(y-t)^{r_2-1} \|u_{(s,t)}\|_C dt ds \\
&+ \frac{\lambda}{\Gamma(r_1)\Gamma(r_2)} \int_0^x \int_0^y (x-s)^{r_1-1}(y-t)^{r_2-1} \\
&\times \left| g\left(s,t,\frac{u_{(s,t)}}{\lambda}\right) - g(s,t,0) \right| dt ds \\
&+ \frac{\lambda}{\Gamma(r_1)\Gamma(r_2)} \int_0^x \int_0^y (x-s)^{r_1-1}(y-t)^{r_2-1} |g(s,t,0)| dt ds \\
&\le |\mu(x,y)| + \frac{\|p\|_\infty}{\Gamma(r_1)\Gamma(r_2)} \int_0^x \int_0^y (x-s)^{r_1-1}(y-t)^{r_2-1} dt ds \\
&+ \frac{\|q\|_\infty}{\Gamma(r_1)\Gamma(r_2)} \int_0^x \int_0^y (x-s)^{r_1-1}(y-t)^{r_2-1} \|u_{(s,t)}\|_C dt ds \\
&+ \frac{k}{\Gamma(r_1)\Gamma(r_2)} \int_0^x \int_0^y (x-s)^{r_1-1}(y-t)^{r_2-1} \|u_{(s,t)}\|_C dt ds \\
&+ \frac{a^{r_1} b^{r_2} g^*}{\Gamma(r_1+1)\Gamma(r_2+1)},
\end{aligned}
$$

where $g^* = \sup_{(s,t)\in J} \|g(s,t,0)\|$. We consider the function τ defined by

$$
\tau(x,y) = \sup\{|u(s,t)| : 0 \le s \le x,\ 0 \le t \le y,\ 0 \le x \le a,\ 0 \le y \le b\}.
$$

Let $(x^*, y^*) \in [-\alpha, x] \times [-\beta, y]$ be such that $\tau(x,y) = |u(x^*, y^*)|$. If $(x^*, y^*) \in J$, then by the previous inequality, we have for $(x,y) \in J$

$$
\begin{aligned}
\tau(x,y) &\le |\mu(x,y)| + \frac{a^{r_1} b^{r_2}(\|p\|_\infty + g^*)}{\Gamma(r_1+1)\Gamma(r_2+1)} \\
&+ \frac{\|q\|_\infty + k}{\Gamma(r_1)\Gamma(r_2)} \int_0^x \int_0^y (x-s)^{r_1-1}(y-t)^{r_2-1} \tau(s,t) dt ds.
\end{aligned}
$$

If $(x^*, y^*) \in \tilde{J}$, then $\tau(x, y) = \|\phi\|_C$ and the previous inequality holds. If $(x, y) \in J$, Lemma 2.43 implies that there exists $\widetilde{k} = \widetilde{k}(r_1, r_2)$ such that

$$\tau(x, y) \leq \left(|\mu(x, y)| + \frac{a^{r_1} b^{r_2}(\|p\|_\infty + g^*)}{\Gamma(r_1 + 1)\Gamma(r_2 + 1)}\right)$$

$$\times \left(1 + \widetilde{k}\frac{\|q\|_\infty + k}{\Gamma(r_1)\Gamma(r_2)} \int_0^x \int_0^y (x - s)^{r_1 - 1}(y - t)^{r_2 - 1} dt ds\right)$$

$$\leq \left(|\mu(x, y)| + \frac{a^{r_1} b^{r_2}(\|p\|_\infty + g^*)}{\Gamma(r_1+1)\Gamma(r_2+1)}\right)\left(1 + \widetilde{k}\frac{(\|q\|_\infty + k)a^{r_1} b^{r_2}}{\Gamma(r_1+1)\Gamma(r_2+1)}\right) := \widetilde{R}.$$

Since for every $(x, y) \in J$, $\|u_{(x,y)}\|_C \leq \tau(x, y)$, we have

$$\|u\|_\infty \leq \max(\|\phi\|_C, \widetilde{R}) := A.$$

This shows that the set $\mathcal{E}$ is bounded. As a consequence of Theorem 2.36, we deduce that $F + G$ has a fixed point u which is a solution to problem (3.13)–(3.15). □

Now we present (without proof) two existence results for the nonlocal problem (3.16)–(3.18).

Definition 3.9. A function $u \in C_{(a,b)}$ is said to be a solution of (3.16)–(3.18) if u satisfies equations (3.16) and (3.18) on J and the condition (3.17) on $\tilde{J}$.

Theorem 3.10. *Assume (3.8.1)–(3.8.3) and the following conditions*

(3.10.1) There exists $\tilde{k} > 0$ such that

$$|Q(u) - Q(v)| \leq \tilde{k}\|u - v\|_\infty, \textit{ for any } u, v \in C(J, \mathbb{R}),$$

(3.10.2) There exists $k^* > 0$ such that

$$|K(u) - K(v)| \leq k^*\|u - v\|_\infty, \textit{ for any } u, v \in C(J, \mathbb{R})$$

hold. If

$$\tilde{k} + k^* + \frac{k a^{r_1} b^{r_2}}{\Gamma(r_1 + 1)\Gamma(r_2 + 1)} < 1,$$

then there exists at least one solution for IVP (3.16)–(3.18) on $[-\alpha, a] \times [-\beta, b]$.

Theorem 3.11. *Assume (3.8.1)–(3.8.3) and the following conditions*

(3.11.1) There exists $\tilde{d} > 0$ such that

$$|Q(u)| \leq \tilde{d}(1 + \|u\|_\infty), \textit{ for any } u \in C(J, \mathbb{R}),$$

(3.11.2) There exists $d^* > 0$ *such that*

$$|K(u)| \leq d^*(1 + \|u\|_\infty), \text{ for any } u \in C(J, \mathbb{R})$$

hold. If (3.19) is satisfied, then there exists at least one solution for IVP (3.16)–(3.18) on $[-\alpha, a] \times [-\beta, b]$.

3.3.3 An Example

As an application of our results we consider the following partial perturbed functional differential equations of the form

$$({}^c D_0^r u)(x, y) = \frac{1 + 3e^{x+y+2}(|u(x-1, y-2)| + 2)}{3e^{x+y+2}(1 + |u(x-1, y-2)|)}, \quad \text{if } (x, y) \in [0, 1] \times [0, 1], \tag{3.20}$$

$$u(x, 0) = x, \; u(0, y) = y^2; \; x, y \in [0, 1], \tag{3.21}$$

$$u(x, y) = x + y^2, \; (x, y) \in [-1, 1] \times [-2, 1] \backslash (0, 1] \times (0, 1]. \tag{3.22}$$

Set

$$f(x, y, u_{(x,y)}) = \frac{|u(x-1, y-2)| + 2}{1 + |u(x-1, y-2)|}, \quad (x, y) \in [0, 1] \times [0, 1],$$

and

$$g(x, y, u_{(x,y)}) = \frac{1}{3e^{x+y+2}(1 + |u(x-1, y-2)|)}, \quad (x, y) \in [0, 1] \times [0, 1].$$

For each $u, \overline{u} \in \mathbb{R}$ and $(x, y) \in [0, 1] \times [0, 1]$ we have

$$|g(x, y, u_{(x,y)}) - g(x, y, \overline{u}_{(x,y)})| \leq \frac{1}{3e^2} \|u - \overline{u}\|_C.$$

Hence condition (3.8.2) is satisfied with $k = \dfrac{1}{3e^2}$. We shall show that condition (3.19) holds with $a = b = 1$. Indeed

$$\frac{k a^{r_1} b^{r_2}}{\Gamma(r_1 + 1)\Gamma(r_2 + 1)} = \frac{1}{3e^2 \Gamma(r_1 + 1)\Gamma(r_2 + 1)} < 1,$$

which is satisfied for each $(r_1, r_2) \in (0, 1] \times (0, 1]$. Also, the function f is continuous on $[0, 1] \times [0, 1] \times [0, \infty)$ and

$$|f(x, y, w)| \leq |w| + 2, \quad \text{for each } (x, y, w) \in [0, 1] \times [0, 1] \times [0, \infty).$$

Thus conditions (3.8.1) and (3.8.3) hold. Consequently, Theorem 3.8 implies that problems (3.20)–(3.22) have at least one solution defined on $[-1,1]\times[-2,1]$.

3.4 Neutral Partial Differential Equations

3.4.1 Introduction

This section is concerned with the existence of solutions of fractional order IVP for the system

$$^{c}D_0^r\Big[u(x,y)-g(x,y,u_{(x,y)})\Big]=f(x,y,u_{(x,y)});\ \text{if } (x,y)\in J:=[0,a]\times[0,b], \tag{3.23}$$

$$u(x,y)=\phi(x,y);\ \text{if } (x,y)\in\tilde{J}:=[-\alpha,a]\times[-\beta,b]\backslash(0,a]\times(0,b], \tag{3.24}$$

$$u(x,0)=\varphi(x),\ x\in[0,a],\ u(0,y)=\psi(y),\ y\in[0,b], \tag{3.25}$$

where $a,b,\alpha,\beta>0,\ f,g:J\times C\to\mathbb{R}^n$ are given functions, ϕ,φ,ψ are as in problem (3.1)–(3.3) and $C:=C([-\alpha,0]\times[-\beta,0],\mathbb{R}^n)$. We present some existing results for the problem (3.23)–(3.25) using Krasnoselskii's fixed point theorem.

3.4.2 Existence of Solutions

Let us start by defining what we mean by a solution of the problem (3.23)–(3.25).

Definition 3.12. A function $u\in C_{(a,b)}=C([-\alpha,a]\times[-\beta,b],\mathbb{R}^n)$ is said to be a solution of (3.1)–(3.3) if u satisfies equations (3.23) and (3.25) on J and the condition (3.24) on $\tilde{J}$.

For each $\delta>0$ we consider the following set,

$$A_\delta=\left\{u\in C_{(a,b)}:u|_{\tilde{J}}=\phi\ \text{and}\ \sup_{(x,y)\in J}\|u(x,y)-\mu(x,y)\|\le\delta\right\}.$$

Theorem 3.13. *Assume that the following hypotheses*

(3.13.1) *The function* $(x,y)\mapsto f(x,y,u)$ *is measurable on* J *for each* $u\in C$,
(3.13.2) *The function* $u\mapsto f(x,y,u)$ *is continuous on* C *for a.e.* $(x,y)\in J$,
(3.13.3) *There exist* $0<r_3<\min\{r_1,r_2\}$ *and a real-valued function* $m(x,y)\in L^{\frac{1}{r_3}}(J)$ *such that*

$$\|f(x,y,u_{(x,y)})\|\le m(x,y),\ \textit{for any } u\in A_\delta \textit{ and for } (x,y)\in J,$$

(3.13.4) *For any* $u \in A_\delta$; $g(x, y, u_{(x,y)}) = g_1(x, y, u_{(x,y)}) + g_2(x, y, u_{(x,y)})$,
(3.13.5) g_1 *is continuous and for any* $u, v \in A_\delta$ *and* $(x, y) \in J$, *we have*

$$\|g_1(x, y, u) - g_1(x, y, v)\| \leq l\|u - v\|_\infty, \quad \text{where } 0 < l < \frac{1}{4},$$

(3.13.6) g_2 *is completely continuous and for any bounded set* $\Lambda \in A_\delta$, *the set* $\{(x, y) \to g_2(x, y, u_{(x,y)}), u \in \Lambda\}$ *is equicontinuous in* $C(J, \mathbb{R}^n)$

are satisfied. Then the IVP (3.23)–(3.25) have at least one solution on $[-\alpha, a] \times [-\beta, b]$.

Proof. First, it is easy to obtain that f is Lebesgue measurable on J according to conditions (3.13.1) and (3.13.2). A direct calculation gives that $(x - s)^{r_1-1}(y - t)^{r_2-1} \in L^{\frac{1}{1-r_3}}([0, x] \times [0, y])$, for $(x, y) \in J$. In the light of the Hölder inequality and (3.13.3), we obtain that $(x-s)^{r_1-1}(y-t)^{r_2-1}f(s, t, u_{(s,t)})$ is Lebesgue integrable with respect to $(s, t) \in [0, x] \times [0, y]$ for all $(x, y) \in J$ and $u \in A_\delta$, and

$$\int_0^x\int_0^y (x - s)^{r_1-1}(y - t)^{r_2-1}\|f(s, t, u_{(s,t)})\|dtds$$
$$\leq \|(x - s)^{r_1-1}(y - t)^{r_2-1}\|_{L^{\frac{1}{1-r_3}}([0,x]\times[0,y])}\|m\|_{L^{\frac{1}{r_3}}(J)}.$$

Let $\widetilde{\phi} \in A_\delta$ be defined as $\widetilde{\phi}|_{\tilde{J}} = \phi$, $\widetilde{\phi}(x, y) = \mu(x, y)$ for all $(x, y) \in J$. If u is a solution of the IVP (3.23)–(3.25), let $u(x, y) = \widetilde{\phi}(x, y) + v(x, y)$, $(x, y) \in [-\alpha, a] \times [-\beta, b]$, then we have $u_{(x,y)} = \widetilde{\phi}_{(x,y)} + v_{(x,y)}$, $(x, y) \in [-\alpha, a] \times [-\beta, b]$. Thus according to (3.13.4), the function v satisfies the equation

$$\begin{aligned} v(x, y) = {} & \frac{1}{\Gamma(r_1)\Gamma(r_2)}\int_0^x\int_0^y (x - s)^{r_1-1}(y - t)^{r_2-1} \\ & \times f(s, t, \widetilde{\phi}_{(s,t)} + v_{(s,t)})dtds \\ & + g_1(x, y, \widetilde{\phi}_{(x,y)} + v_{(x,y)}) + g_2(x, y, \widetilde{\phi}_{(x,y)} + v_{(x,y)}) \\ & - g_1(x, 0, \widetilde{\phi}_{(x,0)} + v_{(x,0)}) - g_2(x, 0, \widetilde{\phi}_{(x,0)} + v_{(x,0)}) \\ & - g_1(0, y, \widetilde{\phi}_{(0,y)} + v_{(0,y)}) - g_2(0, y, \widetilde{\phi}_{(0,y)} + v_{(0,y)}) \\ & + g_1(0, 0, \phi(0, 0) + v_{(0,0)}) + g_2(0, 0, \phi(0, 0) + v_{(0,0)}); \ (x, y) \in J. \end{aligned} \tag{3.26}$$

Since g_1, g_2 are continuous and u is continuous in (x, y) there exists $a', b' > 0$, when $(x, y) \in [0, a') \times [0, b')$,

$$\|g_1(x, y, \widetilde{\phi}_{(x,y)} + v_{(x,y)}) - g_1(0, 0, \phi(0, 0) + v_{(0,0)})\| \leq \frac{\delta}{7}, \tag{3.27}$$

and

$$\|g_2(x,y,\widetilde{\phi}_{(x,y)}+v_{(x,y)})-g_2(0,0,\phi(0,0)+v_{(0,0)})\|\leq\frac{\delta}{7}. \tag{3.28}$$

Choose

$$(\eta,\iota)=\Big\{(a,b),(a',b'),(a'',b'')\Big\},$$

where

$$(a'',b'')=\left(\left(\frac{\delta\Gamma(r_1)\Gamma(r_2)(1+\rho_1)^{1-r_3}}{7M}\right)^{\frac{1}{(1+\rho_1)(1-r_3)}},\right.$$

$$\left.\left(\frac{\delta\Gamma(r_1)\Gamma(r_2)(1+\rho_2)^{1-r_3}}{7M}\right)^{\frac{1}{(1+\rho_2)(1-r_3)}}\right),$$

and

$$\rho_1=\frac{r_1-1}{1-r_3},\quad \rho_2=\frac{r_2-1}{1-r_3}\quad\text{and}\quad M=\|m\|_{L^{\frac{1}{r_3}}(J)}.$$

Define E_δ as follows:

$E_\delta=\{v\in C([-\alpha,\eta]\times[-\beta,\iota],\mathbb{R}^n):v(x,y)=0,\ \text{for}\ (x,y)\in\tilde{J}\ \text{and}\ \|v\|\leq\delta\}.$

Then E_δ is a closed bounded and convex subset of $C([-\alpha,\eta]\times[-\beta,\iota],\mathbb{R}^n)$. On E_δ we define the operators U and V as follows:

$$U(v)(x,y)=\begin{cases}0, & (x,y)\in\tilde{J},\\ g_1(x,y,\widetilde{\phi}_{(x,y)}+v_{(x,y)})-g_1(x,0,\widetilde{\phi}_{(x,0)}+v_{(x,0)}) & \\ -g_1(0,y,\widetilde{\phi}_{(0,y)}+v_{(0,y)})+g_1(0,0,\phi(0,0)+v_{(0,0)}) & (x,y)\in[0,\eta]\times[0,\iota],\end{cases}$$

$$V(v)(x,y)=\begin{cases}0, & (x,y)\in\tilde{J},\\ \frac{1}{\Gamma(r_1)\Gamma(r_2)}\int_0^x\int_0^y(x-s)^{r_1-1}(y-t)^{r_2-1} & \\ \times f(s,t,\widetilde{\phi}_{(s,t)}+v_{(s,t)})dt\,ds & \\ +g_2(x,y,\widetilde{\phi}_{(x,y)}+v_{(x,y)})-g_2(x,0,\widetilde{\phi}_{(x,0)}+v_{(x,0)}) & \\ -g_2(0,y,\widetilde{\phi}_{(0,y)}+v_{(0,y)})+g_2(0,0,\phi(0,0)+v_{(0,0)}) & (x,y)\in[0,\eta]\times[0,\iota].\end{cases}$$

It is easy to see that if the operator equation

$$v=Uv+Vv, \tag{3.29}$$

has a solution $v\in E_\delta$ if and only if v is a solution of (3.26). Thus $u(x,y)=\widetilde{\phi}(x,y)+v(x,y)$ is a solution of (3.23)–(3.25) on $[0,\eta]\times[0,\iota]$. Therefore, the existence of a solution of the IVP (3.23)–(3.25) is equivalent to (3.29) having a fixed point in E_δ. Now we show that $U+V$ has a fixed point in E_δ. The proof is divided into three steps.

Step 1: *$Uu+Vv \in E_\delta$ for every $u, v \in E_\delta$.* In fact, for every $u, v \in E_\delta$, $Uu+Vv \in C([-\alpha,\eta]\times[-\beta,\iota],\mathbb{R}^n)$. Also, it is obvious that $(Uu+Vv)(x,y)=0$, $(x,y)\in[-\alpha,\eta]\times[-\beta,\iota]$. Moreover, for $(x,y)\in[0,\eta]\times[0,\iota]$ by (3.26)–(3.28) and the condition (3.13.3) we have

$$\begin{aligned}
&\|Uu(x,y)+Vv(x,y)\|\\
&\le \|g_1(x,y,\widetilde{\phi}_{(x,y)}+u_{(x,y)})-g_1(0,0,\widetilde{\phi}+u)\|\\
&\quad+\|-g_1(x,0,\widetilde{\phi}_{(x,0)}+u_{(x,0)})+g_1(0,0,\phi(0,0)+u_{(0,0)})\|\\
&\quad+\|-g_1(0,y,\widetilde{\phi}_{(0,y)}+u_{(0,y)})+g_1(0,0,\phi(0,0)+u_{(0,0)})\|\\
&\quad+\|g_2(x,y,\widetilde{\phi}_{(x,y)}+v_{(x,y)})-g_2(0,0,\phi(0,0)+v_{(0,0)})\|\\
&\quad+\|-g_2(x,0,\widetilde{\phi}_{(x,0)}+v_{(x,0)})+g_2(0,0,\phi(0,0)+v_{(0,0)})\|\\
&\quad+\|-g_2(0,y,\widetilde{\phi}_{(0,y)}+v_{(0,y)})+g_2(0,0,\phi(0,0)+v_{(0,0)})\|\\
&\quad+\frac{1}{\Gamma(r_1)\Gamma(r_2)}\int_0^x\int_0^y(x-s)^{r_1-1}(y-t)^{r_2-1}\\
&\quad\times\|f(s,t,\widetilde{\phi}_{(s,t)}+v_{(s,t)})\|\mathrm{d}t\mathrm{d}s\\
&\le\frac{1}{\Gamma(r_1)\Gamma(r_2)}\left(\int_0^x\int_0^y(x-s)^{\frac{r_1-1}{1-r_3}}(y-t)^{\frac{r_2-1}{1-r_3}}\mathrm{d}t\mathrm{d}s\right)^{1-r_3}\\
&\quad\times\left(\int_0^x\int_0^y(m(s,t))^{\frac{1}{r_3}}\mathrm{d}t\mathrm{d}s\right)^{r_3}+\frac{6\delta}{7}\\
&\le\frac{1}{\Gamma(r_1)\Gamma(r_2)}\left(\int_0^x\int_0^y(x-s)^{\frac{r_1-1}{1-r_3}}(y-t)^{\frac{r_2-1}{1-r_3}}\mathrm{d}s\mathrm{d}t\right)^{1-r_3}\\
&\quad\times\left(\int_0^a\int_0^b(m(s,t))^{\frac{1}{r_3}}\mathrm{d}t\mathrm{d}s\right)^{r_3}+\frac{6\delta}{7}\\
&\le\frac{6\delta}{7}+\frac{M\eta^{(1+\rho_1)(1-r_3)}\iota^{(1+\rho_2)(1-r_3)}}{\Gamma(r_1)\Gamma(r_2)(1+\rho_1)^{(1-r_3)}(1+\rho_2)^{(1-r_3)}}\\
&\le\delta.
\end{aligned}$$

Therefore,

$$\|Uu+Vv\|=\sup_{(x,y)\in[0,\eta]\times[0,\iota]}\|Uu(x,y)+Vv(x,y)\|\le\delta,$$

which means that $Uu+Vv\in E_\delta$ for every $u,v\in E_\delta$.

Step 2: *U is a contraction on E_δ.* For any $u, \overline{u} \in E_\delta$, $u_{(x,y)} + \widetilde{\phi}_{(x,y)}, \overline{u}_{(x,y)} + \widetilde{\phi}_{(x,y)} \in A_\delta$. So by (3.13.5), we get that

$$\begin{aligned}\|Uu - U\overline{u}\| &= g_1(x,y,\widetilde{\phi}_{(x,y)} + u_{(x,y)}) - g_1(x,y,\widetilde{\phi}_{(x,y)} + \overline{u}_{(x,y)}) \\ &\quad + g_1(x,0,\widetilde{\phi}_{(x,0)} + u_{(x,0)}) - g_1(x,0,\widetilde{\phi}_{(x,0)} + \overline{u}_{(x,0)}) \\ &\quad + g_1(0,y,\widetilde{\phi}_{(0,y)} + u_{(0,y)}) - g_1(0,y,\widetilde{\phi}_{(0,y)} + \overline{u}_{(0,y)}) \\ &\quad + g_1(0,0,\phi(0,0) + u_{(0,0)}) - g_1(0,0,\phi(0,0) + \overline{u}_{(0,0)}) \\ &\le 4l\|u - \overline{u}\|_\infty,\end{aligned}$$

which implies that

$$\|Uu - U\overline{u}\| \le 4l\|u - \overline{u}\|_\infty.$$

In view of $0 < 4l < 1$, U is a contraction on E_δ.

Step 3: *V is a completely continuous operator.* Let

$$V_1(v)(x,y) = \begin{cases} 0, & (x,y) \in \tilde{J}, \\ g_2(x,y,\widetilde{\phi}_{(x,y)} + v_{(x,y)}) - g_2(x,0,\widetilde{\phi}_{(x,0)} + v_{(x,0)}) & \\ -g_2(0,y,\widetilde{\phi}_{(0,y)} + v_{(0,y)}) + g_2(0,0,\phi(0,0) + v_{(0,0)}), & (x,y) \in [0,\eta] \times [0,\iota], \end{cases}$$

$$V_2(v)(x,y) = \begin{cases} 0, & (x,y) \in \tilde{J}, \\ \frac{1}{\Gamma(r_1)\Gamma(r_2)} \int_0^x \int_0^y (x-s)^{r_1-1}(y-t)^{r_2-1} & \\ \times f(s,t,\widetilde{\phi}_{(s,t)} + v_{(s,t)})\mathrm{d}t\mathrm{d}s, & (x,y) \in [0,\eta] \times [0,\iota]. \end{cases}$$

Clearly, $V = V_1 + V_2$. Since g_2 is completely continuous, V_1 is continuous and $\{V_1 v : v \in E_\delta\}$ is uniformly bounded. From the condition that the set $\{(x,y) \to g_2(x,y,u_{(x,y)} : u \in \Lambda\}$ be equicontinuous for any bounded set Λ in A_δ, we can conclude that V_1 is a completely continuous operator. On the other hand, for any $(x,y) \in [0,\eta] \times [0,\iota]$, we have

$$\begin{aligned}\|V_2 v(x,y)\| &\le \frac{1}{\Gamma(r_1)\Gamma(r_2)} \int_0^x \int_0^y (x-s)^{r_1-1}(y-t)^{r_2-1}\|f(s,t,\widetilde{\phi}_{(s,t)} + v_{(s,t)})\|\mathrm{d}t\mathrm{d}s \\ &\le \frac{1}{\Gamma(r_1)\Gamma(r_2)} \left(\int_0^x \int_0^y (x-s)^{\frac{r_1-1}{1-r_3}} (y-t)^{\frac{r_2-1}{1-r_3}} \mathrm{d}t\mathrm{d}s \right)^{1-r_3} \\ &\quad \times \left(\int_0^x \int_0^y (m(s,t))^{\frac{1}{r_3}} \mathrm{d}t\mathrm{d}s \right)^{r_3} \\ &\le \frac{M\eta^{(1+\rho_1)(1-r_3)}\iota^{(1+\rho_2)(1-r_3)}}{\Gamma(r_1)\Gamma(r_2)(1+\rho_1)^{(1-r_3)}(1+\rho_2)^{(1-r_3)}}.\end{aligned}$$

Hence, $\{V_2v : v \in E_\delta\}$ is uniformly bounded. Now, we will prove that $\{V_2v : v \in E_\delta\}$ is equicontinuous. Let $(x_1, y_1), (x_2, y_2) \in (0, \eta] \times (0, \iota]$, $x_1 < x_2, y_1 < y_2$ and let $u \in E_\delta$. Then

$$
\begin{aligned}
\| V_2(v)(x_2, y_2) - V_2(v)(x_1, y_1)\| &\leq \frac{1}{\Gamma(r_1)\Gamma(r_2)} \left\| \int_0^{x_1}\int_0^{y_1} [(x_2 - s)^{r_1-1}(y_2 - t)^{r_2-1} \right. \\
&- (x_1 - s)^{r_1-1}(y_1 - t)^{r_2-1}] f(s, t, \widetilde{\phi}_{(s,t)} + v_{(s,t)}))\mathrm{d}t\mathrm{d}s \\
&+ \left. \frac{1}{\Gamma(r_1)\Gamma(r_2)} \int_{x_1}^{x_2}\int_{y_1}^{y_2} (x_2 - s)^{r_1-1}(y_2 - t)^{r_2-1} f(s, t, \widetilde{\phi}_{(s,t)} + v_{(s,t)})\mathrm{d}t\mathrm{d}s \right\| \\
&+ \frac{1}{\Gamma(r_1)\Gamma(r_2)} \int_0^{x_1}\int_{y_1}^{y_2} (x_2 - s)^{r_1-1}(y_2 - t)^{r_2-1} \| f(s, t, \widetilde{\phi}_{(s,t)} + v_{(s,t)})\|\mathrm{d}t\mathrm{d}s \\
&+ \frac{1}{\Gamma(r_1)\Gamma(r_2)} \int_{x_1}^{x_2}\int_0^{y_1} (x_2 - s)^{r_1-1}(y_2 - t)^{r_2-1} \| f(s, t, \widetilde{\phi}_{(s,t)} + v_{(s,t)})\|\mathrm{d}t\mathrm{d}s \\
&\leq \frac{M}{\Gamma(r_1)\Gamma(r_2)} \left(\int_0^{x_1}\int_0^{y_1} [(x_1 - s)^{\rho_1}(y_1 - t)^{\rho_2} - (x_2 - s)^{\rho_1}(y_2 - t)^{\rho_2}]\mathrm{d}t\mathrm{d}s \right)^{1-r_3} \\
&+ \frac{M}{\Gamma(r_1)\Gamma(r_2)} \left(\int_{x_1}^{x_2}\int_{y_1}^{y_2} (x_2 - s)^{\rho_1}(y_2 - t)^{\rho_2}\mathrm{d}t\mathrm{d}s \right)^{1-r_3} \\
&+ \frac{M}{\Gamma(r_1)\Gamma(r_2)} \left(\int_0^{x_1}\int_{y_1}^{y_2} (x_2 - s)^{\rho_1}(y_2 - t)^{\rho_2}\mathrm{d}t\mathrm{d}s \right)^{1-r_3} \\
&+ \frac{M}{\Gamma(r_1)\Gamma(r_2)} \left(\int_{x_1}^{x_2}\int_0^{y_1} (x_2 - s)^{\rho_1}(y_2 - t)^{\rho_2}\mathrm{d}t\mathrm{d}s \right)^{1-r_3}
\end{aligned}
$$

As $x_1 \to x_2$, $y_1 \to y_2$ the right-hand side of the above inequality tends to zero. The equicontinuity for the cases $x_1 < x_2 < 0$, $y_1 < y_2 < 0$ and $x_1 \leq 0 \leq x_2$, $y_1 \leq 0 \leq y_2$ is obvious, which means that $\{V_2v : v \in E_\delta\}$ is equicontinuous. Moreover, it is clear that V_2 is continuous. So V_2 is a completely continuous operator. Then $V = V_1 + V_2$ is a completely continuous operator. Therefore, Krasnoselskii's fixed point theorem (Theorem 2.37, [44]) shows that $U + V$ has a fixed point on E_δ which is a solution to problem (3.23)–(3.25). □

In the case where $g_1 = 0$, we get the following result.

Theorem 3.14. *Assume that (3.13.1)–(3.13.3) and g satisfy (3.13.6). Then the IVP (3.23)–(3.25) has at least one solution $[-\alpha, a] \times [-\beta, b]$.*

In the case where $g_2 = 0$, we get the following result.

Theorem 3.15. *Assume that (3.13.1)–(3.13.3) and g satisfy (3.13.5). Then the IVP (3.23)–(3.25) have at least one solution $[-\alpha, a] \times [-\beta, b]$.*

3.4.3 An Example

As an application of our results we consider the following partial hyperbolic functional differential equations of the form:

$$
\begin{aligned}
&{}^cD_0^r\left(u(x,y) - e^{x+y}u(x-1,y-2) - \frac{1}{3e^{x+y+2}(1+|u(x-1,y-2)|)}\right)\\
&= \frac{e^{x+y+2}|u(x-1,y-2)|}{1+|u(x-1,y-2)|}, \quad \text{if } (x,y) \in [0,1]\times[0,1], \qquad (3.30)
\end{aligned}
$$

$$u(x,0) = x,\ u(0,y) = y^2;\ x,y \in [0,1], \qquad (3.31)$$

$$u(x,y) = x + y^2,\ (x,y) \in [-1,1]\times[-2,1]\backslash(0,1]\times(0,1]. \qquad (3.32)$$

Set

$$f(x,y,u_{(x,y)}) = \frac{e^{x+y+2}|u(x-1,y-2)|}{1+|u(x-1,y-2)|},\ (x,y)\in[0,1]\times[0,1],$$

$$g(x,y,u_{(x,y)}) = g_1(x,y,u_{(x,y)}) + g_2(x,y,u_{(x,y)}),\ (x,y)\in[0,1]\times[0,1],$$

where

$$g_1(x,y,u_{(x,y)}) = \frac{1}{3e^{x+y+2}(1+|u(x-1,y-2)|)},$$

$$g_2(x,y,u_{(x,y)}) = e^{x+y}u(x-1,y-2).$$

For each $(x,y)\in[0,1]\times[0,1]$ we have

$$|g_1(x,y,u_{(x,y)}) - g_1(x,y,\overline{u}_{(x,y)})| \le \frac{1}{3e^2}\|u-\overline{u}\|_C.$$

Hence conditions (3.13.4) and (3.13.5) are satisfied with $l = \frac{1}{3e^2}$. It is clear that conditions (3.13.1), (3.13.2), (3.13.3), and (3.13.6) hold with $m(x,y) = e^{x+y+2}$. Consequently, Theorem 3.13 implies that problems (3.30)–(3.32) have at least one solution defined on $[-1,1]\times[-2,1]$.

3.5 Discontinuous Partial Differential Equations in Banach Algebras

3.5.1 Introduction

This section is concerned with the existence of solutions of fractional order IVP, for the system

$$^{c}D_0^r\left(\frac{u(x,y)}{f(x,y,u(x,y))}\right) = g(x,y,u(x,y)), \text{ if } (x,y) \in J, \tag{3.33}$$

$$u(x,0) = \varphi(x),\ x \in [0,a],\ u(0,y) = \psi(y),\ y \in [0,b], \tag{3.34}$$

where $J = [0,a] \times [0,b]$, $a,b > 0$, $f : J \times \mathbb{R} \to \mathbb{R}^*$, $g : J \times \mathbb{R} \to \mathbb{R}$ are given functions satisfying suitable conditions and $\varphi : [0,a] \to \mathbb{R}$, $\psi : [0,b] \to \mathbb{R}$ are given absolutely continuous functions with $\varphi(0) = \psi(0)$. We prove existence of extremal solutions under discontinuous nonlinearity under certain Lipschitz and monotonicity conditions.

3.5.2 Existence of Solutions

Let us start by defining what we mean by a solution of the problem (3.33)–(3.34).

Definition 3.16. A function $u \in C(J,\mathbb{R})$ is said to be a solution of (3.33)–(3.34), if

(i) The function $(x,y) \mapsto (\frac{u(x,y)}{f(x,y,u(x,y))})$ is absolutely continuous,
(ii) u satisfies the equations (3.33)–(3.34) on J

Now, we are concerned with the existence of solutions for the problem (3.33)–(3.34). Note that if the function f is continuous on $J \times \mathbb{R}$, then from the continuity of φ and ψ it follows that $\mu \in C(J,\mathbb{R})$. Let

$$h^* = \|h\|_{L^\infty}.$$

Theorem 3.17. *Assume that the hypotheses*

(3.17.1) The function f is continuous on $J \times \mathbb{R}$
(3.17.2) There exists a function $\alpha \in C(J,\mathbb{R}_+)$ such that

$$|f(x,y,z) - f(x,y,\overline{z})| \le \alpha(x,y)|z-\overline{z}|, \quad \textit{for all } (x,y) \in J, \textit{ for all } z,\overline{z} \in \mathbb{R}$$

(3.17.3) The function g is Carathéodory, and there exists $h \in L^\infty(J,\mathbb{R}_+)$ such that

$$|g(x,y,z)| \le h(x,y), \quad \textit{a.e. } (x,y) \in J, \textit{ for all } z \in \mathbb{R}$$

hold. If

$$\|\alpha\|_\infty \left[\|\mu\|_\infty + \frac{a^{r_1} b^{r_2} h^*}{\Gamma(r_1+1)\Gamma(r_2+1)} \right] < 1, \tag{3.35}$$

Then the IVP (3.33)–(3.34) have at least one solution on J.

Proof. Let $X := C(J, \mathbb{R})$. Define two operators A and B on X by

$$Au(x,y) = f(x,y,u(x,y)); \quad (x,y) \in J, \tag{3.36}$$

$$Bu(x,y) = \mu(x,y) + \frac{1}{\Gamma(r_1)\Gamma(r_2)} \int_0^x \int_0^y (x-s)^{r_1-1}(y-t)^{r_2-1} \times g(s,t,u(s,t))\, dt ds; \quad (x,y) \in J. \tag{3.37}$$

Clearly A and B define the operators $A, B : X \to X$. Now solving (3.33)–(3.34) is equivalent to solving (3.46), which is further equivalent to solving the operator equation

$$Au(x,y)\, Bu(x,y) = u(x,y), \quad (x,y) \in J. \tag{3.38}$$

We show that operators A and B satisfy all the assumptions of Theorem 2.35. First we shall show that A is a Lipschitz. Let $u_1, u_2 \in X$. Then by (3.17.2),

$$\begin{aligned} |Au_1(x,y) - Au_2(x,y)| &= |f(x,y,u_1(x,y)) - f(x,y,u_2(x,y))| \\ &\le \alpha(x,y)|u_1(x,y) - u_2(x,y)| \\ &\le \|\alpha\|_\infty \|u_1 - u_2\|_\infty. \end{aligned}$$

Taking the maximum over (x,y) in the above inequality yields

$$\|Au_1 - Au_2\|_\infty \le \|\alpha\|_\infty \|u_1 - u_2\|_\infty,$$

and so A is a Lipschitz with a Lipschitz constant $\|\alpha\|_\infty$. Next, we show that B is a compact operator on X. Let $\{u_n\}$ be a sequence in X. From (3.17.3) it follows that

$$\|Bu_n\|_\infty \le \|\mu\|_\infty + \frac{a^{r_1} b^{r_2} h^*}{\Gamma(r_1+1)\Gamma(r_2+1)}.$$

As a result $\{Bu_n : n \in \mathbb{N}\}$ is a uniformly bounded set in X. Let $(x_1, y_1), (x_2, y_2) \in J$. Then

$$
\begin{aligned}
|Bu_n(x_1,y_1) - Bu_n(x_2,y_2)| &\le |\mu(x_1,y_1) - \mu(x_2,y_2)| \\
&+ \frac{1}{\Gamma(r_1)\Gamma(r_2)}\Big|\int_0^{x_1}\int_0^{y_1}\big[(x_2-s)^{r_1-1}(y_2-t)^{r_2-1} - (x_1-s)^{r_1-1}(y_1-t)^{r_2-1}\big] \\
&\times g(s,t,u_n(s,t))\mathrm{d}t\mathrm{d}s\Big| \\
&+ \frac{1}{\Gamma(r_1)\Gamma(r_2)}\left|\int_{x_1}^{x_2}\int_{y_1}^{y_2}(x_2-s)^{r_1-1}(y_2-t)^{r_2-1}g(s,t,u_n(s,t))\mathrm{d}t\mathrm{d}s\right| \\
&+ \frac{1}{\Gamma(r_1)\Gamma(r_2)}\left|\int_0^{x_1}\int_{y_1}^{y_2}(x_2-s)^{r_1-1}(y_2-t)^{r_2-1}g(s,t,u_n(s,t))\mathrm{d}t\mathrm{d}s\right| \\
&+ \frac{1}{\Gamma(r_1)\Gamma(r_2)}\left|\int_{x_1}^{x_2}\int_0^{y_1}(x_2-s)^{r_1-1}(y_2-t)^{r_2-1}g(s,t,u_n(s,t))\mathrm{d}t\mathrm{d}s\right| \\
&\le |\mu(x_1,y_1) - \mu(x_2,y_2)| \\
&+ \frac{h^*}{\Gamma(r_1+1)\Gamma(r_2+1)}\Big|\int_0^{x_1}\int_0^{y_1}[(x_2-s)^{r_1-1}(y_2-t)^{r_2-1} \\
&- (x_1-s)^{r_1-1}(y_1-t)^{r_2-1}]\mathrm{d}t\mathrm{d}s\Big| \\
&+ \frac{h^*}{\Gamma(r_1)\Gamma(r_2)}\left|\int_{x_1}^{x_2}\int_{y_1}^{y_2}(x_2-s)^{r_1-1}(y_2-t)^{r_2-1}\mathrm{d}t\mathrm{d}s\right| \\
&+ \frac{h^*}{\Gamma(r_1)\Gamma(r_2)}\left|\int_0^{x_1}\int_{y_1}^{y_2}(x_2-s)^{r_1-1}(y_2-t)^{r_2-1}\mathrm{d}t\mathrm{d}s\right| \\
&+ \frac{h^*}{\Gamma(r_1)\Gamma(r_2)}\left|\int_{x_1}^{x_2}\int_0^{y_1}(x_2-s)^{r_1-1}(y_2-t)^{r_2-1}\mathrm{d}t\mathrm{d}s\right| \\
&\le |\mu(x_1,y_1) - \mu(x_2,y_2)| \\
&+ \frac{h^*}{\Gamma(r_1+1)\Gamma(r_2+1)}[2y_2^{r_2}(x_2-x_1)^{r_1} + 2x_2^{r_1}(y_2-y_1)^{r_2} \\
&+ x_1^{r_1}y_1^{r_2} - x_2^{r_1}y_2^{r_2} - 2(x_2-x_1)^{r_1}(y_2-y_1)^{r_2}]. \\
&\to 0, \quad \text{as } (x_1,y_1)\to(x_2,y_2).
\end{aligned}
$$

From this we conclude that $\{Bu_n : n \in \mathbb{N}\}$ is an equicontinuous set in X. Hence $B : X \to X$ is compact by Arzelà–Ascoli theorem. Moreover,

$$\begin{aligned}M &= \|B(X)\| \\ &\leq |\mu(x,y)| + \frac{1}{\Gamma(r_1)\Gamma(r_2)}\int_0^x\int_0^y (x-s)^{r_1-1}(y-t)^{r_2-1}|g(s,t,u(s,t))|\mathrm{d}t\mathrm{d}s \\ &\leq \|\mu\|_\infty + \frac{a^{r_1}b^{r_2}h^*}{\Gamma(r_1+1)\Gamma(r_2+1)},\end{aligned}$$

and so,

$$\alpha M \leq \|\alpha\|_\infty\left(\|\mu\|_\infty + \frac{a^{r_1}b^{r_2}h^*}{\Gamma(r_1+1)\Gamma(r_2+1)}\right) < 1,$$

by assumption (3.35). To finish, it remains to show that either the conclusion (i) or the conclusion (ii) of Theorem 2.35 holds. We now will show that the conclusion (ii) is not possible. Let $u \in X$ be any solution to (3.33)–(3.34). Then, for any $\lambda \in (0,1)$ we have

$$\begin{aligned}u(x,y) &= \lambda[f(x,y,u(x,y))]\Bigg(\mu(x,y) \\ &\quad + \frac{1}{\Gamma(r_1)\Gamma(r_2)}\int_0^x\int_0^y (x-s)^{r_1-1}(y-t)^{r_2-1}g(s,t,u(s,t))\mathrm{d}t\mathrm{d}s\Bigg),\end{aligned}$$

for $(x,y) \in J$. Therefore,

$$\begin{aligned}|u(x,y)| &\leq |f(x,y,u(x,y))|\Bigg(|\mu(x,y)| \\ &\quad + \frac{1}{\Gamma(r_1)\Gamma(r_2)}\int_0^x\int_0^y (x-s)^{r_1-1}(y-t)^{r_2-1}|g(s,t,u(s,t))|\mathrm{d}t\mathrm{d}s\Bigg) \\ &\leq \left[|f(x,y,u(x,y)) - f(x,y,0)| + |f(x,y,0)|\right] \\ &\quad \times\left(|\mu(x,y)| + \frac{1}{\Gamma(r_1)\Gamma(r_2)}\int_0^x\int_0^y (x-s)^{r_1-1}(y-t)^{r_2-1}h(s,t)\,\mathrm{d}t\mathrm{d}s\right) \\ &\leq (\|\alpha\|_\infty\|u\|_\infty + f^*)\left(\|\mu\|_\infty + \frac{a^{r_1}b^{r_2}h^*}{\Gamma(r_1+1)\Gamma(r_2+1)}\right),\end{aligned}$$

where $f^* = \sup_{(x,y)\in J} |f(x, y, 0)|$, and consequently

$$\|u\|_\infty \leq \frac{f^*\left[\|\mu\|_\infty + \frac{a^{r_1}b^{r_2}h^*}{\Gamma(r_1+1)\Gamma(r_2+1)}\right]}{1 - \|\alpha\|_\infty\left[\|\mu\|_\infty + \frac{a^{r_1}b^{r_2}h^*}{\Gamma(r_1+1)\Gamma(r_2+1)}\right]} := M.$$

Thus the conclusion (ii) of Theorem 2.35 does not hold. Therefore the IVP (3.33)–(3.34) has a solution on J. □

3.5.3 Existence of Extremal Solutions

We equip the space $C(J, \mathbb{R})$ with the order relation $\leq$ with the help of the cone defined by

$$K = \{u \in C(J, \mathbb{R}) : u(x, y) \geq 0, \quad \forall (x, y) \in J\}.$$

Thus $u \leq \bar{u}$ if and only if $u(x, y) \leq \bar{u}(x, y)$ for each $(x, y) \in J$. It is well known that the cone K is positive and normal in $C(J, \mathbb{R})$ ([143]). If $\underline{u}, \bar{u} \in C(J, \mathbb{R})$ and $\underline{u} \leq \bar{u}$, we put

$$[\underline{u}, \overline{u}] = \{u \in C(J, \mathbb{R}) : \underline{u} \leq u \leq \bar{u}\}.$$

Definition 3.18. A function $\underline{u}(\cdot, \cdot) \in C(J, \mathbb{R})$ is said to be a lower solution of (3.33)–(3.34) if we have

$$^cD_0^r\Big[\frac{\underline{u}(x, y)}{f(x, y, \underline{u}(x, y))}\Big] \leq g(x, y, \underline{u}(x, y)), \quad (x, y) \in J,$$

$$\underline{u}(x, 0) \leq \varphi(x), \quad \underline{u}(0, y) \leq \psi(y), \ (x, y) \in J.$$

Similarly a function $\bar{u}(\cdot, \cdot) \in C(J, \mathbb{R})$ is said to be an upper solution of (3.33)–(3.34) if we have

$$^cD_0^r\Big[\frac{\bar{u}(x, y)}{f(x, y, \bar{u}(x, y))}\Big] \geq g(x, y, \bar{u}(x, y)), \quad (x, y) \in J,$$

$$\bar{u}(x, 0) \geq \varphi(x), \quad \bar{u}(0, y) \geq \psi(y), \ (x, y) \in J.$$

Definition 3.19. A solution u_M of the problem (3.33)–(3.34) is said to be maximal if for any other solution u to the problem (3.33)–(3.34) one has $u(x, y) \leq u_M(x, y)$, for all $(x, y) \in J$. Again a solution u_m of the problem (3.33)–(3.34) is said to be minimal if $u_m(x, y) \leq u(x, y)$, for all $(x, y) \in J$ where u is any solution of the problem (3.33)–(3.34) on J.

Let

$$\tilde{h}^* = \|\tilde{h}\|_{L^\infty}.$$

Theorem 3.20. *Assume that the hypotheses (3.17.2) and*

(3.20.1) $f : J \times \mathbb{R}_+ \to \mathbb{R}_+^*$, $g : J \times \mathbb{R}_+ \to \mathbb{R}_+$, $\psi(y) \geq 0$ *on* $[0, b]$ *and*

$$\frac{\varphi(x)}{f(x,0,\varphi(x))} \geq \frac{\varphi(0)}{f(0,0,\varphi(0))} \text{ for all } x \in [0,a],$$

(3.20.2) The functions f *and* g *are Chandrabhan,*

(3.20.3) There exists a function $\tilde{h} \in L^\infty(J, \mathbb{R}_+)$ *such that*

$$|g(x,y,z)| \leq \tilde{h}(x,y), \quad a.e.\ (x,y) \in J, \text{ for all } z \in \mathbb{R},$$

(3.20.4) The problem (3.33)–(3.34) has a lower solution $\underline{u}$ *and an upper solution* $\overline{u}$ *with* $\underline{u} \leq \overline{u}$,

hold. If

$$\|\alpha\|_\infty \left[\|\mu\|_\infty + \frac{a^{r_1} b^{r_2} \tilde{h}^*}{\Gamma(r_1+1)\Gamma(r_2+1)} \right] < 1,$$

then the problem (3.33)–(3.34) has a minimal and a maximal positive solution on J.

Proof. Let $X = C(J, \mathbb{R})$ and consider a closed interval $[\underline{u}, \overline{u}]$ in X which is well defined in view of hypothesis (3.20.4). Define two operators $A, B : [\underline{u}, \overline{u}] \to X$ by (3.36) and (3.37), respectively. Clearly A and B define the operators $A, B : [\underline{u}, \overline{u}] \to K$. Now solving (3.33)–(3.34) is equivalent to solving (3.46), which is further equivalent to solving the operator equation

$$Au(x,y)\,Bu(x,y) = u(x,y), \quad (x,y) \in J. \tag{3.39}$$

We show that operators A and B satisfy all the assumptions of Theorem 2.41. As in Theorem 3.17 we can prove that A is Lipschitz with a Lipschitz constant $\|\alpha\|_\infty$ and B is a completely continuous operator on $[\underline{u}, \overline{u}]$. Now hypothesis (3.20.2) implies that A and B are nondecreasing on $[\underline{u}, \overline{u}]$. To see this, let $u_1, u_2 \in [\underline{u}, \overline{u}]$ be such that $u_1 \leq u_2$. Then by (3.20.2), we get

$$Au_1(x,y) = f(x,y,u_1(x,y)) \leq f(x,y,u_2(x,y)) = Au_2(x,y), \quad \forall (x,y) \in J,$$

and

$$\begin{aligned}
Bu_1(x,y) &= \mu(x,y) + \frac{1}{\Gamma(r_1)\Gamma(r_2)} \int_0^x \int_0^y (x-s)^{r_1-1}(y-t)^{r_2-1} g(s,t,u_1(s,t))\,dt\,ds \\
&\leq \mu(x,y) + \frac{1}{\Gamma(r_1)\Gamma(r_2)} \int_0^x \int_0^y (x-s)^{r_1-1}(y-t)^{r_2-1} g(s,t,u_2(s,t))\,dt\,ds \\
&= Bu_2(x,y), \quad \forall (x,y) \in J.
\end{aligned}$$

So A and B are nondecreasing operators on $[\underline{u}, \overline{u}]$. Again hypothesis (3.20.4) implies

$$\begin{aligned}
\underline{u}(x,y) &= [f(x,y,\underline{u}(x,y))]\Bigg(\mu(x,y) \\
&\quad + \frac{1}{\Gamma(r_1)\Gamma(r_2)}\int_0^x\int_0^y (x-s)^{r_1-1}(y-t)^{r_2-1} g(s,t,\underline{u}(s,t))\,dt ds\Bigg) \\
&\le [f(x,y,z(x,y))]\Bigg(\mu(x,y) \\
&\quad + \frac{1}{\Gamma(r_1)\Gamma(r_2)}\int_0^x\int_0^y (x-s)^{r_1-1}(y-t)^{r_2-1} g(s,t,z(s,t))dt ds\Bigg) \\
&\le [f(x,y,\overline{u}(x,y))]\Bigg(\mu(x,y) \\
&\quad + \frac{1}{\Gamma(r_1)\Gamma(r_2)}\int_0^x\int_0^y (x-s)^{r_1-1}(y-t)^{r_2-1} g(s,t,\overline{u}(s,t))dt ds\Bigg) \\
&\le \overline{u}(x,y),
\end{aligned}$$

for all $(x,y) \in J$ and $z \in [\underline{u}, \overline{u}]$. As a result

$$\underline{u}(x,y) \le Az(x,y)Bz(x,y) \le \overline{u}(x,y), \quad \forall (x,y) \in J \text{ and } z \in [\underline{u}, \overline{u}].$$

Hence $Az\,Bz \in [\underline{u}, \overline{u}]$, for all $z \in [\underline{u}, \overline{u}]$.

Notice for any $u \in [\underline{u}, \overline{u}]$,

$$\begin{aligned}
M &= \|B([\underline{u}, \overline{u}])\| \\
&\le |\mu(x,y)| + \frac{1}{\Gamma(r_1)\Gamma(r_2)}\left|\int_0^x\int_0^y (x-s)^{r_1-1}(y-t)^{r_2-1} g(t,s,u(t,s))dt ds\right| \\
&\le \|\mu\|_\infty + \frac{a^{r_1} b^{r_2} \tilde{h}^*}{\Gamma(r_1+1)\Gamma(r_2+1)}.
\end{aligned}$$

and so,

$$\alpha M \le \|\alpha\|_\infty \left(\|\mu\|_\infty + \frac{a^{r_1} b^{r_2} \tilde{h}^*}{\Gamma(r_1+1)\Gamma(r_2+1)}\right) < 1.$$

Thus the operators A and B satisfy all the conditions of Theorem 2.41 and so the operator equation (3.37) has a least and a greatest solution in $[\underline{u}, \overline{u}]$. This further implies that the problem (3.33)–(3.34) has a minimal and a maximal positive solution on J. □

Theorem 3.21. *Assume that hypotheses (3.17.1), (3.20.1)–(3.20.4) hold. Then the problem (3.33)–(3.34) has a minimal and a maximal positive solution on J.*

Proof. Let $X = C(J, \mathbb{R})$. Consider the order interval $[\underline{u}, \overline{u}]$ in X and define two operators A and B on $[\underline{u}, \overline{u}]$ by (3.36) and (3.37), respectively. Then the problem (3.33)–(3.34) is transformed into an operator equation $Au(x,y)\,Bu(x,y) = u(x,y)$, $(x,y) \in J$ in a Banach algebra X. Notice that (3.20.1) implies $A, B : [\underline{u}, \overline{u}] \to K$. Since the cone K in X is normal, $[\underline{u}, \overline{u}]$ is a norm-bounded set in X.

Next we show that A is completely continuous on $[\underline{u}, \overline{u}]$. Now the cone K in X is normal, so the order interval $[\underline{u}, \overline{u}]$ is norm-bounded. Hence there exists a constant $\varrho > 0$ such that $\|u\| \leq \varrho$ for all $u \in [\underline{u}, \overline{u}]$. As f is continuous on the compact set $J \times [-\varrho, \varrho]$, it attains its maximum , say M. Therefore, for any subset S of $[\underline{u}, \overline{u}]$ we have

$$\begin{aligned}\|A(S)\| &= \sup\{|Au| : u \in S\} \\ &= \sup\left\{\sup_{(x,y)\in J} |f(x,y,u(x,y))| : u \in S\right\} \\ &\leq \sup\left\{\sup_{(x,y)\in J} |f(x,y,u)| : u \in [-\varrho, \varrho]\right\} \\ &\leq M.\end{aligned}$$

This shows that $A(S)$ is a uniformly bounded subset of X. We note that the function $f(x,y,u)$ is uniformly continuous on $J \times [-\varrho, \varrho]$. Therefore, for any $(x_1, y_1), (x_2, y_2) \in J$ we have

$$|f(x_1, y_1, u) - f(x_2, y_2, u)| \to 0 \quad \text{as } (x_1, y_1) \to (x_2, y_2),$$

for all $u \in [-\varrho, \varrho]$. Similarly for any $u_1, u_2 \in [-\varrho, \varrho]$

$$|f(x, y, u_1) - f(x, y, u_2)| \to 0 \quad \text{as } u_1 \to u_2,$$

for all $(x, y) \in J$. Hence any $(x_1, y_1), (x_2, y_2) \in J$, and for any $u \in S$ one has

$$\begin{aligned}|Au(x_1, y_1) - Au(x_2, y_2)| &= |f(x_1, y_1, u(x_1, y_1)) - f(x_2, y_2, u(x_2, y_2))| \\ &\leq |f(x_1, y_1, u(x_1, y_1)) - f(x_2, y_2, u(x_1, y_1)| \\ &\quad + |f(x_2, y_2, u(x_1, y_1)) - f(x_2, y_2, u(x_2, y_2))| \\ &\to 0 \quad \text{as } (x_1, y_1) \to (x_2, y_2).\end{aligned}$$

This shows that $A(S)$ is an equicontinuous set in K. Now an application of the Arzelà–Ascoli theorem yields that A is a completely continuous operator on $[\underline{u}, \overline{u}]$.

Next it can be shown as in the proof of Theorem 3.20 that B is a compact operator on $[\underline{u}, \overline{u}]$. Now an application of Theorem 2.40 yields that the problems (3.33)–(3.34) have a minimal and maximal positive solution on J. □

3.5.4 An Example

As an application of our results we consider the following partial hyperbolic functional differential equations of the form:

$$^{c}D_{0}^{r}\left(\frac{u(x,y)}{f(x,y,u(x,y))}\right) = g(x,y,u(x,y)), \quad \text{if } (x,y) \in [0,1]\times[0,1], \tag{3.40}$$

$$u(x,0) = \varphi(x),\ x \in [0,1],\ u(0,y) = \psi(y),\ y \in [0,1], \tag{3.41}$$

where $f, g : [0,1]\times[0,1]\times\mathbb{R} \to \mathbb{R}$ defined by

$$f(x,y,u) = \frac{1}{e^{x+y+10}(1+|u|)} \tag{3.42}$$

and

$$g(x,y,u) = \frac{1}{e^{x+y+8}(1+u^2)}. \tag{3.43}$$

The functions $\varphi, \psi : [0,1] \to \mathbb{R}$ are defined by

$$\varphi(x) = x^2 e^{-10} \quad \text{and} \quad \psi(y) = ye^{-10}, \text{ for all } x, y \in [0,1]. \tag{3.44}$$

We show that the functions φ, ψ, f, and g satisfy all the hypotheses of Theorem 3.17. We have $f : [0,1]\times[0,1]\times\mathbb{R} \to \mathbb{R}_+^*$, $g : [0,1]\times[0,1]\times\mathbb{R} \to \mathbb{R}_+$. Clearly, the function f satisfies (3.17.1) and (3.17.2) with $\alpha(x,y) = \frac{1}{e^{x+y+10}}$ and $\|\alpha\|_\infty = \frac{1}{e^{10}}$. Also, the function g satisfies (3.17.3) with $h(x,y) = \frac{1}{e^{x+y+8}}$ and $h^* = \frac{1}{e^8}$. A simple computation gives

$$\mu(x,y) = e^x x^2(1+x^2) + e^y y(1+|y|),$$

and $\|\mu\|_\infty \le 4e$. We shall show that condition (3.35) holds. Indeed

$$\|\alpha\|_\infty\left[\|\mu\|_\infty + \frac{a^{r_1}b^{r_2}h^*}{\Gamma(r_1+1)\Gamma(r_2+1)}\right] \le \frac{1}{e^{10}}\left[4e + \frac{1}{e^8\Gamma(r_1+1)\Gamma(r_2+1)}\right] < 1,$$

for some $(r_1, r_2) \in (0,1]\times(0,1]$. Hence by Theorem 3.17, the problem (3.40)–(3.41) have a solution defined on $[0,1]\times[0,1]$.

3.6 Upper and Lower Solutions Method for Partial Hyperbolic Differential Equations

3.6.1 Introduction

This section deals with the existence of solutions to the Darboux problem for the fractional order IVP , for the system

$$({}^cD_0^r u)(x,y) = f(x,y,u(x,y)); \text{ if } (x,y) \in J, \tag{3.45}$$

$$u(x,0) = \varphi(x),\ x \in [0,a],\ u(0,y) = \psi(y),\ y \in [0,b], \tag{3.46}$$

where $J = [0,a] \times [0,b],\ a,b > 0,\ f : J \times \mathbb{R} \to \mathbb{R},\ \varphi : [0,a] \to \mathbb{R},\ \psi : [0,b] \to \mathbb{R}$ are given absolutely continuous functions with $\varphi(0) = \psi(0)$.

3.6.2 Main Result

Let us start by defining what we mean by a solution of the problem (3.45)–(3.46).

Definition 3.22. A function $u \in C(J,\mathbb{R})$ whose mixed derivative D_{xy}^2 exists and integrable is said to be a solution of (3.45)–(3.46) if u satisfies (3.45) and (3.46) on J.

Definition 3.23. A function $z \in C(J,\mathbb{R})$ is said to be a lower solution of (3.45)–(3.46) if z satisfies

$$({}^cD_0^r z)(x,y) \le f(x,y,z(x,y)),\ z(x,0) \le \varphi(x),\ z(0,y) \le \psi(y) \text{ on } J,$$
$$\text{and } z(0,0) \le \varphi(0).$$

The function z is said to be an upper solution of (3.45)–(3.46) if the reversed inequalities hold.

Further, we present conditions for the existence of a solution of our problem.

Theorem 3.24. *Assume that the following hypotheses:*

(3.24.1) The function $f : J \times \mathbb{R} \to \mathbb{R}$ is continuous,
(3.24.2) There exist v and $w \in C(J,\mathbb{R})$, lower and upper solutions for the problem (3.45)–(3.46) such that $v \le w$,

hold. Then the problems (3.45)–(3.46) have at least one solution u such that

$$v(x,y) \le u(x,y) \le w(x,y) \text{ for all } (x,y) \in J.$$

Proof. Transform the problems (3.45)–(3.46) into a fixed point problem. Consider the following modified problems:

$$({}^cD_0^r u)(x,y) = g(x,y,u(x,y)), \text{ if } (x,y) \in J, \tag{3.47}$$

$$u(x,0) = \varphi(x),\ u(0,y) = \psi(y),\ x \in [0,a],\ y \in [0,b], \tag{3.48}$$

where

$$g(x,y,u(x,y)) = f(x,y,h(x,y,u(x,y))),$$
$$h(x,y,u(x,y)) = \max\{v(x,y), \min\{u(x,y), w(x,y)\}\},$$

for each $(x,y) \in J$. A solution to (3.47)–(3.48) is a fixed point of the operator $N : C(J,\mathbb{R}) \to C(J,\mathbb{R})$ defined by

$$N(u)(x,y) = \mu(x,y) + \frac{1}{\Gamma(r_1)\Gamma(r_2)} \int_0^x \int_0^y (x-s)^{r_1-1}(y-t)^{r_2-1} g(s,t,u(s,t))\mathrm{d}t\mathrm{d}s.$$

Notice that g is a continuous function, and from (3.24.2) there exists $M > 0$ such that

$$|g(x,y,u)| \le M, \text{ for each } (x,y) \in J, \text{ and } u \in \mathbb{R}. \tag{3.49}$$

Set

$$\eta = \|\mu\|_\infty + \frac{Ma^{r_1}b^{r_2}}{\Gamma(r_1+1)\Gamma(r_2+1)},$$

and

$$D = \{u \in C(J,\mathbb{R}) : \|u\|_\infty \le \eta\}.$$

Clearly D is a closed convex subset of $C(J,\mathbb{R})$ and that N maps D into D. We shall show that N satisfies the assumptions of Schauder's fixed point theorem. The proof will be given in several steps.

Step 1: *N is continuous.* Let $\{u_n\}$ be a sequence such that $u_n \to u$ in D. Then

$$|N(u_n)(x,y) - N(u)(x,y)|$$
$$\le \frac{1}{\Gamma(r_1)\Gamma(r_2)} \int_0^x \int_0^y |x-s|^{r_1-1}|y-t|^{r_2-1}|g(s,t,u_n(s,t)) - g(s,t,u(s,t))|\mathrm{d}t\mathrm{d}s$$
$$\le \frac{\|g(.,.,u_n(.,.)) - g(.,.,u)\|_\infty}{\Gamma(r_1)\Gamma(r_2)} \int_0^a \int_0^b |x-s|^{r_1-1}|y-t|^{r_2-1}\mathrm{d}t\mathrm{d}s.$$

Since g is a continuous function, we have

$$\|N(u_n) - N(u)\|_\infty \le \frac{a^{r_1}b^{r_2}\|g(.,.,u_n(.,.)) - g(.,.,u(.,.))\|_\infty}{\Gamma(r_1+1)\Gamma(r_2+1)} \to 0 \text{ as } n \to \infty.$$

Step 2: *$N(D)$ is bounded.* This is clear since $N(D) \subset D$ and D is bounded.

Step 3: *$N(D)$ is equicontinuous.* Let $(x_1, y_1), (x_2, y_2) \in (0, a] \times (0, b]$, $x_1 < x_2$, $y_1 < y_2$ and $u \in D$. Then

$$\begin{aligned}
&|N(u)(x_2, y_2) - N(u)(x_1, y_1)| \\
&\quad = |\mu(x_1, y_1) - \mu(x_2, y_2)| \\
&\qquad + \Big|\frac{1}{\Gamma(r_1)\Gamma(r_2)}\int_0^{x_1}\int_0^{y_1}[(x_2 - s)^{r_1-1}(y_2 - t)^{r_2-1} - (x_1 - s)^{r_1-1}(y_1 - t)^{r_2-1}] \\
&\qquad \times g(s, t, u(s, t))dtds \\
&\qquad + \frac{1}{\Gamma(r_1)\Gamma(r_2)}\int_{x_1}^{x_2}\int_{y_1}^{y_2}(x_2 - s)^{r_1-1}(y_2 - t)^{r_2-1}g(s, t, u(s, t))dtds \\
&\qquad + \frac{1}{\Gamma(r_1)\Gamma(r_2)}\int_0^{x_1}\int_{y_1}^{y_2}(x_2 - s)^{r_1-1}(y_2 - t)^{r_2-1}g(s, t, u(s, t))dtds \\
&\qquad + \frac{1}{\Gamma(r_1)\Gamma(r_2)}\int_{x_1}^{x_2}\int_0^{y_1}(x_2 - s)^{r_1-1}(y_2 - t)^{r_2-1}g(s, t, u(s, t))dtds\Big| \\
&\quad \leq |\mu(x_1, y_1) - \mu(x_2, y_2)| \\
&\qquad + \frac{M}{\Gamma(r_1 + 1)\Gamma(r_2 + 1)}[2y_2^{r_2}(x_2 - x_1)^{r_1} + 2x_2^{r_1}(y_2 - y_1)^{r_2} \\
&\qquad + x_1^{r_1}y_1^{r_2} - x_2^{r_1}y_2^{r_2} - 2(x_2 - x_1)^{r_1}(y_2 - y_1)^{r_2}].
\end{aligned}$$

As $x_1 \to x_2$, $y_1 \to y_2$ the right-hand side of the above inequality tends to zero. As a consequence of steps 1–3 together with the Arzelá–Ascoli theorem, we can conclude that $N : D \to D$ is continuous and compact. From an application of Schauder's theorem, we deduce that N has a fixed point u which is a solution of the problem (3.47)–(3.48).

Step 4: *The solution u of (3.47)–(3.48) satisfies*

$$v(x, y) \leq u(x, y) \leq w(x, y) \text{ for all } (x, y) \in J.$$

We prove that

$$u(x, y) \leq w(x, y) \text{ for all } (x, y) \in J.$$

Assume that $u - w$ attains a positive maximum on J at $(\overline{x}, \overline{y}) \in J$; i.e.,

$$(u - w)(\overline{x}, \overline{y}) = \max\{u(x, y) - w(x, y) : (x, y) \in J\} > 0.$$

We distinguish the following cases.

Case 1. If $(\overline{x}, \overline{y}) \in (0, a) \times [0, b]$ there exists $(x^*, y^*) \in (0, a) \times [0, b]$ such that

$$[u(x, y^*) - w(x, y^*)] + [u(x^*, y) - w(x^*, y)] - [u(x^*, y^*) - w(x^*, y^*)] \leq 0;$$
$$\text{for all } (x, y) \in ([x^*, \overline{x}] \times \{y^*\}) \cup (\{x^*\} \times [y^*, b]), \tag{3.50}$$

and

$$u(x, y) - w(x, y) > 0, \text{ for all } (x, y) \in (x^*, \overline{x}] \times [y^*, b]. \tag{3.51}$$

By the definition of h one has

$$^{c}D^{r}u(x, y) = f(x, y, w(x, y)) \text{ for all } (x, y) \in [x^*, \overline{x}] \times [y^*, b].$$

An integration on $[x^*, x] \times [y^*, y]$ for each $(x, y) \in [x^*, \overline{x}] \times [y^*, b]$ yields

$$u(x, y) + u(x^*, y^*) - u(x, y^*) - u(x^*, y)$$
$$= \frac{1}{\Gamma(r_1)\Gamma(r_2)} \int_{x^*}^{x} \int_{y^*}^{y} (x - s)^{r_1 - 1}(y - t)^{r_2 - 1} f(s, t, w(s, t)) \mathrm{d}t \mathrm{d}s. \tag{3.52}$$

From (3.52) and using the fact that w is an upper solution to (3.45)–(3.46) we get

$$u(x, y) + u(x^*, y^*) - u(x, y^*) - u(x^*, y) \leq w(x, y) + w(x^*, y^*)$$
$$-w(x, y^*) - w(x^*, y),$$

which gives,

$$[u(x, y) - w(x, y)] \leq [u(x, y^*) - w(x, y^*)] + [u(x^*, y) - w(x^*, y)]$$
$$-[u(x^*, y^*) - w(x^*, y^*)]. \tag{3.53}$$

Thus from (3.50), (3.51), and (3.53) we obtain the contradiction

$$0 < [u(x, y) - w(x, y)] \leq [u(x, y^*) - w(x, y^*)] + [u(x^*, y) - w(x^*, y)]$$
$$-[u(x^*, y^*) - w(x^*, y^*)] \leq 0; \text{ for all } (x, y)$$
$$\in [x^*, \overline{x}] \times [y^*, b].$$

Case 2. If $\overline{x} = 0$, then $w(0, \overline{y}) < u(0, \overline{y}) \leq w(0, \overline{y})$ which is a contradiction. Thus

$$u(x, y) \leq w(x, y) \text{ for all } (x, y) \in J.$$

Analogously, we can prove that $u(x, y) \geq v(x, y)$; for all $(x, y) \in J$. This shows that the problems (3.47)–(3.48) have a solution u satisfying $v \leq u \leq w$ which is solution of (3.45)–(3.46). □

3.7 Partial Functional Differential Equations with Infinite Delay

3.7.1 Introduction

In this section we discuss the existence of solutions to the Darboux problem for the fractional order IVP, for the system

$$({}^cD_0^r u)(x,y) = f(x,y,u_{(x,y)}), \text{ if } (x,y) \in J, \tag{3.54}$$

$$u(x,y) = \phi(x,y), \text{ if } (x,y) \in \tilde{J}', \tag{3.55}$$

$$u(x,0) = \varphi(x),\ x \in [0,a],\ u(0,y) = \psi(y),\ y \in [0,b], \tag{3.56}$$

where $J = [0,a] \times [0,b],\ a,b > 0,\ \tilde{J}' = (-\infty,a] \times (-\infty,b] \backslash (0,a] \times (0,b],\ f : J \times \mathcal{B} \to \mathbb{R}^n,\ \phi : \tilde{J}' \to \mathbb{R}^n$ are given continuous functions, φ, ψ are as in problems (3.1)–(3.3) and $\mathcal{B}$ is called a phase space that will be specified later. We denote by $u_{(x,y)}$ the element of $\mathcal{B}$ defined by

$$u_{(x,y)}(s,t) = u(x+s, y+t);\ (s,t) \in (-\infty,0] \times (-\infty,0],$$

here $u_{(x,y)}(.,.)$ represents the history of the state from time $-\infty$ up to the present time x and from time $-\infty$ up to the present time y.

Next we consider the following initial value problem for partial neutral functional differential equations:

$${}^cD_0^r\Big(u(x,y) - g(x,y,u_{(x,y)})\Big) = f(x,y,u_{(x,y)}), \text{ if } (x,y) \in J, \tag{3.57}$$

$$u(x,y) = \phi(x,y), \text{ if } (x,y) \in \tilde{J}', \tag{3.58}$$

$$u(x,0) = \varphi(x),\ x \in [0,a],\ u(0,y) = \psi(y),\ y \in [0,b], \tag{3.59}$$

where f, ϕ, φ, ψ are as in problems (3.54)–(3.56) and $g : J \times \mathcal{B} \to \mathbb{R}^n$ is a given continuous function.

3.7.2 The Phase Space $\mathcal{B}$

The notation of the phase space $\mathcal{B}$ plays an important role in the study of both qualitative and quantitative theory for functional differential equations. A usual choice is a seminormed space satisfying suitable axioms , which was introduced by Hale and Kato [138]. For further applications see for instance, the books [139, 152, 179] and their references. For any $(x,y) \in J$ denote $E_{(x,y)} := [0,x] \times \{0\} \cup \{0\} \times [0,y]$; furthermore, in case $x = a,\ y = b$ we write simply E.

Consider the space $(\mathcal{B}, \|(.,.)\|_{\mathcal{B}})$, the seminormed linear space of functions mapping $(-\infty, 0]\times(-\infty, 0]$ into $\mathbb{R}^n$, and satisfying the following fundamental axioms which were adapted from those introduced by Hale and Kato for ordinary differential functional equations:

(A_1) If $z : (-\infty, a] \times (-\infty, b] \to \mathbb{R}^n$ is continuous on J and $z_{(x,y)} \in \mathcal{B}$, for all $(x, y) \in E$, then there are constants $H, K, M > 0$ such that for any $(x, y) \in J$ the following conditions hold:

(i) $z_{(x,y)}$ is in $\mathcal{B}$
(ii) $\|z(x, y)\| \leq H\|z_{(x,y)}\|_{\mathcal{B}}$
(iii) $\|z_{(x,y)}\|_{\mathcal{B}} \leq K \sup_{(s,t)\in[0,x]\times[0,y]} \|z(s,t)\| + M \sup_{(s,t)\in E_{(x,y)}} \|z_{(s,t)}\|_{\mathcal{B}}$

(A_2) For the function $z(.,.)$ in (A_1), $z_{(x,y)}$ is a $\mathcal{B}$-valued continuous function on J.
(A_3) The space $\mathcal{B}$ is complete.

Now, we present some examples of phase spaces [97, 98].

Example 3.25. Let $\mathcal{B}$ be the set of all functions $\phi : (-\infty, 0]\times(-\infty, 0] \to \mathbb{R}^n$ which are continuous on $[-\alpha, 0] \times [-\beta, 0]$, $\alpha, \beta \geq 0$, with the seminorm

$$\|\phi\|_{\mathcal{B}} = \sup_{(s,t)\in[-\alpha,0]\times[-\beta,0]} \|\phi(s,t)\|.$$

Then we have $H = K = M = 1$. The quotient space $\widehat{\mathcal{B}} = \mathcal{B}/\|.\|_{\mathcal{B}}$ is isometric to the space $C([-\alpha, 0]\times[-\beta, 0], \mathbb{R}^n)$ of all continuous functions from $[-\alpha, 0]\times[-\beta, 0]$ into $\mathbb{R}^n$ with the supremum norm, this means that partial differential functional equations with finite delay are included in our axiomatic model.

Example 3.26. Let γ be a real constant and Let C_γ be the set of all continuous functions $\phi : (-\infty, 0]\times(-\infty, 0] \to \mathbb{R}^n$ for which a limit $\lim_{\|(s,t)\|\to\infty} e^{\gamma(s+t)}\phi(s,t)$ exists, with the norm

$$\|\phi\|_{C_\gamma} = \sup_{(s,t)\in(-\infty,0]\times(-\infty,0]} e^{\gamma(s+t)}\|\phi(s,t)\|.$$

Then we have $H = 1$ and $K = M = \max\{e^{-\gamma(a+b)}, 1\}$.

Example 3.27. Let $\alpha, \beta, \gamma \geq 0$ and let

$$\|\phi\|_{CL_\gamma} = \sup_{(s,t)\in[-\alpha,0]\times[-\beta,0]} \|\phi(s,t)\| + \int_{-\infty}^{0}\int_{-\infty}^{0} e^{\gamma(s+t)}\|\phi(s,t)\|\mathrm{d}t\,\mathrm{d}s.$$

be the seminorm for the space CL_γ of all functions $\phi : (-\infty, 0] \times (-\infty, 0] \to \mathbb{R}^n$ which are continuous on $[-\alpha, 0] \times [-\beta, 0]$ measurable on $(-\infty, -\alpha] \times (-\infty, 0] \cup (-\infty, 0] \times (-\infty, -\beta]$, and such that $\|\phi\|_{CL_\gamma} < \infty$. Then

$$H = 1, \; K = \int_{-\alpha}^{0}\int_{-\beta}^{0} e^{\gamma(s+t)} dt ds, \; M = 2.$$

3.7.3 Main Results

Let us start by defining what we mean by a solution of the problems (3.54)–(3.56). Let the space

$$\Omega := \{u : (-\infty, a] \times (-\infty, b] \to \mathbb{R}^n : u_{(x,y)} \in \mathcal{B} \text{ for } (x, y) \in E \text{ and } u|_J \in C(J, \mathbb{R}^n)\}.$$

Definition 3.28. A function $u \in \Omega$ is said to be a solution of (3.54)–(3.56) if u satisfies (3.54) and (3.56) on J and the condition (3.55) on $\tilde{J}'$.

Our first existence result for the IVP (3.54)–(3.56) is based on the Banach contraction principle.

Theorem 3.29. *Assume that the following hypotheses hold:*

(3.29.1) There exists $\ell > 0$ such that

$$\|f(x, y, u) - f(x, y, v)\| \le \ell\|u - v\|_{\mathcal{B}}, \text{ for any } u, \; v \in \mathcal{B} \text{ and } (x, y) \in J.$$

If

$$\frac{\ell K a^{r_1} b^{r_2}}{\Gamma(r_1 + 1)\Gamma(r_2 + 1)} < 1, \tag{3.60}$$

then there exists a unique solution for IVP (3.54)–(3.56) on $(-\infty, a] \times (-\infty, b]$.

Proof. Transform the problems (3.54)–(3.56) into a fixed point problem. Consider the operator $N : \Omega \to \Omega$ defined by

$$N(u)(x, y) = \begin{cases} \phi(x, y), & (x, y) \in \tilde{J}', \\ \mu(x, y) + \frac{1}{\Gamma(r_1)\Gamma(r_2)} \int_0^x \int_0^y (x - s)^{r_1 - 1} & \\ \times (y - t)^{r_2 - 1} f(s, t, u_{(s,t)}) dt ds, & (x, y) \in J. \end{cases} \tag{3.61}$$

Let $v(.,.) : (-\infty, a] \times (-\infty, b] \to \mathbb{R}^n$ be the function defined by

$$v(x,y) = \begin{cases} \phi(x,y), & (x,y) \in \tilde{J}', \\ \mu(x,y), & (x,y) \in J. \end{cases}$$

Then $v_{(x,y)} = \phi$ for all $(x,y) \in E$. For each $w \in C(J,\mathbb{R}^n)$ such that $w(x,y) = 0$ for each $(x,y) \in E$. We denote by $\overline{w}$ the function defined by

$$\overline{w}(x,y) = \begin{cases} 0, & (x,y) \in \tilde{J}', \\ w(x,y) & (x,y) \in J. \end{cases}$$

If $u(.,.)$ satisfies the integral equation

$$u(x,y) = \mu(x,y) + \frac{1}{\Gamma(r_1)\Gamma(r_2)} \int_0^x \int_0^y (x-s)^{r_1-1}(y-t)^{r_2-1} f(s,t,u_{(s,t)})dtds,$$

we can decompose $u(.,.)$ as $u(x,y) = \overline{w}(x,y)+v(x,y);\ (x,y) \in J$, which implies $u_{(x,y)} = \overline{w}_{(x,y)} + v_{(x,y)}$, for every $(x,y) \in J$, and the function $w(.,.)$ satisfies

$$w(x,y) = \frac{1}{\Gamma(r_1)\Gamma(r_2)} \int_0^x \int_0^y (x-s)^{r_1-1}(y-t)^{r_2-1} f(s,t,\overline{w}_{(s,t)} + v_{(s,t)})dtds.$$

Set

$$C_0 = \{w \in C(J,\mathbb{R}^n) :\ w(x,y) = 0 \ \text{ for } (x,y) \in E\},$$

and let $\|.\|_{(a,b)}$ be the seminorm in C_0 defined by

$$\|w\|_{(a,b)} = \sup_{(x,y)\in E} \|w_{(x,y)}\|_{\mathcal{B}} + \sup_{(x,y)\in J} \|w(x,y)\| = \sup_{(x,y)\in J} \|w(x,y)\|,\ w \in C_0.$$

C_0 is a Banach space with norm $\|.\|_{(a,b)}$. Let the operator $P :\ C_0 \to C_0$ be defined by

$$(Pw)(x,y) = \frac{1}{\Gamma(r_1)\Gamma(r_2)} \int_0^x \int_0^y (x-s)^{r_1-1}(y-t)^{r_2-1} f(s,t,\overline{w}_{(s,t)} + v_{(s,t)})dtds, \tag{3.62}$$

for each $(x,y) \in J$. The operator N has a fixed point, which is equivalent to P having a fixed point, and so we turn to proving that P has a fixed point. We shall show that $P : C_0 \to C_0$ is a contraction map. Indeed, consider $w, w^* \in C_0$. Then we have for each $(x,y) \in J$

$$\|P(w)(x,y)-P(w^*)(x,y)\|$$

$$\leq \frac{1}{\Gamma(r_1)\Gamma(r_2)}\int_0^x\int_0^y (x-s)^{r_1-1}(y-t)^{r_2-1}$$

$$\times\|f(s,t,\overline{w}_{(s,t)}+v_{(s,t)})-f(s,t,\overline{w^*}_{(s,t)}+v_{(s,t)})\|\mathrm{d}t\mathrm{d}s$$

$$\leq \frac{1}{\Gamma(r_1)\Gamma(r_2)}\int_0^x\int_0^y (x-s)^{r_1-1}(y-t)^{r_2-1}\ell\|\overline{w}_{(s,t)}-\overline{w^*}_{(s,t)}\|$$

$$\leq \frac{1}{\Gamma(r_1)\Gamma(r_2)}\int_0^x\int_0^y (x-s)^{r_1-1}(y-t)^{r_2-1}\ell K$$

$$\times \sup_{(s,t)\in[0,x]\times[0,y]}\|\overline{w}(s,t)-\overline{w^*}(s,t)\|\mathrm{d}t\mathrm{d}s$$

$$\leq \frac{\ell K}{\Gamma(r_1)\Gamma(r_2)}\int_0^x\int_0^y (x-s)^{r_1-1}(y-t)^{r_2-1}\mathrm{d}t\mathrm{d}s\|\overline{w}-\overline{w^*}\|_{(a,b)}.$$

Therefore

$$\|P(w)-P(w^*)\|_{(a,b)}\leq \frac{\ell K a^{r_1}b^{r_2}}{\Gamma(r_1+1)\Gamma(r_2)+1}\|\overline{w}-\overline{w^*}\|_{(a,b)}.$$

and hence P is a contraction. Therefore, P has a unique fixed point by Banach's contraction principle. □

Now we give an existence result based on the nonlinear alternative of Leray–Schauder type [136].

Theorem 3.30. *Assume that the following hypotheses hold:*

(3.30.1) There exist $p,q\in C(J,\mathbb{R}_+)$ *such that*

$$\|f(x,y,u)\|\leq p(x,y)+q(x,y)\|u\|_{\mathcal{B}},\ \textit{for}\ (x,y)\in J\ \textit{and each}\ u\in\mathcal{B}.$$

Then the IVP (3.54)–(3.56) have at least one solution on $(-\infty,a]\times(-\infty,b]$.

Proof. Let $P: C_0\to C_0$ be defined as in (3.62). We shall show that the operator P is continuous and completely continuous.

Step 1: *P is continuous.* Let $\{w_n\}$ be a sequence such that $w_n\to w$ in C_0. Then

$$\|P(w_n)(x,y)-P(w)(x,y)\|\leq \frac{1}{\Gamma(r_1)\Gamma(r_2)}\int_0^x\int_0^y (x-s)^{r_1-1}(y-t)^{r_2-1}$$

$$\times\|f(s,t,\overline{w_{n(s,t)}}+v_{n(s,t)})-f(s,t,\overline{w}_{(s,t)}+v_{(s,t)})\|\mathrm{d}t\mathrm{d}s.$$

Since f is a continuous function, we have

$$\begin{aligned}\|P(w_n) - P(w)\|_\infty &\le \frac{x^{r_1}y^{r_2}\|f(.,.,\overline{w_n}_{(.,.)} + v_{n(.,.)}) - f(.,.,\overline{w}_{(.,.)} + v_{(.,.)})\|_\infty}{\Gamma(r_1+1)\Gamma(r_2+1)} \\ &\le \frac{a^{r_1}b^{r_2}\|f(.,.,\overline{w_n}_{(.,.)} + v_{n(.,.)}) - f(.,.,\overline{w}_{(.,.)} + v_{(.,.)})\|_\infty}{\Gamma(r_1+1)\Gamma(r_2+1)} \\ &\to 0 \text{ as } n \to \infty.\end{aligned}$$

Step 2: *P maps bounded sets into bounded sets in C_0.* Indeed, it is enough to show that, for any $\eta > 0$, there exists a positive constant $\widetilde{\ell}$ such that, for each $w \in B_\eta = \{w \in C_0 : \|w\|_{(a,b)} \le \eta\}$, we have $\|P(w)\|_\infty \le \widetilde{\ell}$. Let $w \in B_\eta$. By (3.30.1) we have for each $(x, y) \in J$,

$$\begin{aligned}\|P(w)(x,y)\| &\le \frac{1}{\Gamma(r_1)\Gamma(r_2)}\int_0^x\int_0^y (x-s)^{r_1-1}(y-t)^{r_2-1}\|f(s,t,\overline{w}_{(s,t)}+v_{(s,t)})\|dtds \\ &\le \frac{\|p\|_\infty}{\Gamma(r_1)\Gamma(r_2)}\int_0^x\int_0^x (x-s)^{r_1-1}(y-t)^{r_2-1}dtds \\ &\quad + \frac{\|q\|_\infty\eta^*}{\Gamma(r_1)\Gamma(r_2)}\int_0^x\int_0^y (x-s)^{r_1-1}(y-t)^{r_2-1}dtds \\ &\le \frac{\|p\|_\infty + \|q\|_\infty\eta^*}{\Gamma(r_1+1)\Gamma(r_2+1)}a^{r_1}b^{r_2} := \ell^*,\end{aligned}$$

where

$$\begin{aligned}\|\overline{w}_{(s,t)} + v_{(s,t)}\|_{\mathcal{B}} &\le \|\overline{w}_{(s,t)}\|_{\mathcal{B}} + \|v_{(s,t)}\|_{\mathcal{B}} \\ &\le K\eta + K\|\phi(0,0)\| + M\|\phi\|_{\mathcal{B}} := \eta^*.\end{aligned}$$

Hence, $\|P(w)\|_\infty \le \ell^*$.

Step 3: *P maps bounded sets into equicontinuous sets in C_0.* Let $(x_1, y_1), (x_2, y_2) \in (0, a] \times (0, b]$, $x_1 < x_2$, $y_1 < y_2$, B_η be a bounded set as in step 2, and let $w \in B_\eta$. Then

$$\|P(w)(x_2,y_2)-P(w)(x_1,y_1)\|$$
$$\le \frac{1}{\Gamma(r_1)\Gamma(r_2)}\left\|\int_0^{x_1}\int_0^{y_1}\left[(x_2-s)^{r_1-1}(y_2-t)^{r_2-1}-(x_1-s)^{r_1-1}(y_1-t)^{r_2-1}\right]\right.$$
$$\times f(s,t,u_{(s,t)})\mathrm{d}t\mathrm{d}s+\frac{1}{\Gamma(r_1)\Gamma(r_2)}\int_{x_1}^{x_2}\int_{y_1}^{y_2}(x_2-s)^{r_1-1}(y_2-t)^{r_2-1}$$
$$\left.\times f(s,t,\overline{w}_{(s,t)}+v_{(s,t)})\mathrm{d}t\mathrm{d}s\right\|$$
$$+\frac{1}{\Gamma(r_1)\Gamma(r_2)}\int_0^{x_1}\int_{y_1}^{y_2}(x_2-s)^{r_1-1}(y_2-t)^{r_2-1}\|f(s,t,\overline{w}_{(s,t)}+v_{(s,t)})\|\mathrm{d}t\mathrm{d}s$$
$$+\frac{1}{\Gamma(r_1)\Gamma(r_2)}\int_{x_1}^{x_2}\int_0^{y_1}(x_2-s)^{r_1-1}(y_2-t)^{r_2-1}\|f(s,t,\overline{w}_{(s,t)}+v_{(s,t)})\|\mathrm{d}t\mathrm{d}s$$
$$\le \frac{\|p\|_\infty+\|q\|_\infty\eta}{\Gamma(r_1+1)\Gamma(r_2+1)}[y_2^{r_2}(x_2-x_1)^{r_1}+x_2^{r_1}(y_2-y_1)^{r_2}$$
$$-(x_2-x_1)^{r_1}(y_2-y_1)^{r_2}+x_1^{r_1}y_1^{r_2}-x_2^{r_1}y_2^{r_2}]$$
$$+\frac{\|p\|_\infty+\|q\|_\infty\eta}{\Gamma(r_1+1)\Gamma(r_2+1)}(x_2-x_1)^{r_1}(y_2-y_1)^{r_2}$$
$$+\frac{\|p\|_\infty+\|q\|_\infty\eta}{\Gamma(r_1+1)\Gamma(r_2+1)}[x_2^{r_1}-(x_2-x_1)^{r_1}](y_2-y_1)^{r_2}$$
$$+\frac{\|p\|_\infty+\|q\|_\infty\eta}{\Gamma(r_1+1)\Gamma(r_2+1)}(x_2-x_1)^{r_1}[y_2^{r_2}-(y_2-y_1)^{r_2}]$$
$$\le \frac{\|p\|_\infty+\|q\|_\infty\eta}{\Gamma(r_1+1)\Gamma(r_2+1)}[2y_2^{r_2}(x_2-x_1)^{r_1}+2x_2^{r_1}(y_2-y_1)^{r_2}$$
$$+x_1^{r_1}y_1^{r_2}-x_2^{r_1}y_2^{r_2}-2(x_2-x_1)^{r_1}(y_2-y_1)^{r_2}].$$

As $x_1\to x_2$, $y_1\to y_2$ the right-hand side of the above inequality tends to zero. The equicontinuity for the cases $x_1<x_2<0$, $y_1<y_2<0$ and $x_1\le 0\le x_2$, $y_1\le 0\le y_2$ is obvious. As a consequence of Steps 1–3, together with the Arzela–Ascoli theorem , we can conclude that $P: C_0\to C_0$ is continuous and completely continuous.

Step 4: *A priori bounds*. We now show that there exists an open set $U\subseteq C_0$ with $w\neq\lambda P(w)$, for $\lambda\in(0,1)$ and $w\in\partial U$. Let $w\in C_0$ and $w=\lambda P(w)$ for some $0<\lambda<1$. Thus for each $(x,y)\in J$,

$$w(x,y)=\frac{\lambda}{\Gamma(r_1)\Gamma(r_2)}\int_0^x\int_0^y(x-s)^{r_1-1}(y-t)^{r_2-1}f(s,t,u_{(s,t)})\mathrm{d}t\mathrm{d}s.$$

This implies by (3.30.1) that, for each $(x, y) \in J$, we have

$$\|w(x,y)\| \leq \frac{1}{\Gamma(r_1)\Gamma(r_2)} \int_0^x \int_0^y (x-s)^{r_1-1}(y-t)^{r_2-1}[p(s,t)$$

$$+q(s,t)\|\overline{w}_{(s,t)} + v_{(s,t)}\|_{\mathcal{B}}]dt ds$$

$$\leq \frac{\|p\|_\infty a^{r_1} b^{r_2}}{\Gamma(r_1+1)\Gamma(r_2+1)} + \frac{1}{\Gamma(r_1)\Gamma(r_2)} \int_0^x \int_0^y (x-s)^{r_1-1}(y-t)^{r_2-1} q(s,t)$$

$$\times \|\overline{w}_{(s,t)} + v_{(s,t)}\|_{\mathcal{B}} dt ds.$$

But

$$\begin{aligned} \|\overline{w}_{(s,t)} + v_{(s,t)}\|_{\mathcal{B}} &\leq \|\overline{w}_{(s,t)}\|_{\mathcal{B}} + \|v_{(s,t)}\|_{\mathcal{B}} \\ &\leq K \sup\{w(\tilde{s},\tilde{t}) : (\tilde{s},\tilde{t}) \in [0,s] \times [0,t]\} \\ &\quad + M\|\phi\|_{\mathcal{B}} + K\|\phi(0,0)\|. \end{aligned} \tag{3.63}$$

If we name $z(s,t)$ the right-hand side of (3.63), then we have

$$\|\overline{w}_{(s,t)} + v_{(s,t)}\|_{\mathcal{B}} \leq z(x,y),$$

and therefore, for each $(x, y) \in J$ we obtain

$$\|w(x,y)\| \leq \frac{\|p\|_\infty a^{r_1} b^{r_2}}{\Gamma(r_1+1)\Gamma(r_2+1)}$$

$$+\frac{1}{\Gamma(r_1)\Gamma(r_2)} \int_0^x \int_0^y (x-s)^{r_1-1}(y-t)^{r_2-1} q(s,t) z(s,t) dt ds. \tag{3.64}$$

Using the above inequality and the definition of z for each $(x, y) \in J$, we have

$$z(x,y) \leq M\|\phi\|_{\mathcal{B}} + K\|\phi(0,0)\| + \frac{K\|p\|_\infty a^{r_1} b^{r_2}}{\Gamma(r_1+1)\Gamma(r_2+1)}$$

$$+\frac{K\|p\|_\infty}{\Gamma(r_1)\Gamma(r_2)} \int_0^x \int_0^y (x-s)^{r_1-1}(y-t)^{r_2-1} z(s,t) dt ds.$$

Then by Lemma 2.43, there exists $\delta = \delta(r_1, r_2)$ such that we have

$$\|z(x,y)\| \leq R + \delta \frac{K\|q\|_\infty}{\Gamma(r_1)\Gamma(r_2)} \int_0^x \int_0^y (x-s)^{r_1-1}(y-t)^{r_2-1} R dt ds,$$

where

$$R = M\|\phi\|_{\mathcal{B}} + K\|\phi(0,0)\| + \frac{K\|p\|_\infty a^{r_1}b^{r_2}}{\Gamma(r_1+1)\Gamma(r_2+1)}.$$

Hence

$$\|z\|_\infty \le R + \frac{R\delta K\|q\|_\infty a^{r_1}b^{r_2}}{\Gamma(r_1+1)\Gamma(r_2+1)} := \widetilde{M}.$$

Then, (3.64) implies that

$$\|w\|_\infty \le \frac{a^{r_1}b^{r_2}}{\Gamma(r_1+1)\Gamma(r_2+1)}(\|p\|_\infty + \widetilde{M}\|q\|_\infty) := M^*.$$

Set

$$U = \{w \in C_0 : \|w\|_{(a,b)} < M^* + 1\}.$$

$P : \overline{U} \to C_0$ is continuous and completely continuous. By our choice of U, there is no $w \in \partial U$ such that $w = \lambda P(w)$, for $\lambda \in (0,1)$. As a consequence of the nonlinear alternative of Leray–Schauder type [136], we deduce that N has a fixed point which is a solution to problem (3.54)–(3.56). □

Now we present two similar existence results for the problem (3.57)–(3.59).

Definition 3.31. A function $u \in \Omega$ is said to be a solution of (3.57)–(3.59) if u satisfies (3.57) and (3.59) on J and the condition (3.58) on $\tilde{J}'$.

Theorem 3.32. *Assume that (3.29.1) holds and moreover*
(3.32.1) There exists a nonnegative constant ℓ' such that

$$\|g(x,y,u) - g(x,y,v)\| \le \ell'\|u-v\|_{\mathcal{B}},\ \textit{for each } (x,y) \in J,\ \textit{and } u,\ v \in \mathcal{B}.$$

If

$$K\left[4\ell' + \frac{\ell a^{r_1}b^{r_2}}{\Gamma(r_1+1)\Gamma(r_2+1)}\right] < 1, \tag{3.65}$$

then there exists a unique solution for IVP (3.57)–(3.59) on $(-\infty,a]\times(-\infty,b]$.

Proof. Consider the operator $N_1 : \Omega \to \Omega$ defined by

$$N_1(u)(x,y) = \begin{cases} \phi(x,y), & (x,y) \in \tilde{J}', \\ \mu(x,y) + g(x,y,u_{(x,y)}) - g(x,0,u_{(x,0)}) & \\ -g(0,y,u_{(0,y)}) + g(0,0,u_{(0,0)}) & \\ +\frac{1}{\Gamma(r_1)\Gamma(r_2)}\int\limits_0^x\int\limits_0^y (x-s)^{r_1-1}(y-t)^{r_2-1} & \\ \times f(s,t,u_{(s,t)})\mathrm{d}t\mathrm{d}s, & (x,y) \in J. \end{cases}$$

In analogy to Theorem 3.29, we consider the operator $P_1 : C_0 \to C_0$ defined by

$$\begin{aligned} P_1(x,y) &= g(x,y,\overline{w}_{(x,y)} + v_{(x,y)}) - g(x,0,\overline{w}_{(x,0)} + v_{(x,0)}) \\ &\quad - g(0,y,\overline{w}_{(0,y)} + v_{(0,y)}) + g(0,0,\overline{w}_{(0,0)} + v_{(0,0)}) \\ &\quad + \frac{1}{\Gamma(r_1)\Gamma(r_2)} \int_0^x \int_0^y (x-s)^{r_1-1}(y-t)^{r_2-1} f(s,t,\overline{w}_{(s,t)} + v_{(s,t)}) \mathrm{d}t\,\mathrm{d}s, \\ &\quad (x,y) \in J. \end{aligned}$$

We shall show that the operator P_1 is a contraction. Let $w,\ w_* \in C_0$, then following the steps of Theorem 3.29, we have

$$\begin{aligned} &\|P_1(w)(x,y) - P_1(w_*)(x,y)\| \\ &\leq \|g(x,y,\overline{w}_{(x,y)} + v_{(x,y)}) - g(x,y,\overline{w_*}_{(x,y)} + v_{(x,y)})\| \\ &\quad + \|g(x,0,\overline{w}_{(x,0)} + v_{(x,0)}) - g(x,0,\overline{w_*}_{(x,0)} + v_{(x,0)})\| \\ &\quad + \|g(0,y,\overline{w}_{(0,y)} + v_{(0,y)}) - g(0,y,\overline{w_*}_{(0,y)} + v_{(0,y)})\| \\ &\quad + \|g(0,0,\overline{w} + v) - g(0,0,\overline{w_*}_{(0,0)} + v_{(0,0)})\| \\ &\quad + \frac{1}{\Gamma(r_1)\Gamma(r_2)} \int_0^x \int_0^y (x-s)^{r_1-1}(y-t)^{r_2-1} \\ &\quad \times \|f(s,t,\overline{w}_{(s,t)} + v_{(s,t)}) - f(s,t,\overline{w_*}_{(s,t)} + v_{(s,t)})\| \mathrm{d}t\,\mathrm{d}s \\ &\leq 4\ell' K \|\overline{w} - \overline{w_*}\|_{(a,b)} + \frac{1}{\Gamma(r_1)\Gamma(r_2)} \int_0^x \int_0^y (x-s)^{r_1-1}(y-t)^{r_2-1} \\ &\quad \times \ell K \|\overline{w} - \overline{w_*}\| \mathrm{d}t\,\mathrm{d}s. \end{aligned}$$

Therefore

$$\|P_1(w) - P_1(w_*)\|_{(a,b)} \leq K \left[4\ell' + \frac{\ell a^{r_1} b^{r_2}}{\Gamma(r_1+1)\Gamma(r_2)+1} \right] \|\overline{w} - \overline{w_*}\|_{(a,b)}.$$

which implies by (3.65) that P_1 is a contraction. Hence P_1 has a unique fixed point by Banach's contraction principle. □

Our last existence result for the IVP (3.57)–(3.59) is based on the nonlinear alternative of Leray–Schauder type.

Theorem 3.33. *Assume (3.29.1)–(3.30.1) and the following conditions:*

(3.33.1) *The function g is continuous and completely continuous, and for any bounded set D in Ω, the set $\{(x,y) \to g(x,y,u_{(x,y)} : u \in D\}$, is equicontinuous in $C(J,\mathbb{R}^n)$,*

(3.33.2) There exist constants $d_1, d_2 \geq 0$ *such that* $0 \leq d_1 K < \frac{1}{4}$ *and*

$$\|g(x, y, u)\| \leq d_1 \|u\|_{\mathcal{B}} + d_2, \quad (x, y) \in J, \quad u \in \mathcal{B}.$$

Then the IVP (3.57)–(3.59) have at least one solution on $(-\infty, a] \times (-\infty, b]$.

Proof. Let $P_1 : C_0 \to C_0$ defined as in Theorem 3.32. We shall show that the operator P_1 is continuous and completely continuous. Using (3.33.1) it suffices to show that the operator P defined in (3.62) is continuous and completely continuous. This was proved in Theorem 3.30. We now show there exists an open set $U \subseteq C_0$ with $w \neq \lambda P_1(w)$, for $\lambda \in (0, 1)$ and $w \in \partial U$. Let $w \in C_0$ and $w = \lambda p_1(w)$ for some $0 < \lambda < 1$. Thus for each $(x, y) \in J$,

$$\begin{aligned} w(x, y) = & \lambda[g(x, y, \overline{w}_{(x,y)} + v_{(x,y)}) - g(x, 0, \overline{w}_{(x,0)} + v_{(x,0)}) \\ & - g(0, y, \overline{w}_{(0,y)} + v_{(0,y)}) + g(0, 0, \overline{w}_{(0,0)} + v_{(0,0)})] \\ & + \frac{\lambda}{\Gamma(r_1)\Gamma(r_2)} \int_0^x \int_0^y (x - s)^{r_1 - 1}(y - t)^{r_2 - 1} f(s, t, \overline{w}_{(s,t)} + v_{(s,t)}) dt ds, \end{aligned}$$

and

$$\begin{aligned} \|w(x, y)\| = & 4d_1 \|\overline{w}_{(x,y)} + v_{(x,y)}\|_{\mathcal{B}} + \frac{\|p\|_\infty a^{r_1} b^{r_2}}{\Gamma(r_1 + 1)\Gamma(r_2 + 1)} \\ & + \frac{1}{\Gamma(r_1)\Gamma(r_2)} \int_0^x \int_0^y (x-s)^{r_1-1}(y-t)^{r_2-1} q(s, t) \|\overline{w}_{(s,t)} + v_{(s,t)}\|_{\mathcal{B}} dt ds. \end{aligned}$$

Using the above inequality and the definition of z we have that

$$\|z\|_\infty \leq R_1 + \frac{R_1 \delta K \|q^*\|_\infty a^{r_1} b^{r_2}}{(1 - 4d_1 K)\Gamma(r_1 + 1)\Gamma(r_2 + 1)} := L,$$

where

$$R_1 = \frac{1}{1 - 4d_1 K} \left[8d_2 K + \frac{K \|p\|_\infty a^{r_1} b^{r_2}}{\Gamma(r_1 + 1)\Gamma(r_2 + 1)} \right],$$

and

$$\|q^*\|_\infty = \frac{\|q\|_\infty}{1 - 4d_1 K}.$$

Then

$$\|w\|_\infty \leq 4d_1 \|\phi\|_{\mathcal{B}} + 8d_2 + 4Ld_1 + \frac{a^{r_1} b^{r_2}}{\Gamma(r_1+1)\Gamma(r_2+1)} (\|p\|_\infty + L\|q\|_\infty) := L^*.$$

Set

$$U_1 = \{w \in C_0 : \|w\|_{(a,b)} < L^* + 1\}.$$

By our choice of U_1, there is no $w \in \partial U$ such that $w = \lambda P_1(w)$, for $\lambda \in (0,1)$. As a consequence of the nonlinear alternative of Leray–Schauder type [136], we deduce that N_1 has a fixed point which is a solution to problem (3.57)–(3.59). □

3.7.4 An Example

As an application of our results we consider the following partial hyperbolic functional differential equations of the form:

$$({}^cD_0^r u)(x,y) = \frac{ce^{x+y-\gamma(x+y)}\|u_{(x,y)}\|}{(e^{x+y}+e^{-x-y})(1+\|u_{(x,y)}\|)}, \quad \text{if } (x,y) \in J := [0,1]\times[0,1], \tag{3.66}$$

$$u(x,0) = x, \; u(0,y) = y^2, \; x \in [0,1], y \in [0,1], \tag{3.67}$$

$$u(x,y) = x + y^2, \; (x,y) \in \tilde{J}', \tag{3.68}$$

where $\tilde{J}' := (-\infty,1]\times(-\infty,1]\backslash(0,1]\times(0,1]$, $c = \frac{2}{\Gamma(r_1+1)\Gamma(r_2+1)}$ and γ a positive real constant.

Let

$$\mathcal{B}_\gamma = \left\{u \in C((-\infty,0]\times(-\infty,0],\mathbb{R}) : \lim_{\|(\theta,\eta)\|\to\infty} e^{\gamma(\theta+\eta)}u(\theta,\eta) \text{ exists in } \mathbb{R}\right\}.$$

The norm of $\mathcal{B}_\gamma$ is given by

$$\|u\|_\gamma = \sup_{(\theta,\eta)\in(-\infty,0]\times(-\infty,0]} e^{\gamma(\theta+\eta)}|u(\theta,\eta)|.$$

Let

$$E := [0,1]\times\{0\}\cup\{0\}\times[0,1],$$

and $u : (-\infty,1]\times(-\infty,1] \to \mathbb{R}$ such that $u_{(x,y)} \in \mathcal{B}_\gamma$ for $(x,y) \in E$, then

$$\begin{aligned}\lim_{\|(\theta,\eta)\|\to\infty} e^{\gamma(\theta+\eta)}u_{(x,y)}(\theta,\eta) &= \lim_{\|(\theta,\eta)\|\to\infty} e^{\gamma(\theta-x+\eta-y)}u(\theta,\eta)\\ &= e^{-\gamma(x+y)}\lim_{\|(\theta,\eta)\|\to\infty} e^{\gamma(\theta+\eta)}u(\theta,\eta) < \infty.\end{aligned}$$

Hence $u_{(x,y)} \in \mathcal{B}_\gamma$. Finally we prove that

$$\|u_{(x,y)}\|_\gamma = K \sup\{|u(s,t)| : (s,t) \in [0,x] \times [0,y]\} + M \sup\{\|u_{(s,t)}\|_\gamma : (s,t) \in E_{(x,y)}\},$$

where $K = M = 1$ and $H = 1$.

If $x + \theta \leq 0, \ y + \eta \leq 0$ we get

$$\|u_{(x,y)}\|_\gamma = \sup\{|u(s,t)| : (s,t) \in (-\infty, 0] \times (-\infty, 0]\},$$

and if $x + \theta \geq 0, \ y + \eta \geq 0$ then we have

$$\|u_{(x,y)}\|_\gamma = \sup\{|u(s,t)| : (s,t) \in [0,x] \times [0,y]\}.$$

Thus for all $(x + \theta, y + \eta) \in [0,1] \times [0,1]$, we get

$$\|u_{(x,y)}\|_\gamma = \sup\{|u(s,t)| : (s,t) \in (-\infty, 0] \times (-\infty, 0]\} + \sup\{|u(s,t)| : (s,t) \in [0,x] \times [0,y]\}.$$

Then

$$\|u_{(x,y)}\|_\gamma = \sup\{\|u_{(s,t)}\|_\gamma : (s,t) \in E\} + \sup\{|u(s,t)| : (s,t) \in [0,x] \times [0,y]\}.$$

$(\mathcal{B}_\gamma, \|.\|_\gamma)$ is a Banach space. We conclude that $\mathcal{B}_\gamma$ is a phase space . Set

$$f(x,y,u_{(x,y)}) = \frac{c e^{x+y-\gamma(x+y)} \|u_{(x,y)}\|}{(e^{x+y} + e^{-x-y})(1 + \|u_{(x,y)})\|}, \quad (x,y) \in [0,1] \times [0,1].$$

For each $u, \ \overline{u} \in \mathcal{B}_\gamma$ and $(x,y) \in [0,1] \times [0,1]$ we have

$$|f(x,y,u_{(x,y)}) - f(x,y,\overline{u}_{(x,y)})| \leq \frac{e^{x+y} \|u - \overline{u}\|_B}{c(e^{x+y} + e^{-x-y})} \leq \frac{1}{c} \|u - \overline{u}\|_B.$$

Hence condition (3.29.1) is satisfied with $\ell = \frac{1}{c}$. Since $a = b = K = 1$ we get

$$\frac{\ell a^{r_1} b^{r_2} K}{\Gamma(r_1 + 1)\Gamma(r_2 + 1)} = \frac{1}{c\Gamma(r_1 + 1)\Gamma(r_2 + 1)} = \frac{1}{2} < 1,$$

for each $(r_1, r_2) \in (0,1] \times (0,1]$. Consequently Theorem 3.29 implies that problem (3.66)–(3.68) has a unique solution defined on $(-\infty, 1] \times (-\infty, 1]$.

3.8 Partial Hyperbolic Differential Equations with State-Dependent Delay

3.8.1 Introduction

The first result of this section deals with the existence and uniqueness of solutions to fractional order IVP, for the system

$$({}^cD_0^r u)(x,y) = f(x,y,u_{(\rho_1(x,y,u_{(x,y)}),\rho_2(x,y,u_{(x,y)}))}), \text{ if } (x,y) \in J, \tag{3.69}$$

$$u(x,y) = \phi(x,y), \text{ if } (x,y) \in \tilde{J} := [-\alpha,a] \times [-\beta,b] \backslash (0,a] \times (0,b], \tag{3.70}$$

$$u(x,0) = \varphi(x),\ x \in [0,a],\ u(0,y) = \psi(y),\ y \in [0,b], \tag{3.71}$$

where $J = [0,a] \times [0,b]$, $a,b,\alpha,\beta > 0$, $\phi \in C(\tilde{J},\mathbb{R}^n)$, $f : J \times C \to \mathbb{R}^n$, $\rho_1 : J \times C \to [-\alpha,a]$, $\rho_2 : J \times C \to [-\beta,b]$ are given functions, φ, ψ are as in problem (3.1)–(3.3) and $C := C([-\alpha,0] \times [-\beta,0],\mathbb{R}^n)$.

Next we consider the following system of partial neutral hyperbolic differential equations of fractional order:

$$\begin{aligned} {}^cD_0^r[u(x,y) - g(x,y,u_{(\rho_1(x,y,u_{(x,y)}),\rho_2(x,y,u_{(x,y)}))})] \\ = f(x,y,u_{(\rho_1(x,y,u_{(x,y)}),\rho_2(x,y,u_{(x,y)}))}), \text{ if } (x,y) \in J, \end{aligned} \tag{3.72}$$

$$u(x,y) = \phi(x,y), \text{ if } (x,y) \in \tilde{J}, \tag{3.73}$$

$$u(x,0) = \varphi(x),\ x \in [0,a],\ u(0,y) = \psi(y),\ y \in [0,b], \tag{3.74}$$

where f, ρ_1, ρ_2, ϕ, φ, ψ are as in problem (3.69)–(3.71) and $g : J \times C \to \mathbb{R}^n$ is a given continuous function.

The third result deals with the existence of solutions to fractional order partial differential equations

$$({}^cD_0^r u)(x,y) = f(x,y,u_{(\rho_1(x,y,u_{(x,y)}),\rho_2(x,y,u_{(x,y)}))}), \text{ if } (x,y) \in J, \tag{3.75}$$

$$u(x,y) = \phi(x,y), \text{ if } (x,y) \in \tilde{J}' := [-\infty,a] \times [-\infty,b] \backslash (0,a] \times (0,b], \tag{3.76}$$

$$u(x,0) = \varphi(x),\ x \in [0,a],\ u(0,y) = \psi(y),\ y \in [0,b], \tag{3.77}$$

where φ, ψ are as in problem (3.69)–(3.71), $f : J \times \mathcal{B} \to \mathbb{R}^n$, $\phi \in C(\tilde{J}',\mathbb{R}^n)$, $\rho_1 : J \times \mathcal{B} \to (-\infty,a]$, $\rho_2 : J \times \mathcal{B} \to (-\infty,b]$ and $\mathcal{B}$ is a phase space .

Finally we consider the following initial value problem for partial neutral functional differential equations:

$$\begin{aligned} {}^cD_0^r[u(x,y) - g(x,y,u_{(\rho_1(x,y,u_{(x,y)}),\rho_2(x,y,u_{(x,y)}))})] \\ = f(x,y,u_{(\rho_1(x,y,u_{(x,y)}),\rho_2(x,y,u_{(x,y)}))}), \text{ if } (x,y) \in J, \end{aligned} \tag{3.78}$$

$$u(x, y) = \phi(x, y), \text{ if } (x, y) \in \tilde{J}', \tag{3.79}$$

$$u(x, 0) = \varphi(x), \ x \in [0, a], \ u(0, y) = \psi(y), \ y \in [0, b], \tag{3.80}$$

where $f, \rho_1, \rho_2, \phi, \varphi, \psi$ are as in problems (3.75)–(3.77) and $g : J \times \mathcal{B} \to \mathbb{R}^n$ is a given continuous function.

3.8.2 Existence of Solutions for Finite Delay

Definition 3.34. A function $u \in C_{(a,b)} := C([-\alpha, a] \times [-\beta, b], \mathbb{R}^n)$ is said to be a solution of (3.69)–(3.71) if u satisfies (3.69) and (3.71) on J and the condition (3.70) on $\tilde{J}$.

Set $\mathcal{R} := \mathcal{R}_{(\rho_1^-, \rho_2^-)}$

$$= \{(\rho_1(s, t, u), \rho_2(s, t, u)) : (s, t, u) \in J \times C, \ \rho_i(s, t, u) \le 0; \ i = 1, 2\}.$$

We always assume that $\rho_1 : J \times C \to [-\alpha, a], \ \rho_2 : J \times C \to [-\beta, b]$ are continuous and the function $(s, t) \longmapsto u_{(s,t)}$ is continuous from $\mathcal{R}$ into C.

Further, we present conditions for the existence and uniqueness of a solution of problem (3.69)–(3.71).

Theorem 3.35. *Let us assume that the following hypotheses hold:*

(3.35.1) The $f : J \times C \to \mathbb{R}^n$ *is continuous.*
(3.35.2) There exists $k > 0$ *such that*

$$\|f(x, y, u) - f(x, y, v)\| \le k\|u - v\|_C, \textit{ for any } u, \ v \in C \textit{ and } (x, y) \in J.$$

If

$$\frac{k a^{r_1} b^{r_2}}{\Gamma(r_1 + 1)\Gamma(r_2 + 1)} < 1, \tag{3.81}$$

then there exists a unique solution for IVP (3.69)–(3.71) on $[-\alpha, a] \times [-\beta, b]$.

Proof. Transform the problem (3.69)–(3.71) into a fixed point problem. Consider the operator $N : C_{(a,b)} \to C_{(a,b)}$ defined by

$$N(x, y) = \begin{cases} \phi(x, y); & (x, y) \in \tilde{J}, \\ \mu(x, y) + \frac{1}{\Gamma(r_1)\Gamma(r_2)} \int_0^x \int_0^y (x - s)^{r_1 - 1}(y - t)^{r_2 - 1} & \\ \times f(s, t, u_{(\rho_1(s,t,u_{(s,t)}), \rho_2(s,t,u_{(s,t)}))}) dt ds; & (x, y) \in J. \end{cases} \tag{3.82}$$

Let $v, w \in C_{(a,b)}$. Then, for $(x, y) \in [-\alpha, a] \times [-\beta, b]$,

$$\begin{aligned}
&\|N(v)(x,y) - N(w)(x,y)\| \\
&\leq \frac{1}{\Gamma(r_1)\Gamma(r_2)} \int_0^x \int_0^y |(x-s)^{r_1-1}||(y-t)^{r_2-1}| \\
&\quad \times \| f(s,t,v_{(\rho_1(s,t,u_{(s,t)}),\rho_2(s,t,u_{(s,t)}))}) - f(s,t,w_{(\rho_1(s,t,u_{(s,t)}),\rho_2(s,t,u_{(s,t)}))})\| \mathrm{d}t\mathrm{d}s \\
&\leq \frac{k}{\Gamma(r_1)\Gamma(r_2)} \int_0^x \int_0^y (x-s)^{r_1-1}(y-t)^{r_2-1} \| v_{(\rho_1(s,t,u_{(s,t)}),\rho_2(s,t,u_{(s,t)}))} \\
&\quad - w_{(\rho_1(s,t,u_{(s,t)}),\rho_2(s,t,u_{(s,t)}))}\|_C \mathrm{d}t\mathrm{d}s \\
&\leq \frac{k}{\Gamma(r_1)\Gamma(r_2)} \|v - w\|_C \int_0^x \int_0^y (x-s)^{r_1-1}(y-t)^{r_2-1} \mathrm{d}t\mathrm{d}s.
\end{aligned}$$

Consequently,

$$\|N(v) - N(w)\|_{C_{(a,b)}} \leq \frac{k a^{r_1} b^{r_2}}{\Gamma(r_1+1)\Gamma(r_2+1)} \|v - w\|_C.$$

By (3.81), N is a contraction, and hence N has a unique fixed point by Banach's contraction principle. □

Theorem 3.36. *Assume (3.35.1) and the following hypothesis hold:*

(3.36.1) There exist $p, q \in C(J, \mathbb{R}_+)$ such that

$$\|f(x,y,u)\| \leq p(x,y) + q(x,y)\|u\|_C, \textit{ for each } u \in C \textit{ and } (x,y) \in J.$$

Then the IVP (3.69)–(3.71) have at least one solution on $[-\alpha, a] \times [-\beta, b]$.

Proof. Transform the problems (3.69)–(3.71) into a fixed point problem. Consider the operator N defined in (3.82). We shall show that the operator N is continuous and completely continuous.

Step 1: *N is continuous.* Let $\{u_n\}$ be a sequence such that $u_n \to u$ in $C_{(a,b)}$. Let $\eta > 0$ be such that $\|u_n\| \leq \eta$. Then

$$\begin{aligned}
&\|N(u_n)(x,y) - N(u)(x,y)\| \\
&\leq \frac{1}{\Gamma(r_1)\Gamma(r_2)} \int_0^x \int_0^y |(x-s)^{r_1-1}(y-t)^{r_2-1}| \\
&\quad \times \| f(s,t,u_{n(\rho_1(s,t,u_{n(s,t)}),\rho_2(s,t,u_{n(s,t)}))}) \\
&\quad - f(s,t,u_{(\rho_1(s,t,u_{(s,t)}),\rho_2(s,t,u_{(s,t)}))})\| dt ds \\
&\leq \frac{1}{\Gamma(r_1)\Gamma(r_2)} \int_0^x \int_0^y (x-s)^{r_1-1}(y-t)^{r_2-1} \\
&\quad \times \sup_{(s,t)\in J} \| f(s,t,u_{n(\rho_1(s,t,u_{n(s,t)}),\rho_2(s,t,u_{n(s,t)}))}) \\
&\quad - f(s,t,u_{(\rho_1(s,t,u_{(s,t)}),\rho_2(s,t,u_{(s,t)}))})\| dt ds \\
&\leq \frac{\|f(.,.,u_{n(.,.)}) - f(.,.,u)\|_\infty}{\Gamma(r_1)\Gamma(r_2)} \int_0^x \int_0^y (x-s)^{r_1-1}(y-t)^{r_2-1} dt ds.
\end{aligned}$$

Since f is a continuous function, we have

$$\|N(u_n) - N(u)\|_\infty \leq \frac{a^{r_1} b^{r_2} \|f(.,.,u_{n(.,.)}) - f(.,.,u_{(.,.)})\|_\infty}{\Gamma(r_1+1)\Gamma(r_2+1)} \to 0 \text{ as } n \to \infty.$$

Step 2: *N maps bounded sets into bounded sets in $C_{(a,b)}$.* Indeed, it is enough to show that, for any $\eta^* > 0$, there exists a positive constant $\widetilde{\ell}$ such that, for each $u \in B_{\eta^*} = \{u \in C_{(a,b)} : \|u\|_\infty \leq \eta^*\}$, we have $\|N(u)\|_\infty \leq \widetilde{\ell}$. By (3.36.1) we have for each $(x,y) \in J$,

$$\begin{aligned}
\|N(u)(x,y)\| &\leq \|\mu(x,y)\| + \frac{1}{\Gamma(r_1)\Gamma(r_2)} \int_0^x \int_0^y (x-s)^{r_1-1}(y-t)^{r_2-1} \\
&\quad \times \| f(s,t,u_{(\rho_1(s,t,u_{(s,t)}),\rho_2(s,t,u_{(s,t)}))})\| dt ds \\
&\leq \|\mu(x,y)\| + \frac{1}{\Gamma(r_1)\Gamma(r_2)} \int_0^x \int_0^y (x-s)^{r_1-1}(y-t)^{r_2-1} p(s,t) dt ds \\
&\quad + \frac{1}{\Gamma(r_1)\Gamma(r_2)} \int_0^x \int_0^y (x-s)^{r_1-1}(y-t)^{r_2-1} q(s,t)
\end{aligned}$$

$$\times \|u_{(\rho_1(s,t,u_{(s,t)}),\rho_2(s,t,u_{(s,t)}))}\|_\infty \mathrm{d}t\mathrm{d}s$$

$$\leq \|\mu(x,y)\| + \frac{\|p\|_\infty}{\Gamma(r_1)\Gamma(r_2)}\int_0^x\int_0^y (x-s)^{r_1-1}(y-t)^{r_2-1}\mathrm{d}t\mathrm{d}s$$

$$+\frac{\|q\|_\infty \eta^*}{\Gamma(r_1)\Gamma(r_2)}\int_0^x\int_0^y (x-s)^{r_1-1}(y-t)^{r_2-1}\mathrm{d}t\mathrm{d}s.$$

Thus

$$\|N(u)\|_\infty \leq \|\mu\|_\infty + \frac{\|p\|_\infty + \|q\|_\infty\eta^*}{\Gamma(r_1+1)\Gamma(r_2+1)}a^{r_1}b^{r_2} := \widetilde{\ell}.$$

Step 3: *N maps bounded sets into equicontinuous sets in $C_{(a,b)}$.* Let $(x_1,y_1),(x_2,y_2) \in (0,a]\times(0,b]$, $x_1<x_2$, $y_1<y_2$, B_{η^*} be a bounded set of $C_{(a,b)}$ as in Step 2, and let $u\in B_{\eta^*}$. Then

$$\|N(u)(x_2,y_2) - N(u)(x_1,y_1)\| \leq \|\mu(x_1,y_1)-\mu(x_2,y_2)\|$$

$$+\frac{1}{\Gamma(r_1)\Gamma(r_2)}\Bigg\|\int_0^{x_1}\int_0^{y_1}\left[(x_2-s)^{r_1-1}(y_2-t)^{r_2-1}-(x_1-s)^{r_1-1}(y_1-t)^{r_2-1}\right]$$

$$\times f(s,t,u_{(\rho_1(s,t,u_{(s,t)}),\rho_2(s,t,u_{(s,t)}))})\mathrm{d}t\mathrm{d}s$$

$$+\frac{1}{\Gamma(r_1)\Gamma(r_2)}\int_{x_1}^{x_2}\int_{y_1}^{y_2}(x_2-s)^{r_1-1}(y_2-t)^{r_2-1}f(s,t,u_{(\rho_1(s,t,u_{(s,t)}),\rho_2(s,t,u_{(s,t)}))})\mathrm{d}t\mathrm{d}s\Bigg\|$$

$$+\frac{1}{\Gamma(r_1)\Gamma(r_2)}\int_0^{x_1}\int_{y_1}^{y_2}(x_2-s)^{r_1-1}(y_2-t)^{r_2-1}\|f(s,t,u_{(\rho_1(s,t,u_{(s,t)}),\rho_2(s,t,u_{(s,t)}))})\|\mathrm{d}t\mathrm{d}s$$

$$+\frac{1}{\Gamma(r_1)\Gamma(r_2)}\int_{x_1}^{x_2}\int_0^{y_1}(x_2-s)^{r_1-1}(y_2-t)^{r_2-1}\|f(s,t,u_{(\rho_1(s,t,u_{(s,t)}),\rho_2(s,t,u_{(s,t)}))})\|\mathrm{d}t\mathrm{d}s$$

$$\leq \|\mu(x_1,y_1)-\mu(x_2,y_2)\|$$

$$+\frac{\|p\|_\infty+\|q\|_\infty\eta^*}{\Gamma(r_1+1)\Gamma(r_2+1)}[2y_2^{r_2}(x_2-x_1)^{r_1}+2x_2^{r_1}(y_2-y_1)^{r_2}$$

$$+x_1^{r_1}y_1^{r_2}-x_2^{r_1}y_2^{r_2}-2(x_2-x_1)^{r_1}(y_2-y_1)^{r_2}].$$

As $x_1\to x_2$, $y_1\to y_2$ the right-hand side of the above inequality tends to zero. The equicontinuity for the cases $x_1<x_2<0$, $y_1<y_2<0$ and $x_1\leq 0\leq x_2$, $y_1\leq 0\leq y_2$ is obvious. As a consequence of Steps 1–3, together with the Arzela–Ascoli theorem, we can conclude that $N : C_{(a,b)} \to C_{(a,b)}$. is continuous and completely continuous.

Step 4: *A priori bounds.* We now show that there exists an open set $U \subseteq C_{(a,b)}$. with $u \neq \lambda N(u)$, for $\lambda \in (0,1)$ and $u \in \partial U$. Let $u \in C_{(a,b)}$.) and $u = \lambda N(u)$ for some $0 < \lambda < 1$. Thus for each $(x,y) \in J$,

$$u(x,y) = \lambda\mu(x,y) + \frac{\lambda}{\Gamma(r_1)\Gamma(r_2)} \int_0^x \int_0^y (x-s)^{r_1-1}(y-t)^{r_2-1} \times f(s,t,u_{(\rho_1(s,t,u_{(s,t)}),\rho_2(s,t,u_{(s,t)}))})\mathrm{d}t\,\mathrm{d}s.$$

This implies by (3.36.1) that, for each $(x,y) \in J$, we have

$$\begin{aligned}\|u(x,y)\| &\leq \|\mu(x,y)\| + \frac{1}{\Gamma(r_1)\Gamma(r_2)} \int_0^x \int_0^y (x-s)^{r_1-1}(y-t)^{r_2-1} \\ &\quad \times [p(s,t) + q(s,t)\|u_{(\rho_1(s,t,u_{(s,t)}),\rho_2(s,t,u_{(s,t)}))}\|_C]\mathrm{d}t\,\mathrm{d}s \\ &\leq \|\mu(x,y)\| + \frac{\|p\|_\infty}{\Gamma(r_1+1)\Gamma(r_2+1)} a^{r_1} b^{r_2} \\ &\quad + \frac{1}{\Gamma(r_1)\Gamma(r_2)} \int_0^x \int_0^y (x-s)^{r_1-1}(y-t)^{r_2-1} q(s,t)\|u_{(s,t)}\|_C \mathrm{d}t\,\mathrm{d}s.\end{aligned}$$

We consider the function τ defined by

$$\tau(x,y) = \sup\{\|u(s,t)\| : -\alpha \leq s \leq x,\ -\beta \leq t \leq y,\ 0 \leq x \leq a,\ 0 \leq y \leq b\}.$$

Let $(x^*, y^*) \in [-\alpha, x] \times [-\beta, y]$ be such that $\tau(x,y) = \|u(x^*, y^*)\|$. If $(x^*, y^*) \in J$, then by the previous inequality, we have for $(x,y) \in J$,

$$\begin{aligned}\tau(x,y) &\leq \|\mu(x,y)\| + \frac{\|p\|_\infty}{\Gamma(r_1+1)\Gamma(r_2+1)} a^{r_1} b^{r_2} \\ &\quad + \frac{1}{\Gamma(r_1)\Gamma(r_2)} \int_0^x \int_0^y (x-s)^{r_1-1}(y-t)^{r_2-1} q(s,t)\tau(s,t)\mathrm{d}t\,\mathrm{d}s \\ &\leq \|\mu(x,y)\| + \frac{\|p\|_\infty}{\Gamma(r_1+1)\Gamma(r_2+1)} a^{r_1} b^{r_2} \\ &\quad + \frac{\|q\|_\infty}{\Gamma(r_1)\Gamma(r_2)} \int_0^x \int_0^y (x-s)^{r_1-1}(y-t)^{r_2-1} \tau(s,t)\mathrm{d}t\,\mathrm{d}s.\end{aligned}$$

If $(x^*, y^*) \in \tilde{J}$, then $\tau(x, y) = \|\phi\|_C$ and the previous inequality holds. By Lemma 2.43 we have

$$\tau(x, y) \le \left(\|\mu(x, y)\| + \frac{a^{r_1} b^{r_2} \|p\|_\infty}{\Gamma(r_1 + 1)\Gamma(r_2 + 1)} \right) \times \left(1 + \frac{\|q\|_\infty}{\Gamma(r_1)\Gamma(r_2)} \int_0^x \int_0^y (x - s)^{r_1 - 1}(y - t)^{r_2 - 1} dt ds \right)$$

$$\le \left(\|\mu(x, y)\| + \frac{a^{r_1} b^{r_2} \|p\|_\infty}{\Gamma(r_1 + 1)\Gamma(r_2 + 1)} \right) \left(\frac{a^{r_1} b^{r_2} \|q\|_\infty}{\Gamma(r_1 + 1)\Gamma(r_2 + 1)} \right) := M.$$

Since for every $(x, y) \in J$, $\|u_{(x,y)}\|_C \le \tau(x, y)$, we have

$$\|u\|_\infty \le \max(\|\phi\|_C, M) := R.$$

Set

$$U = \{u \in C_{(a,b)} : \|u\|_\infty < R + 1\}.$$

By our choice of U, there is no $u \in \partial U$ such that $u = \lambda N(u)$, for $\lambda \in (0, 1)$. As a consequence of the nonlinear alternative of Leray–Schauder type [136], we deduce that N has a fixed point u in $\overline{U}$ which is a solution to problem (3.69)–(3.71). □

Now, we present existence results for the problem (3.72)–(3.74). We give two results, the first one considered by using Banach's contraction principle (Theorem 3.38) and the second result is based on the nonlinear alternative of Leray–Schauder (Theorem 3.39).

Definition 3.37. A function $u \in C_{(a,b)}$ is said to be a solution of (3.72)–(3.74) if u satisfies equations (3.72) and (3.74) on J and the condition (3.73) on $\tilde{J}$.

Theorem 3.38. *Assume (3.35.1) and the following hypotheses hold:*

(3.38.1) The function $g : J \times C \to \mathbb{R}^n$ is continuous.
(3.38.2) There exists a constant $c_1 > 0$ such that for every $(x, y) \in J$

$$\|g(x, y, u) - g(x, y, v)\| \le c_1 \|u - v\|_C, \textit{ for any } u, v \in C$$

(3.38.3) There exists a constant $\ell > 0$ such that for every $(x, y) \in J$,

$$\|f(x, y, u) - f(x, y, v)\| \le \ell \|u - v\|_C, \textit{ for any } u, v \in C.$$

If

$$4c_1 + \frac{\ell a^{r_1} b^{r_2}}{\Gamma(r_1 + 1)\Gamma(r_2 + 1)} < 1, \tag{3.83}$$

then there exists a unique solution for the IVP (3.72)–(3.74) on $[-\alpha, a] \times [-\beta, b]$.

Proof. Consider the operator $N_1 : C_{(a,b)} \to C_{(a,b)}$ defined by

$$N_1(u)(x,y) = \begin{cases} \phi(x,y), & (x,y) \in \tilde{J}, \\ \mu(x,y) + g(x,y,u_{(x,y)}) \\ -g(x,0,u_{(x,0)}) - g(0,y,u_{(0,y)}) + g(0,0,u) \\ +\frac{1}{\Gamma(r_1)\Gamma(r_2)} \int_0^x \int_0^y (x-s)^{r_1-1}(y-t)^{r_2-1} \\ \times f(s,t,u_{(\rho_1(s,t,u_{(s,t)}),\rho_2(s,t,u_{(s,t)}))})dt\,ds, & (x,y) \in J, \end{cases} \tag{3.84}$$

We shall show that the operator N_1 is a contraction. Let $v, w \in C_{(a,b)}$. Then we have

$$\begin{aligned}
&\|N_1(v)(x,y)) - N_1(w)(x,y)\| \\
&\leq \|g(x,y,v_{(x,y)}) - g(x,y,w_{(x,y)})\| + \|g(x,0,v_{(x,0)}) - g(x,0,w_{(x,0)})\| \\
&\quad + \|g(0,y,v_{(0,y)}) - g(0,y,w_{(0,y)})\| + \|g(0,0,v) - g(0,0,w)\| \\
&\quad + \frac{1}{\Gamma(r_1)\Gamma(r_2)} \int_0^x \int_0^y |x-s|^{r_1-1}|y-t|^{r_2-1} \\
&\quad \times \|f(s,t,v_{(\rho_1(s,t,u_{(s,t)}),\rho_2(s,t,u_{(s,t)}))}) - f(s,t,w_{(\rho_1(s,t,u_{(s,t)}),\rho_2(s,t,u_{(s,t)}))})\|dt\,ds \\
&\leq c_1(\|v_{(x,y)} - w_{(x,y)}\|_C + \|v_{(x,0)} - w_{(x,0)}\|_C + \|v_{(0,y)} - w_{(0,y)}\|_C + \|v - w\|_C) \\
&\quad + \frac{1}{\Gamma(r_1)\Gamma(r_2)} \int_0^x \int_0^y [(x-s)^{r_1-1}(y-t)^{r_2-1} \\
&\quad \times \sup_{(s,t)\in J} \|f(s,t,v_{(\rho_1(s,t,u_{(s,t)}),\rho_2(s,t,u_{(s,t)}))}) \\
&\quad - f(s,t,w_{(\rho_1(s,t,u_{(s,t)}),\rho_2(s,t,u_{(s,t)}))})\|]dt\,ds \\
&\leq 4c_1\|v - w\|_{[-\alpha,a]\times[-\beta,b]} \\
&\quad + \frac{\ell}{\Gamma(r_1)\Gamma(r_2)}\|v - w\|_{[-\alpha,a]\times[-\beta,b]} \int_0^x \int_0^y (x-s)^{r_1-1}(y-t)^{r_2-1}dt\,ds \\
&\leq 4c_1\|v - w\|_{[-\alpha,a]\times[-\beta,b]} + \frac{\ell x^{r_1} y^{r_2}}{\Gamma(r_1+1)\Gamma(r_2+1)}\|v - w\|_{[-\alpha,a]\times[-\beta,b]}.
\end{aligned}$$

Consequently

$$\|N_1(v)(x,y) - N_1(w)(x,y)\| \leq \left[4c_1 + \frac{\ell a^{r_1} b^{r_2}}{\Gamma(r_1+1)\Gamma(r_2+1)}\right] \|v - w\|_{[-\alpha,a]\times[-\beta,b]}.$$

By (3.83), N_1 is a contraction , and hence N_1 has a unique fixed point by Banach's contraction principle. □

Our second existence result for IVP (3.72)–(3.74) is based on the nonlinear alternative of Leray–Schauder.

Theorem 3.39. *Assume (3.35.1), (3.38.1) and that the following hypotheses hold:*

(3.39.1) There exist $p, q \in C(J, \mathbb{R}_+)$ such that

$$\|f(x, y, u)\| \leq p(x, y) + q(x, y)\|u\|_C \text{ for } (x, y) \in J \text{ and } u \in C.$$

(3.39.2) The function g is completely continuous , and for any bounded set B in $C_{(a,b)}$, the set $\{(x, y) \to g(x, y, u_{(x,y)}) : u \in B\}$ is equicontinuous in $C(J, \mathbb{R}^n)$, and there exist constants $0 < d_1 < \frac{1}{4}$, $d_2 > 0$ such that

$$\|g(x, y, u)\| \leq d_1\|u\|_C + d_2,\ (x, y) \in J,\ u \in C.$$

Then the IVP (3.72)–(3.74) have at least one solution on $[-\alpha, a] \times [-\beta, b]$.

Proof. Consider the operator N_1 defined in (3.84). We shall show that the operator N_1 is continuous and completely continuous . Using (H'_{55}) it suffices to show that the operator $N_2 : C_{(a,b)} \to C_{(a,b)}$ defined by

$$N_2(u)(x, y) = \begin{cases} \phi(x, y), & (x, y) \in \tilde{J}, \\ \mu(x, y) \\ +\frac{1}{\Gamma(r_1)\Gamma(r_2)} \int_0^x \int_0^y (x - s)^{r_1-1}(y - t)^{r_2-1} \\ \times f(s, t, u_{(\rho_1(s,t,u_{(s,t)}), \rho_2(s,t,u_{(s,t)}))})dt ds, & (x, y) \in J, \end{cases}$$

is continuous and completely continuous . As in Theorem 3.6 [6], we can show that N_2 is continuous and completely continuous. We now show there exists an open set $U \subseteq C_{(a,b)}$ with $u \neq \lambda N_1(u)$, for $\lambda \in (0, 1)$ and $u \in \partial U$.
Let $u \in C_{(a,b)}$ and $u = \lambda N_1(u)$ for some $0 < \lambda < 1$. Then for each $(x, y) \in J$,

$$u(x, y) = \lambda \big[\mu(x, y) + g(x, y, u_{(x,y)}) - g(x, 0, u_{(x,0)}) - g(0, y, u_{(0,y)}) + g(0, 0, u)$$

$$+\frac{1}{\Gamma(r_1)\Gamma(r_2)} \int_0^x \int_0^y (x - s)^{r_1-1}(y - t)^{r_2-1}$$

$$\times\ f(s, t, u_{(\rho_1(s,t,u_{(s,t)}), \rho_2(s,t,u_{(s,t)}))})dt ds\big].$$

This implies by (3.39.1) and (3.39.2) that, for each $(x, y) \in J$, we have

$$\begin{aligned}
\|u(x,y)\| &\le \|\mu(x,y)\| + d_1(\|u_{(x,y)}\|_C + \|u_{(x,0)}\|_C + \|u_{(0,y)}\|_C + \|u\|_C) + 4d_2 \\
&\quad + \frac{1}{\Gamma(r_1)\Gamma(r_2)} \int_0^x \int_0^y (x-s)^{r_1-1}(y-t)^{r_2-1}[p(s,t)+q(s,t) \\
&\quad \times \|u_{(\rho_1(s,t,u_{(s,t)}),\rho_2(s,t,u_{(s,t)}))}\|_C]\mathrm{d}t\mathrm{d}s \\
&\le \|\mu(x,y)\| + 4d_2 + 4d_1\|u\|_{[-\alpha,a]\times[-\beta,b]} + \frac{a^{r_1}b^{r_2}\|p\|_\infty}{\Gamma(r_1+1)\Gamma(r_2+1)} \\
&\quad + \frac{1}{\Gamma(r_1)\Gamma(r_2)} \int_0^x \int_0^y (x-s)^{r_1-1}(y-t)^{r_2-1} q(s,t)\|u_{(s,t)}\|_C \mathrm{d}t\mathrm{d}s.
\end{aligned}$$

We consider the function ω defined by

$$\omega(x,y) = \sup\{u(s,t) : -\alpha \le s \le x, \ -\beta \le t \le y\}, \ 0 \le x \le a, \ 0 \le y \le b.$$

Let $(x^*, y^*) \in [-\alpha, x] \times [-\beta, y]$ such that $\omega(x,y) = \|u(x^*, y^*)\|$. By the previous inequality we have, for $(x, y) \in J$,

$$\begin{aligned}
\omega(x,y) &\le \frac{1}{1-4d_1}\Bigg[\|\mu(x,y)\| + 4d_2 + \frac{a^{r_1}b^{r_2}\|p\|_\infty}{\Gamma(r_1+1)\Gamma(r_2+1)} \\
&\quad + \frac{1}{\Gamma(r_1)\Gamma(r_2)} \int_0^x \int_0^y (x-s)^{r_1-1}(y-t)^{r_2-1} q(s,t)\omega(s,t)\mathrm{d}t\mathrm{d}s\Bigg] \\
&\le \frac{1}{1-4d_1}\Bigg[\|\mu(x,y)\| + 4d_2 + \frac{a^{r_1}b^{r_2}\|p\|_\infty}{\Gamma(r_1+1)\Gamma(r_2+1)} \\
&\quad + \frac{\|q\|_\infty}{\Gamma(r_1)\Gamma(r_2)} \int_0^x \int_0^y (x-s)^{r_1-1}(y-t)^{r_2-1} \omega(s,t)\mathrm{d}t\mathrm{d}s\Bigg].
\end{aligned}$$

If $(x^*, y^*) \in \tilde{J}$, then $\omega(x,y) = \|\phi\|_C$ and the previous inequality holds. Then Lemma 2.43 implies

$$\omega(x,y) \le \frac{1}{1-4d_1}\left(\|\mu(x,y)\| + 4d_2 + \frac{a^{r_1}b^{r_2}\|p\|_\infty}{\Gamma(r_1+1)\Gamma(r_2+1)}\right)$$
$$\times\left(1 + \frac{\|q\|_\infty}{(1-4d_1)\Gamma(r_1)\Gamma(r_2)}\int_0^x\int_0^y (x-s)^{r_1-1}(y-t)^{r_2-1}dtds\right)$$
$$\le \frac{1}{1-4d_1}\left(\|\mu(x,y)\| + 4d_2 + \frac{a^{r_1}b^{r_2}\|p\|_\infty}{\Gamma(r_1+1)\Gamma(r_2+1)}\right)$$
$$\times\left(1 + \frac{a^{r_1}b^{r_2}\|q\|_\infty}{(1-4d_1)\Gamma(r_1+1)\Gamma(r_2+1)}\right) := \overline{M}.$$

Since for every $(x,y) \in J$, $\|u_{(x,y)}\|_C \le \omega(x,y)$, we have

$$\|u\|_\infty \le \max\{\|\phi\|_C, \overline{M}\} := \overline{R}.$$

Set

$$U_1 = \{u \in C_{(a,b)} : \|u\|_\infty < \overline{R} + 1\}.$$

The operator $N_1 : \overline{U_1} \to C_{(a,b)}$ is continuous and completely continuous. From the choice of U there is no $u \in \partial U_1$ such that $u = \lambda N_1(u)$ for $\lambda \in (0,1)$. As a consequence of the nonlinear alternative of Leray–Schauder type [136], we deduce that N_1 has fixed point u in U_1, which is a solution of the IVP (3.72)–(3.74). □

3.8.3 Existence of Solutions for Infinite Delay

Let us start in this section by defining what we mean by a solution of the problem (3.75)–(3.77). Let the space

$$\Omega := \{u : (-\infty, a] \times (-\infty, b] \to \mathbb{R}^n : u_{(x,y)} \in \mathcal{B} \text{ for } (x,y) \in E$$
$$\text{and } u|_J \in C(J, \mathbb{R}^n)\}.$$

Definition 3.40. A function $u \in \Omega$ is said to be a solution of (3.75)–(3.77) if u satisfies (3.75) and (3.77) on J and the condition (3.76) on $\tilde{J}'$.

Set $\mathcal{R}' := \mathcal{R}'_{(\rho_1^-, \rho_2^-)}$

$$= \{(\rho_1(s,t,u), \rho_2(s,t,u)) : (s,t,u) \in J \times \mathcal{B},\ \rho_i(s,t,u) \le 0;\ i = 1,2\}.$$

We always assume that $\rho_1 : J \times \mathcal{B} \to (-\infty, a]$, $\rho_2 : J \times \mathcal{B} \to (-\infty, b]$ are continuous and the function $(s,t) \longmapsto u_{(s,t)}$ is continuous from $\mathcal{R}'$ into $\mathcal{B}$.

We will need to introduce the following hypothesis:

(C_ϕ) There exists a continuous bounded function $L : \mathcal{R}'_{(\rho_1^-,\rho_2^-)} \to (0,\infty)$ such that

$$\|\phi_{(s,t)}\|_{\mathcal{B}} \leq L(s,t)\|\phi\|_{\mathcal{B}}, \text{ for any}(s,t) \in \mathcal{R}'.$$

In the sequel we will make use of the following generalization of a consequence of the phase space axioms ([147], Lemma 2.1).

Lemma 3.41. *If* $u \in \Omega$, *then*

$$\|u_{(s,t)}\|_{\mathcal{B}} = (M + L')\|\phi\|_{\mathcal{B}} + K \sup_{(\theta,\eta)\in[0,\max\{0,s\}]\times[0,\max\{0,t\}]} \|u(\theta,\eta)\|,$$

where

$$L' = \sup_{(s,t)\in\mathcal{R}'} L(s,t).$$

Our first existence result for the IVP (3.75)–(3.77) is based on the Banach contraction principle.

Theorem 3.42. *Assume that the following hypothesis holds:*

(3.42.1) There exists $\ell' > 0$ *such that*

$$\|f(x,y,u) - f(x,y,v)\| \leq \ell'\|u-v\|_{\mathcal{B}}, \textit{ for any } u,\ v \in \mathcal{B} \textit{ and } (x,y) \in J.$$

If

$$\frac{\ell' K a^{r_1} b^{r_2}}{\Gamma(r_1+1)\Gamma(r_2+1)} < 1, \tag{3.85}$$

then there exists a unique solution for IVP (3.75)–(3.77) on $(-\infty,a]\times(-\infty,b]$.

Proof. Transform the problem (3.75)–(3.77) into a fixed point problem. Consider the operator $N : \Omega \to \Omega$ defined by

$$N(u)(x,y) = \begin{cases} \phi(x,y), & (x,y) \in \tilde{J}', \\ \mu(x,y) + \frac{1}{\Gamma(r_1)\Gamma(r_2)} \int_0^x \int_0^y (x-s)^{r_1-1}(y-t)^{r_2-1} & \\ \times f(s,t,u_{(\rho_1(s,t,u_{(s,t)}),\rho_2(s,t,u_{(s,t)}))})\mathrm{d}t\mathrm{d}s, & (x,y) \in J. \end{cases} \tag{3.86}$$

Let $v(.,.) : (-\infty,a]\times(-\infty,b] \to \mathbb{R}^n$ be a function defined by

$$v(x,y) = \begin{cases} \phi(x,y), & (x,y) \in \tilde{J}', \\ \mu(x,y), & (x,y) \in J. \end{cases}$$

Then $v_{(x,y)} = \phi$ for all $(x,y) \in E$. For each $w \in C(J, \mathbb{R}^n)$ with $w(x,y) = 0$ for each $(x,y) \in E$ we denote by $\overline{w}$ the function defined by

$$\overline{w}(x,y) = \begin{cases} 0, & (x,y) \in \tilde{J}', \\ w(x,y) & (x,y) \in J. \end{cases}$$

If $u(.,.)$ satisfies the integral equation

$$u(x,y) = \mu(x,y) + \frac{1}{\Gamma(r_1)\Gamma(r_2)} \int_0^x \int_0^y (x-s)^{r_1-1}(y-t)^{r_2-1} \times f(s,t,u_{(\rho_1(s,t,u_{(s,t)}),\rho_2(s,t,u_{(s,t)}))})\mathrm{d}t\,\mathrm{d}s,$$

we can decompose $u(.,.)$ as $u(x,y) = \overline{w}(x,y) + v(x,y)$; $(x,y) \in J$, which implies $u_{(x,y)} = \overline{w}_{(x,y)} + v_{(x,y)}$, for every $(x,y) \in J$, and the function $w(.,.)$ satisfies

$$w(x,y) = \frac{1}{\Gamma(r_1)\Gamma(r_2)} \int_0^x \int_0^y (x-s)^{r_1-1}(y-t)^{r_2-1} \times f(s,t,\overline{w}_{(\rho_1(s,t,u_{(s,t)}),\rho_2(s,t,u_{(s,t)}))} + v_{(\rho_1(s,t,u_{(s,t)}),\rho_2(s,t,u_{(s,t)}))})\mathrm{d}t\,\mathrm{d}s.$$

Set

$$C_0 = \{w \in \Omega : w(x,y) = 0 \text{ for } (x,y) \in E\},$$

and let $\|.\|_{(a,b)}$ be the seminorm in C_0 defined by

$$\|w\|_{(a,b)} = \sup_{(x,y)\in E} \|w_{(x,y)}\|_{\mathcal{B}} + \sup_{(x,y)\in J} \|w(x,y)\| = \sup_{(x,y)\in J} \|w(x,y)\|, \; w \in C_0.$$

C_0 is a Banach space with norm $\|.\|_{(a,b)}$. Let the operator $P : C_0 \to C_0$ be defined by

$$(Pw)(x,y) = \frac{1}{\Gamma(r_1)\Gamma(r_2)} \int_0^x \int_0^y (x-s)^{r_1-1}(y-t)^{r_2-1} \times f(s,t,\overline{w}_{(\rho_1(s,t,u_{(s,t)}),\rho_2(s,t,u_{(s,t)}))} + v_{(\rho_1(s,t,u_{(s,t)}),\rho_2(s,t,u_{(s,t)}))})\mathrm{d}t\,\mathrm{d}s, \tag{3.87}$$

for each $(x,y) \in J$. The operator N has a fixed point if and only if P has a fixed point, and so we turn to proving that P has a fixed point. We shall show that $P : C_0 \to C_0$ is a contraction map. Indeed, consider $w, w^* \in C_0$. Then we have for each $(x,y) \in J$

$$\|P(w)(x,y) - P(w^*)(x,y)\|$$
$$\leq \frac{1}{\Gamma(r_1)\Gamma(r_2)} \int_0^x \int_0^y (x-s)^{r_1-1}(y-t)^{r_2-1}$$
$$\times \| f(s,t,\overline{w}_{(\rho_1(s,t,u_{(s,t)}),\rho_2(s,t,u_{(s,t)}))} + v_{(\rho_1(s,t,u_{(s,t)}),\rho_2(s,t,u_{(s,t)}))})$$
$$- f(s,t,\overline{w^*}_{(\rho_1(s,t,u_{(s,t)}),\rho_2(s,t,u_{(s,t)}))} + v_{(\rho_1(s,t,u_{(s,t)}),\rho_2(s,t,u_{(s,t)}))})\| \mathrm{d}t\mathrm{d}s$$
$$\leq \frac{1}{\Gamma(r_1)\Gamma(r_2)} \int_0^x \int_0^y (x-s)^{r_1-1}(y-t)^{r_2-1}\ell'$$
$$\times \|\overline{w}_{(\rho_1(s,t,u_{(s,t)}),\rho_2(s,t,u_{(s,t)}))} - \overline{w^*}_{(\rho_1(s,t,u_{(s,t)}),\rho_2(s,t,u_{(s,t)}))}\|_{\mathcal{B}} \mathrm{d}t\mathrm{d}s$$
$$\leq \frac{\ell'}{\Gamma(r_1)\Gamma(r_2)} \int_0^x \int_0^y (x-s)^{r_1-1}(y-t)^{r_2-1} \|\overline{w}_{(s,t)} - \overline{w^*}_{(s,t)}\|_{\mathcal{B}} \mathrm{d}t\mathrm{d}s$$
$$\leq \frac{\ell' K}{\Gamma(r_1)\Gamma(r_2)} \int_0^x \int_0^y (x-s)^{r_1-1}(y-t)^{r_2-1}$$
$$\times \sup_{(s,t)\in[0,x]\times[0,y]} \|\overline{w}(s,t) - \overline{w^*}(s,t)\| \mathrm{d}t\mathrm{d}s$$
$$\leq \frac{\ell' K}{\Gamma(r_1)\Gamma(r_2)} \int_0^x \int_0^y (x-s)^{r_1-1}(y-t)^{r_2-1} \mathrm{d}t\mathrm{d}s \|\overline{w} - \overline{w^*}\|_{(a,b)} \mathrm{d}t\mathrm{d}s.$$

Therefore

$$\|P(w) - P(w^*)\|_{(a,b)} \leq \frac{\ell' K a^{r_1} b^{r_2}}{\Gamma(r_1+1)\Gamma(r_2)+1} \|\overline{w} - \overline{w^*}\|_{(a,b)}.$$

and hence P is a contraction. Therefore, P has a unique fixed point by Banach's contraction principle. □

Now we give an existence result based on the nonlinear alternative of Leray–Schauder type [136].

Theorem 3.43. *Assume* (C_ϕ) *and the following hypothesis:*

(3.43.1) There exist $p, q \in C(J, \mathbb{R}_+)$ *such that*

$$\|f(x,y,u)\| \leq p(x,y) + q(x,y)\|u\|_{\mathcal{B}} \textit{ for } (x,y) \in J \textit{ and each } u \in \mathcal{B}.$$

Then the IVP (3.75)–(3.77) have at least one solution on $(-\infty, a] \times (-\infty, b]$.

Proof. Let $P : C_0 \to C_0$ defined as in (3.87). We shall show that the operator P is continuous and completely continuous.

Step 1: *P is continuous.* Let $\{w_n\}$ be a sequence such that $w_n \to w$ in C_0. Then

$$\begin{aligned}\|P(w_n)(x,y) - P(w)(x,y)\| &\le \frac{1}{\Gamma(r_1)\Gamma(r_2)}\int_0^x\int_0^y (x-s)^{r_1-1}(y-t)^{r_2-1}\\ &\quad\times\|f(s,t,\overline{w}_{n(\rho_1(s,t,u_{n(s,t)}),\rho_2(s,t,u_{n(s,t)}))}+v_{n(\rho_1(s,t,u_{n(s,t)}),\rho_2(s,t,u_{n(s,t)}))})\\ &\quad-f(s,t,\overline{w}_{(\rho_1(s,t,u_{(s,t)}),\rho_2(s,t,u_{(s,t)}))}+v_{(\rho_1(s,t,u_{(s,t)}),\rho_2(s,t,u_{(s,t)}))})\|dtds\\ &\le \frac{1}{\Gamma(r_1)\Gamma(r_2)}\int_0^x\int_0^y (x-s)^{r_1-1}(y-t)^{r_2-1}\\ &\quad\times\|f(s,t,\overline{w}_{n(s,t)}+v_{n((s,t)}-f(s,t,\overline{w}_{(s,t)}+v_{(s,t)})\|dtds.\end{aligned}$$

Since f is a continuous function, we have

$$\begin{aligned}\|P(w_n)-P(w)\|_\infty &\le \frac{x^{r_1}y^{r_2}\|f(.,.,\overline{w_n}_{(.,.)}+v_{n(.,.)})-f(.,.,\overline{w}_{(.,.)}+v_{(.,.)})\|_\infty}{\Gamma(r_1+1)\Gamma(r_2+1)}\\ &\le \frac{a^{r_1}b^{r_2}\|f(.,.,\overline{w_n}_{(.,.)}+v_{n(.,.)})-f(.,.,\overline{w}_{(.,.)}+v_{(.,.)})\|_\infty}{\Gamma(r_1+1)\Gamma(r_2+1)}\\ &\to 0 \text{ as } n\to\infty.\end{aligned}$$

Step 2: *P maps bounded sets into bounded sets in C_0.* Indeed, it is enough to show that, for any $\eta > 0$, there exists a positive constant $\widetilde{\ell}$ such that, for each $w \in B_\eta = \{w \in C_0 : \|w\|_{(a,b)} \le \eta\}$, we have $\|P(w)\|_\infty \le \widetilde{\ell}$.
Lemma 3.41 implies that

$$\begin{aligned}\|\overline{w}_{(s,t)}+v_{(s,t)}\|_{\mathcal{B}} &\le \|\overline{w}_{(s,t)}\|_{\mathcal{B}}+\|v_{(s,t)}\|_{\mathcal{B}}\\ &\le K\eta + K\|\phi(0,0)\| + (M+L')\|\phi\|_{\mathcal{B}}.\end{aligned}$$

Set

$$\eta^* := K\eta + K\|\phi(0,0)\| + (M+L')\|\phi\|_{\mathcal{B}}.$$

Let $w \in B_\eta$. By $(C2)$ we have for each $(x,y)\in J$,

$$\begin{aligned}
\|P(w)(x,y)\| &\leq \frac{1}{\Gamma(r_1)\Gamma(r_2)}\int_0^x\int_0^y (x-s)^{r_1-1}(y-t)^{r_2-1} \\
&\quad \times \|f(s,t,\overline{w}_{(\rho_1(s,t,u_{(s,t)}),\rho_2(s,t,u_{(s,t)}))}+v_{(\rho_1(s,t,u_{(s,t)}),\rho_2(s,t,u_{(s,t)}))})\|\mathrm{d}t\mathrm{d}s \\
&\leq \frac{1}{\Gamma(r_1)\Gamma(r_2)}\left\|\int_0^x\int_0^y (x-s)^{r_1-1}(y-t)^{r_2-1}p(s,t)\mathrm{d}t\mathrm{d}s\right\| \\
&\quad +\frac{1}{\Gamma(r_1)\Gamma(r_2)}\int_0^x\int_0^y (x-s)^{r_1-1}(y-t)^{r_2-1}q(s,t) \\
&\quad \times \|\overline{w}_{(\rho_1(s,t,u_{(s,t)}),\rho_2(s,t,u_{(s,t)}))}+v_{(\rho_1(s,t,u_{(s,t)}),\rho_2(s,t,u_{(s,t)}))}\|_{\mathcal{B}}\mathrm{d}t\mathrm{d}s \\
&\leq \frac{\|p\|_\infty}{\Gamma(r_1)\Gamma(r_2)}\int_0^x\int_0^x (x-s)^{r_1-1}(y-t)^{r_2-1}\mathrm{d}t\mathrm{d}s \\
&\quad +\frac{\|q\|_\infty\eta^*}{\Gamma(r_1)\Gamma(r_2)}\int_0^x\int_0^y (x-s)^{r_1-1}(y-t)^{r_2-1}\mathrm{d}t\mathrm{d}s \\
&\leq \frac{\|p\|_\infty+\|q\|_\infty\eta^*}{\Gamma(r_1+1)\Gamma(r_2+1)}a^{r_1}b^{r_2} := \ell^*.
\end{aligned}$$

Step 3: *P maps bounded sets into equicontinuous sets in C_0.* Let $(x_1,y_1),(x_2,y_2)\in(0,a]\times(0,b]$, $x_1<x_2$, $y_1<y_2$, B_η be a bounded set as in step 2, and let $w\in B_\eta$. Then

$$\begin{aligned}
&\|P(w)(x_2,y_2)-P(w)(x_1,y_1)\| \\
&\leq \frac{1}{\Gamma(r_1)\Gamma(r_2)}\left\|\int_0^{x_1}\int_0^{y_1}[(x_2-s)^{r_1-1}(y_2-t)^{r_2-1}-(x_1-s)^{r_1-1}(y_1-t)^{r_2-1}]\right. \\
&\quad \times f(s,t,u_{(\rho_1(s,t,u_{(s,t)}),\rho_2(s,t,u_{(s,t)}))})\mathrm{d}t\mathrm{d}s \\
&\quad +\frac{1}{\Gamma(r_1)\Gamma(r_2)}\int_{x_1}^{x_2}\int_{y_1}^{y_2}(x_2-s)^{r_1-1}(y_2-t)^{r_2-1} \\
&\quad \times f(s,t,\overline{w}_{(\rho_1(s,t,u_{(s,t)}),\rho_2(s,t,u_{(s,t)}))}+v_{(\rho_1(s,t,u_{(s,t)}),\rho_2(s,t,u_{(s,t)}))})\mathrm{d}t\mathrm{d}s\Big\| \\
&\quad +\frac{1}{\Gamma(r_1)\Gamma(r_2)}\int_0^{x_1}\int_{y_1}^{y_2}(x_2-s)^{r_1-1}(y_2-t)^{r_2-1}
\end{aligned}$$

$$\times \|f(s,t,\overline{w}_{(\rho_1(s,t,u_{(s,t)}),\rho_2(s,t,u_{(s,t)}))} + v_{(\rho_1(s,t,u_{(s,t)}),\rho_2(s,t,u_{(s,t)}))})\|dtds$$

$$+ \frac{1}{\Gamma(r_1)\Gamma(r_2)} \int_{x_1}^{x_2}\int_{0}^{y_1} (x_2-s)^{r_1-1}(y_2-t)^{r_2-1}$$

$$\times \|f(s,t,\overline{w}_{(\rho_1(s,t,u_{(s,t)}),\rho_2(s,t,u_{(s,t)}))} + v_{(\rho_1(s,t,u_{(s,t)}),\rho_2(s,t,u_{(s,t)}))})\|dtds$$

$$\leq \frac{\|p\|_\infty + \|q\|_\infty \eta}{\Gamma(r_1)\Gamma(r_2)} \int_{0}^{x_1}\int_{0}^{y_1} [(x_1-s)^{r_1-1}(y_1-t)^{r_2-1}$$

$$-(x_2-s)^{r_1-1}(y_2-t)^{r_2-1}]dtds$$

$$+ \frac{\|p\|_\infty + \|q\|_\infty \eta}{\Gamma(r_1)\Gamma(r_2)} \int_{x_1}^{x_2}\int_{y_1}^{y_2} (x_2-s)^{r_1-1}(y_2-t)^{r_2-1}dtds$$

$$+ \frac{\|p\|_\infty + \|q\|_\infty \eta}{\Gamma(r_1)\Gamma(r_2)} \int_{0}^{x_1}\int_{y_1}^{y_2} (x_2-s)^{r_1-1}(y_2-t)^{r_2-1}dtds$$

$$+ \frac{\|p\|_\infty + \|q\|_\infty \eta}{\Gamma(r_1)\Gamma(r_2)} \int_{x_1}^{x_2}\int_{0}^{y_1} (x_2-s)^{r_1-1}(y_2-t)^{r_2-1}dtds$$

$$\leq \frac{\|p\|_\infty + \|q\|_\infty \eta}{\Gamma(r_1+1)\Gamma(r_2+1)}[2y_2^{r_2}(x_2-x_1)^{r_1} + 2x_2^{r_1}(y_2-y_1)^{r_2}$$

$$+ x_1^{r_1}y_1^{r_2} - x_2^{r_1}y_2^{r_2} - 2(x_2-x_1)^{r_1}(y_2-y_1)^{r_2}].$$

As $x_1 \to x_2$, $y_1 \to y_2$ the right-hand side of the above inequality tends to zero. The equicontinuity for the cases $x_1 < x_2 < 0$, $y_1 < y_2 < 0$ and $x_1 \leq 0 \leq x_2$, $y_1 \leq 0 \leq y_2$ is obvious. As a consequence of Steps 1–3, together with the Arzela–Ascoli theorem, we can conclude that $P : C_0 \to C_0$ is continuous and completely continuous.

Step 4: *A priori bounds*. We now show there exists an open set $U \subseteq C_0$ with $w \neq \lambda P(w)$, for $\lambda \in (0,1)$ and $w \in \partial U$. Let $w \in C_0$ and $w = \lambda P(w)$ for some $0 < \lambda < 1$. Then for each $(x,y) \in J$,

$$w(x,y) = \frac{\lambda}{\Gamma(r_1)\Gamma(r_2)} \int_{0}^{x}\int_{0}^{y} (x-s)^{r_1-1}(y-t)^{r_2-1}$$

$$\times f(s,t,u_{(\rho_1(s,t,u_{(s,t)}),\rho_2(s,t,u_{(s,t)}))})dtds.$$

This implies by (3.43.1) that, for each $(x, y) \in J$, we have

$$\|w(x,y)\| \leq \frac{1}{\Gamma(r_1)\Gamma(r_2)}\int_0^x\int_0^y (x-s)^{r_1-1}(y-t)^{r_2-1}[p(s,t)$$
$$+q(s,t)\|\overline{w}_{(\rho_1(s,t,u_{(s,t)}),\rho_2(s,t,u_{(s,t)}))} + v_{(\rho_1(s,t,u_{(s,t)}),\rho_2(s,t,u_{(s,t)}))}\|_{\mathcal{B}}]\mathrm{d}t\mathrm{d}s$$
$$\leq \frac{\|p\|_\infty a^{r_1}b^{r_2}}{\Gamma(r_1+1)\Gamma(r_2+1)}$$
$$+\frac{1}{\Gamma(r_1)\Gamma(r_2)}\int_0^x\int_0^y (x-s)^{r_1-1}(y-t)^{r_2-1}q(s,t)\|\overline{w}_{(s,t)} + v_{(s,t)}\|_{\mathcal{B}}\mathrm{d}t\mathrm{d}s.$$

But

$$\|\overline{w}_{(s,t)} + v_{(s,t)}\|_{\mathcal{B}} \leq \|\overline{w}_{(s,t)}\|_{\mathcal{B}} + \|v_{(s,t)}\|_{\mathcal{B}}$$
$$\leq K\sup\{w(\tilde{s},\tilde{t}) : (\tilde{s},\tilde{t}) \in [0,s]\times[0,t]\}$$
$$+(M+L')\|\phi\|_{\mathcal{B}} + K\|\phi(0,0)\|. \tag{3.88}$$

If we name $z(s,t)$ the right-hand side of (3.88), then we have

$$\|\overline{w}_{(s,t)} + v_{(s,t)}\|_{\mathcal{B}} \leq z(x,y),$$

and therefore, for each $(x, y) \in J$, we obtain

$$\|w(x,y)\| \leq \frac{\|p\|_\infty a^{r_1}b^{r_2}}{\Gamma(r_1+1)\Gamma(r_2+1)}$$
$$+\frac{1}{\Gamma(r_1)\Gamma(r_2)}\int_0^x\int_0^y (x-s)^{r_1-1}(y-t)^{r_2-1}q(s,t)z(s,t)\mathrm{d}t\mathrm{d}s. \tag{3.89}$$

Using the above inequality and the definition of z for each $(x, y) \in J$ we have

$$z(x,y) \leq (M+L')\|\phi\|_{\mathcal{B}} + K\|\phi(0,0)\| + \frac{K\|p\|_\infty a^{r_1}b^{r_2}}{\Gamma(r_1+1)\Gamma(r_2+1)}$$
$$+\frac{K\|p\|_\infty}{\Gamma(r_1)\Gamma(r_2)}\int_0^x\int_0^y (x-s)^{r_1-1}(y-t)^{r_2-1}z(s,t)\mathrm{d}t\mathrm{d}s.$$

Then by Lemma 2.43, there exists $\delta = \delta(r_1, r_2)$ such that we have

$$\|z(x,y)\| \leq R + \delta \frac{K\|q\|_\infty}{\Gamma(r_1)\Gamma(r_2)} \int_0^x \int_0^y (x-s)^{r_1-1}(y-t)^{r_2-1} R dt ds,$$

where

$$R = (M + L')\|\phi\|_{\mathcal{B}} + K\|\phi(0,0)\| + \frac{K\|p\|_\infty a^{r_1} b^{r_2}}{\Gamma(r_1+1)\Gamma(r_2+1)}.$$

Hence

$$\|z\|_\infty \leq R + \frac{R\delta K\|q\|_\infty a^{r_1} b^{r_2}}{\Gamma(r_1+1)\Gamma(r_2+1)} := \widetilde{M}.$$

Then, (3.89) implies that

$$\|w\|_\infty \leq \frac{a^{r_1} b^{r_2}}{\Gamma(r_1+1)\Gamma(r_2+1)}(\|p\|_\infty + \widetilde{M}\|q\|_\infty) := M^*.$$

Set

$$U = \{w \in C_0 : \|w\|_{(a,b)} < M^* + 1\}.$$

$P : \overline{U} \to C_0$ is continuous and completely continuous. By our choice of U, there is no $w \in \partial U$ such that $w = \lambda P(w)$, for $\lambda \in (0,1)$. As a consequence of the nonlinear alternative of Leray–Schauder type [136], we deduce that N has a fixed point which is a solution to problem (3.75)–(3.77). □

Now we present two similar existence results for the problems (3.78)–(3.80).

Definition 3.44. A function $u \in \Omega$ is said to be a solution of (3.78)–(3.80) if u satisfies (3.78) and (3.80) on J and the condition (3.79) on $\tilde{J}'$.

Theorem 3.45. *Assume that (3.42.1) holds and moreover (3.45.1) There exists a nonnegative constant c_2 such that*

$$\|g(x,y,u) - g(x,y,v)\| \leq c_2\|u-v\|_{\mathcal{B}}, \text{ for each } (x,y) \in J, \text{ and } u, v \in \mathcal{B}.$$

If

$$K\left[4c_2 + \frac{\ell' a^{r_1} b^{r_2}}{\Gamma(r_1+1)\Gamma(r_2+1)}\right] < 1, \tag{3.90}$$

then there exists a unique solution for IVP (3.78)–(3.80) on $(-\infty, a] \times (-\infty, b]$.

Proof. Consider the operator $N_1 : \Omega \to \Omega$ defined by

$$N_1(u)(x,y) = \begin{cases} \phi(x,y), & (x,y) \in \tilde{J}', \\ \mu(x,y) + g(x,y,u_{(x,y)}) - g(x,0,u_{(x,0)}) \\ -g(0,y,u_{(0,y)}) + g(0,0,u_{(0,0)}) \\ +\frac{1}{\Gamma(r_1)\Gamma(r_2)} \int\limits_0^x \int\limits_0^y (x-s)^{r_1-1}(y-t)^{r_2-1} \\ f(s,t,u_{(\rho_1(s,t,u_{(s,t)}),\rho_2(s,t,u_{(s,t)}))})dtds, & (x,y) \in J. \end{cases}$$

In analogy to Theorem 3.42, we consider the operator $P_1 : C_0 \to C_0$ defined by

$$\begin{aligned} P_1(x,y) = {} & g(x,y,\overline{w}_{(x,y)} + v_{(x,y)}) - g(x,0,\overline{w}_{(x,0)} + v_{(x,0)}) \\ & -g(0,y,\overline{w}_{(0,y)} + v_{(0,y)}) + g(0,0,\overline{w}_{(0,0)} + v_{(0,0)}) \\ & +\frac{1}{\Gamma(r_1)\Gamma(r_2)} \int\limits_0^x \int\limits_0^y (x-s)^{r_1-1}(y-t)^{r_2-1} \\ & \times f(s,t,\overline{w}_{(\rho_1(s,t,u_{(s,t)}),\rho_2(s,t,u_{(s,t)}))} \\ & +v_{(\rho_1(s,t,u_{(s,t)}),\rho_2(s,t,u_{(s,t)}))})dtds, \ (x,y) \in J. \end{aligned}$$

We shall show that the operator P_1 is a contraction. Let $w, \ w_* \in C_0$, then following the steps of Theorem 3.42, we have

$$\begin{aligned} & \|P_1(w)(x,y) - P_1(w_*)(x,y)\| \\ & \le \|g(x,y,\overline{w}_{(x,y)} + v_{(x,y)}) - g(x,y,\overline{w_*}_{(x,y)} + v_{(x,y)})\| \\ & \quad +\|g(x,0,\overline{w}_{(x,0)} + v_{(x,0)}) - g(x,0,\overline{w_*}_{(x,0)} + v_{(x,0)})\| \\ & \quad +\|g(0,y,\overline{w}_{(0,y)} + v_{(0,y)}) - g(0,y,\overline{w_*}_{(0,y)} + v_{(0,y)})\| \\ & \quad +\|g(0,0,\overline{w} + v) - g(0,0,\overline{w_*}_{(0,0)} + v_{(0,0)})\| \\ & \quad +\frac{1}{\Gamma(r_1)\Gamma(r_2)} \int\limits_0^x \int\limits_0^y (x-s)^{r_1-1}(y-t)^{r_2-1} \\ & \quad \times\|f(s,t,\overline{w}_{(\rho_1(s,t,u_{(s,t)}),\rho_2(s,t,u_{(s,t)}))} + v_{(\rho_1(s,t,u_{(s,t)}),\rho_2(s,t,u_{(s,t)}))}) \\ & \quad -f(s,t,\overline{w_*}_{(\rho_1(s,t,u_{(s,t)}),\rho_2(s,t,u_{(s,t)}))} + v_{(\rho_1(s,t,u_{(s,t)}),\rho_2(s,t,u_{(s,t)}))})\|dtds \\ & \le 4c_2K\|\overline{w} - \overline{w_*}\|_{(a,b)} + \frac{1}{\Gamma(r_1)\Gamma(r_2)} \int\limits_0^x \int\limits_0^y (x-s)^{r_1-1}(y-t)^{r_2-1} \\ & \quad \times\ell' K\|\overline{w} - \overline{w_*}\|dtds. \end{aligned}$$

Therefore

$$\| P_1(w) - P_1(w_*)\|_{(a,b)} \leq K \left[4c_2 + \frac{\ell' a^{r_1} b^{r_2}}{\Gamma(r_1+1)\Gamma(r_2)+1} \right] \|\overline{w} - \overline{w_*}\|_{(a,b)}.$$

which implies by (3.90) that P_1 is a contraction. Hence P_1 has a unique fixed point by Banach's contraction principle. □

Our last existence result for the IVP (3.78)–(3.80) is based on the nonlinear alternative of Leray–Schauder type.

Theorem 3.46. *Assume* (C_ϕ), *(3.42.1), (3.43.1), and the following conditions:*

(3.46.1) The function g *is continuous and completely continuous, and for any bounded set* D *in* Ω, *the set* $\{(x,y) \to g(x,y,u_{(x,y)}) : u \in D\}$, *is equicontinuous in* $C(J,\mathbb{R}^n)$,

(3.46.2) There exist constants $d_1, d_2 \geq 0$ *such that* $0 \leq d_1 K < \frac{1}{4}$ *and*

$$\|g(x,y,u)\| \leq d_1 \|u\|_{\mathcal{B}} + d_2, \quad (x,y) \in J, \quad u \in \mathcal{B}.$$

Then the IVP (3.78)–(3.80) have at least one solution on $(-\infty,a] \times (-\infty,b]$.

Proof. Let $P_1 : C_0 \to C_0$ defined as in Theorem 3.45. We shall show that the operator P_1 is continuous and completely continuous. Using (3.46.1) it suffices to show that the operator $P_2 : C_0 \to C_0$ defined by

$$\begin{aligned} P_2(w)(x,y) &= g(x,y,\overline{w}_{(x,y)} + v_{(x,y)}) - g(x,0,\overline{w}_{(x,0)} + v_{(x,0)}) \\ &\quad - g(0,y,\overline{w}_{(0,y)} + v_{(0,y)}) + g(0,0,\overline{w}_{(0,0)} + v_{(0,0)}) \\ &\quad + \frac{1}{\Gamma(r_1)\Gamma(r_2)} \int_0^x \int_0^y (x-s)^{r_1-1}(y-t)^{r_2-1} \\ &\quad \times f(s,t,\overline{w}_{(\rho_1(s,t,u_{(s,t)}),\rho_2(s,t,u_{(s,t)}))} + v_{(\rho_1(s,t,u_{(s,t)}),\rho_2(s,t,u_{(s,t)}))})dtds, \end{aligned}$$

is continuous and completely continuous . This was proved in Theorem 3.36. We now show there exists an open set $U \subseteq C_0$ with $w \neq \lambda P_2(w)$, for $\lambda \in (0,1)$ and $w \in \partial U$. Let $w \in C_0$ and $w = \lambda p_2(w)$ for some $0 < \lambda < 1$. Then for each $(x,y) \in J$,

$$\begin{aligned} w(x,y) &= \lambda[g(x,y,\overline{w}_{(x,y)} + v_{(x,y)}) - g(x,0,\overline{w}_{(x,0)} + v_{(x,0)}) \\ &\quad - g(0,y,\overline{w}_{(0,y)} + v_{(0,y)}) + g(0,0,\overline{w}_{(0,0)} + v_{(0,0)})] \\ &\quad + \frac{\lambda}{\Gamma(r_1)\Gamma(r_2)} \int_0^x \int_0^y (x-s)^{r_1-1}(y-t)^{r_2-1} \\ &\quad \times f(s,t,\overline{w}_{(\rho_1(s,t,u_{(s,t)}),\rho_2(s,t,u_{(s,t)}))} + v_{(\rho_1(s,t,u_{(s,t)}),\rho_2(s,t,u_{(s,t)}))})dtds, \end{aligned}$$

and

$$\|w(x,y)\| = 4d_1\|\overline{w}_{(x,y)} + v_{(x,y)}\|_B + \frac{\|p\|_\infty a^{r_1} b^{r_2}}{\Gamma(r_1+1)\Gamma(r_2+1)}$$

$$+\frac{1}{\Gamma(r_1)\Gamma(r_2)}\int_0^x\int_0^y (x-s)^{r_1-1}(y-t)^{r_2-1}q(s,t)\|\overline{w}_{(s,t)}+v_{(s,t)}\|_B dt ds.$$

Using the above inequality and the definition of z we have that

$$\|z\|_\infty \le R_1 + \frac{R_1\delta K\|q^*\|_\infty a^{r_1}b^{r_2}}{(1-4d_1K)\Gamma(r_1+1)\Gamma(r_2+1)} := L,$$

where

$$R_1 = \frac{1}{1-4d_1K}\left[8d_2K + \frac{K\|p\|_\infty a^{r_1}b^{r_2}}{\Gamma(r_1+1)\Gamma(r_2+1)}\right],$$

and

$$\|q^*\|_\infty = \frac{\|q\|_\infty}{1-4d_1K}.$$

Then

$$\|w\|_\infty \le 4d_1\|\phi\|_B + 8d_2 + 4Ld_1 + \frac{a^{r_1}b^{r_2}}{\Gamma(r_1+1)\Gamma(r_2+1)}(\|p\|_\infty + L\|q\|_\infty) := \widetilde{L}.$$

Set

$$U_1 = \{w \in C_0 : \|w\|_{(a,b)} < \widetilde{L} + 1\}.$$

By our choice of U_1, there is no $w \in \partial U$ such that $w = \lambda P_2(w)$, for $\lambda \in (0,1)$. As a consequence of the nonlinear alternative of Leray–Schauder type [136], we deduce that N_1 has a fixed point which is a solution to problems (3.78)–(3.80). □

3.8.4 Examples

3.8.4.1 Example 1

As an application of our results we consider the following fractional order hyperbolic partial functional differential equations of the form:

$$({}^cD_0^r u)(x,y) = \frac{|u(x-\sigma_1(u(x,y)), y-\sigma_2(u(x,y)))|q+2}{10e^{x+y+4}(1+|u(x-\sigma_1(u(x,y)), y-\sigma_2(u(x,y)))|)},$$
$$\text{if } (x,y) \in [0,1]\times[0,1], \tag{3.91}$$

$$u(x,0) = x, \; u(0,y) = y^2, \; x \in [0,1], \; y \in [0,1], \tag{3.92}$$

$$u(x,y) = x + y^2, \; (x,y) \in [-1,1] \times [-2,1] \backslash (0,1] \times (0,1], \tag{3.93}$$

where $\sigma_1 \in C(\mathbb{R},[0,1])$, $\sigma_2 \in C(\mathbb{R},[0,2])$. Set

$$\rho_1(x,y,\varphi) = x - \sigma_1(\varphi(0,0)), \;\; (x,y,\varphi) \in J \times C([-1,0] \times [-2,0],\mathbb{R}),$$

$$\rho_2(x,y,\varphi) = y - \sigma_2(\varphi(0,0)), \;\; (x,y,\varphi) \in J \times C([-1,0] \times [-2,0],\mathbb{R}),$$

$$f(x,y,\varphi) = \frac{|\varphi|+2}{(10e^{x+y+4})(1+|\varphi|)}, \; (x,y) \in [0,1] \times [0,1], \; \varphi \in C([-1,0] \times [-2,0],\mathbb{R}).$$

For each $\varphi, \overline{\varphi} \in C([-1,0] \times [-2,0],\mathbb{R})$ and $(x,y) \in [0,1] \times [0,1]$ we have

$$|f(x,y,\varphi) - f(x,y,\overline{\varphi})| \leq \frac{1}{10e^4}\|\varphi - \overline{\varphi}\|_C.$$

Hence conditions (3.35.1) and (3.35.2) are satisfied with $k = \frac{1}{10e^4}$. We shall show that condition (3.81) holds with $a = b = 1$. Indeed

$$\frac{ka^{r_1}b^{r_2}}{\Gamma(r_1+1)\Gamma(r_2+1)} = \frac{1}{10e^4\Gamma(r_1+1)\Gamma(r_2+1)} < 1,$$

which is satisfied for each $(r_1,r_2) \in (0,1] \times (0,1]$. Consequently, Theorem 3.35 implies that problems (3.91)–(3.93) have a unique solution defined on $[-1,1] \times [-2,1]$.

3.8.4.2 Example 2

Consider the following fractional order hyperbolic partial functional differential equations of the form

$$({}^cD_0^r u)(x,y) = \frac{|u(x - \sigma_1(u(x,y)), y - \sigma_2(u(x,y)))|}{10e^{x+y+4}(1 + |u(x - \sigma_1(u(x,y)), y - \sigma_2(u(x,y)))|)}, \quad \text{if } (x,y) \in [0,1] \times [0,1], \tag{3.94}$$

$$u(x,0) = x, \; u(0,y) = y^2, \; x \in [0,1], \; y \in [0,1], \tag{3.95}$$

$$u(x,y) = x + y^2, \; (x,y) \in [-1,1] \times [-2,1] \backslash (0,1] \times (0,1], \tag{3.96}$$

where $\sigma_1 \in C(\mathbb{R},[0,1])$, $\sigma_2 \in C(\mathbb{R},[0,2])$. Theorem 3.36 implies that problem (3.94)–(3.95) has at least one solution defined on $[-1,1] \times [-2,1]$.

3.8.4.3 Example 3

Consider now the following fractional order hyperbolic partial neutral functional differential equations of the form

$$\begin{aligned}
&{}^cD_0^r\left(u(x,y)-\frac{1}{4e^{x+y+5}(1+|u(x-\sigma_1(u(x,y)),y-\sigma_2(u(x,y)))|)}\right)\\
&=\frac{|u(x-\sigma_1(u(x,y)),y-\sigma_2(u(x,y)))|+2}{10e^{x+y+4}(1+|u(x-\sigma_1(u(x,y)),y-\sigma_2(u(x,y)))|)},\\
&\text{if } (x,y)\in[0,1]\times[0,1], \qquad (3.97)
\end{aligned}$$

$$u(x,0)=x,\ u(0,y)=y^2,\ x\in[0,1],\ y\in[0,1], \qquad (3.98)$$

$$u(x,y)=x+y^2,\ (x,y)\in[-1,1]\times[-2,1]\backslash(0,1]\times(0,1], \qquad (3.99)$$

where $\sigma_1\in C(\mathbb{R},[0,1])$, $\sigma_2\in C(\mathbb{R},[0,2])$. We can easily show that hypotheses $(H'_{51})-(H'_{53})$ are satisfied and condition (3.83) holds with $a=b=1$. Consequently, Theorem 3.38 implies that problems (3.97)–(3.99) have a unique solution defined on $[-1,1]\times[-2,1]$.

3.8.4.4 Example 4

We consider now the following fractional order partial hyperbolic differential equations with infinite delay of the form:

$$\begin{aligned}
({}^cD_0^r u)(x,y)&=\frac{ce^{x+y-\gamma(x+y)}}{e^{x+y}+e^{-x-y}}\\
&\quad\times\frac{|u(x-\sigma_1(u(x,y)),y-\sigma_2(u(x,y)))|}{1+|u(x-\sigma_1(u(x,y)),y-\sigma_2(u(x,y)))|},\\
&\quad\text{if } (x,y)\in[0,1]\times[0,1], \qquad (3.100)
\end{aligned}$$

$$u(x,0)=x,\ u(0,y)=y^2,\ x\in[0,1],\ y\in[0,1], \qquad (3.101)$$

$$u(x,y)=x+y^2,\ (x,y)\in(-\infty,1]\times(-\infty,1]\backslash(0,1]\times(0,1], \qquad (3.102)$$

where $J:=[0,1]\times[0,1]$, $c=\frac{2}{\Gamma(r_1+1)\Gamma(r_2+1)}$, γ a positive real constant, and $\sigma_1,\sigma_2\in C(\mathbb{R},[0,\infty))$. Let $\mathcal{B}_\gamma$ be the phase space defined in the Example of Sect. 3.7. Set

$$\rho_1(x,y,\varphi)=x-\sigma_1(\varphi(0,0)),\ (x,y,\varphi)\in J\times\mathcal{B}_\gamma,$$

$$\rho_2(x,y,\varphi)=y-\sigma_2(\varphi(0,0)),\ (x,y,\varphi)\in J\times\mathcal{B}_\gamma,$$

$$f(x,y,\varphi) = \frac{ce^{x+y-\gamma(x+y)}|\varphi|}{(e^{x+y}+e^{-x-y})(1+|\varphi|)}, \ (x,y) \in [0,1]\times[0,1], \ \varphi \in \mathcal{B}_\gamma.$$

For each $\varphi, \ \overline{\varphi} \in \mathcal{B}_\gamma$ and $(x,y) \in [0,1]\times[0,1]$ we have

$$|f(x,y,\varphi) - f(x,y,\overline{\varphi})| \le \frac{1}{c}\|\varphi - \overline{\varphi}\|_\gamma.$$

Hence condition (3.42.1) is satisfied with $\ell' = \frac{1}{c}$. Since $a = b = K = 1$ we get

$$\frac{K\ell' a^{r_1} b^{r_2} K}{\Gamma(r_1+1)\Gamma(r_2+1)} = \frac{1}{c\Gamma(r_1+1)\Gamma(r_2+1)} = \frac{1}{2} < 1,$$

for each $(r_1, r_2) \in (0,1]\times(0,1]$. Consequently, Theorem 3.42 implies that problems (3.100)–(3.102) have a unique solution defined on $(-\infty,1]\times(-\infty,1]$.

3.8.4.5 Example 5

Consider the following partial neutral functional differential equations of the form:

$$\begin{aligned} &{}^cD_0^r\left(u(x,y) - \frac{1}{(3e^{x+y+2})(1+|u(x-\sigma_1(u(x,y)), y-\sigma_2(u(x,y)))|}\right) \\ &= \frac{e^{x+y+2}|u(x-\sigma_1(u(x,y)), y-\sigma_2(u(x,y)))|}{1+|u(x-\sigma_1(u(x,y)), y-\sigma_2(u(x,y)))|}, \\ &\text{if } (x,y) \in [0,1]\times[0,1], \end{aligned} \tag{3.103}$$

$$u(x,0) = x, \ u(0,y) = y^2, \ x \in [0,1], \ y \in [0,1], \tag{3.104}$$

$$u(x,y) = x + y^2, \ (x,y) \in (-\infty,1]\times(-\infty,1]\backslash(0,1]\times(0,1], \tag{3.105}$$

where $\sigma_1, \sigma_2 \in C(\mathbb{R},[0,\infty))$. Let $\mathcal{B} := \mathcal{B}_{\gamma=0}$. Set

$$\rho_1(x,y,\varphi) = x - \sigma_1(\varphi(0,0)), \ (x,y,\varphi) \in J \times \mathcal{B},$$

$$\rho_2(x,y,\varphi) = y - \sigma_2(\varphi(0,0)), \ (x,y,\varphi) \in J \times \mathcal{B},$$

$$f(x,y,\varphi) = \frac{e^{x+y+2}|\varphi|}{1+|\varphi|}, \ (x,y) \in [0,1]\times[0,1], \ \varphi \in \mathcal{B}$$

and

$$g(x,y,\varphi) = \frac{1}{(3e^{x+y+2})(1+|\varphi|)}.$$

For each $\varphi,\ \overline{\varphi} \in \mathcal{B}$ and $(x, y) \in [0, 1] \times [0, 1]$ we have

$$|g(x, y, \varphi) - g(x, y, \overline{\varphi})| \leq \frac{1}{3e^2}\|\varphi - \overline{\varphi}\|_{\mathcal{B}}.$$

Hence condition (3.45.1) holds with $c_2 = \dfrac{1}{3e^2}$.

It is clear that condition (3.42.1) holds with $\ell' = e^2$. Since $a = b = K = 1$ we get

$$K[4c_2 + \frac{\ell' a^{r_1} b^{r_2} K}{\Gamma(r_1+1)\Gamma(r_2+1)}] < 1,$$

for each $(r_1, r_2) \in (0, 1] \times (0, 1]$. Consequently, Theorem 3.45 implies that problems (3.103)–(3.105) have a unique solution defined on $(-\infty, 1] \times (-\infty, 1]$.

3.9 Global Uniqueness Results for Partial Hyperbolic Differential Equations

3.9.1 Introduction

In this section, we provide sufficient conditions for the global existence and uniqueness of some classes of fractional order partial hyperbolic differential equations. Firstly, we present a global existence and uniqueness of solutions to fractional order IVP , for the system

$$({}^cD_0^r u)(x, y) = f(x, y, u_{(x,y)}),\ \text{if } (x, y) \in J_\infty, \tag{3.106}$$

$$u(x, y) = \phi(x, y),\ \text{if } (x, y) \in \tilde{J}_\infty, \tag{3.107}$$

$$u(x, 0) = \varphi(x),\ x \in [0, \infty),\ u(0, y) = \psi(y),\ y \in [0, \infty), \tag{3.108}$$

where $J_\infty = [0, \infty) \times [0, \infty)$, $\alpha, \beta > 0$, $\tilde{J}_\infty := [-\alpha, \infty) \times [-\beta, \infty)\backslash(0, \infty) \times (0, \infty)$, $f : J_\infty \times C \to \mathbb{R}^n$, $\phi \in C(\tilde{J}_\infty, \mathbb{R}^n)$, $\varphi, \psi : [0, \infty) \to \mathbb{R}^n$, are given absolutely continuous functions with $\varphi(x) = \phi(x, 0)$ and $\psi(y) = \phi(0, y)$ for each $x, y \in [0, \infty)$ and $C := C([-\alpha, 0] \times [-\beta, 0])$. If $u \in C([-\alpha, \infty) \times [-\beta, \infty), \mathbb{R}^n)$, then for any $(x, y) \in J_\infty$ define $u_{(x,y)}$ by

$$u_{(x,y)}(s, t) = u(x + s, y + t),\ \text{for } (s, t) \in [-\alpha, 0] \times [-\beta, 0].$$

Next we consider the following IVP for partial neutral functional differential equations with finite delay of the form:

$${}^cD_0^r\Big(u(x, y) - g(x, y, u_{(x,y)})\Big) = f(x, y, u_{(x,y)}),\ \text{if } (x, y) \in J_\infty, \tag{3.109}$$

$$u(x, y) = \phi(x, y), \text{ if } (x, y) \in \tilde{J}_\infty, \tag{3.110}$$

$$u(x, 0) = \varphi(x), \ x \in [0, \infty), \ u(0, y) = \psi(y), \ y \in [0, \infty), \tag{3.111}$$

where f, ϕ, φ, ψ are as in problems (3.106)–(3.108) and $g : J_\infty \times C \to \mathbb{R}^n$ is a given function. The third result deals with the existence of solutions to fractional order partial hyperbolic functional differential equations with infinite delay of the form

$$({}^c D_0^r u)(x, y) = f(x, y, u_{(x,y)}); \text{ if } (x, y) \in J_\infty, \tag{3.112}$$

$$u(x, y) = \phi(x, y); \text{ if } (x, y) \in \tilde{J}'_\infty, \tag{3.113}$$

$$u(x, 0) = \varphi(x), \ x \in [0, \infty), \ u(0, y) = \psi(y), \ y \in [0, \infty), \tag{3.114}$$

where φ, ψ are as in problems (3.106)–(3.108) and $\tilde{J}'_\infty = \mathbb{R}^2 \backslash (0, \infty) \times (0, \infty)$, $f : J_\infty \times \mathcal{B} \to \mathbb{R}^n$, $\phi \in C(\tilde{J}'_\infty, \mathbb{R}^n)$ and $\mathcal{B}$ is a phase space. We denote by $u_{(x,y)}$ the element of $\mathcal{B}$ defined by

$$u_{(x,y)}(s, t) = u(x + s, y + t); \ (s, t) \in (-\infty, 0] \times (-\infty, 0].$$

Finally we consider the following initial value problem for partial neutral functional differential equations with infinite delay :

$${}^c D_0^r \Big(u(x, y) - g(x, y, u_{(x,y)}) \Big) = f(x, y, u_{(x,y)}), \text{ if } (x, y) \in J_\infty, \tag{3.115}$$

$$u(x, y) = \phi(x, y), \text{ if } (x, y) \in \tilde{J}'_\infty, \tag{3.116}$$

$$u(x, 0) = \varphi(x), \ x \in [0, \infty), \ u(0, y) = \psi(y), \ y \in [0, \infty), \tag{3.117}$$

where f, ϕ, φ, ψ are as in problem (3.112)–(3.114) and $g : J_\infty \times \mathcal{B} \to \mathbb{R}^n$ is a given continuous function.

Let X be a Fréchet space with a family of seminorms $\{\|\cdot\|_n\}_{n \in \mathbb{N}}$. We assume that the family of seminorms $\{\|\cdot\|_n\}$ verifies:

$$\|x\|_1 \le \|x\|_2 \le \|x\|_3 \le \dots \quad \textit{for every } x \in X.$$

Let $Y \subset X$, we say that Y is bounded if for every $n \in \mathbb{N}$, there exists $\overline{M}_n > 0$ such that

$$\|y\|_n \le \overline{M}_n \quad \textit{for all } y \in Y.$$

To X we associate a sequence of Banach spaces $\{(X^n, \|\cdot\|_n)\}$ as follows : For every $n \in \mathbb{N}$, we consider the equivalence relation $\sim_n$ defined by : $x \sim_n y$ if and only if $\|x - y\|_n = 0$ for $x, y \in X$. We denote $X^n = (X|_{\sim_n}, \|\cdot\|_n)$ the quotient space, the completion of X^n with respect to $\|\cdot\|_n$. To every $Y \subset X$, we associate a sequence $\{Y^n\}$ of subsets $Y^n \subset X^n$ as follows : For every $x \in X$, we denote $[x]_n$ the equivalence class of x of subset X^n and we defined $Y^n = \{[x]_n : x \in Y\}$. We denote $\overline{Y^n}$, $int_n(Y^n)$ and $\partial_n Y^n$, respectively, the closure, the interior, and the boundary of Y^n with respect to $\|\cdot\|_n$ in X^n. For more information about this subject see [127].

Definition 3.47. Let X be a Fréchet space . A function $N : X \longrightarrow X$ is said to be a contraction if for each $n \in \mathbb{N}$ there exists $k_n \in (0,1)$ such that

$$\|N(u) - N(v)\|_n \le k_n \|u - v\|_n \quad \text{for all } u, v \in X.$$

Theorem 3.48. *[127] Let X be a Fréchet space and $Y \subset X$ a closed subset in X. Let $N : Y \longrightarrow X$ be a contraction such that $N(Y)$ is bounded. Then one of the following statements holds:*

(a) The operator N has a unique fixed point.
(b) There exists $\lambda \in [0,1)$, $n \in \mathbb{N}$ and $u \in \partial_n Y^n$ such that $\|u - \lambda N(u)\|_n = 0$.

3.9.2 Global Result for Finite Delay

For each $p \in \mathbb{N}$ we consider following set $C_p = C([-\alpha, p] \times [-\beta, p], \mathbb{R}^n)$ and we define in $C_\infty := C([-\alpha, \infty) \times [-\beta, \infty), \mathbb{R}^n)$ the seminorms by

$$\|u\|_p = \sup\{\|u(x,y)\| : -\alpha \le x \le p, \; -\beta \le y \le p\}.$$

Then C_∞ is a Fréchet space with the family of seminorms $\{\|u\|_p\}_{p\in\mathbb{N}}$.

Let us start by defining what we mean by a solution of the problem (3.106)–(3.108).

Definition 3.49. A function $u \in C_\infty$ whose mixed derivative D^2_{xy} exists and is integrable is said to be a solution of (3.106)–(3.108) if u satisfies (3.106) and (3.108) on J_∞ and the condition (3.107) on $\tilde{J}_\infty$.

Further, we present conditions for the existence and uniqueness of a solution of problems (3.106)–(3.108).

Theorem 3.50. *Let $J_0 := [0, p] \times [0, p]$; $p \in \mathbb{N}$. Assume that*

(3.50.1) The function $f : J_\infty \times C \to \mathbb{R}^n$ is continuous
(3.50.2) For each $p \in \mathbb{N}$, there exists $l_p \in C(J_0, \mathbb{R}_+)$ such that for each $(x,y) \in J_0$

$$\|f(x,y,u) - f(x,y,v)\| \le l_p(x,y)\|u - v\|_C, \textit{ for each } u, \; v \in C$$

If

$$\frac{l_p^* p^{r_1+r_2}}{\Gamma(r_1+1)\Gamma(r_2+1)} < 1, \tag{3.118}$$

where

$$l_p^* = \sup_{(x,y)\in J_0} l_p(x,y),$$

then there exists a unique solution for IVP (3.106)–(3.108) on $[-\alpha, \infty) \times [-\beta, \infty)$.

Proof. Transform the problems (3.106)–(3.108) into a fixed point problem. Consider the operator $N : C_\infty \to C_\infty$ defined by

$$N(u)(x,y) = \begin{cases} \phi(x,y), & (x,y) \in \tilde{J}_\infty, \\ \mu(x,y) \\ +\frac{1}{\Gamma(r_1)\Gamma(r_2)} \int_0^x \int_0^y (x-s)^{r_1-1}(y-t)^{r_2-1} f(s,t,u_{(s,t)}) dt ds, & (x,y) \in J_\infty. \end{cases}$$

Let u be a possible solution of the problem $u = \lambda N(u)$ for some $0 < \lambda < 1$. This implies that for each $(x,y) \in J_0$, we have

$$u(x,y) = \lambda \mu(x,y) + \frac{\lambda}{\Gamma(r_1)\Gamma(r_2)} \int_0^x \int_0^y (x-s)^{r_1-1}(y-t)^{r_2-1} f(s,t,u_{(s,t)}) dt ds.$$

This implies by (3.50.2) that

$$\|u(x,y)\| \le \|\mu(x,y)\| + \frac{f^* p^{r_1+r_2}}{\Gamma(r_1+1)\Gamma(r_2+1)} + \frac{1}{\Gamma(r_1)\Gamma(r_2)} \int_0^x \int_0^y (x-s)^{r_1-1}(y-t)^{r_2-1} l_p(s,t) \|u_{(s,t)}\|_C dt ds,$$

where

$$f^* = \sup_{(x,y)\in J_0} \|f(x,y,0)\|.$$

We consider the function τ defined by

$$\tau(x,y) = \sup\{\|u(s,t)\| : -\alpha \le s \le x, \ -\beta \le t \le y, \ x,y \in [0,p]\}.$$

Let $(x^*, y^*) \in [-\alpha, x] \times [-\beta, y]$ be such that $\tau(x,y) = \|u(x^*, y^*)\|$. If $(x^*, y^*) \in J_0$, then by the previous inequality, we have for $(x,y) \in J_0$,

$$\|u(x,y)\| \le \|\mu(x,y)\| + \frac{f^* p^{r_1+r_2}}{\Gamma(r_1+1)\Gamma(r_2+1)} + \frac{1}{\Gamma(r_1)\Gamma(r_2)} \int_0^x \int_0^y (x-s)^{r_1-1}(y-t)^{r_2-1} l_p(s,t) \tau(s,t) dt ds. \tag{3.119}$$

If $(x^*, y^*) \in \tilde{J}_\infty$, then $\tau(x,y) = \|\phi\|_C$ and the previous inequality holds. By (3.119) we obtain that

$$\begin{aligned}\tau(x,y) &\leq \|\mu(x,y)\| + \frac{f^* p^{r_1+r_2}}{\Gamma(r_1+1)\Gamma(r_2+1)} \\ &\quad + \frac{1}{\Gamma(r_1)\Gamma(r_2)} \int_0^x \int_0^y (x-s)^{r_1-1}(y-t)^{r_2-1} l_p(s,t)\tau(s,t)\mathrm{d}t\mathrm{d}s \\ &\leq \|\mu(x,y)\| + \frac{f^* p^{r_1+r_2}}{\Gamma(r_1+1)\Gamma(r_2+1)} \\ &\quad + \frac{l_p^*}{\Gamma(r_1)\Gamma(r_2)} \int_0^x \int_0^y (x-s)^{r_1-1}(y-t)^{r_2-1} \tau(s,t)\mathrm{d}t\mathrm{d}s,\end{aligned}$$

and Lemma 2.43 implies that there exists a constant $\delta = \delta(r_1, r_2)$ such that

$$\begin{aligned}\tau(x,y) &\leq \left(\|\mu\|_p + \frac{f^* p^{r_1+r_2}}{\Gamma(r_1+1)\Gamma(r_2+1)}\right)\left(1 + \frac{\delta l_p^*}{\Gamma(r_1+1)\Gamma(r_2+1)}\right) \\ &:= M_p.\end{aligned}$$

Then from (3.119) we have

$$\begin{aligned}\|u\|_p &\leq \|\mu\|_p + \frac{f^* p^{r_1+r_2}}{\Gamma(r_1+1)\Gamma(r_2+1)} + \frac{M_p l_p^*}{\Gamma(r_1+1)\Gamma(r_2+1)} \\ &:= M_p^*.\end{aligned}$$

Since for every $(x,y) \in J_0$, $\|u_{(x,y)}\|_C \leq \tau(x,y)$, we have

$$\|u\|_p \leq \max(\|\phi\|_C, M_p^*) := R_p.$$

Set

$$U = \{u \in C_\infty : \|u\|_p \leq R_p + 1 \text{ for all } p \in \mathbb{N}\}.$$

We shall show that $N : U \longrightarrow C_p$ is a contraction map. Indeed, consider $v, w \in U$. Then for each $(x,y) \in J_0$, we have

$$\begin{aligned}&\|N(v)(x,y) - N(w)(x,y)\| \\ &\leq \frac{1}{\Gamma(r_1)\Gamma(r_2)} \int_0^x \int_0^y (x-s)^{r_1-1}(y-t)^{r_2-1} \|f(s,t,v_{(s,t)}) - f(s,t,w_{(s,t)})\|\mathrm{d}t\mathrm{d}s\end{aligned}$$

$$
\begin{aligned}
&\leq \frac{1}{\Gamma(r_1)\Gamma(r_2)} \int_0^x \int_0^y (x-s)^{r_1-1}(y-t)^{r_2-1} l_{(p,q)}(s,t) \|v_{(s,t)} - w_{(s,t)}\|_C \, dt ds \\
&\leq \frac{l_p^* p^{r_1+r_2}}{\Gamma(r_1+1)\Gamma(r_2+1)} \|v-w\|_p.
\end{aligned}
$$

Thus

$$
\|N(v) - N(w)\|_p \leq \frac{l_p^* p^{r_1+r_2}}{\Gamma(r_1+1)\Gamma(r_2+1)} \|v-w\|_p.
$$

Hence by (3.118), $N : U \longrightarrow C_p$ is a contraction. By our choice of U, there is no $u \in \partial_n U^n$ such that $u = \lambda N(u)$, for $\lambda \in (0,1)$. As a consequence of Theorem 3.48, we deduce that N has a unique fixed point u in U which is a solution to problem (3.106)–(3.108). □

Now we present a global existence and uniqueness result for the problems (3.109)–(3.111).

Definition 3.51. A function $u \in C([-\alpha, \infty) \times [-\beta, \infty), \mathbb{R}^n)$ is said to be a global solution of (3.109)–(3.111) if u satisfies (3.109) and (3.111) on J_∞ and the condition (3.110) on $\tilde{J}_\infty$.

Theorem 3.52. *Assume that (3.50.1), (3.50.2): and the following condition hold*

(3.52.1) *For each $p = 1, 2, \ldots$, there exists a constant c_p with $0 < c_p < \frac{1}{4}$ such that for each $(x, y) \in J_0$ we have*

$$
\|g(x,y,u) - g(x,y,v)\| \leq c_p \|u - v\|, \text{ for each } u, v \in C.
$$

If

$$
4c_p + \frac{l_p^* p^{r_1+r_2}}{\Gamma(r_1+1)\Gamma(r_2+1)} < 1,
$$

then there exists a unique solution for IVP (3.109)–(3.111) on $[-\alpha, \infty) \times [-\beta, \infty)$.

Proof. Transform the problems (3.109)–(3.111) into a fixed point problem. Consider the operator $N_1 : C_\infty \to C_\infty$ defined by

$$
N_1(u)(x,y) = \begin{cases}
\phi(x,y), & (x,y) \in \tilde{J}_\infty, \\
\mu(x,y) + g(x,y,u_{(x,y)}) - g(x,0,u_{(x,0)}) & \\
-g(0,y,u_{(0,y)}) + g(0,0,u_{(0,0)}) & \\
+\frac{1}{\Gamma(r_1)\Gamma(r_2)} \int_0^x \int_0^y (x-s)^{r_1-1}(y-t)^{r_2-1} & \\
\times f(s,t,u_{(s,t)}) dt ds, & (x,y) \in J_\infty,
\end{cases}
$$

In order to use the nonlinear alternative, we shall obtain a priori estimates for the solutions of the integral equation

$$u(x,y) = \lambda(\mu(x,y)+g(x,y,u_{(x,y)})-g(x,0,u_{(x,0)})-g(0,y,u_{(0,y)})+g(0,0,u_{(0,0)})$$
$$+\frac{\lambda}{\Gamma(r_1)\Gamma(r_2)}\int_0^x\int_0^y (x-s)^{r_1-1}(y-t)^{r_2-1} f(s,t,u_{(s,t)})\mathrm{d}t\,\mathrm{d}s,$$

for some $\lambda \in (0,1)$. Then using (3.50.1), (3.50.2), (3.52.1), and (3.119) we get

$$\|u(x,y)\| \le \|\mu(x,y)\| + \frac{f^* p^{r_1+r_2}}{\Gamma(r_1+1)\Gamma(r_2+1)}$$
$$+\|g(x,y,u_{(x,y)})\|+\|g(x,0,u_{(x,0)})\|+\|g(0,y,u_{(0,y)})\|$$
$$+\|g(0,0,u_{(0,0)}))\|$$
$$+\frac{1}{\Gamma(r_1)\Gamma(r_2)}\int_0^x\int_0^y (x-s)^{r_1-1}(y-t)^{r_2-1} l_p(s,t)\tau(s,t)\mathrm{d}t\,\mathrm{d}s,$$

then we obtain

$$\|u(x,y)\| \le \|\mu(x,y)\| + \frac{f^* p^{r_1+r_2}}{\Gamma(r_1+1)\Gamma(r_2+1)}$$
$$+4c_p\tau(x,y)+\|g(x,y,0)\|+\|g(x,0,0)\|+\|g(0,y,0)\|+\|g(0,0,0)\|$$
$$+\frac{l_p^*}{\Gamma(r_1)\Gamma(r_2)}\int_0^x\int_0^y (x-s)^{r_1-1}(y-t)^{r_2-1}\tau(s,t)\mathrm{d}t\,\mathrm{d}s. \qquad (3.120)$$

Replacing (3.120) in the definition of $\tau(x,y)$ we get

$$\tau(x,y) \le \frac{1}{1-4c_p}\left[\|\mu(x,y)\| + \frac{f^* p^{r_1+r_2}}{\Gamma(r_1+1)\Gamma(r_2+1)} + 4g^*\right]$$
$$+\frac{\widetilde{l}_p^*}{\Gamma(r_1)\Gamma(r_2)}\int_0^x\int_0^y (x-s)^{r_1-1}(y-t)^{r_2-1}\tau(s,t)\mathrm{d}t\,\mathrm{d}s,$$

where $\widetilde{l}_p^* = \frac{l_p^*}{1-4c_p}$ and $g_p^* = \sup_{(x,y)\in J_0} \|g(x,y,0)\|$. By Lemma 2.43, there exists a constant $\delta = \delta(r_1, r_2)$ such that

$$\|\tau\|_p \leq \frac{1}{1-4c_p}\left[\|\mu\|_p + \frac{f^* p^{r_1+r_2}}{\Gamma(r_1+1)\Gamma(r_2+1)} + 4g_p^*\right] \times \left[1 + \frac{\delta \widetilde{l}_p^*}{\Gamma(r_1+1)\Gamma(r_2+1)}\right] := D_p. \tag{3.121}$$

Then from (3.120) and (3.121) we get

$$\|u\|_p \leq \|\mu\|_p + \frac{f^* p^{r_1+r_2}}{\Gamma(r_1+1)\Gamma(r_2+1)} + 4g_p^* + 4c_p D_p + \frac{D_p l_p^*}{\Gamma(r_1+1)\Gamma(r_2+1)} := D_p^*.$$

Since for every $(x,y) \in J_0$, $\|u_{(x,y)}\|_C \leq \tau(x,y)$, we have

$$\|u\|_p \leq \max(\|\phi\|_C, D_p^*) := R_p^*.$$

Set

$$U_1 = \{u \in C_\infty : \|u\|_p \leq R_p^* + 1 \text{ for all } p = 1, 2, \ldots\}.$$

Clearly, U_1 is a closed subset of C_0. As in Theorem 3.50, we can show that $N_1 : U_1 \longrightarrow C_p$ is a contraction operator. Indeed

$$\|N_1(v) - N_1(w)\|_p \leq \left(4c_p + \frac{l_p^* p^{r_1+r_2}}{\Gamma(r_1+1)\Gamma(r_2+1)}\right)\|v - w\|_p,$$

for each $v, w \in U_1$, and $(x,y) \in J_0$. From the choice of U_1, there is no $u \in \partial_n U_1^n$ such that $u = \lambda N_1(u)$, for some $\lambda \in (0,1)$. As a consequence of Theorem 3.48, we deduce that N_1 has a unique fixed point u in U_1 which is a solution to problem (3.109)–(3.111). □

3.9.3 *Global Result for Infinite Delay*

In this section we present a global existence and uniqueness result for problem (3.112)–(3.114). Let the space

$$\Omega := \{u : \mathbb{R}^2 \to \mathbb{R}^n : u_{(x,y)} \in \mathcal{B} \text{ for } (x,y) \in E_\infty \text{ and } u|_{J_\infty} \in C(J_\infty, \mathbb{R}^n)\},$$

where $E_\infty = [0,\infty) \times \{0\} \cup \{0\} \times [0,\infty)$.

Definition 3.53. A function $u \in \Omega$ is said to be a solution of (3.112)–(3.114) if u satisfies (3.13) and (3.114) on J_∞ and the condition (3.113) on $\tilde{J}'_\infty$.

For each $p \in \mathbb{N}$ we consider the following set;

$$C'_p = \{u : (-\infty, p] \times (-\infty, p] \to \mathbb{R}^n : u \in \mathcal{B} \cap C(J_0, \mathbb{R}^n),\ u_{(x,y)} = 0$$
$$\text{for } (x, y) \in E_{(p,p)}\},$$

and we define in

$$C'_\infty := \{u : \mathbb{R}^2 \to \mathbb{R}^n : u \in \mathcal{B} \cap C([0, \infty) \times [0, \infty), \mathbb{R}^n),\ u_{(x,y)} = 0$$
$$\text{for } (x, y) \in E_\infty\}$$

the seminorms by

$$\begin{aligned} \|u\|_{p'} &= \sup_{(x,y)\in E_p} \|u_{(x,y)}\|_{\mathcal{B}} + \sup_{(x,y)\in J_0} \|u(x, y)\| \\ &= \sup_{(x,y)\in J_0} \|u(x, y)\|,\ u \in C'_p. \end{aligned}$$

Then C'_∞ is a Fréchet space with the family of seminorms $\{\|u\|_{p'}\}$.

Further, we present conditions for the existence and uniqueness of a solution of problems (3.112)–(3.114).

Theorem 3.54. *(3.54.1) The function $f : J_\infty \times \mathcal{B} \longrightarrow \mathbb{R}^n$ is continuous,*
(3.54.2) For each $p \in \mathbb{N}$ there exists $l'_p \in C(J_0, \mathbb{R}^n)$ such that for $(x, y) \in J_0$, we have

$$\|f(x, y, u) - f(x, y, v)\| \le l'_p(x, y)\|u - v\|_{\mathcal{B}},\ \textit{for each } u,\ v \in \mathcal{B}.$$

If

$$\frac{K l'^*_p p^{r_1+r_2}}{\Gamma(r_1 + 1)\Gamma(r_2 + 1)} < 1, \tag{3.122}$$

where

$$l'^*_p = \sup_{(x,y)\in J_0} l'_p(x, y),$$

then there exists a unique solution for IVP (3.112)–(3.114) on $\mathbb{R}^2$.

Proof. Transform the problems (3.112)–(3.114) into a fixed point problem. Consider the operator $N' : \Omega \to \Omega$ defined by

$$N'(u)(x,y) = \begin{cases} \phi(x,y); & (x,y) \in \tilde{J}'_\infty, \\ \mu(x,y) \\ +\frac{1}{\Gamma(r_1)\Gamma(r_2)} \int_0^x \int_0^y (x-s)^{r_1-1}(y-t)^{r_2-1} \\ \times f(s,t,u_{(s,t)})\mathrm{d}t\mathrm{d}s; & (x,y) \in J_\infty. \end{cases}$$

Let $v(.,.) : \mathbb{R}^2 \to \mathbb{R}^n$. be a function defined by

$$v(x,y) = \begin{cases} \phi(x,y); \ (x,y) \in \tilde{J}'_\infty, \\ \mu(x,y); \ (x,y) \in J_\infty. \end{cases}$$

Then $v_{(x,y)} = \phi$ for all $(x,y) \in E_\infty$. For each $w \in C(J,\mathbb{R}^n)$ with $w(x,y) = 0$, for all $(x,y) \in E_\infty$, we denote by $\overline{w}$ the function defined by

$$\overline{w}(x,y) = \begin{cases} 0; & (x,y) \in \tilde{J}'_\infty, \\ w(x,y); & (x,y) \in J_\infty. \end{cases}$$

If $u(.,.)$ satisfies the integral equation

$$u(x,y) = \mu(x,y) + \frac{1}{\Gamma(r_1)\Gamma(r_2)} \int_0^x \int_0^y (x-s)^{r_1-1}(y-t)^{r_2-1} f(s,t,u_{(s,t)})\mathrm{d}t\mathrm{d}s,$$

we can decompose $u(.,.)$ as $u(x,y) = \overline{w}(x,y) + v(x,y), \ x,y \geq 0$, which implies that $u_{(x,y)} = \overline{w}_{(x,y)} + v_{(x,y)}$, for every $x,y \geq 0$, and the function $w(.,.)$ satisfies

$$w(x,y) = \frac{1}{\Gamma(r_1)\Gamma(r_2)} \int_0^x \int_0^y (x-s)^{r_1-1}(y-t)^{r_2-1} f(s,t,\overline{w}_{(s,t)} + v_{(s,t)})\mathrm{d}t\mathrm{d}s.$$

Let the operator $P' : C'_\infty \to C'_\infty$ be defined by

$$(P'w)(x,y) = \frac{1}{\Gamma(r_1)\Gamma(r_2)} \int_0^x \int_0^y (x-s)^{r_1-1}(y-t)^{r_2-1} \\ \times f(s,t,\overline{w}_{(s,t)} + v_{(s,t)})\mathrm{d}t\mathrm{d}s; \ (x,y) \in J_\infty. \tag{3.123}$$

The operator N' has a fixed point if and only if P' has a fixed point, and so we turn to prove that P' has a fixed point. We shall use the alternative to prove that P' has a fixed point. Let w be a possible solution of the problem $w = P'(w)$ for some $0 < \lambda < 1$. This implies that for each $(x,y) \in J_0$, we have

$$w(x,y) = \frac{\lambda}{\Gamma(r_1)\Gamma(r_2)} \int_0^x \int_0^y (x-s)^{r_1-1}(y-t)^{r_2-1} f(s,t,\overline{w}_{(s,t)} + v_{(s,t)}))\mathrm{d}t\mathrm{d}s.$$

This implies by (3.54.1) that

$$\|w(x,y)\| \leq \frac{f_p^* p^{r_1+r_2}}{\Gamma(r_1+1)\Gamma(r_2+1)} + \frac{1}{\Gamma(r_1)\Gamma(r_2)} \int_0^x \int_0^y (x-s)^{r_1-1}(y-t)^{r_2-1}$$

$$l'_p(s,t)\|\overline{w}_{(s,t)} + v_{(s,t)})\|_{\mathcal{B}}\mathrm{d}t\mathrm{d}s, \tag{3.124}$$

where

$$f_p^* = \sup\{\|f(x,y,0)\| : (x,y) \in J_0\}.$$

But

$$\begin{aligned} \|\overline{w}_{(s,t)} + v_{(s,t)}\|_{\mathcal{B}} &\leq \|\overline{w}_{(s,t)}\|_{\mathcal{B}} + \|v_{(s,t)}\|_{\mathcal{B}} \\ &\leq K \sup\{u(\tilde{s},\tilde{t}) : (\tilde{s},\tilde{t}) \in [0,s]\times[0,t]\} \\ &\quad + M\|\phi\|_{\mathcal{B}} + K\|\phi(0,0)\|. \end{aligned} \tag{3.125}$$

If we name $z(s,t)$ the right-hand side of (3.125), then we have

$$\|\overline{w}_{(s,t)} + v_{(s,t)}\|_{\mathcal{B}} \leq z(s,t). \tag{3.126}$$

Therefore, from (3.124) and (3.126) we get

$$\begin{aligned} \|w(x,y)\| &\leq \frac{f_p^* p^{r_1+r_2}}{\Gamma(r_1+1)\Gamma(r_2+1)} \\ &\quad + \frac{1}{\Gamma(r_1)\Gamma(r_2)} \int_0^x \int_0^y (x-s)^{r_1-1}(y-t)^{r_2-1} l'_p(s,t) z(s,t)\mathrm{d}t\mathrm{d}s. \end{aligned} \tag{3.127}$$

Replacing (3.127) in the definition of w, we have that

$$\begin{aligned} \|z(x,y)\| &\leq \frac{K f_p^* p^{r_1+r_2}}{\Gamma(r_1+1)\Gamma(r_2+1)} + M\|\phi\|_{\mathcal{B}} \\ &\quad + \frac{K l'^*_p}{\Gamma(r_1)\Gamma(r_2)} \int_0^x \int_0^y (x-s)^{r_1-1}(y-t)^{r_2-1} z(s,t)\mathrm{d}t\mathrm{d}s. \end{aligned}$$

By Lemma 2.43, there exists a constant $\delta = \delta(r_1, r_2)$ such that

$$\|z\|_{p'} \leq \left(\frac{Kf_p^* p^{r_1+r_2}}{\Gamma(r_1+1)\Gamma(r_2+1)} + M\|\phi\|_{\mathcal{B}}\right) \times \left(1 + \frac{\delta K l_p'^*}{\Gamma(r_1+1)\Gamma(r_2+1)}\right) := \widetilde{M}.$$

Then from (3.127) we have

$$\|w\|_{p'} \leq \widetilde{M}\frac{l_p'^* p^{r_1+r_2}}{\Gamma(r_1+1)\Gamma(r_2+1)} + \frac{f_p^* p^{r_1+r_2}}{\Gamma(r_1+1)\Gamma(r_2+1)} := \widetilde{M}^*.$$

Set

$$U' = \{w \in C'_\infty : \|w\|_{p'} \leq \widetilde{M}^* + 1 \text{ for all } p \in \mathbb{N}\}.$$

We shall show that $P' : U' \longrightarrow C'_p$ is a contraction map. Indeed, consider $w, w^* \in U'$. Then for each $(x, y) \in J_0$, we have

$$\begin{aligned}
&\|P'(w)(x,y) - P'(w^*)(x,y)\| \\
&\leq \frac{1}{\Gamma(r_1)\Gamma(r_2)}\int_0^x\int_0^y (x-s)^{r_1-1}(y-t)^{r_2-1} \\
&\quad \times \|f(s,t,\overline{w}_{(s,t)} + v_{(s,t)}) - f(s,t,\overline{w^*}_{(s,t)} + v_{(s,t)})\|dtds \\
&\leq \frac{1}{\Gamma(r_1)\Gamma(r_2)}\int_0^x\int_0^y (x-s)^{r_1-1}(y-t)^{r_2-1} l_p'(s,t)\|\overline{w}_{(s,t)} - \overline{w^*}_{(s,t)}\|_{\mathcal{B}}dtds \\
&\leq K\frac{l_p'^* p^{r_1+r_2}}{\Gamma(r_1+1)\Gamma(r_2+1)}\|w - w^*\|_{p'}.
\end{aligned}$$

Thus

$$\|P'(w) - P'(w^*)\|_{p'} \leq \frac{K l_p'^* p^{r_1+r_2}}{\Gamma(r_1+1)\Gamma(r_2+1)}\|w - w^*\|_{p'}.$$

Hence by (3.122), $P' : U' \longrightarrow C'_p$ is a contraction. By our choice of U', there is no $w \in \partial_n(U')^n$ such that $w = \lambda P'(w)$, for $\lambda \in (0, 1)$. As a consequence of Theorem 3.48, we deduce that N' has a unique fixed point which is a solution to problems (3.112)–(3.114). □

Now we present an existence result for the problems (3.115)–(3.117).

Definition 3.55. A function $u \in \Omega$ is said to be a global solution of (3.115)–(3.117) if u satisfies (3.115) and (3.117) on J_∞ and the condition (3.116) on $\tilde{J}'_\infty$.

Theorem 3.56. *Let $f, g : J_\infty \times \mathcal{B} \longrightarrow \mathbb{R}^n$ be continuous functions. Assume that (3.54.1), (3.54.2), and the following condition hold:*

(3.56.1) For each $p = 1, 2, \dots$, there exists a constant c'_p with $0 < Kc'_p < \frac{1}{4}$ such that for any $(x, y) \in J_0$, we have

$$\|g(x, y, u) - g(x, y, v)\| \le c'_p \|u - v\|_{\mathcal{B}}, \quad \textit{for any } u, v \in \mathcal{B}.$$

If

$$4c'_p + \frac{K l'^{*}_p p^{r_1 + r_2}}{\Gamma(r_1 + 1)\Gamma(r_2 + 1)} < 1, \textit{ for each } p \in \mathbb{N},$$

then there exists a unique solution for IVP (3.115)–(3.117) on $\mathbb{R}^2$.

Proof. Consider the operator $N'_1 : \Omega \to \Omega$ defined by

$$N'_1(u)(x, y) = \begin{cases} \phi(x, y), & (x, y) \in \tilde{J}'_\infty, \\ \mu(x, y) + g(x, y, u_{(x,y)}) - g(x, 0, u_{(x,0)}) & \\ -g(0, y, u_{(0,y)}) + g(0, 0, u_{(0,0)}) & \\ +\frac{1}{\Gamma(r_1)\Gamma(r_2)} \int_0^x \int_0^y (x - s)^{r_1 - 1} (y - t)^{r_2 - 1} & \\ \times f(s, t, u_{(s,t)}) \mathrm{d}t \mathrm{d}s, & (x, y) \in J_\infty. \end{cases}$$

In analogy to Theorem 3.54, we consider the operator $P'_1 : C'_\infty \to C'_\infty$ defined by

$$\begin{aligned} P'_1(w)(x, y) &= g(x, y, \overline{w}_{(x,y)} + v_{(x,y)}) - g(x, 0, \overline{w}_{(x,0)} + v_{(x,0)}) \\ &\quad -g(0, y, \overline{w}_{(0,y)} + v_{(0,y)}) + g(0, 0, \overline{w}_{(0,0)} + v_{(0,0)}) \\ &\quad +\frac{1}{\Gamma(r_1)\Gamma(r_2)} \int_0^x \int_0^y (x - s)^{r_1 - 1} (y - t)^{r_2 - 1} \\ &\quad \times f(s, t, \overline{w}_{(s,t)} + v_{(s,t)}) \mathrm{d}t \mathrm{d}s, \ (x, y) \in J. \end{aligned}$$

In order to use the nonlinear alternative, we shall obtain a priori estimates for the solutions of the integral equation

$$\begin{aligned} w(x, y) &= \lambda(g(x, y, \overline{w}_{(x,y)} + v_{(x,y)}) - g(x, 0, \overline{w}_{(x,0)} + v_{(x,0)}) \\ &\quad -g(0, y, \overline{w}_{(0,y)} + v_{(0,y)}) + g(0, 0, \overline{w}_{(0,0)} + v_{(0,0)})) \\ &\quad +\frac{\lambda}{\Gamma(r_1)\Gamma(r_2)} \int_0^x \int_0^y (x - s)^{r_1 - 1} (y - t)^{r_2 - 1} f(s, t, \overline{w}_{(s,t)} + v_{(s,t)}) \mathrm{d}t \mathrm{d}s, \end{aligned}$$

for some $\lambda \in (0,1)$. Then from (3.54.1), (3.54.2), (3.56.1), (3.124) and (3.126) we get

$$\begin{aligned}\|w(x,y)\| \leq & \frac{f_p^* p^{r_1+r_2}}{\Gamma(r_1+1)\Gamma(r_2+1)} + 4c'_p z(x,y) \\ & + \|g(x,y,0)\| + \|g(x,0,0)\| + \|g(0,y,0)\| + \|g(0,0,0)\| \\ & + \frac{1}{\Gamma(r_1)\Gamma(r_2)} \int_0^x \int_0^y (x-s)^{r_1-1}(y-t)^{r_2-1} l'_p(s,t) z(s,t) dt ds. \quad (3.128)\end{aligned}$$

Replacing (3.128) in the definition of $z(x,y)$ we get

$$\begin{aligned}z(x,y) \leq & \frac{1}{1-4Kc_p}\Bigg[M\|\phi\|_{\mathcal{B}} + 4K\|\phi(0,0)\| + 4K\|g(0,0,\phi(0,0))\| \\ & + 4Kg_p^* + \frac{Kf_p^* p^{r_1+r_2}}{\Gamma(r_1+1)\Gamma(r_2+1)}\Bigg] \\ & + \frac{\tilde{l}'^*_p}{\Gamma(r_1)\Gamma(r_2)} \int_0^x \int_0^y (x-s)^{r_1-1}(y-t)^{r_2-1} z(s,t) dt ds,\end{aligned}$$

where $\tilde{l}'^*_p(x,y) = \frac{l'^*_p}{1-4Kc_p}$, $g_p^* = \sup\{\|g(x,y,0)\| : (x,y) \in J_0\}$.

By (3.128) and Lemma 2.43, there exists a constant $\delta = \delta(r_1,r_2)$ such that

$$\begin{aligned}z(x,y) \leq & \frac{1}{1-4Kc_p}\Bigg[M\|\phi\|_{\mathcal{B}} + 4K\|\phi(0,0)\| + 4K\|g(0,0,\phi(0,0))\| \\ & + 4Kg_p^* + \frac{Kf_p^* p^{r_1+r_2}}{\Gamma(r_1+1)\Gamma(r_2+1)}\Bigg] \\ & \times \left[1 + \frac{\delta \tilde{l}'^*_p}{\Gamma(r_1+1)\Gamma(r_2+1)}\right] := D'. \quad (3.129)\end{aligned}$$

Then from (3.128) and (3.129) we get

$$\begin{aligned}\|w\|_{p'} \leq & \frac{(D' l'^*_p + f_p^*) p^{r_1+r_2}}{\Gamma(r_1+1)\Gamma(r_2+1)} + 4c'_p D' + 4g_p^* \\ := & D'^*.\end{aligned}$$

Since for every $(x, y) \in J_0$, $\|w_{(x,y)}\|_{\mathcal{B}} \leq z(x, y)$, we have

$$\|w\|_{p'} \leq \max(\|\phi\|_{\mathcal{B}}, D'^*) := R'^*.$$

Set

$$U_1' = \{w \in C_\infty' : \|w\|_{p'} \leq R'^* + 1\}.$$

Clearly, U_1' is a closed subset of C_∞'. As in Theorem 3.54, we can show that $P_1' : U_1' \longrightarrow C_\infty'$ is a contraction operator. From the choice of U_1', there is no $w \in \partial_n U_1^n$ such that $w = \lambda P_1'(w)$, for some $\lambda \in (0, 1)$. As a consequence of Theorem 3.48, we deduce that N_1' has a unique fixed point which is a solution to problem (3.115)–(3.117). □

3.9.4 Examples

3.9.4.1 Example 1

As an application of our results we consider the following partial hyperbolic functional differential equations with finite delay of the form

$$^cD_0^r u(x, y) = \frac{c_p}{e^{x+y+2}(1 + |u(x-1, y-2)|)}, \quad \text{if } (x, y) \in [0, \infty) \times [0, \infty), \tag{3.130}$$

$$u(x, 0) = x, \; u(0, y) = y^2, \; x, y \in [0, \infty), \tag{3.131}$$

$$u(x, y) = x + y^2, \; (x, y) \in [-1, \infty) \times [-2, \infty) \backslash (0, \infty) \times (0, \infty), \tag{3.132}$$

where

$$c_p = \frac{\Gamma(r_1 + 1)\Gamma(r_2 + 1)}{p^{r_1+r_2}}; \; p \in \mathbb{N}^*.$$

Set

$$f(x, y, u_{(x,y)}) = \frac{c_p}{e^{x+y+2}(1 + |u(x-1, y-2)|)}; \; (x, y) \in [0, \infty) \times [0, \infty).$$

For each $p \in \mathbb{N}^*$ and $(x, y) \in J_0$ we have

$$|f(x, y, u_{(x,y)}) - f(x, y, \overline{u}_{(x,y)})| \leq \frac{c_p}{e^2} \|u - \overline{u}\|_C.$$

Hence conditions (3.50.1) and (3.50.2) are satisfied with $l_p^* = \frac{c_p}{e^2}$. We shall show that condition (3.118) holds for all $p \in \mathbb{N}^*$. Indeed

$$\frac{l_p^* p^{r_1+r_2}}{\Gamma(r_1 + 1)\Gamma(r_2 + 1)} = \frac{1}{e^2} < 1,$$

which is satisfied for each $(r_1, r_2) \in (0,1] \times (0,1]$. Consequently, Theorem 3.50 implies that problems (3.130)–(3.132) have at least one solution defined on $[-1,\infty) \times[-2,\infty)$.

3.9.4.2 Example 2

We consider now the following partial hyperbolic functional differential equations with infinite delay of the form:

$$({}^cD_0^r u)(x,y) = \frac{4e^{x+y}}{c_p\pi^2(e^{x+y}+e^{-x-y})} \times \int_{-\infty}^{-x}\int_{-\infty}^{-y} \frac{e^{\gamma(\theta+\eta)}u(x+\theta,y+\eta)d\eta d\theta}{(1+(x+\theta)^2)(1+(y+\eta)^2)};$$

$$\text{if } (x,y)\in[0,\infty)\times[0,\infty), \tag{3.133}$$

$$u(x,y) = x+y^2;\ (x,y)\in\mathbb{R}^2\backslash(0,\infty)\times(0,\infty), \tag{3.134}$$

$$u(x,0)=x,\ u(0,y)=y^2,\ x\in[0,\infty),\ y\in[0,\infty), \tag{3.135}$$

where $c_p = \frac{3p^{r_1+r_2}}{\Gamma(r_1+1)\Gamma(r_2+1)}$, $p\in\mathbf{N}^*$, and γ a positive real constant. Let $\mathcal{B}_\gamma$ be the phase space defined in the Example of Sect. 3.7 and let

$$f(x,y,u) = \frac{4e^{x+y}}{c_p\pi^2(e^{x+y}+e^{-x-y})}\int_{-\infty}^{-x}\int_{-\infty}^{-y}\frac{e^{\gamma(\theta+\eta)}u(x+\theta,y+\eta)}{(1+(x+\theta)^2)(1+(y+\eta)^2)}d\eta d\theta;$$

for each $(x,y,u)\in J_\infty\times\mathcal{B}_\gamma$. Then for each $u,v\in\mathcal{B}_\gamma$, we have

$$\begin{aligned}
&|f(x,y,u)-f(x,y,v)| \\
&= \left|\frac{4e^{x+y}}{c_p\pi^2(e^{x+y}+e^{-x-y})}\int_{-\infty}^{-x}\int_{-\infty}^{-y}\frac{e^{\gamma(\theta+\eta)}u(x+\theta,y+\eta)}{(1+(x+\theta)^2)(1+(y+\eta)^2)}d\eta d\theta\right. \\
&\quad \left. -\frac{4e^{x+y}}{c_p\pi^2(e^{x+y}+e^{-x-y})}\int_{-\infty}^{-x}\int_{-\infty}^{-y}\frac{e^{\gamma(\theta+\eta)}v(x+\theta,y+\eta)}{(1+(x+\theta)^2)(1+(y+\eta)^2)}d\eta d\theta\right| \\
&\le \frac{4e^{x+y}}{c_p\pi^2(e^{x+y}+e^{-x-y})}\int_{-\infty}^{-x}\int_{-\infty}^{-y}\frac{e^{\gamma(\theta+\eta)}|u(x+\theta,y+\eta)-v(x+\theta,y+\eta)|}{(1+(x+\theta)^2)(1+(y+\eta)^2)}d\eta d\theta
\end{aligned}$$

$$\leq \frac{4e^{x+y}}{c_p\pi^2(e^{x+y}+e^{-x-y})}\int_{-\infty}^{0}\int_{-\infty}^{0}\frac{e^{\gamma(\theta+\eta)}|u(\theta,\eta)-v(\theta,\eta)|}{(1+\theta^2)(1+\eta^2)}d\eta d\theta$$

$$\leq \frac{4e^{x+y}}{c_p\pi^2(e^{x+y}+e^{-x-y})}\int_{0}^{\infty}\int_{0}^{\infty}\frac{1}{(1+\theta^2)(1+\eta^2)}d\eta d\theta\,\|u-v\|_\gamma$$

$$\leq \frac{e^{x+y}}{c_p(e^{x+y}+e^{-x-y})}\|u-v\|_\gamma.$$

Hence condition (3.54.2) is satisfied with $l'_p(x,y) = \frac{e^{x+y}}{c_p(e^{x+y}+e^{-x-y})}$. Since

$$l'^* = \sup\left\{\frac{e^{x+y}}{c_p(e^{x+y}+e^{-x-y})} : (x,y)\in[0,\infty)\times[0,\infty)\right\}\leq\frac{1}{c_p}$$

and $K = 1$, we have

$$\frac{Kl'^*p^{r_1+r_2}}{\Gamma(r_1+1)\Gamma(r_2+1)} = \frac{1}{3} < 1.$$

Hence condition (3.122) holds for each $(r_1,r_2)\in(0,1]\times(0,1]$ and all $p\in\mathbb{N}^*$. Consequently, Theorem 3.54 implies that problems (3.133)–(3.135) have a unique solution defined on $\mathbb{R}^2$.

3.10 Notes and Remarks

The results of Chap. 3 are taken from Abbas and Benchohra [5–7, 10, 14] and Abbas et al. [3, 30, 32, 35, 37]. Other results may be found in [41, 44, 66, 71, 110, 244, 245, 247, 249, 251–255].

Chapter 4
Partial Hyperbolic Functional Differential Inclusions

4.1 Introduction

In this chapter, we shall present existence results for some classes of initial value problems for partial hyperbolic differential inclusions with fractional order involving the Caputo fractional derivative, when the right-hand side is convex as well as nonconvex valued. Some results rely on the nonlinear alternative of Leray–Schauder type. In other results, we shall use the fixed-point theorem for contraction multivalued maps due to Covitz and Nadler.

4.2 Partial Hyperbolic Differential Inclusions

4.2.1 Introduction

This section deals with the existence of solutions to fractional order IVP, for the system

$$\left({}^{c}D_{0}^{r}u\right)(x,y) \in F\left(x,y,u_{(x,y)}\right), \text{ if } (x,y) \in J := [0,a] \times [0,b], \tag{4.1}$$

$$u(x,y) = \phi(x,y), \text{ if } (x,y) \in \tilde{J} := [-\alpha,a] \times [-\beta,b] \backslash (0,a] \times (0,b], \tag{4.2}$$

$$u(x,0) = \varphi(x),\ x \in [0,a],\ u(0,y) = \psi(y),\ y \in [0,b], \tag{4.3}$$

where $a,b,\alpha,\beta > 0,\ F : J \times C \to \mathcal{P}(\mathbb{R})$ is a compact-valued multivalued map, $\mathcal{P}(\mathbb{R})$ is the family of all subsets of $\mathbb{R}$, $\phi \in C(\tilde{J},\mathbb{R})$, $\varphi : [0,a] \to \mathbb{R}$, $\psi : [0,b] \to \mathbb{R}$ are given absolutely continuous functions with $\varphi(x) = \phi(x,0)$, $\psi(y) = \phi(0,y)$ for each $x \in [0,a],\ y \in [0,b]$, and $C := C([-\alpha,0] \times [-\beta,0],\mathbb{R})$.

S. Abbas et al., *Topics in Fractional Differential Equations*, Developments in Mathematics 27, DOI 10.1007/978-1-4614-4036-9_4,

4.2.2 The Convex Case

In this section, we are concerned with the existence of solutions for the problems (4.1)–(4.3) when the right-hand side is compact and convex valued.

Definition 4.1. A function $u \in C_{(a,b)} := C([-\alpha, a] \times [-\beta, b], \mathbb{R})$ such that its mixed derivative D^2_{xy} exists on J is said to be a solution of (4.1)–(4.3), if there exists a function $f \in L^1(J, \mathbb{R})$ with $f(x,y) \in F(x, y, u_{(x,y)})$, for a.e. $(x,y) \in J$, such that $({}^cD^r_0 u)(x,y) = f(x,y)$ for a.e. $(x,y) \in J$, and u satisfies (4.3) on J and the condition (4.2) on $\tilde{J}$.

Theorem 4.2. *Assume the following hypotheses hold:*

(4.2.1) $F : J \times \mathbb{R} \longrightarrow \mathcal{P}_{cp,cv}(\mathbb{R})$ *is a Carathéodory multivalued map.*

(4.2.2) There exist $p \in C(J, \mathbb{R}_+)$ *and* $\Psi : [0,\infty) \to (0,\infty)$ *continuous and nondecreasing such that*

$$\|F(x,y,u)\|_{\mathcal{P}} \le p(x,y)\Psi(|u|) \text{ for } (x,y) \in J \text{ and each } u \in \mathbb{R},$$

(4.2.3) There exists $l \in C(J, \mathbb{R}_+)$ *such that*

$$H_d(F(x,y,u), F(x,y,\overline{u})) \le l(x,y)|u - \overline{u}| \text{ for every } u, \overline{u} \in \mathbb{R},$$

and

$$d(0, F(x,y,0)) \le l(x,y), \text{ a.e. } (x,y) \in J,$$

(4.2.4) There exists a number $M > 0$ *such that*

$$\frac{M}{\|\mu\|_\infty + \dfrac{\Psi(M)p^* a^{r_1} b^{r_2}}{\Gamma(r_1+1)\Gamma(r_2+1)}} > 1, \tag{4.4}$$

where $p^* = \sup_{(x,y)\in J} p(x,y)$*. Then the IVP (4.1)–(4.3) have at least one solution on* $[-\alpha, a] \times [-\beta, b]$.

Proof. Transform the problems (4.1)–(4.3) into a fixed-point problem. Consider the multivalued operator $N : C_{(a,b)} \to P(C_{(a,b)})$ defined by $N(u) = \{h \in C_{(a,b)}\}$ with

$$h(x,y) = \begin{cases} \phi(x,y), & (x,y) \in \tilde{J}, \\ \mu(x,y) \\ +\dfrac{1}{\Gamma(r_1)\Gamma(r_2)} \displaystyle\int_0^x \int_0^y (x-s)^{r_1-1}(y-t)^{r_2-1} f(s,t)\,\mathrm{d}t\,\mathrm{d}s, & (x,y) \in J, \end{cases}$$

where $f \in S_{F,u}$.

Remark 4.3. Clearly, from Lemma 2.12, the fixed points of N are solutions to (4.1)–(4.3).

We shall show that N satisfies the assumptions of the nonlinear alternative of Leray–Schauder type . The proof will be given in several steps.

Step 1: *$N(u)$ is convex for each $u \in C_{(a,b)}$.* Indeed, if $h_1,\ h_2$ belong to $N(u)$, then there exist $f_1,\ f_2 \in S_{F,u}$ such that for each $(x,y) \in J$ we have

$$h_i(x,y) = \mu(x,y) + \frac{1}{\Gamma(r_1)\Gamma(r_2)} \int_0^x \int_0^y (x-s)^{r_1-1}(y-t)^{r_2-1} f_i(s,t)\mathrm{d}t\mathrm{d}s, \quad i = 1,2.$$

Let $0 \le d \le 1$. Then, for each $(x,y) \in J$, we have

$$(dh_1 + (1-d)h_2)(x,y) = \mu(x,y) + \frac{1}{\Gamma(r_1)\Gamma(r_2)} \int_0^x \int_0^y (x-s)^{r_1-1}(y-t)^{r_2-1} \times [df_1(s,t) + (1-d)f_2(s,t)]\mathrm{d}t\mathrm{d}s,$$

and for each $(x,y) \in \tilde{J}$, we have

$$(dh_1 + (1-d)h_2)(x,y) = \phi(x,y).$$

Since $S_{F,y}$ is convex (because F has convex values), we have

$$dh_1 + (1-d)h_2 \in N(u).$$

Step 2: *N maps bounded sets into bounded sets in $C_{(a,b)}$.* Let $B_{\eta^*} = \{u \in C_{(a,b)} : \|u\|_\infty \le \eta^*\}$ be a bounded set in $C_{(a,b)}$ and $u \in B_{\eta^*}$. Then for each $h \in N(u)$, there exists $f \in S_{F,u}$ such that

$$h(x,y) = \mu(x,y) + \frac{1}{\Gamma(r_1)\Gamma(r_2)} \int_0^x \int_0^y (x-s)^{r_1-1}(y-t)^{r_2-1} f(s,t)\mathrm{d}t\mathrm{d}s.$$

By (4.2.2) we have for each $(x,y) \in J$,

$$\begin{aligned}|h(x,y)| &\le |\mu(x,y)| + \frac{1}{\Gamma(r_1)\Gamma(r_2)} \int_0^x \int_0^y (x-s)^{r_1-1}(y-t)^{r_2-1} |f(s,t)|\mathrm{d}t\mathrm{d}s \\ &\le |\mu(x,y)| + \frac{1}{\Gamma(r_1)\Gamma(r_2)} \int_0^x \int_0^y (x-s)^{r_1-1}(y-t)^{r_2-1} p(s,t)\Psi(\|u_{(s,t)}\|)\mathrm{d}t\mathrm{d}s.\end{aligned}$$

Then

$$\|h\|_\infty \leq \|\mu\|_\infty + \frac{\Psi(\eta^*)a^{r_1}b^{r_2}p^*}{\Gamma(r_1+1)\Gamma(r_2+1)} := \ell_1.$$

On the other hand, for each $(x,y) \in \tilde{J}$,

$$\|h\|_\infty \leq \|\phi\|_\infty := \ell_2.$$

Thus, for each $(x,y) \in [-\alpha, a] \times [-\beta, b]$,

$$\|h\|_\infty \leq \min\{\ell_1, \ell_2\} := \ell.$$

Step 3: *N maps bounded sets into equicontinuous sets of* $C_{(a,b)}$. Let (x_1, y_1), $(x_2, y_2) \in J$, $x_1 < x_2$ and $y_1 < y_2$, B_{η^*} be a bounded set of $C_{(a,b)}$ as in Step 2, let $u \in B_{\eta^*}$ and $h \in N(u)$, then

$$\begin{aligned}
&|h(x_2,y_2) - h(x_1,y_1)| \\
&= \Bigg|\mu(x_2,y_2) - \mu(x_1,y_1) + \frac{1}{\Gamma(r_1)\Gamma(r_2)}\int_0^{x_1}\int_0^{y_1}[(x_2-s)^{r_1-1}(y_2-t)^{r_2-1} \\
&\quad -(x_1-s)^{r_1-1}(y_1-t)^{r_2-1}]f(s,t)\mathrm{d}t\mathrm{d}s \\
&\quad +\frac{1}{\Gamma(r_1)\Gamma(r_2)}\int_{x_1}^{x_2}\int_{y_1}^{y_2}(x_2-s)^{r_1-1}(y_2-t)^{r_2-1}f(s,t)\mathrm{d}t\mathrm{d}s \\
&\quad +\frac{1}{\Gamma(r_1)\Gamma(r_2)}\int_{0}^{x_1}\int_{y_1}^{y_2}(x_2-s)^{r_1-1}(y_2-t)^{r_2-1}f(s,t)\mathrm{d}t\mathrm{d}s \\
&\quad +\frac{1}{\Gamma(r_1)\Gamma(r_2)}\int_{x_1}^{x_2}\int_{0}^{y_1}(x_2-s)^{r_1-1}(y_2-t)^{r_2-1}f(s,t)\mathrm{d}t\mathrm{d}s\Bigg| \\
&\leq |\mu(x_1,y_1) - \mu(x_2,y_2)| \\
&\quad +\frac{p^*\Psi(\eta^*)}{\Gamma(r_1+1)\Gamma(r_2+1)}\left[2y_2^{r_2}(x_2-x_1)^{r_1} + 2x_2^{r_1}(y_2-y_1)^{r_2}\right. \\
&\quad \left.+x_1^{r_1}y_1^{r_2} - x_2^{r_1}y_2^{r_2} - 2(x_2-x_1)^{r_1}(y_2-y_1)^{r_2}\right].
\end{aligned}$$

As $x_1 \longrightarrow x_2$ and $y_1 \longrightarrow y_2$, the right-hand side of the above inequality tends to zero. The equicontinuity for the cases $x_1 < x_2 < 0$, $y_1 < y_2 < 0$ and $x_1 \leq 0 \leq x_2$, $y_1 \leq 0 \leq y_2$ is obvious. As a consequence of Steps 1–3 together with the Arzelá–Ascoli theorem, we can conclude that N is completely continuous.

Step 4: *N has a closed graph.* Let $u_n \to u_*$, $h_n \in N(u_n)$, and $h_n \to h_*$. We need to show that $h_* \in N(u_*)$.

$h_n \in N(u_n)$ means that there exists $f_n \in S_{F,u_n}$ such that, for each $(x,y) \in J$,

$$h_n(x,y) = \mu(x,y) + \frac{1}{\Gamma(r_1)\Gamma(r_2)} \int_0^x \int_0^y (x-s)^{r_1-1}(y-t)^{r_2-1} f_n(s,t)\mathrm{d}t\mathrm{d}s,$$

and for $(x,y) \in \tilde{J}$, $h_n(x,y) = \phi(x,y)$.

We must show that there exists $f_* \in S_{F,u*}$ such that, for each $(x,y) \in J$,

$$h_*(x,y) = \mu(x,y) + \frac{1}{\Gamma(r_1)\Gamma(r_2)} \int_0^x \int_0^y (x-s)^{r_1-1}(y-t)^{r_2-1} f_*(s,t)\mathrm{d}t\mathrm{d}s,$$

and for $(x,y) \in \tilde{J}$, $h_*(x,y) = \phi(x,y)$.

Since $F(x,y,\cdot)$ is upper semicontinuous, then for every $\varepsilon > 0$, there exist $n_0(\epsilon) \geq 0$ such that for every $n \geq n_0$, we have

$$f_n(x,y) \in F(x,y,u_{n(x,y)}) \subset F(x,y,u_{*(x,y)}) + \varepsilon B(0,1), \text{ a.e. } (x,y) \in J.$$

Since $F(.,.,.)$ has compact values, then there exists a subsequence f_{n_m} such that

$$f_{n_m}(\cdot,\cdot) \to f_*(\cdot,\cdot) \text{ as } m \to \infty$$

and

$$f_*(x,y) \in F(x,y,u_{*(x,y)}), \text{ a.e. } (x,y) \in J.$$

For every $w \in F(x,y,u_{*(x,y)})$, we have

$$|f_{n_m}(x,y) - u_*(x,y)| \leq |f_{n_m}(x,y) - w| + |w - f_*(x,y)|.$$

Then

$$|f_{n_m}(x,y) - f_*(x,y)| \leq d\Big(f_{n_m}(x,y), F(x,y,u_{*(x,y)})\Big).$$

By an analogous relation, obtained by interchanging the roles of f_{n_m} and f_*, it follows that

$$\begin{aligned}|f_{n_m}(x,y) - u_*(x,y)| &\leq H_d\Big(F(x,y,u_{n(x,y)}), F(x,y,u_{*(x,y)})\Big) \\ &\leq l(x,y)\|u_n - u_*\|_\infty.\end{aligned}$$

Then

$$
\begin{aligned}
|h_{n(x,y)} - h_{*(x,y)}| &\leq \frac{1}{\Gamma(r_1)\Gamma(r_2)} \int_0^x \int_0^y (x-s)^{r_1-1}(y-t)^{r_2-1}|f_m(s,t) \\
&\quad - f_*(s,t)|\mathrm{d}t\mathrm{d}s \\
&\leq \frac{l^*\|u_{n_m} - u_*\|_\infty}{\Gamma(r_1)\Gamma(r_2)} \int_0^x \int_0^y (x-s)^{r_1-1}(y-t)^{r_2-1}\mathrm{d}t\mathrm{d}s,
\end{aligned}
$$

where $l^* = \sup_{(x,y)\in J} l(x,y)$. Hence

$$
\|h_{n_m} - h_*\|_\infty \leq \frac{a^{r_1}b^{r_2}l^*}{\Gamma(r_1+1)\Gamma(r_2+1)}|u_{n_m} - u_*\|_\infty \to 0 \text{ as } m \to \infty.
$$

Step 5: *A priori bounds on solutions.* Let u be a possible solution of the problems (4.1)–(4.3). Then, there exists $f \in S_{F,u}$ such that, for each $(x,y) \in J$,

$$
\begin{aligned}
|u(x,y)| &\leq |\mu(x,y)| + \frac{1}{\Gamma(r_1)\Gamma(r_2)} \int_0^x \int_0^y (x-s)^{r_1-1}(y-t)^{r_2-1}|f(s,t)|\mathrm{d}t\mathrm{d}s \\
&\leq |\mu(x,y)| + \frac{1}{\Gamma(r_1)\Gamma(r_2)} \int_0^x \int_0^y (x-s)^{r_1-1}(y-t)^{r_2-1}p(s,t)\psi(\|u_{(s,t)}\|)\mathrm{d}t\mathrm{d}s \\
&\leq |\mu(x,y)| + \frac{\Psi(\|u_{(s,t)}\|)}{\Gamma(r_1)\Gamma(r_2)} \int_0^x \int_0^y (x-s)^{r_1-1}(y-t)^{r_2-1}p(s,t)\mathrm{d}t\mathrm{d}s \\
&\leq \|\mu\|_\infty + \frac{\Psi(\|u\|_\infty)p^*a^{r_1}b^{r_2}}{\Gamma(r_1+1)\Gamma(r_2+1)},
\end{aligned}
$$

and for each $(x,y) \in \tilde{J}$, $|u(x,y)| = |\phi(x,y)|$. This implies by (4.2.2) that, for each $(x,y) \in J$, we have

$$
\frac{\|u\|_\infty}{\|\mu\|_\infty + \dfrac{\Psi(\|u\|_\infty)p^*a^{r_1}b^{r_2}}{\Gamma(r_1+1)\Gamma(r_2+1)}} < 1
$$

Then by condition (4.4), there exists M such that $\|u\|_\infty \neq M$.

Let

$$
U = \{u \in C_{(a,b)} : \|u\|_\infty < M^*\},
$$

where $M^* = \min\{M, \|\phi\|_C\}$. The operator $N : \overline{U} \to \mathcal{P}(C_{(a,b)})$ is upper semicontinuous and completely continuous. From the choice of U, there is no $u \in \partial U$ such that $u \in \lambda N(u)$ for some $\lambda \in (0, 1)$. As a consequence of the nonlinear alternative of Leray–Schauder type (Lemma 2.33, [136]), we deduce that N has a fixed point u in $\overline{U}$ which is a solution of the problems (4.1)–(4.3). □

We present another existence result for problems (4.1)–(4.3) using Bohnenblust–Karlin fixed-point theorem [81].

Theorem 4.4. *Assume (4.2.1), (4.2.2) and the following hypothesis holds:*

(4.4.1) There exist $p \in C(J, \mathbb{R}_+)$ such that

$$\|F(x, y, u)\|_{\mathcal{P}} \leq p(x, y)(|u| + 1) \text{ for } (x, y) \in J \text{ and each } u \in \mathbb{R}.$$

If

$$\Gamma(r_1 + 1)\Gamma(r_2 + 1) > p^* a^{r_1} b^{r_2}, \tag{4.5}$$

then problems (4.1)–(4.3) have at least one solution.

Proof. We shall show that the operator N defined in Theorem 4.2 satisfies fixed-point theorem. Let $\rho > 0$ be such that

$$\rho > \frac{\Gamma(r_1 + 1)\Gamma(r_2 + 1)\|\mu\|_\infty + p^* a^{r_1} b^{r_2}}{\Gamma(r_1 + 1)\Gamma(r_2 + 1) - p^* a^{r_1} b^{r_2}},$$

and consider the subset

$$D_\rho = \{u \in C_{(a,b)} : \|u\|_\infty \leq \rho\}.$$

Clearly, the subset D_ρ is closed, bounded, and convex. From (4.5) we have $N(D_\rho) \subseteq D_\rho$. As before the multivalued operator $N : D_\rho \to \mathcal{P}(D_\rho)$ is upper semicontinuous and completely continuous. Hence Lemma 2.38 implies that N has a fixed point which is a solution to problems (4.1)–(4.3). □

4.2.3 *The Nonconvex Case*

We present now a result for the problems (4.1)–(4.3) with a nonconvex-valued right-hand side. Our considerations are based on the fixed-point theorem for contraction multivalued maps given by Covitz and Nadler [96].

Theorem 4.5. *Assume (4.2.3) and the following hypothesis holds:*

(4.5.1) $F : J \times \mathbb{R} \longrightarrow \mathcal{P}_{cp}(\mathbb{R})$ has the property that $F(\cdot, u) : J \to \mathcal{P}_{cp}(\mathbb{R})$ is measurable for each $u \in \mathbb{R}$.

If

$$\frac{l^* a^{r_1} b^{r_2}}{\Gamma(r_1+1)\Gamma(r_2+1)} < 1, \tag{4.6}$$

then the IVP (4.1)–(4.3) has at least one solution on J.

Remark 4.6. For each $u \in C_{(a,b)}$, the set $S_{F,u}$ is nonempty since by (4.5.1), F has a measurable selection (see [93], Theorem III.6).

Proof. We shall show that N satisfies the assumptions of Lemma 2.39. The proof will be given in two steps.

Step 1: $N(u) \in \mathcal{P}_{cl}(C_{(a,b)})$ *for each* $u \in C_{(a,b)}$. Indeed, let $(u_n)_{n\geq 0} \in N(u)$ such that $u_n \longrightarrow \tilde{u}$ in $C_{(a,b)}$. Then, $\tilde{u} \in C_{(a,b)}$ and there exists $f_n(.,.) \in S_{F,u}$ such that, for each $(x,y) \in J$,

$$u_n(x,y) = \mu(x,y) + \frac{1}{\Gamma(r_1)\Gamma(r_2)} \int_0^x \int_0^y (x-s)^{r_1-1}(y-t)^{r_2-1} f_n(s,t)\mathrm{d}t\mathrm{d}s,$$

and for each $(x,y) \in \tilde{J}$,

$$u_n(x,y) = \phi(x,y).$$

Using the fact that F has compact values and from (4.2.3), we may pass to a subsequence if necessary to get that $f_n(.,.)$ converges weakly to f in $L^1_w(J,\mathbb{R})$ (the space endowed with the weak topology). A standard argument shows that $f_n(.,.)$ converges strongly to f and hence $f \in S_{F,u}$. Then, for each $(x,y) \in J$,

$$u_n(x,y) \longrightarrow \tilde{u}(x,y) = \mu(x,y) + \frac{1}{\Gamma(r_1)\Gamma(r_2)} \int_0^x \int_0^y (x-s)^{r_1-1}(y-t)^{r_2-1} f(s,t)\mathrm{d}t\mathrm{d}s,$$

and for each $(x,y) \in \tilde{J}$, $u_n(x,y) \longrightarrow \tilde{u}(x,y) = \phi(x,y)$.

So, $\tilde{u} \in N(u)$.

Step 2: *There exists* $\gamma < 1$ *such that*

$$H_d(N(u), N(\overline{u})) \leq \gamma \|u - \overline{u}\|_\infty \textit{ for each } u, \overline{u} \in C_{(a,b)}.$$

Let $u, \overline{u} \in C_{(a,b)}$ and $h \in N(u)$. Then, there exists $f(x,y) \in F(x,y,u_{(x,y)})$ such that that for each $(x,y) \in J$

$$h(x,y) = \mu(x,y) + \frac{1}{\Gamma(r_1)\Gamma(r_2)} \int_0^x \int_0^y (x-s)^{r_1-1}(y-t)^{r_2-1} f(s,t)\mathrm{d}t\mathrm{d}s,$$

and for each $(x,y) \in \tilde{J}$, $h(x,y) = \phi(x,y)$.

From (4.2.3) it follows that

$$H_d(F(x,y,u_{(x,y)}),F(x,y,\overline{u}_{(x,y)}))\le l(x,y)|u_{(x,y)}-\overline{u}_{(x,y)}|.$$

Hence, there exists $w\in F(x,y,\overline{u}_{(x,y)})$ such that

$$|f(x,y)-w|\le l(x,y)|u_{(x,y)}-\overline{u}_{(x,y)}|,\ (x,y)\in J.$$

Consider $U:J\to\mathcal{P}(\mathbb{R})$ given by

$$U(x,y)=\{w\in\mathbb{R}:|f(x,y)-w|\le l(x,y)|u_{(x,y)}-\overline{u}_{(x,y)}|\}.$$

Since the multivalued operator $U_1(x,y)=U(x,y)\cap F(x,y,\overline{u}_{(x,y)})$ is measurable (see Proposition III.4 in [93]), there exists a function $\overline{f}(x,y)$ which is a measurable selection for U_1. So, $\overline{f}(x,y)\in F(x,y,\overline{u}_{(x,y)})$, and for each $(x,y)\in J$,

$$|f(x,y)-\overline{f}(x,y)|\le l(x,y)|u_{(x,y)}-\overline{u}_{(x,y)}|.$$

Let us define for each $(x,y)\in J$

$$\overline{h}(x,y)=\mu(x,y)+\frac{1}{\Gamma(r_1)\Gamma(r_2)}\int_0^x\int_0^y(x-s)^{r_1-1}(y-t)^{r_2-1}\overline{f}(s,t)\mathrm{d}t\mathrm{d}s,$$

and for each $(x,y)\in\tilde{J}$, $\overline{h}(x,y)=\phi(x,y)$. Then for each $(x,y)\in\tilde{J}$, $\|h-\overline{h}\|_\infty=0$, and for each $(x,y)\in J$

$$\begin{aligned}|h(x,y)-\overline{h}(x,y)|&\le\frac{1}{\Gamma(r_1)\Gamma(r_2)}\int_0^x\int_0^y(x-s)^{r_1-1}(y-t)^{r_2-1}|f(s,t)\\&\quad-\overline{f}(s,t)|\mathrm{d}t\mathrm{d}s\\&\le\frac{1}{\Gamma(r_1)\Gamma(r_2)}\int_0^x\int_0^y(x-s)^{r_1-1}(y-t)^{r_2-1}l(s,t)\|u_{(s,t)}\\&\quad-\overline{u}_{(s,t)}\|\mathrm{d}t\mathrm{d}s\\&\le\frac{1}{\Gamma(r_1)\Gamma(r_2)}\int_0^x\int_0^y(x-s)^{r_1-1}(y-t)^{r_2-1}l(s,t)\|u_{(s,t)}\\&\quad-\overline{u}_{(s,t)}\|\mathrm{d}t\mathrm{d}s\\&\le\frac{\|u_{(.,.)})-\overline{u}_{(.,.)})\|_\infty}{\Gamma(r_1)\Gamma(r_2)}\int_0^x\int_0^y(x-s)^{r_1-1}(y-t)^{r_2-1}l(s,t)\mathrm{d}t\mathrm{d}s.\end{aligned}$$

Thus, for each $(x, y) \in [-\alpha, a] \times [-\beta, b]$

$$\|h - \overline{h}\|_\infty \leq \frac{l^* a^{r_1} b^{r_2}}{\Gamma(r_1 + 1)\Gamma(r_2 + 1)} \|u - \overline{u}\|_\infty.$$

By an analogous relation, obtained by interchanging the roles of u and $\overline{u}$, it follows that

$$H_d(N(u), N(\overline{u})) \leq \frac{l^* a^{r_1} b^{r_2}}{\Gamma(r_1 + 1)\Gamma(r_2 + 1)} \|u - \overline{u}\|_\infty.$$

So by (4.6), N is a contraction and thus, by Lemma 2.39, N has a fixed point u which is solution to (4.1)–(4.3). □

4.2.4 An Example

As an application of the main results, we consider the fractional differential inclusion

$$^c D^r u(x, y) \in F(x, y, u_{(x,y)}), \text{ a.e. } (x, y) \in J = [0, 1] \times [0, 1], \tag{4.7}$$

$$u(x, y) = x + y^2, \text{ a.e. } (x, y) \in ([-1, 1] \times [-2, 1]) \setminus (0, 1] \times (0, 1], \tag{4.8}$$

$$u(x, 0) = x, \; u(0, y) = y^2, \; x \in [0, 1], \; y \in [0, 1], \tag{4.9}$$

where $r = (r_1, r_2) \in (0, 1] \times (0, 1]$. Set

$$F(x, y, u_{(x,y)}) = \left\{u \in \mathbb{R} : f_1(x, y, u_{(x,y)}) \leq u \leq f_2(x, y, u_{(x,y)})\right\},$$

where $f_1, f_2 : [0, 1] \times [0, 1] \times C([-1, 0] \times [-2, 0], \mathbb{R}) \to \mathbb{R}$. We assume that for each $(x, y) \in J$, $f_1(x, y, .)$ is lower semicontinuous (i.e., the set $\{z \in C([-1, 0] \times [-2, 0], \mathbb{R}) : f_1(x, y, z) > \mu\}$ is open for each $\mu \in \mathbb{R}$), and assume that for each $(x, y) \in J$, $f_2(x, y, .)$ is upper semicontinuous (i.e., the set $\{z \in C([-1, 0] \times [-2, 0], \mathbb{R}) : f_2(x, y, z) < \mu\}$ is open for each $\mu \in \mathbb{R}$). Assume that there are $p \in C(J, \mathbb{R}_+)$ and $\psi^* : [0, \infty) \to (0, \infty)$ continuous and nondecreasing such that

$$\max\left(|f_1(x, y, z)|, \; |f_2(x, y, z)|\right) \leq p(x, y)\psi^*(\|z\|),$$

for each $(x, y) \in J$ and all $z \in C([-1, 0] \times [-2, 0], \mathbb{R})$.

It is clear that F is compact and convex valued, and it is upper semi-continuous (see [104]). Since all the conditions of Theorem 4.2 are satisfied, problem (4.7)–(4.9) has at least one solution u on $[-1, 1] \times [-2, 1]$.

4.3 Existence Results for Partial Hyperbolic Differential Inclusions

4.3.1 Introduction

In this section we study the following fractional order IVP, for the system:

$$({}^cD_0^r u)(x,y) \in F(x,y,u(x,y)), \text{ if } (x,y) \in J, \tag{4.10}$$

$$u(x,0) = \varphi(x),\ x \in [0,a],\ u(0,y) = \psi(y),\ y \in [0,b], \tag{4.11}$$

where $J = [0,a] \times [0,b],\ a,b > 0,\ F : J \times \mathbb{R}^n \to \mathcal{P}(\mathbb{R}^n)$ is a compact-valued multivalued map, $\mathcal{P}(\mathbb{R}^n)$ is the family of all subsets of $\mathbb{R}^n$, $\varphi : [0,a] \to \mathbb{R}^n$, $\psi : [0,b] \to \mathbb{R}^n$ are given absolutely continuous functions with $\varphi(0) = \psi(0)$.

4.3.2 Existence of Solutions

Let us start by defining what we mean by a solution of the problems (4.10) and (4.11).

Definition 4.7. A function $u \in C(J,\mathbb{R}^n)$ with its mixed derivative D_{xy}^2 exists and is integrable is said to be a solution of (4.10) and (4.11), if there exists a function $f \in L^1(J,\mathbb{R}^n)$ with $f(x,y) \in F(x,y,u(x,y))$ such that $({}^cD_0^r u)(x,y) = f(x,y)$, and u satisfies (4.11) on J.

Definition 4.8. Let (X,d) be a metric space and consider a set-valued map T on X with nonempty closed values in X. T is said to be a λ-contraction if there exists $0 < \lambda < 1$ such that for each $u, v \in X$,

$$H_d(T(u),T(v)) \leq \lambda d(u,v).$$

Remark 4.9. If X is complete, then every set-valued contraction has a fixed point.

We denote by $\text{Fix}(T)$ the set of all fixed points of the set-valued map T. Obviously, $\text{Fix}(T)$ is closed.

Lemma 4.10 ([184]). *Let X be a complete metric space and suppose that T_1, T_2 are λ-contractions with closed values in X. Then*

$$H_d(\text{Fix}(T_1),\text{Fix}(T_2)) \leq \frac{1}{1-\lambda} \sup_{z \in X} d(T_1(z),T_2(z)).$$

Theorem 4.11. *Assume that the following hypotheses:*

(4.11.1) $F : J \times \mathbb{R}^n \longrightarrow \mathcal{P}_{cl}(\mathbb{R}^n)$ *has the property that*

$$F(\cdot, \cdot, u) : J \to P_{cl}(\mathbb{R}^n) \text{ is measurable for each } u \in \mathbb{R}^n,$$

(4.11.2) There exists $l \in C(J, \mathbb{R}_+)$ *such that*

$$H_d(F(x,y,u), F(x,y,\overline{u})) \leq l(x,y)\|u - \overline{u}\| \text{ for every } u, \overline{u} \in \mathbb{R}^n,$$

and

$$d(0, F(x,y,0)) \leq l(x,y), \text{ a.e. } (x,y) \in J,$$

are satisfied and let $v(.,.) \in L^1(J, \mathbb{R}^n)$ *be such that there exists* $q(.,.) \in L^1(J, \mathbb{R})$ *with*

$$d({}^cD_0^r v(x,y), F(x,y,v(x,y))) \leq q(x,y) \text{ for each } (x,y) \in J.$$

If

$$\frac{a^{r_1}b^{r_2}l^*}{\Gamma(r_1+1)\Gamma(r_2+1)} < 1, \tag{4.12}$$

where $l^* = \sup\{l(x,y) : (x,y) \in J\}$, *then for every* $\epsilon > 0$ *there exists* $u(.,.)$ *a solution of (4.10) and (4.11) satisfying for all* $(x,y) \in J$

$$\|u - v\|_1 \leq \frac{a^{r_1}b^{r_2}\|q\|_1}{\Gamma(r_1+1)\Gamma(r_2+1) - a^{r_1}b^{r_2}l^*} + \epsilon. \tag{4.13}$$

Proof. For $w(.,.) \in L^1(J, \mathbb{R}^n)$ define the following set-valued maps:

$$M_w(x,y) = F\Big(x, y, \mu(x,y) + (I_0^r w)(x,y)\Big); \ (x,y) \in J, \tag{4.14}$$

$$T(w) = \{\phi(.,.) \in L^1(J, \mathbb{R}^n); \ \phi(x,y) \in M_w(x,y)\}, \ (x,y) \in J. \tag{4.15}$$

It follows from Lemma 2.12 that $u(.,.)$ is a solution of (4.10) and (4.11) if and only if ${}^cDu(.,.)$ is a fixed point of T. We shall prove first that $T(w)$ is nonempty and closed for every $w \in L^1(J, \mathbb{R}^n)$. The fact that the set-valued map $M_w(.,.)$ is measurable is well known. For example, the map

$$(x,y) \to \mu(x,y) + (I_0^r w)(x,y)$$

can be approximated by step functions and we can apply Theorem III. 40 in [93]. Since the values of F are closed with the measurable selection theorem (Theorem III.6 in [93]) we infer that $M_w(.,.)$ admits a measurable selection ϕ. First (4.14) and (4.15) imply that for each $(x,y) \in J$,

$$\begin{aligned}\|\phi(x,y)\| &\le d\Big(0, F(x,y,0)\Big) + d_H\Big(F(x,y,0), F(x,y,\mu(x,y) + (I_0^r w)(x,y))\Big) \\ &\le l(x,y)(1 + \|\mu\|_\infty + (I_0^r w)(x,y)),\end{aligned}$$

which shows that $\phi \in L^1(J,\mathbb{R}^n)$ and $T(w)$ is nonempty. On the other hand, the set $T(w)$ is also closed. Indeed, if $\phi_n \in T(w)$ and $\|\phi_n - \phi\|_1 \to 0$ then we can pass to a subsequence $\phi_{n_k} \in T(w)$ such that $\phi_{n_k}(x,y) \to \phi(x,y)$ for a.e. $(x,y) \in J$, and we find that $\phi \in T(w)$. We show next that $T(.)$ is a contraction on $L^1(J,\mathbb{R}^n)$. Let $z, \overline{z} \in L^1(J,\mathbb{R}^n)$ be given, $\phi \in T(z)$ and let $\delta > 0$. Consider the following set-valued map:

$$G(x,y) = M_{\overline{z}}(x,y) \cap \left\{u \in \mathbb{R}^n : \|\phi(x,y) - u\| \le \frac{\delta}{ab} + l(x,y)(I_0^r(z-\overline{z}))(x,y)\right\}.$$

From Proposition III.4 in [93], G is measurable and from (4.11.1) and (4.11.2), G has nonempty closed values. Therefore, there exists $\gamma(.,.)$ a measurable selection of $G(.,.)$. It follows that $\gamma \in T(w)$ and according to the definition of the norm we have for each $(x,y) \in J$,

$$\begin{aligned}\|\phi - \gamma\|_1 &= \int_0^a\int_0^b \|\phi(x,y) - \gamma(x,y)\| \mathrm{d}y\mathrm{d}x \\ &\le \int_0^a\int_0^b \frac{\delta}{ab}\mathrm{d}y\mathrm{d}x + \int_0^a\int_0^b l(x,y)(I_0^r(z-\overline{z}))(x,y)\mathrm{d}y\mathrm{d}x \\ &\le \delta + \frac{1}{\Gamma(r_1)\Gamma(r_2)}\int_0^a\int_0^b\left[\int_0^a\int_0^b l(x,y)(x-s)^{r_1-1}(y-t)^{r_2-1}\mathrm{d}y\mathrm{d}x\right] \\ &\quad \times \|z(s,t) - \overline{z}(s,t)\|\mathrm{d}t\mathrm{d}s \\ &\le \delta + \frac{a^{r_1}b^{r_2}l^*}{\Gamma(r_1+1)\Gamma(r_2+1)}\|z - \overline{z}\|_1,\end{aligned}$$

hence

$$\|\phi - \gamma\|_1 \le \delta + \frac{a^{r_1}b^{r_2}l^*}{\Gamma(r_1+1)\Gamma(r_2+1)}\|z - \overline{z}\|_1.$$

Since $\delta > 0$ was chosen arbitrary, we deduce that

$$d(\phi, T(w)) \le \frac{a^{r_1}b^{r_2}l^*}{\Gamma(r_1+1)\Gamma(r_2+1)}\|z - \overline{z}\|_1.$$

Replacing z by $\overline{z}$ we obtain

$$d_H(T(z), T(\overline{z})) \leq \frac{a^{r_1} b^{r_2} l^*}{\Gamma(r_1+1)\Gamma(r_2+1)} \|z - \overline{z}\|_1,$$

thus by (4.12), T is a contraction on $L^1(J, \mathbb{R}^n)$. We consider next the following set-valued maps:

$$\overline{F}(x,y,u) = F(x,y,u) + q(x,y)[-1,1];\ (x,y,u) \in J \times \mathbb{R}^n,$$
$$\overline{M}_w(x,y) = F_1\Big(x,y,\mu(x,y) + (I_0^r w)(x,y)\Big);\ (x,y) \in J,$$
$$\overline{T}(w) = \{\phi(.,.) \in L^1(J,\mathbb{R}^n);\ \phi(x,y) \in \overline{M}_w(x,y)\},\ (x,y) \in J.$$

Obviously, $\overline{F}(.,.,.)$ satisfies hypotheses (4.11.1) and (4.11.2). Repeating the previous step of the proof we obtain that $\overline{T}$ is also a contraction on $L^1(J,\mathbb{R}^n)$ with closed nonempty values. We prove next the following estimate:

$$d_H(T(w), \overline{T}(w)) \leq \int_0^a \int_0^b q(x,y)\mathrm{d}y\mathrm{d}x. \tag{4.16}$$

Let $\phi \in T(w),\ \delta > 0$ and define

$$\overline{G}(x,y) = \overline{M}_w(x,y) \cap \left\{ z \in \mathbb{R}^n : \|\phi(x,y) - z\| \leq q(x,y) + \frac{\delta}{ab} \right\}.$$

With the same arguments used for the set-valued map $G(.,.)$, we deduce that $\overline{G}(.,.)$ is measurable with nonempty closed values. Hence let $\psi(.,.)$ be a measurable selection of $\overline{G}(.,.)$. It follows that $\psi \in \overline{T}(w)$ and one has for each $(x,y) \in J$,

$$\begin{aligned}
\|\phi - \psi\|_1 &= \int_0^a \int_0^b \|\phi(x,y) - \psi(x,y)\|\mathrm{d}y\mathrm{d}x \\
&\leq \int_0^a \int_0^b \left(q(x,y) + \frac{\delta}{ab} \right) \mathrm{d}y\mathrm{d}x \\
&\leq \delta + \int_0^a \int_0^b q(x,y)\mathrm{d}y\mathrm{d}x.
\end{aligned}$$

Since δ is arbitrary, as above we obtain (4.16). We apply Lemma 4.10 and we infer that

$$d_H(\mathrm{Fix}(T), \mathrm{Fix}(\overline{T})) \leq \frac{1}{1 - \frac{a^{r_1} b^{r_2} l^*}{\Gamma(r_1+1)\Gamma(r_2+1)}} \int_0^a \int_0^b q(x, y) \mathrm{d}y \mathrm{d}x$$

$$= \frac{\Gamma(r_1+1)\Gamma(r_2+1)}{\Gamma(r_1+1)\Gamma(r_2+1) - a^{r_1} b^{r_2} l^*} \int_0^a \int_0^b q(x, y) \mathrm{d}y \mathrm{d}x.$$

Since ${}^c D_0^r v(.,.) \in \mathrm{Fix}(\overline{T})$ it follows that there exists $w \in \mathrm{Fix}(T)$ such that for any $\epsilon > 0$ we have

$$\|{}^c D_0^r v - w\|_1 \leq \frac{\Gamma(r_1+1)\Gamma(r_2+1)}{\Gamma(r_1+1)\Gamma(r_2+1) - a^{r_1} b^{r_2} l^*} \int_0^a \int_0^b q(x, y) \mathrm{d}y \mathrm{d}x + \epsilon \frac{\Gamma(r_1+1)\Gamma(r_2+1)}{a^{r_1} b^{r_2}}. \tag{4.17}$$

We define

$$u(x, y) = \mu(x, y) + (I_0^r w)(x, y); \; (x, y) \in J.$$

By (4.17) we get for each $(x, y) \in J$,

$$\begin{aligned}
\|u - v\|_1 &= \int_0^a \int_0^b \|u(x, y) - v(x, y)\| \mathrm{d}y \mathrm{d}x \\
&\leq \frac{1}{\Gamma(r_1)\Gamma(r_2)} \int_0^a \int_0^b \left[\int_0^a \int_0^b (x - s)^{r_1 - 1} (y - t)^{r_2 - 1} \mathrm{d}y \mathrm{d}x \right] \\
&\quad \times \|w(s, t) - {}^c D_0^r v(s, t)\| \mathrm{d}t \mathrm{d}s \\
&\leq \frac{a^{r_1} b^{r_2}}{\Gamma(r_1+1)\Gamma(r_2+1)} \|w - {}^c D_0^r v\|_1 \\
&\leq \frac{a^{r_1} b^{r_2} \|q\|_1}{\Gamma(r_1+1)\Gamma(r_2+1) - a^{r_1} b^{r_2} l^*} + \epsilon.
\end{aligned}$$

Hence, for all $(x, y) \in J$,

$$\|u - v\|_1 \leq \frac{a^{r_1} b^{r_2} \|q\|_1}{\Gamma(r_1+1)\Gamma(r_2+1) - a^{r_1} b^{r_2} l^*} + \epsilon.$$

□

Remark 4.12. Suppose that the assumptions in Theorem 4.11 are satisfied with $v \equiv 0$, then for every $\epsilon > 0$, problems (4.10) and (4.11) have at least one solution u on J such that

$$\|u\|_1 \leq \frac{a^{r_1} b^{r_2} \|q\|_1}{\Gamma(r_1+1)\Gamma(r_2+1) - a^{r_1} b^{r_2} l^*} + \epsilon.$$

4.3.3 *Qualitative Properties and Topological Structure of the Solution Set*

4.3.3.1 Topological Structure of the Solution Set

Now, we present a result on the topological structure of the set of solutions of (4.10) and (4.11).

Theorem 4.13. *Assume that (4.11.1), (4.11.2), and the following hypothesis hold:*

(4.13.1) There exists $p(.,.) \in C(J, \mathbb{R}_+)$ such that

$$\|F(x,y,z)\|_{\mathcal{P}} \leq p(x,y) \text{ for } (x,y) \in J \text{ and each } z \in \mathbb{R}^n.$$

Then the solution set of (4.10) and (4.11) in not empty and compact in $C(J, \mathbb{R}^n)$.

Proof. Let

$$S = \{u \in C(J, \mathbb{R}^n) : u \text{ is solution of } (4.10) \text{ and } (4.11)\}.$$

From Theorem 4.11, $S \neq \emptyset$. Now, we can prove that S is compact. Let $(u_n)_{n \in \mathbf{N}} \in S$, then there exists $f_n \in S_{F, u_n}$ such that for $(x,y) \in J$,

$$u_n(x,y) = \mu(x,y) + (I_0^r f_n)(x,y).$$

From (4.13.1) we can prove that there exists a constant $M > 0$ such that

$$\|u_n\|_\infty \leq M, \quad \text{for every } n \geq 1.$$

We can easily show that the set $\{u_n : n \geq 1\}$ is equicontinuous in $C(J, \mathbb{R}^n)$; hence by Arzéla–Ascoli theorem we can conclude that there exists a subsequence (denoted again by $\{u_n\}$) of $\{u_n\}$ such that u_n converges to u in $C(J, \mathbb{R}^n)$. We shall show that there exist $f(.,.) \in F(.,.,u_.)$ such that

$$u(x,y) = \mu(x,y) + (I_0^r f)(x,y).$$

Since $F(x,y,.)$ is upper semicontinuous, then for every $\varepsilon > 0$, there exists $n_0(\varepsilon) \geq 0$ such that for every $n \geq n_0$, we have

$$f_n(x,y) \in F(x,y,u_n(x,y)) \subset F(x,y,u(x,y)) + \varepsilon B(0,1), \text{ a.e. } (x,y) \in J.$$

Since F has compact values, there exists subsequence $f_{n_m}(.)$ such that

$$f_{n_m}(.,.) \to f(.,.) \text{ as } m \to \infty$$

and

$$f(x,y) \in F(x,y,u(x,y)), \text{ a.e. } (x,y) \in J.$$

It is clear that

$$\|f_{n_m}(x,y)\| \leq p(x,y), \text{ a.e. } (x,y) \in J.$$

By Lebesgue's dominated convergence theorem, we conclude that $f \in L^1(J,\mathbb{R}^n)$ which implies that $f \in S_{F,u}$. Thus

$$u(x,y) = \mu(x,y) + (I_0^r f)(x,y);\ (x,y) \in J.$$

Hence $S \in \mathcal{P}_{cp}(C(J,\mathbb{R}^n))$. □

4.3.3.2 On the Set of Solutions for the Darboux Problem for Fractional Order Partial Hyperbolic Differential Inclusions

Now, we prove the arcwise connectedness of the solution set for the IVP, for the system

$$^cD_0^r u(x,y) \in F\Big(x,y,u(x,y),G\big(x,y,u(x,y)\big)\Big);\ \text{if } (x,y) \in J := [0,a]\times[0,b], \tag{4.18}$$

$$\begin{cases} u(x,0) = \varphi(x);\ x \in [0,a], \\ u(0,y) = \psi(y);\ y \in [0,b], \\ \varphi(0) = \psi(0), \end{cases} \tag{4.19}$$

where $a,b > 0$, $F : J \times \mathbb{R}^n \times \mathbb{R}^n \to \mathcal{P}(\mathbb{R}^n)$, $G : J \times \mathbb{R}^n \to \mathcal{P}(\mathbb{R}^n)$ are given multifunctions, $\varphi \in AC([0,a],\mathbb{R}^n)$ and $\psi \in AC([0,b],\mathbb{R}^n)$.

Given a continuous function $d : J \to (0,\infty)$, we denote by L^1 the Banach space of all (equivalence class of) Lebesgue measurable functions $w : J \to \mathbb{R}^n$, endowed with the norm

$$\|w\|_{L^1} = \int_0^a\int_0^b d(x,y)\|w(x,y)\|dydx. \tag{4.20}$$

Definition 4.14. A space X is said to be arcwise connected if any two distinct points can be joined by an arc, i.e., a path f which is a homeomorphism between the unit interval $[0, 1]$ and its image $f([0, 1])$. A metric space Z is called absolute retract if, for any metric space X and any $X_0 \in \mathcal{P}_{cl}(X)$, every continuous function $g : X_0 \to Z$ has a continuous extension $g : X \to Z$ over X.

Every continuous image of an absolute retract is an arcwise connected space.

Let S be a separable Banach space. Set $E := L^1(J)$.

Lemma 4.15 ([190]). *Assume that $G : S \times E \to \mathcal{P}_{cl}(E)$ and $F : S \times E \times E \to \mathcal{P}_{cl}(E)$ are Hausdorff continuous multifunctions with decomposable values, satisfying the following conditions:*

(a) There exists $L \in [0, 1)$ such that, for every $s \in S$ and every $u, u' \in E$,

$$H_d(G(s, u), G(s, u')) \leq L\|u - u'\|_1.$$

(b) There exists $M \in [0, 1)$ such that $L + M < 1$ and for every $s \in S$ and every $u, v, w, z \in E$.

$$H_d(F(s, u, z), F(s, v, w))\| \leq M(\|u - v\|_1 + \|z - w\|_1).$$

Set $\mathrm{Fix}(\Gamma(s, .)) = \{u \in E : u \in \Gamma(s, u)\}$, *where*

$$\Gamma(s, u) = F(s, u, G(s, u)); \ (s, u) \in S \times E.$$

Then

1. *For every $s \in S$ the set* $\mathrm{Fix}(\Gamma(s, .))$ *is nonempty and arcwise connected.*
2. *For any $s_i \in S$, and any $u_i \in$* $\mathrm{Fix}(\Gamma(s, .))$*; $i = 1, \dots, p$, there exists a continuous function $\gamma(s) \in$* $\mathrm{Fix}(\Gamma(s, .))$ *for all $s \in S$ and $\gamma(s_i) = u_i$; $i = 1, \dots, p$.*

Lemma 4.16 ([190]). *Let $U : J \to \mathcal{P}_{cl}(\mathbb{R}^n)$ and $V : J \times \mathbb{R}^n \to \mathcal{P}_{cl}(\mathbb{R}^n)$ be two multifunctions, satisfying the following conditions:*

(a) U is measurable and there exists $\rho \in E$ such that $H_d(U(x, y)), \{0\}) \leq \rho(x, y)$ for almost all $(x, y) \in J$.
(b) The multifunction $(x, y) \to V(x, y, u)$ is measurable for every $u \in E$.
(c) The multifunction $u \to V(x, y, u)$ is Hausdorff continuous for all $(x, y) \in J$.

Let $v : J \to \mathbb{R}^n$ be a measurable selection from $(x, y) \to V(x, y, U(x, y))$. Then there exists a selection $u \in E$ such that $v(x, y) \in V(x, y, u(x, y))$; $(x, y) \in J$.

Lemma 4.17 ([80]). *Let $\zeta \in (0, 1)$ and let $N : J \to [0, \infty)$ be an integrable function. Then there exists a continuous function $d : J \to (0, \infty)$ which, for every $(x, y) \in J$, satisfies*

$$\int_x^a \int_y^b N(s,t)d(s,t)\mathrm{d}t\mathrm{d}s = \zeta(d(x,y)-1).$$

Theorem 4.18. *Let $G : J \times \mathbb{R}^n \to \mathcal{P}_{cl}(\mathbb{R}^n)$ and $F : J \times \mathbb{R}^n \times \mathbb{R}^n \to \mathcal{P}_{cl}(\mathbb{R}^n)$ be two set-valued maps, satisfying the following assumptions:*

(4.18.1) The set-valued maps $(x,y) \to F(x,y,u,v)$ and $(x,y) \to G(x,y,u)$ are measurable for all $u, v \in \mathbb{R}^n$.

(4.18.2) There exists a positive integrable function $l : J \to \mathbb{R}$ such that, for every $u, u' \in \mathbb{R}^n$,

$$H_d(G(x,y,u), G(x,y,u')) \le l(x,y)\|u-u'\|; \ a.e.\ (x,y) \in J.$$

(4.18.3) There exists a positive integrable function $m : J \to \mathbb{R}$ and $\eta \in [0,1]$ such that, for every $u, v, u', v' \in \mathbb{R}^n$,

$$H_d(F(x,y,u,v), G(x,y,u',v)) \le m(x,y)\|u-u'\| + \eta(\|v-v'\|; \ a.e.\ (x,y) \in J.$$

(4.18.4) There exist positive integrable functions $f, g : J \to \mathbb{R}$ such that

$$H_d(\{0\}, F(x,y,\{0\},\{0\})) \le f(x,y), \ a.e.\ (x,y) \in J$$

and

$$H_d(\{0\}, G(x,y,\{0\})) \le g(x,y); \ a.e.\ (x,y) \in J.$$

Then,

1. *For every $\mu \in \mathcal{M}$, the solution set $S(\mu)$ of problems (4.18) and (4.19) is nonempty and arcwise connected in the space $C(J)$.*
2. *For any $\mu_i \in \mathcal{M}$ and any $u_i \in S(\mu)$; $i = 1, \ldots, p$, there exists a continuous function $s : \mathcal{M} \to C(J)$ such that $s(\mu) \in S(\mu)$ for any $\mu \in \mathcal{M}$ and $s(\mu_i) = u_i$; $i = 1, \ldots, p$.*
3. *The set $S = \cup_{\mu \in \mathcal{M}} S(\mu)$ is arcwise connected in the space $C(J)$.*

Proof. In what follows $N(x,y) = \max\{l(x,y), m(x,y);\ (x,y) \in J\}$ and take $\zeta \in (0,1)$ such that $2\zeta + \eta < 1$ and $d : J \to (0,\infty)$ in (4.20) is the corresponding mapping found in Lemma 4.17.

1. For $\mu \in \mathcal{M}$ and $u \in L^1$, set

$$u_\mu(x,y) = \mu(x,y) + (I_\theta^r u)(x,y); \ (x,y) \in J.$$

Define the multifunctions $\alpha : \mathcal{M} \times L^1 \to \mathcal{P}(L^1)$ and $\beta : \mathcal{M} \times L^1 \times L^1 \to \mathcal{P}(L^1)$ by

$$\alpha(\mu,u) = \{v \in L^1 : v(x,y) \in G(x,y,u_\mu(x,y)); \ a.e.\ (x,y) \in J\},$$

$$\beta(\mu, u, v) = \{w \in L^1 : w(x, y) \in F(x, y, u_\mu(x, y), v(x, y));\ a.e.\ (x, y) \in J\},$$

where $\mu \in \mathcal{M}$ and $u, v \in L^1$. We prove that α and β satisfy the hypotheses of Lemma 4.15.

Since u_μ is measurable and G satisfies hypotheses (4.18.1) and (4.18.2), the multifunction $G_\mu : (x, y) \to G(x, y, u_\mu(x, y))$ is measurable and $G_\mu \in \mathcal{P}_{cl}(L^1)$, G_μ has a measurable selection. Therefore due to hypothesis (4.18.4), we get $\alpha(\mu, u) \neq \emptyset$. Also, by simple computation, it follows that the set $\alpha(\mu, u)$ is closed and decomposable. In the same way we obtain that $\alpha(\mu, u) \in \mathcal{P}_{cl}(L^1)$ is a decomposable set.

Set $d := \int_0^a \int_0^b d(x, y)\mathrm{d}y\mathrm{d}x$. Pick $(\mu, u), (\mu_1, u_1) \in \mathcal{M} \times L^1$ and choose $v \in \alpha(\mu, u)$. For each $\epsilon > 0$ there exists $v_1 \in \alpha(\mu_1, u_1)$ such that, for every $(x, y) \in J$ one has

$$\|v(x, y) - v_1(x, y)\| \leq H_d\Big(G(x, y, u_\mu(x, y)), G(x, y, u_{\mu_1}(x, y))\Big) + \epsilon$$

$$\leq N(x, y)\Big(\|\mu(x, y) - \mu_1(x, y)\| + \frac{1}{\Gamma(r_1)\Gamma(r_2)} \int_0^x \int_0^y (x - s)^{r_1 - 1}(y - t)^{r_2 - 1}$$

$$\|u(s, t) - u_1(s, t)\|\mathrm{d}t\mathrm{d}s\Big) + \epsilon.$$

Hence, for any $\epsilon > 0$,

$$\|v - v_1\|_{L^1} \leq \|\mu - \mu_1\|_\infty \int_0^a \int_0^b d(x, y)N(x, y)\mathrm{d}y\mathrm{d}x$$

$$+ \frac{1}{\Gamma(r_1)\Gamma(r_2)} \int_0^a \int_0^b d(x, y)N(x, y)$$

$$\times \left(\int_0^x \int_0^y (x - s)^{r_1 - 1}(y - t)^{r_2 - 1} \|u(s, t) - u_1(s, t)\|\mathrm{d}t\mathrm{d}s \right)\mathrm{d}y\mathrm{d}x + \epsilon d$$

$$\leq \zeta(d(a, b) - 1)\|\mu - \mu_1\|_\infty + \frac{1}{\Gamma(r_1)\Gamma(r_2)} \int_0^a \int_0^b \|u(s, t) - u_1(s, t)\|$$

$$\times \left(\int_s^a \int_t^b (x - s)^{r_1 - 1}(y - t)^{r_2 - 1} d(x, y)N(x, y)\mathrm{d}y\mathrm{d}x \right)\mathrm{d}t\mathrm{d}s + \epsilon d$$

$$\leq \zeta(d(a,b)-1)\|\mu-\mu_1\|_\infty$$
$$+\frac{1}{\Gamma(r_1)\Gamma(r_2)}\int_0^a\int_0^b \|u(s,t)-u_1(s,t)\|$$
$$\times\Big(\int_s^a\int_t^b (x-s)^{\frac{r_1-1}{1-r_3}}(y-t)^{\frac{r_2-1}{1-r_3}}dydx\Big)^{1-r_3}$$
$$\times\Big(\int_s^a\int_t^b d^{\frac{1}{r_3}}(x,y)N^{\frac{1}{r_3}}(x,y)dydx\Big)^{r_3}dtds+\epsilon d,$$

where $0<r_3<\min\{r_1,r_2\}$. Then, for any $\epsilon>0$,

$$\|v-v_1\|_{L^1}\leq \zeta(d(a,b)-1)\|\mu-\mu_1\|_\infty+\frac{a^{(\omega_1+1)(1-r_3)}b^{(\omega_2+1)(1-r_3)}}{(\omega_1+1)^{(1-r_3)}(\omega_2+1)^{(1-r_3)}\Gamma(r_1)\Gamma(r_2)}$$
$$\times\int_0^a\int_0^b \zeta^{r_3}\Big(d^{\frac{1}{r_3}}(s,t)N^{\frac{1-r_3}{r_3}}(s,t)-1\Big)^{r_3}\|u(s,t)-u_1(s,t)\|dtds+\epsilon d,$$

where $\omega_1=\frac{r_1-1}{1-r_3}$, $\omega_2=\frac{r_2-1}{1-r_3}$. Thus, for any $\epsilon>0$,

$$\|v-v_1\|_{L^1}\leq \zeta(d(a,b)-1)\|\mu-\mu_1\|_\infty+\frac{a^{(\omega_1+1)(1-r_3)}b^{(\omega_2+1)(1-r_3)}}{(\omega_1+1)^{(1-r_3)}(\omega_2+1)^{(1-r_3)}\Gamma(r_1)\Gamma(r_2)}$$
$$\times\zeta^{r_3}N^*\|u-u_1\|_{L^1}+\epsilon d,$$

where $N^*=\sup_{(x,y)\in J} N^{\frac{1-r_3}{r_3}}(x,y)$. Hence, for any $\epsilon>0$,

$$\|v-v_1\|_{L^1}\leq \zeta(d(a,b)-1)\|\mu-\mu_1\|_\infty$$
$$+\frac{\zeta^{r_3}N^*a^{(\omega_1+1)(1-r_3)}b^{(\omega_2+1)(1-r_3)}}{(\omega_1+1)^{(1-r_3)}(\omega_2+1)^{(1-r_3)}\Gamma(r_1)\Gamma(r_2)}\|u-u_1\|_{L^1}+\epsilon d.$$

This implies that,

$$d_{L^1}(v,\alpha(\mu_1,u_1))\leq \zeta(d(a,b)-1)\|\mu-\mu_1\|_\infty$$
$$+\frac{\zeta^{r_3}N^*a^{(\omega_1+1)(1-r_3)}b^{(\omega_2+1)(1-r_3)}}{(\omega_1+1)^{(1-r_3)}(\omega_2+1)^{(1-r_3)}\Gamma(r_1)\Gamma(r_2)}\|u-u_1\|_{L^1},$$

for all $v\in\alpha(\mu,u)$. Therefore,

$$d_{L^1}(\alpha(\mu,u),\alpha(\mu_1,u_1))\leq \zeta(d(a,b)-1)\|\mu-\mu_1\|_\infty$$
$$+\frac{\zeta^{r_3}N^*a^{(\omega_1+1)(1-r_3)}b^{(\omega_2+1)(1-r_3)}}{(\omega_1+1)^{(1-r_3)}(\omega_2+1)^{(1-r_3)}\Gamma(r_1)\Gamma(r_2)}.$$

Consequently,

$$H_d(\alpha(\mu,u),\alpha(\mu_1,u_1)) \leq \zeta(d(a,b)-1)\|\mu-\mu_1\|_\infty + \frac{\zeta^{r_3}N^* a^{(\omega_1+1)(1-r_3)} b^{(\omega_2+1)(1-r_3)}}{(\omega_1+1)^{(1-r_3)}(\omega_2+1)^{(1-r_3)}\Gamma(r_1)\Gamma(r_2)},$$

which shows that α is Hausdorff continuous and satisfies the assumptions of Lemma 4.15. Also, by the same method, we obtain that the multifunction β is Hausdorff continuous and satisfies the assumptions of Lemma 4.15. Define $\Gamma(\mu,u) = \beta(\mu,u,\alpha(\mu,u))$; $(\mu,u) \in \mathcal{M} \times L^1$. According to Lemma 4.15, the set $\text{Fix}(\Gamma(s,.) = \{u \in E : u \in \Gamma(s,u)\}$ is nonempty and arcwise connected in L^1. Moreover, for fixed $\mu_i \in \mathcal{M}$ and $u_i \in \text{Fix}(\Gamma(\mu_i,.))$; $i = 1,\ldots,p$, there exists a continuous function $\gamma : \mathcal{M} \to L^1$ such that

$$\gamma(\mu) \in \text{Fix}(\Gamma(\mu,.); \text{ for all } \mu \in \mathcal{M}, \tag{4.21}$$

$$\gamma(\mu) = u_i; \; i = 1,\ldots,p. \tag{4.22}$$

We shall prove that

$$\text{Fix}(\Gamma(\mu,.)) = \big\{u \in L^1 : u(x,y) \in F(x,y,u_\mu(x,y),G(x,y,u_\mu(x,y)));$$
$$\times a.e.\ (x,y) \in J\big\}. \tag{4.23}$$

Denote by $A(\mu)$ the right-hand side of (4.23). If $u \in \text{Fix}(\Gamma(\mu,.))$ then there is $v \in \alpha(\mu,v)$ such that $u \in \beta(\mu,u,v)$. Therefore, $v(x,y) \in G(x,y,u_\mu(x,y))$ and

$$u(x,y) \in F(x,y,u_\mu(x,y),v(x,y)) \subset F(x,y,u_\mu(x,y),G(x,y,u_\mu(x,y)));$$
$$\times\ a.e.\ (x,y) \in J,$$

so that $\text{Fix}(\Gamma(\mu,.)) \subset A(\mu)$. Let now $u \in A(\mu)$. By Lemma 4.16, there exists a selection $v \in L^1$ of the multifunction $(x,y) \to G(x,y,u_\mu(x,y))$ satisfying

$$u(x,y) \in F(x,y,u_\mu(x,y),v(x,y)); \text{ a.e. } (x,y) \in J.$$

Hence $v \in \alpha(\mu,v)$ and $u \in \beta(\mu,u,v)$ and thus $u \in \Gamma(\mu,u)$, which implies that $A(\mu) \subset \text{Fix}(\Gamma(\mu,.))$ and so that (4.23).

We next note that the function $T : L^1 \to C(J)$,

$$T(u)(x,y) = I_\theta^r u(x,y); \; (x,y) \in J$$

is continuous and one has

$$S(\mu) = \mu + T(\text{Fix}(\Gamma(\mu,.))); \; \mu \in \mathcal{M}. \tag{4.24}$$

Since $\text{Fix}(\Gamma(\mu,.))$ is nonempty and arcwise connected in L^1, the set $S(\mu)$ has the same properties in $C(J)$.

2. Let $\mu_i \in \mathcal{M}$ and let $u_i \in S(\mu_i)$; $i = 1, \ldots, p$ be fixed. By (4.24) there exists $v_i \in \text{Fix}(\Gamma(\mu_i, .))$ such that

$$u_i = \mu_i + T(v_i);\ i = 1, \ldots, p.$$

If $\gamma : \mathcal{M} \to L^1$ is a continuous function satisfying (4.21) and (4.22) we define, for every $\mu \in \mathcal{M}$,

$$S(\mu) = \mu + T(\gamma(\mu)).$$

Obviously, the function $s : \mathcal{M} \to C(J)$ is continuous, $s(\mu) \in S(\mu)$ for all $\mu \in \mathcal{M}$, and

$$S(\mu_i) = \mu_i + T(\gamma(\mu_i)) = \mu_i + T(v_i) = u_i;\ i = 1, \ldots, p.$$

3. Let $u_1, u_2 \in S = \bigcup_{\mu \in \mathcal{M}} S(\mu)$ and choose $\mu_i \in \mathcal{M}$; $i = 1, 2$ such that $u_i \in S(\mu_i)$; $i = 1, 2$. From the conclusion of 2 we deduce the existence of a continuous function $s : \mathcal{M} \to S(J)$ satisfying $s(\mu_i) = u_i$; $i = 1, 2$ and $s(\mu) \in S(\mu)$; $\mu \in \mathcal{M}$. Let $h : [0, 1] \to \mathcal{M}$ be a continuous function such that $h(0) = \mu_1$ and $h(1) = \mu_2$. Then the function $s \circ h : [0, 1] \to C(J)$ is continuous and verifies

$$s \circ h(0) = u_1,\ s \circ h(1) = u_2,$$
$$s \circ h(\mu) \in S(h(\mu)) \subset S;\ \mu \in \mathcal{M}. \qquad \square$$

4.3.3.3 On the Set of Solutions of Fractional Order Riemann–Liouville Integral Inclusions

Now, we prove the arcwise connectedness of the solution set for the following Fredholm-type fractional-order Riemann–Liouville integral inclusion:

$$u(x, y) = \mu(x, y) + \frac{1}{\Gamma(r_1)\Gamma(r_2)} \int_0^a \int_0^b (a - s)^{r_1 - 1}(b - t)^{r_2 - 1} f(x, y, s, t, \xi(s, t)) dt ds, \tag{4.25}$$

$$\xi(x, y) \in F\Big(x, y, u(x, y), G\big(x, y, u(x, y)\big)\Big);\ \text{if } (x, y) \in J := [0, a] \times [0, b], \tag{4.26}$$

where $a, b, r_1, r_2 \in (0, \infty)$, $\mu : J \to \mathbb{R}^n$, $f : J \times J \times \mathbb{R}^n \to \mathbb{R}^n$ are given functions, $F : J \times \mathbb{R}^n \times \mathbb{R}^n \to \mathcal{P}(\mathbb{R}^n)$, $G : J \times \mathbb{R}^n \to \mathcal{P}(\mathbb{R}^n)$ are given multifunctions.

Theorem 4.19. *Let $G : J \times \mathbb{R}^n \to \mathcal{P}_{cl}(\mathbb{R}^n)$ and $F : J \times \mathbb{R}^n \times \mathbb{R}^n \to \mathcal{P}_{cl}(\mathbb{R}^n)$ be two set-valued maps, satisfying the following assumptions:*

(4.19.1) *The set-valued maps* $(x, y) \to F(x, y, u, v)$ *and* $(x, y) \to G(x, y, u)$ *are measurable for all* $u, v \in \mathbb{R}^n$.

(4.19.2) *There exists a positive integrable function* $l : J \to \mathbb{R}$ *such that, for every* $u, u' \in \mathbb{R}^n$,

$$H_d(G(x, y, u), G(x, y, u')) \le l(x, y)\|u - u'\|; \text{ a.e. } (x, y) \in J.$$

(4.19.3) *There exists a positive integrable function* $m : J \to \mathbb{R}$ *and* $\eta \in [0, 1]$ *such that, for every* $u, v, u', v' \in \mathbb{R}^n$,

$$H_d(F(x, y, u, v), G(x, y, u', v)) \le m(x, y)\|u - u'\| + \eta\|v - v'\|;$$

we deduce the existence of a continuous $(x, y) \in J$.

(4.19.4) *There exist positive integrable functions* $f_1, f_2 : J \to \mathbb{R}$ *such that*

$$H_d(\{0\}, F(x, y, \{0\}, \{0\})) \le f_1(x, y), \text{ a.e. } (x, y) \in J$$

$$H_d(\{0\}, G(x, y, \{0\})) \le f_2(x, y);$$

we deduce the existence of a continuous $(x, y) \in J$.

(4.19.5) *The function* $f : J \times J \times \mathbb{R}^n \to \mathbb{R}^n$ *is continuous and bounded and there exists a constant* $M > 0$ *such that*

$$\|f(x, y, s, t, \xi_1) - f(x, y, s, t, \xi_2)\| \le M\|\xi_1 - \xi_2\|; \ \xi_1, \xi_2 \in \mathbb{R}^n.$$

Then,

1. *For every* $\mu \in C(J)$, *the solution set* $S(\mu)$ *of problems (4.25) and (4.26) is nonempty and arcwise connected in the space* $C(J)$.
2. *For any* $\mu_i \in C(J)$ *and any* $u_i \in S(\mu)$; $i = 1, \ldots, p$, *there exists a continuous function* $s : C(J) \to C(J)$ *such that* $s(\mu) \in S(\mu)$ *for any* $\mu \in C(J)$ *and* $s(\mu_i) = u_i$; $i = 1, \ldots, p$.
3. *The set* $S = \cup_{\mu \in C(J)} S(\mu)$ *is arcwise connected in the space* $C(J)$.

Proof. In what follows $N(x, y) = \max\{l(x, y), m(x, y); \ (x, y) \in J\}$ and $\zeta \in (0, 1)$ will be taken such that $2\zeta + \eta < 1$ and $d : J \to (0, \infty)$ in (4.20) is the corresponding mapping found in Lemma 4.17.

1. For $\mu \in C(J)$ and $u \in L^1$, set

$$u_\mu(x, y) = \mu(x, y) + \frac{1}{\Gamma(r_1)\Gamma(r_2)} \int_0^a \int_0^b (a - s)^{r_1 - 1}(b - t)^{r_2 - 1} \times f(x, y, s, t, \xi(s, t))dtds; \ (x, y) \in J.$$

Define the multifunctions $\alpha : C(J) \times L^1 \to \mathcal{P}(L^1)$ and $\beta : C(J) \times L^1 \times L^1 \to \mathcal{P}(L^1)$ by

$$\alpha(\mu, \xi) = \{v \in L^1 : v(x, y) \in G(x, y, u_\mu(x, y)); \text{ a.e. } (x, y) \in J,$$

$$\beta(\mu, \xi, v) = \{w \in L^1 : w(x, y) \in F(x, y, u_\mu(x, y), v(x, y)); \text{ a.e. } (x, y) \in J,$$

where $\mu \in C(J)$ and $\xi, v \in L^1$. We prove that α and β satisfy the hypothesis of Lemma 4.15.

Since u_μ is measurable and G satisfies hypothesis (4.19.1) and (4.19.2), the multifunction $G_\mu : (x, y) \to G(x, y, u_\mu(x, y))$ is measurable and $G_\mu \in \mathcal{P}_{cl}(L^1)$, G_μ has a measurable selection. Therefore due to hypothesis (4.19.4), we get $\alpha(\mu, \xi) \neq \emptyset$. Also, by simple computation, it follows that the set $\alpha(\mu, \xi)$ is closed and decomposable. In the same way we obtain that $\alpha(\mu, \xi) \in \mathcal{P}_{cl}(L^1)$ is a decomposable set.

Set $d := \int_0^a \int_0^b d(x, y)\mathrm{d}y\mathrm{d}x$. Pick $(\mu, \xi), (\mu_1, \xi_1) \in C(J) \times L^1$ and choose $v \in \alpha(\mu, \xi)$. For each $\epsilon > 0$ there exists $v_1 \in \alpha(\mu_1, \xi_1)$ such that, for every $(x, y) \in J$, one has

$$\begin{aligned}
\|v(x, y) - v_1(x, y)\| &\leq H_d\Big(G(x, y, u_\mu(x, y)), G(x, y, u_{\mu_1}(x, y))\Big) + \epsilon \\
&\leq N(x, y)\Big(\|\mu(x, y) - \mu_1(x, y)\| \\
&\quad + \frac{1}{\Gamma(r_1)\Gamma(r_2)} \int_0^x \int_0^y (x - s)^{r_1 - 1}(y - t)^{r_2 - 1} M \|\xi(s, t) \\
&\quad - \xi_1(s, t)\|\mathrm{d}t\mathrm{d}s\Big) + \epsilon.
\end{aligned}$$

Thus, for any $\epsilon > 0$,

$$\begin{aligned}
\|v - v_1\|_{L^1} &\leq \|\mu - \mu_1\|_\infty \int_0^a \int_0^b d(x, y)N(x, y)\mathrm{d}y\mathrm{d}x \\
&\quad + \frac{M}{\Gamma(r_1)\Gamma(r_2)} \int_0^a \int_0^b d(x, y)N(x, y) \\
&\quad \times \left(\int_0^a \int_0^b (a - s)^{r_1 - 1}(b - t)^{r_2 - 1} \|\xi(s, t) - \xi_1(s, t)\|\mathrm{d}t\mathrm{d}s\right)\mathrm{d}y\mathrm{d}x + \epsilon d
\end{aligned}$$

$$\leq \zeta(d(a,b)-1)\|\mu-\mu_1\|_\infty + \frac{M}{\Gamma(r_1)\Gamma(r_2)}\int_0^a\int_0^b \|\xi(s,t)-\xi_1(s,t)\|$$

$$\times\left(\int_s^a\int_t^b (a-s)^{r_1-1}(b-t)^{r_2-1}d(x,y)N(x,y)\mathrm{d}y\mathrm{d}x\right)\mathrm{d}t\mathrm{d}s+\epsilon d$$

$$\leq \zeta(d(a,b)-1)\|\mu-\mu_1\|_\infty$$

$$+\frac{M}{\Gamma(r_1)\Gamma(r_2)}\int_0^a\int_0^b \|\xi(s,t)-\xi_1(s,t)\|$$

$$\times\left(\int_s^a\int_t^b (a-s)^{\frac{r_1-1}{1-r_3}}(b-t)^{\frac{r_2-1}{1-r_3}}\mathrm{d}y\mathrm{d}x\right)^{1-r_3}$$

$$\times\left(\int_s^a\int_t^b d^{\frac{1}{r_3}}(x,y)N^{\frac{1}{r_3}}(x,y)\mathrm{d}y\mathrm{d}x\right)^{r_3}\mathrm{d}t\mathrm{d}s+\epsilon d,$$

where $0<r_3<\min\{r_1,r_2\}$. Then, for any $\epsilon>0$,

$$\|v-v_1\|_{L^1}\leq \zeta(d(a,b)-1)\|\mu-\mu_1\|_\infty$$
$$+\frac{Ma^{(\omega_1+1)(1-r_3)}b^{(\omega_2+1)(1-r_3)}}{(\omega_1+1)^{(1-r_3)}(\omega_2+1)^{(1-r_3)}\Gamma(r_1)\Gamma(r_2)}$$
$$\times\int_0^a\int_0^b \zeta^{r_3}\left(d^{\frac{1}{r_3}}(s,t)N^{\frac{1-r_3}{r_3}}(s,t)-1\right)^{r_3}\|\xi(s,t)$$
$$-\xi_1(s,t)\|\mathrm{d}t\mathrm{d}s+\epsilon d,$$

where $\omega_1=\frac{r_1-1}{1-r_3}$, $\omega_2=\frac{r_2-1}{1-r_3}$. Then, for any $\epsilon>0$,

$$\|v-v_1\|_{L^1}\leq \zeta(d(a,b)-1)\|\mu-\mu_1\|_\infty$$
$$+\frac{Ma^{(\omega_1+1)(1-r_3)}b^{(\omega_2+1)(1-r_3)}}{(\omega_1+1)^{(1-r_3)}(\omega_2+1)^{(1-r_3)}\Gamma(r_1)\Gamma(r_2)}$$
$$\times\zeta^{r_3}N^*\|\xi-\xi_1\|_{L^1}+\epsilon d,$$

where $N^* = \sup_{(x,y)\in J} N^{\frac{1-r_3}{r_3}}(x,y)$. Hence, for any $\epsilon > 0$, we have

$$\|v - v_1\|_{L^1} \leq \zeta(d(a,b)-1)\|\mu - \mu_1\|_\infty + \frac{M\zeta^{r_3}N^* a^{(\omega_1+1)(1-r_3)} b^{(\omega_2+1)(1-r_3)}}{(\omega_1+1)^{(1-r_3)}(\omega_2+1)^{(1-r_3)}\Gamma(r_1)\Gamma(r_2)}\|\xi - \xi_1\|_{L^1} + \epsilon d.$$

This implies that

$$d_{L^1}(v, \alpha(\mu_1, \xi_1)) \leq \zeta(d(a,b)-1)\|\mu - \mu_1\|_\infty + \frac{M\zeta^{r_3}N^* a^{(\omega_1+1)(1-r_3)} b^{(\omega_2+1)(1-r_3)}}{(\omega_1+1)^{(1-r_3)}(\omega_2+1)^{(1-r_3)}\Gamma(r_1)\Gamma(r_2)}\|\xi - \xi_1\|_{L^1},$$

for all $v \in \alpha(\mu, u)$. Therefore,

$$d_{L^1}(\alpha(\mu,\xi), \alpha(\mu_1, \xi_1)) \leq \zeta(d(a,b)-1)\|\mu - \mu_1\|_\infty + \frac{M\zeta^{r_3}N^* a^{(\omega_1+1)(1-r_3)} b^{(\omega_2+1)(1-r_3)}}{(\omega_1+1)^{(1-r_3)}(\omega_2+1)^{(1-r_3)}\Gamma(r_1)\Gamma(r_2)}\|\xi - \xi_1\|_{L^1}.$$

Consequently,

$$H_d(\alpha(\mu,\xi), \alpha(\mu_1, \xi_1)) \leq \zeta(d(a,b)-1)\|\mu - \mu_1\|_\infty + \frac{M\zeta^{r_3}N^* a^{(\omega_1+1)(1-r_3)} b^{(\omega_2+1)(1-r_3)}}{(\omega_1+1)^{(1-r_3)}(\omega_2+1)^{(1-r_3)}\Gamma(r_1)\Gamma(r_2)}\|\xi - \xi_1\|_{L^1},$$

which shows that α is Hausdorff continuous and satisfies the assumptions of Lemma 4.15. Also, by the same method, we obtain that the multifunction β is Hausdorff continuous and satisfies the assumptions of Lemma 4.15. Define $\Gamma(\mu,\xi) = \beta(\mu,\xi,\alpha(\mu,\xi))$; $(\mu,\xi) \in C(J) \times L^1$. According to Lemma 4.15, the set $\mathrm{Fix}(\Gamma(s,.) = \{\xi \in E : \xi \in \Gamma(s,\xi)\}$ is nonempty and arcwise connected in L^1. Moreover, for fixed $\mu_i \in C(J)$ and $\xi_i \in \mathrm{Fix}(\Gamma(\mu_i,.))$; $i = 1,\ldots,p$, there exists a continuous function $\gamma : C(J) \to L^1$ such that

$$\gamma(\mu) \in \mathrm{Fix}(\Gamma(\mu,.); \text{ for all } \mu \in C(J), \tag{4.27}$$

$$\gamma(\mu) = \xi_i;\ i = 1,\ldots,p. \tag{4.28}$$

We shall prove that

$$\mathrm{Fix}(\Gamma(\mu,.)) = \{\xi \in L^1 : \xi(x,y) \in F(x,y,u_\mu(x,y), G(x,y,u_\mu(x,y)));\ \text{a.e. } (x,y) \in J\}. \tag{4.29}$$

Denote by $A(\mu)$ the right-hand side of (4.29). If $\xi \in \operatorname{Fix}(\Gamma(\mu,.))$ then there is $v \in \alpha(\mu, v)$ such that $\xi \in \beta(\mu, \xi, v)$. Therefore, $v(x,y) \in G(x,y,u_\mu(x,y))$ and

$$\begin{aligned}\xi(x,y) \in F(x,y,u_\mu(x,y),v(x,y)) \subset F(x,y,u_\mu(x,y),\\ \times G(x,y,u_\mu(x,y))); \text{a.e. } (x,y) \in J,\end{aligned}$$

so that $\operatorname{Fix}(\Gamma(\mu,.)) \subset A(\mu)$. Let now $\xi \in A(\mu)$. By Lemma 4.16, there exists a selection $v \in L^1$ of the multifunction $(x,y) \to G(x,y,u_\mu(x,y))$ satisfying

$$\xi(x,y) \in F(x,y,u_\mu(x,y),v(x,y)); \text{ a.e. } (x,y) \in J.$$

Hence $v \in \alpha(\mu, v)$ and $\xi \in \beta(\mu, \xi, v)$ and thus $\xi \in \Gamma(\mu, u)$, which implies that $A(\mu) \subset \operatorname{Fix}(\Gamma(\mu,.))$ and so that (4.29).

We next note that by (4.19.5), the function $T : L^1 \to C(J)$,

$$\begin{aligned}T(\xi)(x,y) = \frac{1}{\Gamma(r_1)\Gamma(r_2)} \int_0^a \int_0^b (a-s)^{r_1-1}(b-t)^{r_2-1}\\ \times f(x,y,s,t,\xi(s,t))dtds; \ (x,y) \in J,\end{aligned}$$

is continuous and one has

$$S(\mu) = \mu + T(\operatorname{Fix}(\Gamma(\mu,.))); \ \mu \in C(J). \tag{4.30}$$

Since $\operatorname{Fix}(\Gamma(\mu,.))$ is nonempty and arcwise connected in L^1, the set $S(\mu)$ has the same properties in $C(J)$.

2. Let $\mu_i \in C(J)$ and let $u_i \in S(\mu_i)$; $i = 1, \ldots, p$ be fixed. By (4.30) there exists $v_i \in \operatorname{Fix}(\Gamma(\mu_i,.))$ such that

$$u_i = \mu_i + T(v_i); \ i = 1, \ldots, p.$$

If $\gamma : C(J) \to L^1$ is a continuous function satisfying (4.27) and (4.28) we define, for every $\mu \in C(J)$,

$$S(\mu) = \mu + T(\gamma(\mu)).$$

Obviously, the function $s : C(J) \to C(J)$ is continuous, $s(\mu) \in S(\mu)$ for all $\mu \in C(J)$, and

$$S(\mu_i) = \mu_i + T(\gamma(\mu_i)) = \mu_i + T(v_i) = u_i; \ i = 1, \ldots, p.$$

3. Let $u_1, u_2 \in S = \bigcup_{\mu \in \mathcal{M}} S(\mu)$ and choose $\mu_i \in \mathcal{M}$; $i = 1, 2$ such that $u_i \in S(\mu_i)$; $i = 1, 2$. From the conclusion of 2 we deduce the existence of a continuous function $s : C(J) \to S(J)$ satisfying $s(\mu_i) = u_i$; $i = 1, 2$ and

$s(\mu) \in S(\mu)$; $\mu \in C(J)$. Let $h : [0,1] \to C(J)$ be a continuous function such that $h(0) = \mu_1$ and $h(1) = \mu_2$. Then the function $s \circ h : [0,1] \to C(J)$ is continuous and verifies

$$s \circ h(0) = u_1, \; s \circ h(1) = u_2,$$
$$s \circ h(\mu) \in S(h(\mu)) \subset S; \; \mu \in C(J).$$

□

Corollary 4.20. *Consider the following Fredholm-type integral inclusion:*

$$u(x,y) = \mu(x,y) + \int_0^a \int_0^b f(x,y,s,t,\xi(s,t))dtds, \tag{4.31}$$

$$\xi(x,y) \in F\Big(x,y,u(x,y),G\Big(x,y,u(x,y)\Big)\Big); \; \textit{if } (x,y) \in J := [0,a] \times [0,b]. \tag{4.32}$$

Suppose that Hypotheses (4.19.1)–(4.19.5) hold. Then,

1. *For every $\mu \in C(J)$, the solution set $S'(\mu)$ of problems (4.31) and (4.32) is nonempty and arcwise connected in the space $C(J)$.*
2. *For any $\mu_i \in C(J)$ and any $u_i \in S'(\mu)$; $i = 1,\ldots,p$, there exists a continuous function $s' : C(J) \to C(J)$ such that $s'(\mu) \in S'(\mu)$ for any $\mu \in C(J)$ and $s'(\mu_i) = u_i$; $i = 1,\ldots,p$.*
3. *The set $S' = \cup_{\mu \in C(J)} S'(\mu)$ is arcwise connected in the space $C(J)$.*

4.4 Upper and Lower Solutions Method for Partial Differential Inclusions

4.4.1 Introduction

This section deals with the existence of solutions to the Darboux problem for the fractional order hyperbolic differential inclusion

$$({}^cD_0^r u)(x,y) \in F(x,y,u(x,y)), \text{ if } (x,y) \in J, \tag{4.33}$$

$$u(x,0) = \varphi(x), \; x \in [0,a], \; u(0,y) = \psi(y), \; y \in [0,b], \tag{4.34}$$

where $J = [0,a] \times [0,b]$, $a,b > 0$, $F : J \times \mathbb{R}^n \to \mathcal{P}(\mathbb{R}^n)$ is a compact-valued multivalued map and φ, ψ are as in problems (4.10) and (4.11)

4.4.2 Main Result

Let us start by defining what we mean by a solution of problem (4.33) and (4.34).

Definition 4.21. A function $u \in C(J, \mathbb{R}^n)$ is said to be a solution of (4.33) and (4.34), if there exists a function $f \in L^1(J, \mathbb{R}^n)$ with $f(x, y) \in F(x, y, u(x, y))$ such that $({}^cD_0^r u)(x, y) = f(x, y)$, and u satisfies the (4.34) on J.

Let $z,\ \bar{z} \in C(J, \mathbb{R}^n)$ be such that

$$z(x, y) = (z_1(x, y), z_2(x, y), \ldots, z_n(x, y)),\ (x, y) \in J$$

and

$$\bar{z}(x, y) = (\bar{z}_1(x, y), \bar{z}_2(x, y), \ldots, \bar{z}_n(x, y)),\ (x, y) \in J.$$

The notation $z \leq \bar{z}$ means that

$$z_i(x, y) \leq \bar{z}_i(x, y),\ i = 1, \ldots, n.$$

Theorem 4.22. *Assume that the following hypotheses:*

(4.22.1) $F : J \times \mathbb{R}^n \longrightarrow \mathcal{P}_{cp,cv}(\mathbb{R}^n)$ *is* L^1*-Carathéodory*
(4.22.2) There exists $l \in C(J, \mathbb{R}_+)$ *such that*

$$H_d(F(x, y, u), F(x, y, \overline{u})) \leq l(x, y)\|u - \overline{u}\| \text{ for every } u, \overline{u} \in \mathbb{R}^n$$

and

$$d(0, F(x, y, 0)) \leq l(x, y),\ a.e.\ (x, y) \in J$$

(4.22.3) There exist v *and* $w \in C(J, \mathbb{R}^n)$*, lower and upper solutions for the problems (4.33) and (4.34) such that* $v(x, y) \leq w(x, y)$ *for each* $(x, y) \in J$

hold. Then problem (4.33) and (4.34) has at least one solution u *such that*

$$v(x, y) \leq u(x, y) \leq w(x, y) \text{ for all } (x, y) \in J.$$

Proof. Transform problem (4.33) and (4.34) into a fixed-point problem. Consider the following modified problem:

$$({}^cD_0^r u)(x, y) \in F(x, y, g(u(x, y))),\ \text{if } (x, y) \in J, \tag{4.35}$$

$$u(x, 0) = \varphi(x),\ u(0, y) = \psi(y),\ x \in [0, a] \text{ and } y \in [0, b], \tag{4.36}$$

where $g : C(J, \mathbb{R}^n) \longrightarrow C(J, \mathbb{R}^n)$ be the truncation operator defined by

$$(gu)(x,y)=\begin{cases} v(x,y), & u(x,y)<v(x,y),\\ u(x,y), & v(x,y)\le u(x,y)\le w(x,y),\\ w(x,y), & u(x,y)>w(x,y).\end{cases}$$

A solution to (4.35) and (4.36) is a fixed point of the operator $G : C(J,\mathbb{R}^n) \longrightarrow \mathcal{P}(C(J,\mathbb{R}^n))$ defined by

$$G(u)=\begin{cases} h\in C(J,\mathbb{R}^n):\\ h(x,y)=\mu(x,y)\\ +\dfrac{1}{\Gamma(r_1)\Gamma(r_2)}\displaystyle\int_0^x\int_0^y (x-s)^{r_1-1}(y-t)^{r_2-1}f(s,t)\mathrm{d}t\mathrm{d}s,\ (x,y)\in J,\end{cases}$$

where $f\in \tilde{S}_{F,g(u)}=\{f\in S_{F,g(u)} : f(x,y)\ge f_1(x,y)$ on A_1 and $f(x,y)\le f_2(x,y)$ on $A_2\}$, $A_1=\{(x,y)\in J : u(x,y)<v(x,y)\le w(x,y)\}$, $A_2=\{(x,y)\in J : u(x,y)\le w(x,y)<u(x,y)\}$.

Remark 4.23. (A) For each $u\in C(J,\mathbb{R}^n)$, the set $\tilde{S}_{F,g(u)}$ is nonempty. In fact, (4.15.1) implies that there exists $f_3\in S_{F,g(u)}$, so we set

$$f=f_1\chi_{A_1}+f_2\chi_{A_2}+f_3\chi_{A_3},$$

where χ_{A_i} is the characteristic function of A_i; $i=1,2$, and

$$A_3=\{(x,y)\in J : v(x,y)\le u(x,y)\le w(x,y)\}.$$

Then, by decomposability, $f\in \tilde{S}_{F,g(u)}$.

(B) By the definition of g it is clear that $F(.,.,g(u)(.,.))$ is an L^1-Carathéodory multivalued map with compact convex values and there exists $\phi_1\in C(J,\mathbb{R}_+)$ such that

$$\|F(x,y,g(u(x,y)))\|_{\mathcal{P}}\le \phi_1(x,y) \text{ for each } u\in\mathbb{R}^n.$$

Set

$$\eta=\|\mu\|_\infty+\frac{a^{r_1}b^{r_2}\phi_1^*}{\Gamma(r_1+1)\Gamma(r_2+1)},$$

where

$$\phi_1^*=\sup\{\phi_1(x,y) : (x,y)\in J\}$$

and

$$D=\{u\in C(J,\mathbb{R}^n) : \|u\|_\infty\le\eta\}.$$

Clearly D is a closed convex subset of $C(J,\mathbb{R}^n)$ and that G maps D into D. We shall show that D satisfies the assumptions of Lemma 2.38. The proof will be given in several steps.

Step 1: *$G(u)$ is convex for each $u \in D$.* Indeed, if $h_1,\ h_2$ belong to $G(u)$, then there exist $u_1,\ u_2 \in S_{F,g(u)}$ such that for each $(x,y) \in J$ we have

$$h_i(x,y) = \mu(x,y) + \frac{1}{\Gamma(r_1)\Gamma(r_2)}\int_0^x\int_0^y (x-s)^{r_1-1}(y-t)^{r_2-1}u_i(s,t)\mathrm{d}t\mathrm{d}s,\quad i=1,2.$$

Let $0 \le \xi \le 1$. Then, for each $(x,y) \in J$, we have

$$(\xi h_1 + (1-\xi)h_2)(x,y) = \mu(x,y) + \frac{1}{\Gamma(r_1)\Gamma(r_2)}\int_0^x\int_0^y (x-s)^{r_1-1}(y-t)^{r_2-1} \times[\xi u_1(s,t) + (1-\xi)u_2(s,t)]\mathrm{d}t\mathrm{d}s.$$

Since $\tilde{S}_{F,g(u)}$ is convex (because F has convex values), we have

$$\xi h_1 + (1-\xi)h_2 \in G(u).$$

Step 2: *$G(D)$ is bounded.* This is clear since $G(D) \subset D$ and D is bounded.

Step 3: *$G(D)$ is equicontinuous.* Let $(x_1,y_1),(x_2,y_2) \in J,\ \ x_1 < x_2$ and $y_1 < y_2$, let $u \in D$ and $h \in G(u)$, then there exists $z \in S_{F,g(u)}$ such that for each $(x,y) \in J$ we have

$$\begin{aligned}
\|h(x_2,y_2) - h(x_1,y_1)\| &= \Bigg\| \mu(x_2,y_2) - \mu(x_1,y_1) + \frac{1}{\Gamma(r_1)\Gamma(r_2)}\int_0^{x_1}\int_0^{y_1} [(x_2-s)^{r_1-1}(y_2-t)^{r_2-1} \\
&\quad -(x_1-s)^{r_1-1}(y_1-t)^{r_2-1}]z(s,t)\mathrm{d}s\mathrm{d}t \\
&\quad + \frac{1}{\Gamma(r_1)\Gamma(r_2)}\int_0^{x_1}\int_{y_1}^{y_2} (x_2-s)^{r_1-1}(y_2-t)^{r_2-1}z(s,t)\mathrm{d}t\mathrm{d}s \\
&\quad + \frac{1}{\Gamma(r_1)\Gamma(r_2)}\int_{x_1}^{x_2}\int_0^{y_1} (x_2-s)^{r_1-1}(y_2-t)^{r_2-1}z(s,t)\mathrm{d}t\mathrm{d}s
\end{aligned}$$

$$+\frac{1}{\Gamma(r_1)\Gamma(r_2)}\int_{x_1}^{x_2}\int_{y_1}^{y_2}(x_2-s)^{r_1-1}(y_2-t)^{r_2-1}z(s,t)\mathrm{d}t\mathrm{d}s\Bigg\|$$

$$\leq \|\mu(x_1,y_1)-\mu(x_2,y_2)\|$$

$$+\frac{\phi_1^*}{\Gamma(r_1+1)\Gamma(r_2+1)}[2y_2^{r_2}(x_2-x_1)^{r_1}+2x_2^{r_1}(y_2-y_1)^{r_2}$$

$$+x_1^{r_1}y_1^{r_2}-x_2^{r_1}y_2^{r_2}-2(x_2-x_1)^{r_1}(y_2-y_1)^{r_2}].$$

As $x_1 \longrightarrow x_2$ and $y_1 \longrightarrow y_2$, the right-hand side of the above inequality tends to zero. As a consequence of Steps 1–3 together with the Arzelá–Ascoli theorem, we can conclude that $G : D \longrightarrow \mathcal{P}(D)$ is compact.

Step 4: *G has a closed graph.* Let $u_n \to u_*$, $h_n \in G(u_n)$, and $h_n \to h_*$. We need to show that $h_* \in G(u_*)$.
$h_n \in G(u_n)$ means that there exists $z_n \in \tilde{S}_{F,g(u_n)}$ such that, for each $(x,y) \in J$,

$$h_n(x,y)=\mu(x,y)+\frac{1}{\Gamma(r_1)\Gamma(r_2)}\int_0^x\int_0^y(x-s)^{r_1-1}(y-t)^{r_2-1}z_n(s,t)\mathrm{d}t\mathrm{d}s.$$

We must show that there exists $z_* \in \tilde{S}_{F,g(u_*)}$ such that, for each $(x,y) \in J$,

$$h_*(x,y)=\mu(x,y)+\frac{1}{\Gamma(r_1)\Gamma(r_2)}\int_0^x\int_0^y(x-s)^{r_1-1}(y-t)^{r_2-1}z_*(s,t)\mathrm{d}t\mathrm{d}s.$$

Since $F(x,y,\cdot)$ is upper semicontinuous, then for every $\varepsilon > 0$, there exist $n_0(\epsilon) \geq 0$ such that for every $n \geq n_0$, we have

$$f_n(x,y) \in F(x,y,u_n(x,y)) \subset F(x,y,u_*(x,y)) + \varepsilon B(0,1), \text{ a.e. } (x,y) \in J.$$

Since $F(.,.,.)$ has compact values, then there exists a subsequence f_{n_m} such that

$$f_{n_m}(\cdot,\cdot) \to f_*(\cdot,\cdot) \text{ as } m \to \infty$$

and

$$f_*(x,y) \in F(x,y,u_*)(x,y), \text{ a.e. } (x,y) \in J.$$

For every $w \in F(x,y,u_*(x,y))$, we have

$$\|f_{n_m}(x,y)-u_*(x,y)\| \leq \|f_{n_m}(x,y)-w\| + \|w-f_*(x,y)\|.$$

Then

$$\|f_{n_m}(x,y) - f_*(x,y)\| \leq d\Big(f_{n_m}(x,y), F(x,y,u_*(x,y))\Big).$$

By an analogous relation, obtained by interchanging the roles of f_{n_m} and f_*, it follows that

$$\begin{aligned}\|f_{n_m}(x,y) - u_*(x,y)\| &\leq H_d(F(x,y,u_n(x,y)), F(x,y,u_*(x,y)) \\ &\leq l(x,y)\|u_n - u_*\|_\infty.\end{aligned}$$

Then

$$\begin{aligned}\|h_{n_m}(x,y) - h_*(x,y)\| &\leq \frac{1}{\Gamma(r_1)\Gamma(r_2)}\int_0^x\int_0^y (x-s)^{r_1-1}(y-t)^{r_2-1}\|f_{n_m}(s,t) \\ &\quad - f_*(s,t)\|\mathrm{d}t\mathrm{d}s \\ &\leq \frac{\|u_{n_m} - u_*\|_\infty}{\Gamma(r_1)\Gamma(r_2)}\int_0^x\int_0^y (x-s)^{r_1-1}(y-t)^{r_2-1}l(s,t)\mathrm{d}t\mathrm{d}s \\ &\leq \frac{a^{r_1}b^{r_2}l^*}{\Gamma(r_1+1)\Gamma(r_2+1)}\|u_{n_m} - u_*\|_\infty,\end{aligned}$$

where

$$l^* = \sup\{l(x,y) : (x,y) \in J\}.$$

Hence

$$\|h_{n_m} - h_*\|_\infty \leq \frac{a^{r_1}b^{r_2}l^*}{\Gamma(r_1+1)\Gamma(r_2+1)}\|u_{n_m} - u_*\|_\infty \to 0 \text{ as } m \to \infty.$$

Step 5: *The solution u of (4.35) and (4.36) satisfies*

$$v(x,y) \leq u(x,y) \leq w(x,y) \text{ for all } (x,y) \in J.$$

We prove that

$$u(x,y) \leq w(x,y) \text{ for all } (x,y) \in J.$$

Assume that $u - w$ attains a positive maximum on J at $(\overline{x}, \overline{y}) \in J$; i.e.,

$$(u-w)(\overline{x}, \overline{y}) = \max\{u(x,y) - w(x,y) : (x,y) \in J\} > 0.$$

We distinguish the following cases.

Case 1 If $(\overline{x}, \overline{y}) \in (0, a) \times [0, b]$ there exists $(x^*, y^*) \in (0, a) \times [0, b]$ such that

$$[u(x, y^*) - w(x, y^*)] + [u(x^*, y) - w(x^*, y)]$$
$$- [u(x^*, y^*) - w(x^*, y^*)] \leq 0; \text{ for all } (x, y) \in ([x^*, \overline{x}] \times \{y^*\}) \cup (\{x^*\} \times [y^*, b]) \tag{4.37}$$

and

$$u(x, y) - w(x, y) > 0, \text{ for all } (x, y) \in (x^*, \overline{x}] \times [y^*, b]. \tag{4.38}$$

By the definition of g one has

$${}^cD_0^r u(x, y) \in F(x, y, g(u(x, y))) \text{ for all } (x, y) \in [x^*, \overline{x}] \times [y^*, b],$$

then there exists $f \in F(x, y, g(u(x, y)))$ such that

$${}^cD_0^r u(x, y) = f(x, y) \text{ for all } (x, y) \in [x^*, \overline{x}] \times [y^*, b].$$

An integration on $[x^*, x] \times [y^*, y]$ for each $(x, y) \in [x^*, \overline{x}] \times [y^*, b]$ yields

$$u(x, y) + u(x^*, y^*) - u(x, y^*) - u(x^*, y)$$
$$= \frac{1}{\Gamma(r_1)\Gamma(r_2)} \int_{x^*}^{x} \int_{y^*}^{y} (x - s)^{r_1 - 1}(y - t)^{r_2 - 1} f(s, t) dt ds. \tag{4.39}$$

From (4.39) and using the fact that w is an upper solution to (3.1)–(3.2) we get

$$u(x, y)+u(x^*, y^*)-u(x, y^*)-u(x^*, y) \leq w(x, y)+w(x^*, y^*)-w(x, y^*)-w(x^*, y),$$

which gives

$$[u(x, y) - w(x, y)] \leq [u(x, y^*) - w(x, y^*)] + [u(x^*, y) - w(x^*, y)]$$
$$-[u(x^*, y^*) - w(x^*, y^*)]. \tag{4.40}$$

Thus from (4.37), (4.38) and (4.40) we obtain the contradiction

$$0 < [u(x, y) - w(x, y)]$$
$$\leq [u(x, y^*) - w(x, y^*)] + [u(x^*, y) - w(x^*, y)] - [u(x^*, y^*) - w(x^*, y^*)] \leq 0;$$
$$\text{for all } (x, y) \in [x^*, \overline{x}] \times [y^*, b].$$

case 2 If $\overline{x} = 0$, then $w(0, \overline{y}) < u(0, \overline{y}) \leq w(0, \overline{y})$ which is a contradiction. Thus

$$u(x, y) \leq w(x, y) \text{ for all } (x, y) \in J.$$

Analogously, we can prove that $u(x, y) \geq v(x, y)$, for all $(x, y) \in J$. This shows that the problems (4.35) and (4.36) have a solution u satisfying $v \leq u \leq w$ which is a solution of (4.33) and (4.34). □

4.5 Partial Functional Differential Inclusions with Infinite Delay

4.5.1 *Introduction*

This section deals with the existence of solutions to fractional order IVP , for the system

$$({}^cD_0^r u)(x, y) \in F(x, y, u_{(x,y)}), \text{ if } (x, y) \in J, \tag{4.41}$$

$$u(x, y) = \phi(x, y), \text{ if } (x, y) \in \tilde{J}', \tag{4.42}$$

$$u(x, 0) = \varphi(x), \ x \in [0, a], \ u(0, y) = \psi(y), \ y \in [0, b], \tag{4.43}$$

where $J = [0, a] \times [0, b]$, $a, b > 0$, $\tilde{J}' = (-\infty, a] \times (-\infty, b] \backslash (0, a] \times (0, b]$, $F : J \times \mathcal{B} \to \mathcal{P}(\mathbb{R}^n)$ is a multivalued map with compact, convex values, $\mathcal{P}(\mathbb{R}^n)$ is the family of all subsets of $\mathbb{R}^n$, $\phi : \tilde{J}' \to \mathbb{R}^n$ is a given continuous function, φ, ψ are as in problem (3.1)–(3.3) and $\mathcal{B}$ is a phase space. Next we consider the following IVP for partial neutral functional differential inclusions:

$${}^cD_0^r[u(x, y) - g(x, y, u_{(x,y)})] \in F(x, y, u_{(x,y)}), \text{ if } (x, y) \in J, \tag{4.44}$$

$$u(x, y) = \phi(x, y), \text{ if } (x, y) \in \tilde{J}', \tag{4.45}$$

$$u(x, 0) = \varphi(x), \ x \in [0, a], \ u(0, y) = \psi(y), \ y \in [0, b], \tag{4.46}$$

where F, ϕ, φ, ψ are as in problem (4.41)–(4.43) and $g : J \times \mathcal{B} \to \mathcal{P}(\mathbb{R}^n)$ is a given continuous function.

4.5.2 *Main Results*

Let us start by defining what we mean by a solution of the problem (4.41)–(4.43). Let the space

$$\Omega := \{u : (-\infty, a] \times (-\infty, b] \to \mathbb{R}^n : u_{(x,y)} \in \mathcal{B} \text{ for } (x, y) \in E \text{ and } u|_J \text{is continuous}\}.$$

Definition 4.24. A function $u \in \Omega$ is said to be a solution of (4.41)–(4.43) if there exists a function $f \in L^1(J, \mathbb{R}^n)$ with $f(x, y) \in F(x, y, u_{(x,y)})$ such that $({}^cD_0^r u)(x, y) = f(x, y)$ and u satisfies (4.43) on J and the condition (4.42) on $\tilde{J}'$.

Theorem 4.25. *Assume that*

(4.25.1) $F : J \times \mathcal{B} \to \mathcal{P}_{cp,cv}(\mathbb{R}^n)$ *is a Carathéodory multivalued map*

(4.25.2) There exists $l \in L^\infty(J, \mathbb{R})$ *such that*

$$H_d(F(x, y, u), F(x, y, \overline{u})) \leq l(x, y)\|u - \overline{u}\|_{\mathcal{B}} \textit{ for every } u, \overline{u} \in \mathcal{B}$$

and

$$d(0, F(x, y, 0)) \leq l(x, y), \textit{ a.e. } (x, y) \in J.$$

Then the IVP (4.41)–(4.43) has at least one solution on $(-\infty, a] \times (-\infty, b]$.

Proof. Let $l^* = \|l\|_{L^\infty}$. Transform the problems (4.41)–(4.43) into a fixed-point problem. Consider the multivalued operator $N : \Omega \to \mathcal{P}(\Omega)$ defined by

$$N(x, y) = \{h \in \Omega\},$$

such that

$$h(x, y) = \begin{cases} \phi(x, y), & (x, y) \in \tilde{J}', \\ \mu(x, y) \\ \quad + \frac{1}{\Gamma(r_1)\Gamma(r_2)} \int_0^x \int_0^y (x - s)^{r_1 - 1}(y - t)^{r_2 - 1} f(s, t) \mathrm{d}t \mathrm{d}s, \ f \in S_{F,u}, & (x, y) \in J. \end{cases}$$

Let $v(.,.) : (-\infty, a] \times (-\infty, b] \to \mathbb{R}^n$. be a function defined by,

$$v(x, y) = \begin{cases} \phi(x, y), \ (x, y) \in \tilde{J}', \\ \mu(x, y), \ (x, y) \in J. \end{cases}$$

Then $v_{(x,y)} = \phi$ for all $(x, y) \in E$. For each $w \in C(J, \mathbb{R}^n)$ with $w(0, 0) = 0$, we denote by $\overline{w}$ the function defined by

$$\overline{w}(x, y) = \begin{cases} 0, & (x, y) \in \tilde{J}', \\ w(x, y) & (x, y) \in J. \end{cases}$$

If $u(.,.)$ satisfies the integral equation

$$u(x, y) = \mu(x, y) + \frac{1}{\Gamma(r_1)\Gamma(r_2)} \int_0^x \int_0^y (x - s)^{r_1 - 1}(y - t)^{r_2 - 1} f(s, t) \mathrm{d}t \mathrm{d}s,$$

we can decompose $u(.,.)$ as $u(x, y) = \overline{w}(x, y) + v(x, y);\ (x, y) \in J$, which implies $u_{(x,y)} = \overline{w}_{(x,y)} + v_{(x,y)}$, for every $(x, y) \in J$, and the function $w(.,.)$ satisfies

$$w(x,y) = \frac{1}{\Gamma(r_1)\Gamma(r_2)} \int_0^x \int_0^y (x-s)^{r_1-1}(y-t)^{r_2-1} f(s,t)\mathrm{d}t\mathrm{d}s,$$

where $f \in S_{F,\overline{w}_{(s,t)}+v_{(s,t)}}$. Set

$$C_0 = \{w \in C(J,\mathbb{R}^n) : w(x,y) = 0 \text{ for } (x,y) \in E\},$$

and let $\|.\|_{(a,b)}$ be the seminorm in C_0 defined by

$$\|w\|_{(a,b)} = \sup_{(x,y)\in E} \|w_{(x,y)}\|_{\mathcal{B}} + \sup_{(x,y)\in J} \|w(x,y)\| = \sup_{(x,y)\in J} \|w(x,y)\|, \ w \in C_0.$$

C_0 is a Banach space with norm $\|.\|_{(a,b)}$. Let the operator $P : C_0 \to \mathcal{P}(C_0)$ be defined by

$$(Pw)(x,y) = \{h \in C_0\},$$

such that

$$h(x,y) = \frac{1}{\Gamma(r_1)\Gamma(r_2)} \int_0^x \int_0^y (x-s)^{r_1-1}(y-t)^{r_2-1} f(s,t)\mathrm{d}t\mathrm{d}s, \ (x,y) \in J,$$

where $f \in S_{F,\overline{w}_{(s,t)}+v_{(s,t)}}$. Obviously, that the operator N has a fixed point is equivalent to P has a fixed point .

Step 1: *$P(u)$ is convex for each $u \in C_0$.* Indeed, if h_1, h_2 belong to $P(u)$, then there exist $f_1, f_2 \in S_{F,\overline{w}_{(s,t)}+v_{(s,t)}}$ such that for each $(x,y) \in J$ we have

$$h_i(x,y) = \frac{1}{\Gamma(r_1)\Gamma(r_2)} \int_0^x \int_0^y (x-s)^{r_1-1}(y-t)^{r_2-1} f_i(s,t)\mathrm{d}t\mathrm{d}s, \ i = 1,2.$$

Let $0 \le \xi \le 1$. Then, for each $(x,y) \in J$, we have

$$(\xi h_1 + (1-\xi)h_2)(x,y) = \frac{1}{\Gamma(r_1)\Gamma(r_2)} \int_0^x \int_0^y (x-s)^{r_1-1}(y-t)^{r_2-1} \times[\xi f_1(s,t) + (1-\xi) f_2(s,t)]\mathrm{d}t\mathrm{d}s.$$

Since $\tilde{S}_{F,\overline{w}_{(s,t)}+v_{(s,t)}}$ is convex (because F has convex values), we have

$$\xi h_1 + (1-\xi)h_2 \in P(u).$$

Step 2: *P maps bounded sets into bounded sets in C_0.* Indeed, it is enough to show that there exists a positive constant ℓ such that, for each $z \in B_\rho = \{u \in C_0 : \|z\| \leq \rho\}$, one has $\|P(z)\| \leq \ell$. Let $z \in B_\rho$ and $h \in P(z)$. Then there exists $f \in S_{F,\overline{w}_{(s,t)}+v_{(s,t)}}$, such that, for each $(x, y) \in J$, we have

$$h(x, y) = \frac{1}{\Gamma(r_1)\Gamma(r_2)} \int_0^x \int_0^y (x - s)^{r_1-1}(y - t)^{r_2-1} f(s, t) \mathrm{d}t \mathrm{d}s.$$

Then, for each $(x, y) \in J$,

$$\begin{aligned}
\|h(x, y)\| &\leq \frac{1}{\Gamma(r_1)\Gamma(r_2)} \int_0^x \int_0^y (x - s)^{r_1-1}(y - t)^{r_2-1} \|f(s, t)\| \mathrm{d}t \mathrm{d}s \\
&\leq \frac{1}{\Gamma(r_1)\Gamma(r_2)} \int_0^x \int_0^y (x - s)^{r_1-1}(y - t)^{r_2-1} l(s, t) \\
&\quad \times (1 + \|\overline{w}_{(s,t)} + v_{(s,t)}\|_B) \mathrm{d}t \mathrm{d}s \\
&\leq \frac{l^*(1 + \rho^*)}{\Gamma(r_1)\Gamma(r_2)} \int_0^x \int_0^y (x - s)^{r_1-1}(y - t)^{r_2-1} \mathrm{d}t \mathrm{d}s \\
&\leq \frac{a^{r_1} b^{r_2} l^*(1 + \rho^*)}{\Gamma(r_1 + 1)\Gamma(r_2 + 1)},
\end{aligned}$$

where

$$\begin{aligned}
\|\overline{w}_{(s,t)} + v_{(s,t)}\|_{\mathcal{B}} &\leq \|\overline{w}_{(s,t)}\|_{\mathcal{B}} + \|v_{(s,t)}\|_{\mathcal{B}} \\
&\leq K\rho + M\|\phi\|_{\mathcal{B}} = \rho^*.
\end{aligned}$$

Step 3: *$P(B_\rho)$ is equicontinuous.* Let $P(B_\rho)$ as in Step 2 and let $(x_1, y_1), (x_2, y_2) \in J$, $x_1 < x_2$ and $y_1 < y_2$, let $u \in B_\rho$ and $h \in P(u)$, then there exists $f \in S_{F,\overline{w}_{(s,t)}+v_{(s,t)}}$ such that for each $(x, y) \in J$ we have

$$\begin{aligned}
&\|h(x_2, y_2) - h(x_1, y_1)\| \\
&= \Bigg\| \frac{1}{\Gamma(r_1)\Gamma(r_2)} \int_0^{x_1} \int_0^{y_1} [(x_2 - s)^{r_1-1}(y_2 - t)^{r_2-1} - (x_1 - s)^{r_1-1}(y_1 - t)^{r_2-1}] \\
&\quad \times f(s, t) \mathrm{d}t \mathrm{d}s + \frac{1}{\Gamma(r_1)\Gamma(r_2)} \int_{x_1}^{x_2} \int_{y_1}^{y_2} (x_2 - s)^{r_1-1}(y_2 - t)^{r_2-1} f(s, t) \mathrm{d}t \mathrm{d}s
\end{aligned}$$

$$+\frac{1}{\Gamma(r_1)\Gamma(r_2)}\int_0^{x_1}\int_{y_1}^{y_2}(x_2-s)^{r_1-1}(y_2-t)^{r_2-1}f(s,t)\mathrm{d}t\mathrm{d}s$$

$$+\frac{1}{\Gamma(r_1)\Gamma(r_2)}\int_{x_1}^{x_2}\int_0^{y_1}(x_2-s)^{r_1-1}(y_2-t)^{r_2-1}f(s,t)\mathrm{d}t\mathrm{d}s\Bigg\|$$

$$\leq \frac{l^*(1+\rho^*)}{\Gamma(r_1+1)\Gamma(r_2+1)}[2y_2^{r_2}(x_2-x_1)^{r_1}+2x_2^{r_1}(y_2-y_1)^{r_2}$$

$$+x_1^{r_1}y_1^{r_2}-x_2^{r_1}y_2^{r_2}-2(x_2-x_1)^{r_1}(y_2-y_1)^{r_2}].$$

As $x_1 \longrightarrow x_2$ and $y_1 \longrightarrow y_2$, the right-hand side of the above inequality tends to zero. As a consequence of Steps 1–3 together with the Arzelá–Ascoli theorem, we can conclude that $N : C_0 \longrightarrow \mathcal{P}_{cp}(C_0)$ is completely continuous.

Step 4: *P has a closed graph.* Let $u_n \to u_*$, $h_n \in P(u_n)$, and $h_n \to h_*$. We need to show that $h_* \in P(u_*)$. $h_n \in P(u_n)$ means that there exists $f_n \in S_{F,\overline{w}_{(s,t)}+v_{(s,t)}}$ such that, for each $(x,y) \in J$,

$$h_n(x,y) = \frac{1}{\Gamma(r_1)\Gamma(r_2)}\int_0^x\int_0^y(x-s)^{r_1-1}(y-t)^{r_2-1}f_n(s,t)\mathrm{d}t\mathrm{d}s.$$

We must show that there exists $f_* \in \tilde{S}_{F,\overline{w}_{(s,t)}+v_{(s,t)})}$ such that, for each $(x,y) \in J$,

$$h_*(x,y) = \frac{1}{\Gamma(r_1)\Gamma(r_2)}\int_0^x\int_0^y(x-s)^{r_1-1}(y-t)^{r_2-1}f_*(s,t)\mathrm{d}t\mathrm{d}s.$$

Since $F(x,y,\cdot)$ is upper semicontinuous, then for every $\varepsilon > 0$, there exist $n_0(\epsilon) \geq 0$ such that for every $n \geq n_0$, we have

$$f_n(x,y) \in F(x,y,\overline{w}_n(x,y)+v_{(x,y)}) \subset F(x,y,\overline{w}*_{(x,y)})+\varepsilon B(0,1), \text{ a.e. } (x,y) \in J.$$

Since $F(.,.,.)$ has compact values, then there exists a subsequence f_{n_m} such that

$$f_{n_m}(\cdot,\cdot) \to f_*(\cdot,\cdot) \text{ as } m \to \infty$$

and

$$f_*(x,y) \in F(x,y,\overline{w}_*(x,y)+v_{(x,y)}), \text{ a.e. } (x,y) \in J.$$

Then for every $w \in F(x,y,\overline{w}(x,y)+v_{(x,y)})$, we have

$$\|f_{n_m}(x,y)-f_*(x,y)\| \leq \|f_{n_m}(x,y)-w\| + \|w-f_*(x,y)\|.$$

Then

$$\| f_{n_m}(x,y) - f_*(x,y)\| \le d\Big(f_{n_m}(x,y), F(x,y,\overline{w}_*(x,y) + v_{(x,y)})\Big).$$

By an analogous relation, obtained by interchanging the roles of f_{n_m} and f_*, it follows that

$$\begin{aligned}\| f_{n_m}(x,y) - f_*(x,y)\| &\le H_d\Big(F(x,y,\overline{w}_n(x,y)+v_{(x,y)}), F(x,y,\overline{w}_*(x,y)+v_{(x,y)})\Big)\\ &\le l(x,y)\|\overline{w}_n - \overline{w}_*\|_{\mathcal{B}}.\end{aligned}$$

Then

$$\begin{aligned}\|h_{n_m}(x,y) - h_*(x,y)\| &\le \frac{1}{\Gamma(r_1)\Gamma(r_2)}\int_0^x\int_0^y (x-s)^{r_1-1}(y-t)^{r_2-1}\|\overline{w}_{n_m}(s,t)\\ &\quad -\overline{w}_*(s,t)\|\mathrm{d}t\mathrm{d}s\\ &\le \frac{\|\overline{w}_{n_m} - \overline{w}_*\|_{(a,b)}}{\Gamma(r_1)\Gamma(r_2)}\int_0^x\int_0^y (x-s)^{r_1-1}(y-t)^{r_2-1} l(s,t)\mathrm{d}t\mathrm{d}s\\ &\le \frac{a^{r_1}b^{r_2}l^*}{\Gamma(r_1+1)\Gamma(r_2+1)}\|\overline{w}_{n_m} - \overline{w}_*\|_{(a,b)}.\end{aligned}$$

Hence

$$\|h_{n_m} - h_*\|_{(a,b)} \le \frac{a^{r_1}b^{r_2}l^*}{\Gamma(r_1+1)\Gamma(r_2+1)}\|\overline{w}_{n_m} - \overline{w}_*\|_{(a,b)} \to 0 \text{ as } m \to \infty.$$

Step 5: (A priori bounds). We now show there exists an open set $U \subseteq C_0$ with $w \in \lambda P(w)$, for $\lambda \in (0,1)$ and $w \in \partial U$. Let $w \in C_0$ and $w \in \lambda P(w)$ for some $0 < \lambda < 1$. Thus there exists $f \in S_{F,\overline{w}_{(s,t)}+v_{(s,t)}}$ such that, for each $(x,y) \in J$,

$$w(x,y) = \frac{\lambda}{\Gamma(r_1)\Gamma(r_2)}\int_0^x\int_0^y (x-s)^{r_1-1}(y-t)^{r_2-1} f(s,t)\mathrm{d}t\mathrm{d}s.$$

This implies by (4.25.2) that, for each $(x,y) \in J$, we have

$$\begin{aligned}\|w(x,y)\| \le &\frac{1}{\Gamma(r_1)\Gamma(r_2)}\int_0^x\int_0^y (x-s)^{r_1-1}(y-t)^{r_2-1}\\ &\times l(s,t)(1 + \|\overline{w}_{(s,t)} + v_{(s,t)}\|_{\mathcal{B}})\mathrm{d}t\mathrm{d}s\end{aligned}$$

$$\leq \frac{l^* a^{r_1} b^{r_2}}{\Gamma(r_1+1)\Gamma(r_2+1)} + \frac{l^*}{\Gamma(r_1)\Gamma(r_2)} \int_0^x \int_0^y (x-s)^{r_1-1}(y-t)^{r_2-1} \|\overline{w}_{(s,t)} + v_{(s,t)}\|_{\mathcal{B}} dt ds.$$

But

$$\begin{aligned} \|\overline{w}_{(s,t)} + v_{(s,t)}\|_{\mathcal{B}} &\leq \|\overline{w}_{(s,t)}\|_{\mathcal{B}} + \|v_{(s,t)}\|_{\mathcal{B}} \\ &\leq K \sup\{w(\tilde{s},\tilde{t}) : (\tilde{s},\tilde{t}) \in [0,s]\times[0,t]\} + M\|\phi\|_{\mathcal{B}}. \end{aligned} \quad (4.47)$$

If we name $z(s,t)$ the right-hand side of (4.47), then we have

$$\|\overline{w}_{(s,t)} + v_{(s,t)}\|_{\mathcal{B}} \leq z(x,y),$$

and therefore, for each $(x,y) \in J$, we obtain

$$\|w(x,y)\| \leq \frac{l^* a^{r_1} b^{r_2}}{\Gamma(r_1+1)\Gamma(r_2+1)} + \frac{l^*}{\Gamma(r_1)\Gamma(r_2)} \int_0^x \int_0^y (x-s)^{r_1-1}(y-t)^{r_2-1} z(s,t) dt ds. \quad (4.48)$$

Using the above inequality and the definition of z we have that

$$z(x,y) \leq M\|\phi\|_{\mathcal{B}} + \frac{K l^* a^{r_1} b^{r_2}}{\Gamma(r_1+1)\Gamma(r_2+1)} + \frac{K l^*}{\Gamma(r_1)\Gamma(r_2)} \int_0^x \int_0^y (x-s)^{r_1-1}(y-t)^{r_2-1} z(s,t) dt ds,$$

for each $(x,y) \in J$. Then, Lemma 2.43 implies there exists $\delta = \delta(r_1, r_2)$

$$\|z(x,y)\| \leq R + \delta \frac{R K l^*}{\Gamma(r_1)\Gamma(r_2)} \int_0^x \int_0^y (x-s)^{r_1-1}(y-t)^{r_2-1} dt ds,$$

where

$$R = M\|\phi\|_{\mathcal{B}} + \frac{K l^* a^{r_1} b^{r_2}}{\Gamma(r_1+1)\Gamma(r_2+1)}.$$

Hence

$$\|z\|_\infty \leq R + \frac{R\delta K l^* a^{r_1} b^{r_2}}{\Gamma(r_1+1)\Gamma(r_2+1)} := \widetilde{M}.$$

Then, (4.48) implies that

$$\|w\|_\infty \leq \frac{l^* a^{r_1} b^{r_2}}{\Gamma(r_1+1)\Gamma(r_2+1)}(1+\widetilde{M}) := M^*.$$

Set

$$U = \{w \in C_0 : \|w\|_{(a,b)} < M^* + 1\}.$$

$P : \overline{U} \to C_0$ is continuous and completely continuous. By Theorem 2.33 and our choice of U, there is no $w \in \partial U$ such that $w \in \lambda P(w)$, for $\lambda \in (0,1)$. As a consequence of the nonlinear alternative of Leray–Schauder type [136], we deduce that N has a fixed point which is a solution to problems (4.41)–(4.43). □

Now we present a similar existence result for the problems (4.44)–(4.46).

Definition 4.26. A function $u \in \Omega$ is said to be a solution of (4.44)–(4.46) if there exists $f \in F(x,y,u_{(x,y)})$ such that u satisfies the equations ${}^cD_0^r[u(x,y) - g(x,y,u_{(x,y)})] = f(x,y)$ and (4.46) on J and the condition (4.45) on $\tilde{J}'$.

Theorem 4.27. *Assume (4.25.1) and (4.25.2) and the following condition holds:*

(4.27.1) The function g is continuous and completely continuous, and for any bounded set B in Ω, the set $\{(x,y) \to g(x,y,u) : u \in B\}$, is equicontinuous in $C(J,\mathbb{R}^n)$, and there exist constants $d_1, d_2 \geq 0$ such that $0 \leq Kd_1 < \frac{1}{4}$ and

$$\|g(x,y,u)\| \leq d_1\|u\|_{\mathcal{B}} + d_2,\ (x,y) \in J,\ u \in \mathcal{B}.$$

Then the IVP (4.44)–(4.46) have at least one solution on $(-\infty,a] \times (-\infty,b]$.

Proof. Consider the operator $N_1 : \Omega \to \mathcal{P}(\Omega)$ defined by

$$(N_1u)(x,y) = \left\{ h \in \Omega : h(x,y) = \begin{cases} \phi(x,y), & (x,y) \in \tilde{J}', \\ \mu(x,y) + g(x,y,u_{(x,y)}) & \\ -g(x,0,u_{(x,0)}) - g(0,y,u_{(0,y)}) & \\ +g(0,0,u_{(0,0)}) & \\ +\frac{1}{\Gamma(r_1)\Gamma(r_2)}\int_0^x\int_0^y (x-s)^{r_1-1} & \\ (y-t)^{r_2-1} f(s,t)\,dt\,ds, & (x,y) \in J, \end{cases} \right\},$$

where $f \in S_{F,u}$. In analogy to Theorem 4.25, we consider the operator $P_1 : C_0 \to \mathcal{P}(C_0)$ defined by

$$(P_1u)(x,y) = \left\{ h \in \Omega : h(x,y) = \begin{cases} 0, & (x,y) \in \tilde{J}', \\ g(x,y,\overline{w}_{(x,y)} + v_{(x,y)}) & \\ -g(x,0,\overline{w}_{(x,0)} + v_{(x,0)}) & \\ -g(0,y,\overline{w}_{(0,y)} & \\ +v_{(0,y)}) + g(0,0,\overline{w}_{(0,0)} + v_{(0,0)}) & \\ +\frac{1}{\Gamma(r_1)\Gamma(r_2)}\int_0^x\int_0^y (x-s)^{r_1-1} & \\ (y-t)^{r_2-1} f(s,t)\,dt\,ds, & (x,y) \in J, \end{cases} \right\},$$

where $f \in S_{F,\overline{w}+v}$. We shall show that the operator P_1 is continuous and completely continuous. Using (4.27.1) it suffices to show that the operator $P_2 : C_0 \to \mathcal{P}(C_0)$ defined by

$$(P_2u)(x,y) = \left\{ h \in \Omega : h(x,y) = \begin{cases} 0, & (x,y) \in \tilde{J}', \\ \frac{1}{\Gamma(r_1)\Gamma(r_2)} \int_0^x \int_0^y (x-s)^{r_1-1} & \\ (y-t)^{r_2-1} f(s,t)\mathrm{d}t\mathrm{d}s, & (x,y) \in J, \end{cases} \right\},$$

is continuous and completely continuous. This was proved in Theorem 4.25. We now show that there exists an open set $U \subseteq C_0$ with $w \in \lambda P_1(w)$, for $\lambda \in (0,1)$ and $w \in \partial U$. Let $w \in C_0$ and $w \in \lambda P_1(w)$ for some $0 < \lambda < 1$. Thus for each $(x,y) \in J$,

$$\begin{aligned} w(x,y) &= \lambda[g(x,y,\overline{w}_{(x,y)} + v_{(x,y)}) - g(x,0,\overline{w}_{(x,0)} + v_{(x,0)}) \\ &\quad - g(0,y,\overline{w}_{(0,y)} + v_{(0,y)}) + g(0,0,\overline{w}_{(0,0)} + v_{(0,0)})] \\ &\quad + \frac{\lambda}{\Gamma(r_1)\Gamma(r_2)} \int_0^x \int_0^y (x-s)^{r_1-1}(y-t)^{r_2-1} f(s,t)\mathrm{d}t\mathrm{d}s, \end{aligned}$$

where $f \in F(x,y,\overline{w}+v)$. Then

$$\begin{aligned} \|w(x,y)\| &\le 4d_1\|\overline{w}_{(x,y)} + v_{(x,y)}\|_{\mathcal{B}} + \frac{l^* a^{r_1} b^{r_2}}{\Gamma(r_1+1)\Gamma(r_2+1)} \\ &\quad + \frac{1}{\Gamma(r_1)\Gamma(r_2)} \int_0^x \int_0^y (x-s)^{r_1-1}(y-t)^{r_2-1} l(s,t)\|\overline{w}_{(s,t)} + v_{(s,t)}\|_{\mathcal{B}}\mathrm{d}t\mathrm{d}s. \end{aligned}$$

Using the above inequality and the definition of z we have that

$$\|z\|_\infty \le R_1 + \frac{R_1 \delta(r_1,r_2) K l^{**} a^{r_1} b^{r_2}}{(1-4d_1K)\Gamma(r_1+1)\Gamma(r_2+1)} := L,$$

where

$$R_1 = \frac{1}{1-4d_1K}\left[8d_2K + \frac{K l^* a^{r_1} b^{r_2}}{\Gamma(r_1+1)\Gamma(r_2+1)}\right]$$

and

$$l^{**} = \frac{l^*}{1-4d_1K}.$$

Then

$$\|w\|_\infty \le 4d_1\|\phi\|_{\mathcal{B}} + 8d_2 + 4Ld_1 + \frac{a^{r_1} b^{r_2} l^*(1+L)}{\Gamma(r_1+1)\Gamma(r_2+1)} := L^*.$$

Set

$$U_1 = \{w \in C_0 : \|w\|_{(a,b)} < L^* + 1\}.$$

By Theorem 2.33 and our choice of U_1, there is no $w \in \partial U$ such that $w \in \lambda P_2(w)$, for $\lambda \in (0,1)$. As a consequence of the nonlinear alternative of Leray–Schauder type [136], we deduce that N_1 has a fixed point which is a solution to problems (4.44)–(4.46). □

4.5.3 *An Example*

As an application of our results we consider the following partial hyperbolic functional differential inclusion of the form:

$$({}^cD_0^r u)(x,y) \in F(x,y,u_{(x,y)}), \text{ if } (x,y) \in J := [0,1] \times [0,1], \tag{4.49}$$

$$u(x,y) = \phi(x,y),\ (x,y) \in (-\infty,1] \times (-\infty,1] \backslash (0,1] \times (0,1], \tag{4.50}$$

Let $\gamma \geq 0$ and consider the phase space $\mathcal{B}_\gamma$ defined in the example of Sect. 3.7. Set

$$F(x,y,u_{(x,y)}) = \{u \in \mathbb{R} : f_1(x,y,u_{(x,y)}) \leq u \leq f_2(x,y,u_{(x,y)})\},$$

where $f_1, f_2 : [0,1] \times [0,1] \times \mathcal{B}_\gamma \to \mathbb{R}$. We assume that for each $(x,y) \in J$, $f_1(x,y,.)$ is lower semicontinuous (i.e., the set $\{z \in \mathcal{B}_\gamma : f_1(x,y,z) > \nu\}$ is open for each $\nu \in \mathbb{R}$), and assume that for each $(x,y) \in J$, $f_2(x,y,.)$ is upper semicontinuous (i.e., the set $\{z \in \mathcal{B}_\gamma : f_2(x,y,z) < \nu\}$ is open for each $\nu \in \mathbb{R}$). Assume that there are $l \in L^\infty(J,\mathbb{R}_+)$ and $\Psi : [0,\infty) \to (0,\infty)$ continuous and nondecreasing such that

$$\max(|f_1(x,y,z)|, |f_2(x,y,z)|) \leq l(x,y)\Psi(|z|), \text{ for a.e. } (x,y) \in J \text{ and all } z \in \mathcal{B}_\gamma.$$

It is clear that F is compact and convex valued, and it is upper semi-continuous (see [104]). Since all the conditions of Theorem 4.25 are satisfied, problems (4.49) and (4.50) have at least one solution defined on $(-\infty,1] \times (-\infty,1]$.

4.6 Fractional Order Riemann–Liouville Integral Inclusions with two Independent Variables and Multiple Time Delay

4.6.1 *Introduction*

This section deals with the existence and uniqueness of solutions for the following system of fractional integral inclusions:

$$u(x,y) - \sum_{i=1}^{m} b_i(x,y)u(x-\xi_i, y-\mu_i) \in I_0^r F(x,y,u(x,y)); \text{ if } (x,y) \in J := [0,a] \times [0,b], \quad (4.51)$$

$$u(x,y) = \Phi(x,y); \quad \text{if } (x,y) \in \tilde{J} := [-\xi, a] \times [-\mu, b] \backslash (0,a] \times (0,b], \quad (4.52)$$

where $a, b > 0$, $\xi_i, \mu_i \geq 0$; $i = 1, \dots, m$, $\xi = \max_{i=1,\dots,m} \{\xi_i\}$, $\mu = \max_{i=1,\dots,m} \{\mu_i\}$, $F : J \times \mathbb{R}^n \to \mathcal{P}(\mathbb{R}^n)$ is a set-valued function with nonempty values in $\mathbb{R}^n$, $\mathcal{P}(\mathbb{R}^n)$ is the family of all nonempty subsets of $\mathbb{R}^n$, $I_0^r F(x,y,u(x,y))$ is the definite integral for the set-valued functions F of order $r = (r_1, r_2) \in (0,\infty) \times (0,\infty)$, $b_i : J \to \mathbb{R}^n$; $i = 1 \cdots m$, and $\Phi : \tilde{J} \to \mathbb{R}^n$ are given continuous functions such that

$$\Phi(x,0) = \sum_{i=1}^{m} b_i(x,0)\Phi(x-\xi_i, -\mu_i); \; x \in [0,a]$$

and

$$\Phi(0,y) = \sum_{i=1}^{m} b_i(0,y)\Phi(-\xi_i, y-\mu_i); \; y \in [0,b].$$

We establish the existence results for the problems (4.51) and (4.52) when the right-hand side is convex as well as when it is nonconvex valued. Our approach is based on appropriate fixed-point theorems, namely, Bohnenblust–Karlin fixed-point theorem for the convex case and Covitz–Nadler for the nonconvex case. This approach is now standard; however, its utilization is new in the framework of the considered integral inclusions.

4.6.2 Existence of Solutions

Let $C := C([-\xi, a] \times [-\mu, b], \mathbb{R}^n)$ be the Banach space of all continuous functions from $[-\xi, a] \times [-\mu, b]$ into $\mathbb{R}^n$ with the norm

$$\|w\|_C = \sup_{(x,y) \in [-\xi,a] \times [-\mu,b]} \|w(x,y)\|.$$

Let us start by defining what we mean by a solution of the problems (4.51) and (4.52).

Definition 4.28. A function $u \in C$ is said to be a solution of (4.51) and (4.52) if there exists $f \in S_{F,u}$ such that u satisfies the equation

$$u(x,y) = \sum_{i=1}^{m} b_i(x,y)u(x-x_i, y-y_i) + I_\theta^r f(x,y)$$

on J and condition (4.52) on $\tilde{J}$.

Lemma 4.29 ([180]). *Let X be a Banach space. Let $F : J \times X \longrightarrow \mathcal{P}_{cp,cv}(X)$ be an L^1-Carathéodory multivalued map and let Λ be a linear continuous mapping from $L^1(J, X)$ to $C(J, X)$, then the operator*

$$\begin{aligned} \Lambda \circ S_F : C(J, X) &\longrightarrow \mathcal{P}_{cp,cv}(C(J, X)), \\ u &\longmapsto (\Lambda \circ S_F)(u) := \Lambda(S_{F,u}) \end{aligned}$$

is a closed graph operator in $C(J, X) \times C(J, X)$.

Let $F : J \times \mathbb{R}^n \to \mathcal{P}(\mathbb{R}^n)$ be a set-valued function with nonempty values in $\mathbb{R}^n$. $I_0^r F(x, y, u(x, y))$ are the definite integral for the set-valued functions F of order $r = (r_1, r_2) \in (0, \infty) \times (0, \infty)$ which is defined as

$$\begin{aligned} &I_0^r F(x, y, u(x, y)) \\ &= \left\{ \frac{1}{\Gamma(r_1)\Gamma(r_2)} \int_0^x \int_0^y (x-s)^{r_1-1}(y-t)^{r_2-1} f(s,t)\,dt\,ds : f(x, y) \in S_{F,u} \right\}. \end{aligned}$$

Set

$$B = \max_{i=1\cdots m} \left\{ \sup_{(x,y)\in J} \|b_i(x, y)\| \right\}.$$

Theorem 4.30 (Convex case). *Assume*

(4.30.1) The multifunction F is Carathéodory,
(4.30.2) There exist positive functions $h, l \in L^\infty(J)$ such that

$$\|F(x, y, u)\|_{\mathcal{P}} \le h(x, y) + l(x, y)\|u\|, \text{ a.e. } (x, y) \in J \text{ for all } u \in \mathbb{R}^n.$$

If

$$mB + \frac{a^{r_1} b^{r_2} l^*}{\Gamma(1 + r_1)\Gamma(1 + r_2)} < 1,$$

where $l^* = \|l\|_{L^\infty}$, then the problems (4.51) and (4.52) have at least one solution u on $[-\xi, a] \times [-\mu, b]$.

Proof. Transform the problems (4.51) and (4.52) into a fixed-point problem. Consider the multivalued operator $N : C \to \mathcal{P}(C)$ defined by

$$N(u)(x,y) = \left\{ g \in C : g(x,y) = \begin{cases} \Phi(x, y); & (x, y) \in \tilde{J}, \\ \sum_{i=1}^{m} b_i(x, y)u(x - \xi_i, y - \mu_i) & \\ +I_\theta^r f(x, y);\ f \in S_{F,u}; & (x, y) \in J. \end{cases} \right\} \tag{4.53}$$

Clearly, the fixed points of N are solutions to (4.51) and (4.52). Let

$$B_R = \{u \in C : \|u\|_C \leq R\}$$

be a closed bounded and convex subset of C, where

$$R = \text{Max}||\Phi||, \frac{a^{r_1} b^{r_2} h^*}{(1 - mB)\Gamma(1 + r_1)\Gamma(1 + r_2) - a^{r_1} b^{r_2} l^*}$$

and

$$h^* = \|h\|_{L^\infty}.$$

We shall show that N satisfies the assumptions of Lemma 2.38. The proof will be given by several steps.

Step 1: $N(B_R) \subset B_R$. Let $u \in B_R$. We must show that $N(u) \in B_R$. For each $g \in N(u)$, there exists $f \in S_{F,u}$ such that, for each $(x, y) \in J$, we have

$$\begin{aligned}
\|g(x,y)\| &\leq \sum_{i=1}^{m} \|b_i(x,y)\| \|u(x - \xi_i, y - \mu_i)\| \\
&\quad + \frac{1}{\Gamma(r_1)\Gamma(r_2)} \int_0^x \int_0^y (x-s)^{r_1-1}(y-t)^{r_2-1} \|f(s,t)\| \mathrm{d}t\mathrm{d}s \\
&\leq mB\|u\| + \frac{1}{\Gamma(r_1)\Gamma(r_2)} \int_0^x \int_0^y (x-s)^{r_1-1}(y-t)^{r_2-1} (h(s,t) \\
&\quad + l(s,t)\|u\|)\mathrm{d}t\mathrm{d}s \\
&\leq mBR + \frac{a^{r_1} b^{r_2}}{\Gamma(1+r_1)\Gamma(1+r_2)} (h^* + l^* R) = R,
\end{aligned}$$

and, for all $(x, y) \in \tilde{J}$ and $g(x, y) \in N(u)$, we have

$$\|g(x,y)\| = \|\Phi(x,y)\| \leq R.$$

Thus, we get for all $(x, y) \in [-\xi, a] \times [-\mu, b]$ and $g(x, y) \in N(u)$, we have

$$\|g(x,y)\| \leq R.$$

Step 2: $N(B_R)$ *is a relatively compact set.* We must show that N is a compact operator. Since B_R is a bounded closed and convex set and $N(B_R) \subset B_R$, it follows that $N(B_R)$ is a bounded closed and convex set. Moreover, for $0 \leq x_1 \leq x_2 \leq a$

and $0 \leq y_1 \leq y_2 \leq b$ and $u \in B_R$, then for each $g \in N(u)$, there exists $f \in S_{F,u}$ such that, for each $(x, y) \in J$, we have

$$\begin{aligned}
&\|g(x_1, y_1) - g(x_2, y_2)\| \\
&\leq \sum_{i=1}^{m} \|b_i(x_1, y_1)u(x_1 - \xi_i, y_1 - \mu_i) - b_i(x_2, y_2)u(x_2 - \xi_i, y_2 - \mu_i)\| \\
&\quad + \frac{1}{\Gamma(r_1)\Gamma(r_2)} \int_0^{x_1}\int_0^{y_1} [(x_2 - s)^{r_1-1}(y_2 - t)^{r_2-1} \\
&\quad -(x_1 - s)^{r_1-1}(y_1 - t)^{r_2-1}]\|f(s,t)\|dt ds \\
&\quad + \frac{1}{\Gamma(r_1)\Gamma(r_2)} \int_{x_1}^{x_2}\int_{y_1}^{y_2} (x_2 - s)^{r_1-1}(y_2 - t)^{r_2-1}\|f(s,t)\|dt ds \\
&\quad + \frac{1}{\Gamma(r_1)\Gamma(r_2)} \int_0^{x_1}\int_{y_1}^{y_2} (x_2 - s)^{r_1-1}(y_2 - t)^{r_2-1}\|f(s,t)\|dt ds \\
&\quad + \frac{1}{\Gamma(r_1)\Gamma(r_2)} \int_{x_1}^{x_2}\int_0^{y_1} (x_2 - s)^{r_1-1}(y_2 - t)^{r_2-1}\|f(s,t)\|dt ds \\
&\leq B\sum_{i=1}^{m} \|u(x_1 - \xi_i, y_1 - \mu_i) - u(x_2 - \xi_i, y_2 - \mu_i)\| \\
&\quad + \frac{h^*}{\Gamma(r_1)\Gamma(r_2)}[2y_2^{r_2}(x_2 - x_1)^{r_1} + 2x_2^{r_1}(y_2 - y_1)^{r_2} \\
&\quad + x_1^{r_1}y_1^{r_2} - x_2^{r_1}y_2^{r_2} - 2(x_2 - x_1)^{r_1}(y_2 - y_1)^{r_2}].
\end{aligned}$$

As $x_1 \longrightarrow x_2$ and $y_1 \longrightarrow y_2$, the right-hand side of the above inequality tends to zero. The equicontinuity for the cases $x_1 < x_2 < 0$, $y_1 < y_2 < 0$ and $x_1 \leq 0 \leq x_2$, $y_1 \leq 0 \leq y_2$ is obvious. An application of Arzela–Ascoli theorem yields that N maps B_R into C a compact set in C, i.e., $N : B_R \to \mathcal{P}(C)$ is a compact operator. Thus $N(B_R)$ is relatively compact.

Step 3: *N is upper semicontinuous on B_R.* Let $u_n \to u_*$, $h_n \in N(u_n)$ and $h_n \to h_*$. We need to show that $h_* \in N(u_*)$.

$h_n \in N(u_n)$ means that there exists $f_n \in S_{F,u_n}$ such that

$$\begin{cases} h_n(x,y) = \Phi(x,y) & (x,y) \in \tilde{J}, \\ h_n(x,y) = \sum_{i=1}^{m} b_i(x,y)u_n(x-\xi_i, y-\mu_i) & \\ +\frac{1}{\Gamma(r_1)\Gamma(r_2)}\int_0^x\int_0^y (x-s)^{r_1-1}(y-t)^{r_2-1} f_n(s,t)\mathrm{d}t\mathrm{d}s & (x,y) \in J. \end{cases}$$

We must show that there exists $f_* \in S_{F,u_*}$ such that

$$\begin{cases} h_*(x,y) = \Phi(x,y) & (x,y) \in \tilde{J}, \\ h_*(x,y) = \sum_{i=1}^{m} b_i(x,y)u_*(x-\xi_i, y-\mu_i) & \\ +\frac{1}{\Gamma(r_1)\Gamma(r_2)}\int_0^x\int_0^y (x-s)^{r_1-1}(y-t)^{r_2-1} f_*(s,t)\mathrm{d}t\mathrm{d}s & (x,y) \in J. \end{cases}$$

Now, we consider the linear continuous operator

$$\begin{aligned} \Lambda : L^1([-\xi,a]\times[-\mu,b]) &\longrightarrow C, \\ f &\longmapsto \Lambda(f)(x,y) \end{aligned}$$

such that

$$\begin{cases} \Lambda(f)(x,y) = \Phi(x,y) & (x,y) \in \tilde{J}, \\ \Lambda(f)(x,y) = \sum_{i=1}^{m} b_i(x,y)u_*(x-\xi_i, y-\mu_i) & \\ +\frac{1}{\Gamma(r_1)\Gamma(r_2)}\int_0^x\int_0^y (x-s)^{r_1-1}(y-t)^{r_2-1} f(s,t)\mathrm{d}t\mathrm{d}s & (x,y) \in J. \end{cases}$$

From Lemma 4.29, it follows that $\Lambda \circ S_F$ is a closed graph operator. Clearly, for each $(x,y) \in J$, we have

$$\left\| \left[h_n(x,y) - \sum_{i=1}^{m} b_i(x,y)u_n(x-\xi_i, y-\mu_i) \right] - \left[h_*(x,y) - \sum_{i=1}^{m} b_i(x,y)u_*(x-\xi_i, y-\mu_i) \right] \right\| \to 0 \text{ as } n \to \infty.$$

Moreover, from the definition of Λ, we have

$$\left[h_n(x,y) - \sum_{i=1}^{m} b_i(x,y)u_n(x-\xi_i, y-\mu_i) \right] \in \Lambda(S_{F,u_n}).$$

Since $u_n \to u_*$, it follows from Lemma 4.29 that, for some $f_* \in \Lambda(S_{F,u_*})$, we have

$$h_*(x,y) - \sum_{i=1}^{m} b_i(x,y)u_*(x-\xi_i, y-\mu_i)$$
$$= \frac{1}{\Gamma(r_1)\Gamma(r_2)}\int_0^x\int_0^y (x-s)^{r_1-1}(y-t)^{r_2-1} f_*(s,t)\mathrm{d}t\mathrm{d}s.$$

From Lemma 2.21, we can conclude that N is u.s.c.

Step 4: *N has convex values.* Let $u \in C$ and $g_1, g_2 \in N(u)$, then there exist $f_1, f_2 \in S_{F,u}$ such that

$$g_k(x,y) = \sum_{i=1}^{m} b_i(x,y)u(x-\xi_i, y-\mu_i) + I_\theta^r f_k(x,y);\ k = 1,2.$$

Let $0 \le \zeta \le 1$, then for each $(x,y) \in J$, we have

$$\begin{aligned}[\zeta g_1 + (1-\zeta)g_2](x,y) &= \sum_{i=1}^{m} b_i(x,y)[\zeta u(x-\xi_i, y-\mu_i) \\ &\quad +(1-\zeta)u(x-\xi_i, y-\mu_i)] \\ &\quad +I_\theta^r[\zeta f_1 + (1-\zeta)f_2](x,y) \\ &= \sum_{i=1}^{m} b_i(x,y)u(x-\xi_i, y-\mu_i) + I_\theta^r[\zeta f_1 + (1-\zeta)f_2](x,y),\end{aligned}$$

and for each $(x,y) \in \tilde{J}$, we have $[\zeta g_1 + (1-\zeta)g_2](x,y) = \Phi(x,y)$.

The convexity of $S_{F,u}$ and $F(x,y,u)$ implies that $[\zeta g_1 + (1-\zeta)g_2] \in N(u)$. Hence $N(u)$ is convex for each $u \in C$. As a consequence of Lemma 2.38, we deduce that N has a fixed point which is a solution for the problems (4.51) and (4.52). □

Theorem 4.31 (Non-convex case). *Assume*

(4.31.1) The multifunction $F : J \times \mathbb{R}^n \to \mathcal{P}_{cp}(\mathbb{R}^n)$ has the property that

$$F(\cdot,\cdot,u) : J \to \mathcal{P}_{cp}(\mathbb{R}^n) \text{ is measurable for each } u \in \mathbb{R}^n,$$

(4.31.2) There exists a positive function $m \in L^\infty(J)$ such that

$$H_d(F(x,y,u), F(x,y,v)) \le m(x,y)\|u-v\| \text{ for every } u, v \in \mathbb{R}^n,$$

and

$$d(0, F(x,y,0)) \le m(x,y), \text{ a.e. } (x,y) \in J.$$

If

$$mB + \frac{m^* a^{r_1} b^{r_2}}{\Gamma(1+r_1)\Gamma(1+r_2)} < 1, \tag{4.54}$$

where $m^* = \|m\|_{L^\infty}$, *then problems (4.51) and (4.52) have at least one solution on* $[-\xi, a] \times [-\mu, b]$.

Proof. For each $u \in C$ the set $S_{F,u}$ is nonempty since by (4.31.1), F has a nonempty measurable selection (see [93], Theorem III.6). We shall show that N defined in Theorem 4.30 satisfies the assumptions of Lemma 2.39. The proof will be given in two steps.

Step 1: $N(u) \in \mathcal{P}_{cl}(C)$ *for each* $u \in C$. Indeed, let $(g_n)_{n\geq 0} \in N(u)$ such that $g_n \longrightarrow g$. Then, $g \in C$ and there exists $f_n \in S_{F,u}$ such that, for each $(x,y) \in J$,

$$\begin{cases} g_n(x,y) = \Phi(x,y) & (x,y) \in \tilde{J}, \\ g_n(x,y) = \sum\limits_{i=1}^{m} b_i(x,y)u(x-\xi_i, y-\mu_i) & \\ +\frac{1}{\Gamma(r_1)\Gamma(r_2)} \int\limits_0^x \int\limits_0^y (x-s)^{r_1-1}(y-t)^{r_2-1} f_n(s,t)\mathrm{d}t\mathrm{d}s & (x,y) \in J. \end{cases}$$

Using the fact that F has compact values and from (4.31.2), we may pass to a subsequence if necessary to get that $f_n(.,.)$ converges weakly to f in $L^1_w(J)$ (the space endowed with the weak topology). An application of a standard argument shows that $f_n(.,.)$ converges strongly to f and hence $f \in S_{F,u}$. Then, for each $(x,y) \in [-\xi, a] \times [-\mu, b]$, $g_n(x,y) \to g(x,y)$, where

$$\begin{cases} g(x,y) = \Phi(x,y) & (x,y) \in \tilde{J}, \\ g(x,y) = \sum\limits_{i=1}^{m} b_i(x,y)u(x-\xi_i, y-\mu_i) & \\ +\frac{1}{\Gamma(r_1)\Gamma(r_2)} \int\limits_0^x \int\limits_0^y (x-s)^{r_1-1}(y-t)^{r_2-1} f(s,t)\mathrm{d}t\mathrm{d}s & (x,y) \in J. \end{cases}$$

So, $g \in N(u)$.

Step 2: *There exists* $\gamma < 1$ *such that* $H_d(N(u), N(v)) \leq \gamma \|u - v\|_C$ *for each* $u, v \in C$. Let $u, v \in C$ and $g \in N(u)$. Then, there exists $f \in S_{F,u}$ such that

$$\begin{cases} g(x,y) = \Phi(x,y) & (x,y) \in \tilde{J}, \\ g(x,y) = \sum\limits_{i=1}^{m} b_i(x,y)u(x-\xi_i, y-\mu_i) & \\ +\frac{1}{\Gamma(r_1)\Gamma(r_2)} \int\limits_0^x \int\limits_0^y (x-s)^{r_1-1}(y-t)^{r_2-1} f(s,t)\mathrm{d}t\mathrm{d}s & (x,y) \in J. \end{cases}$$

From (4.31.2) it follows that

$$H_d(F(x,y,u(x,y)),F(x,y,v(x,y))) \le m(x,y)\|u(x,y)-v(x,y)\|.$$

Hence, there exists $w \in S_{F,u}$ such that

$$\|f(x,y)-w(x,y)\| \le m(x,y)\|u(x,y)-v(x,y)\|;\ (x,y)\in J.$$

Consider $U : J \to \mathcal{P}(\mathbb{R}^n)$ given by

$$U(x,y)=\{w\in C : \|f(x,y)-w(x,y)\| \le m(x,y)\|u(x,y)-v(x,y)\|\}.$$

Since the multivalued operator $u(x,y)=U(x,y)\cap F(x,y,v(x,y))$ is measurable (see Proposition III.4 in [93]), there exists a function $u_2(x,y)$ which is a measurable selection for u. So, $\overline{f}(x,y)\in S_{F,v}$, and for each $(x,y)\in J$,

$$\|f(x,y)-\overline{f}(x,y)\| \le m(x,y)\|u(x,y)-v(x,y)\|.$$

Let us define

$$\begin{cases} \overline{g}(x,y)=\Phi(x,y) & (x,y)\in \tilde{J},\\ \overline{g}(x,y)=\sum\limits_{i=1}^{m} b_i(x,y)v(x-\xi_i,y-\mu_i) & \\ +\frac{1}{\Gamma(r_1)\Gamma(r_2)}\int\limits_0^x\int\limits_0^y (x-s)^{r_1-1}(y-t)^{r_2-1}\overline{f}(s,t)\mathrm{d}t\mathrm{d}s & (x,y)\in J. \end{cases}$$

Then, for each $(x,y)\in[-\xi,a]\times[-\mu,b]$, we get

$$\begin{aligned} \|g(x,y)-\overline{g}(x,y)\| &\le \sum_{i=1}^{m}\|b_i(x,y)\|\|u(x-\xi_i,y-\mu_i)-v(x-\xi_i,y-\mu_i)\|\\ &\quad+\frac{1}{\Gamma(r_1)\Gamma(r_2)}\int\limits_0^x\int\limits_0^y (x-s)^{r_1-1}(y-t)^{r_2-1}\\ &\quad\times\|f(s,t)-\overline{f}(s,t)\|\mathrm{d}t\mathrm{d}s\\ &\le mB\|u-v\|_C\\ &\quad+\frac{1}{\Gamma(r_1)\Gamma(r_2)}\int\limits_0^x\int\limits_0^y (x-s)^{r_1-1}(y-t)^{r_2-1}\\ &\quad\times m(s,t)\|u-v\|_C\mathrm{d}t\mathrm{d}s\\ &\le mB\|u-v\|_C+m^*\frac{a^{r_1}b^{r_2}}{\Gamma(1+r_1)\Gamma(1+r_2)}\|u-v\|_C\\ &=\left(mB+\frac{m^*a^{r_1}b^{r_2}}{\Gamma(1+r_1)\Gamma(1+r_2)}\right)\|u-v\|_C. \end{aligned}$$

Thus, for each $(x, y) \in [-\xi, a] \times [-\mu, b]$, we get

$$\|g - \overline{g}\|_C \le \left(mB + \frac{m^* a^{r_1} b^{r_2}}{\Gamma(1 + r_1)\Gamma(1 + r_2)}\right)\|u - v\|_C.$$

By an analogous relation, obtained by interchanging the roles of u and v, it follows that

$$H_d(N(u), N(v)) \le \left(mB + \frac{m^* a^{r_1} b^{r_2}}{\Gamma(1 + r_1)\Gamma(1 + r_2)}\right)\|u - v\|_C.$$

So by (4.54), N is a contraction and thus, by Lemma 2.39, N has a fixed point u which is solution to (4.51) and (4.52) on $[-\xi, a] \times [-\mu, b]$. □

4.6.3 An Example

As an application of our results we consider the following system of fractional integral inclusions of the form:

$$u(x, y) - \frac{x^3 y}{8} u(x - 1, y - 3) + \frac{x^4 y^2}{12} u\left(x - 2, y - \frac{1}{4}\right)$$

$$+ \frac{1}{14} u\left(x - \frac{3}{2}, y - 2\right) \in I_\theta^r F(x, y, u); \text{ if } (x, y) \in J := [0, 1] \times [0, 1], \quad (4.55)$$

$$u(x, y) = \Phi(x, y); \text{ if } (x, y) \in \tilde{J} := [-2, 1] \times [-3, 1] \backslash (0, 1] \times (0, 1], \quad (4.56)$$

where $m = 3,\ r = \left(\frac{1}{2}, \frac{1}{5}\right)$,

$$F(x, y, u) = \left[\frac{e^{x+y-\frac{|u|}{2}}}{1 + |u|}, \frac{e^{x+y}(1 + e^{-10}|u|)}{1 + |u|}\right]; \text{ a.a. } (x, y) \in J \text{ and for all } u \in \mathbb{R},$$

and $\Phi : \tilde{J} \to \mathbb{R}$ is a continuous function satisfying

$$\Phi(x, 0) = \frac{1}{14}\Phi\left(x - \frac{3}{2}, -2\right),\ \Phi(0, y) = \frac{1}{14}\Phi\left(-\frac{3}{2}, y - 2\right);\ x, y \in [0, 1]. \quad (4.57)$$

Notice that condition (4.57) is satisfied by $\Phi \equiv 0$.

Set

$$b_1(x, y) = \frac{x^3 y}{8},\ b_2(x, y) = \frac{x^4 y^2}{12},\ b_3(x, y) = \frac{1}{14}.$$

Then, $B = \frac{1}{8}$ and

$$\|F(x, y, u)\| \le e^{x+y}(1 + e^{-10}|u|) \text{ for a.e. } (x, y) \in J \text{ and all } u \in \mathbb{R}.$$

It is clear that F is Carathéodory; hence condition (4.30.1) is satisfied. We shall show that condition (4.30.2) holds with $a = b = 1, h^* = e^2,\ l^* = e^{-8}$. A simple computation shows that

$$mB + \frac{a^{r_1} b^{r_2} l^*}{\Gamma(1+r_1)\Gamma(1+r_2)} = \frac{3}{8} + \frac{1}{e^8 \Gamma(\frac{3}{2})\Gamma(\frac{6}{5})} < 1.$$

In view of Theorem 4.30, the problem (4.55) and (4.56) have a solution defined on $[-2,1] \times [-3,1]$.

4.7 Notes and Remarks

The results of Chap. 3 are taken from Abbas and Benchohra [8, 9, 12, 15, 18, 19, 24] and Abbas et al. [4, 36]. Other results may be found in [49, 242, 243].

Chapter 5
Impulsive Partial Hyperbolic Functional Differential Equations

5.1 Introduction

In this chapter, we shall present existence results for some classes of initial value problems for fractional order partial hyperbolic differential equations with impulses at fixed or variable times impulses.

5.2 Impulsive Partial Hyperbolic Functional Differential Equations

5.2.1 Introduction

This section concerns the existence results to fractional order IVP , for the system

$$({}^cD^r_{x_k}u)(x,y) = f(x,y,u(x,y)), \text{ if } (x,y) \in J_k;\ k = 0,\dots,m, \tag{5.1}$$

$$u(x_k^+, y) = u(x_k^-, y) + I_k(u(x_k^-, y));\ \text{ if } y \in [0,b],\ k = 1,\dots,m, \tag{5.2}$$

$$u(x,0) = \varphi(x),\ u(0,y) = \psi(y),\ \text{ if } x \in [0,a] \text{ and } y \in [0,b], \tag{5.3}$$

where $J_0 = [0,x_1]\times[0,b]$, $J_k = (x_k, x_{k+1}]\times[0,b]$; $k = 1,\dots,m$, $a,b > 0$, $0 = x_0 < x_1 < \cdots < x_m < x_{m+1} = a$, $f : J\times\mathbb{R}^n \to \mathbb{R}^n$, $J = [0,a]\times[0,b]$, and $I_k : \mathbb{R}^n \to \mathbb{R}^n$, $k = 0,1,\dots,m$ are given functions, $\varphi : [0,a] \to \mathbb{R}^n$, $\psi : [0,b] \to \mathbb{R}^n$ are absolutely continuous functions with $\varphi(0) = \psi(0)$. Next we consider the following nonlocal IVP , for the system:

$$({}^cD^r_{x_k}u)(x,y) = f(x,y,u(x,y));\ \text{ if } (x,y) \in J_k;\ k = 0,\dots,m \tag{5.4}$$

$$u(x_k^+, y) = u(x_k^-, y) + I_k(u(x_k^-, y));\ \text{ if } y \in [0,b],\ k = 1,\dots,m, \tag{5.5}$$

S. Abbas et al., *Topics in Fractional Differential Equations*, Developments in Mathematics 27, DOI 10.1007/978-1-4614-4036-9_5,

$$u(x,0)+Q(u)=\varphi(x),\ u(0,y)+K(u)=\psi(y),\quad \text{if } x\in[0,a] \text{ and } y\in[0,b], \tag{5.6}$$

where f, φ, ψ, I_k; $k = 1,\dots m$, are as in problems (5.1–5.3) and $Q, K : PC(J,\mathbb{R}^n) \to \mathbb{R}^n$ are continuous functions. $PC(J,\mathbb{R}^n)$ is a Banach space to be specified later.

5.2.2 Existence of Solutions

To define the solutions of problem (5.1)–(5.3), we shall consider the space

$$\begin{aligned} PC(J,\mathbb{R}^n) =\{u : J \to \mathbb{R}^n : u \in C(J_k,\mathbb{R}^n);\ k = 0,1,\dots,m, \text{ and there}\\ \text{exist } u(x_k^-,y) \text{ and } u(x_k^+,y);\ k=1,\dots,m,\\ \text{with } u(x_k^-,y)=u(x_k,y) \text{ for each } y\in[0,b]\}. \end{aligned}$$

This set is a Banach space with the norm

$$\|u\|_{PC} = \sup_{(x,y)\in J} \|u(x,y)\|.$$

Definition 5.1. A function $u \in PC(J_k,\mathbb{R}^n)$; $k = 0,\dots,m$ whose r-derivative exists on J_k; $k = 0,\dots,m$ is said to be a solution of (5.1)–(5.3) if u satisfies $({}^cD^r_{x_k}u)(x,y) = f(x,y,u(x,y))$ on J_k; $k = 0,\dots,m$, and conditions (5.2), (5.3) are satisfied.

Our first result is based on the Banach fixed-point theorem.

Theorem 5.2. *Assume that*

(5.2.1) There exists a constant $l > 0$ such that

$$\begin{aligned} &\|f(x,y,u) - f(x,y,\overline{u})\|\\ &\quad \le l\|u-\overline{u}\|, \textit{ for each } (x,y)\in J, \textit{and each } u,\overline{u}\in\mathbb{R}^n. \end{aligned}$$

(5.2.2) There exists a constant $l^* > 0$ such that

$$\|I_k(u) - I_k(\overline{u})\| \le l^*\|u-\overline{u}\|, \textit{ for each } u,\overline{u}\in\mathbb{R}^n,\ k=1,\dots,m.$$

If

$$2ml^* + \frac{2la^{r_1}b^{r_2}}{\Gamma(r_1+1)\Gamma(r_2+1)} < 1, \tag{5.7}$$

then (5.1)–(5.3) *have a unique solution on* J.

Proof. We transform the problem (5.1)–(5.3) into a fixed-point problem. Consider the operator $F : PC(J, \mathbb{R}^n) \to PC(J, \mathbb{R}^n)$ defined by

$$F(u)(x,y) = \mu(x,y) + \sum_{0<x_k<x} (I_k(u(x_k^-, y)) - I_k(u(x_k^-, 0)))$$
$$+ \frac{1}{\Gamma(r_1)\Gamma(r_2)} \sum_{0<x_k<x} \int_{x_{k-1}}^{x_k} \int_0^y (x_k - s)^{r_1-1}(y-t)^{r_2-1}$$
$$\times f(s,t,u(s,t))\mathrm{d}t\mathrm{d}s$$
$$+ \frac{1}{\Gamma(r_1)\Gamma(r_2)} \int_{x_k}^{x} \int_0^y (x-s)^{r_1-1}(y-t)^{r_2-1} f(s,t,u(s,t))\mathrm{d}t\mathrm{d}s.$$

Clearly that by Lemma 2.15, the fixed points of the operator F are solution of the problem (5.1)–(5.3). We shall use the Banach contraction principle to prove that F has a fixed point. We shall show that F *is a contraction*. Let $u, v \in PC(J, \mathbb{R}^n)$. Then, for each $(x,y) \in J$, we have

$$\|F(u)(x,y) - F(v)(x,y)\|$$
$$\leq \sum_{k=1}^{m} (\|I_k(u(x_k^-, y)) - I_k(v(x_k^-, y))\| + \|I_k(u(x_k^-, 0)) - I_k(v(x_k^-, 0))\|)$$
$$+ \frac{1}{\Gamma(r_1)\Gamma(r_2)} \sum_{k=1}^{m} \int_{x_{k-1}}^{x_k} \int_0^y (x_k - s)^{r_1-1}(y-t)^{r_2-1}$$
$$\times \|f(s,t,u(s,t)) - f(s,t,v(s,t))\|\mathrm{d}t\mathrm{d}s$$
$$+ \frac{1}{\Gamma(r_1)\Gamma(r_2)} \int_{x_k}^{x} \int_0^y (x-s)^{r_1-1}(y-t)^{r_2-1}$$
$$\times \|f(s,t,u(s,t)) - f(s,t,v(s,t))\|\mathrm{d}t\mathrm{d}s$$
$$\leq \sum_{k=1}^{m} l^*(\|u(x_k^-, y) - v(x_k^-, y)\| + \|u(x_k^-, 0) - v(x_k^-, 0)\|)$$
$$+ \frac{l}{\Gamma(r_1)\Gamma(r_2)} \sum_{k=1}^{m} \int_{x_{k-1}}^{x_k} \int_0^y (x_k - s)^{r_1-1}(y-t)^{r_2-1}\|u(s,t) - v(s,t)\|\mathrm{d}t\mathrm{d}s$$
$$+ \frac{l}{\Gamma(r_1)\Gamma(r_2)} \int_{x_k}^{x} \int_0^y (x-s)^{r_1-1}(y-t)^{r_2-1}\|u(s,t) - v(s,t)\|\mathrm{d}t\mathrm{d}s$$

$$\leq \left[2ml^* + \frac{la^{r_1}b^{r_2}}{\Gamma(r_1+1)\Gamma(r_2+1)} + \frac{la^{r_1}b^{r_2}}{\Gamma(r_1+1)\Gamma(r_2+1)}\right]\|u-v\|_\infty$$

$$\leq \left[2ml^* + \frac{2la^{r_1}b^{r_2}}{\Gamma(r_1+1)\Gamma(r_2+1)}\right]\|u-v\|_\infty.$$

By the condition (5.7), we conclude that F is a contraction . As a consequence of Banach fixed-point theorem, we deduce that F has a fixed point which is a solution of the problems (5.1)–(5.3). □

In the following theorem we give an existence result for the problems (5.1)–(5.3) by applying the nonlinear alternative of Leray–Schauder type.

Theorem 5.3. *Let $f(\cdot,\cdot,u) \in PC(J,\mathbb{R}^n)$ for each $u \in \mathbb{R}^n$. Assume that the following conditions hold:*

(5.3.1) There exists $\phi_f \in C(J,\mathbb{R}_+)$ and $\psi_* : [0,\infty) \to (0,\infty)$ continuous and nondecreasing such that

$$\|f(x,y,u)\| \leq \phi_f(x,y)\psi_*(\|u\|) \quad \textit{for all } (x,y) \in J,\ u \in \mathbb{R}^n.$$

(5.3.2) There exists $\psi^* : [0,\infty) \to (0,\infty)$ continuous and nondecreasing such that

$$\|I_k(u)\| \leq \psi^*(\|u\|) \quad \textit{for all } u \in \mathbb{R}^n,$$

(5.3.3) There exists a number $\overline{M} > 0$ such that

$$\frac{\overline{M}}{\|\mu\|_\infty + 2m\psi^*(\overline{M}) + \frac{2a^{r_1}b^{r_2}\phi_f^0\psi_*(\overline{M})}{\Gamma(r_1+1)\Gamma(r_2+1)}} > 1,$$

where $\phi_f^0 = \sup\{\phi_f(x,y) : (x,y) \in J\}$.

Then (5.1)–(5.3) *has at least one solution on J.*

Proof. Consider the operator F defined in Theorem 5.2.

Step 1: *F is continuous.* Let $\{u_n\}$ be a sequence such that $u_n \to u$ in $PC(J,\mathbb{R}^n)$. There exists $\eta > 0$ such that $\|u_n\| \leq \eta$. Then for each $(x,y) \in J$, we have

$$\|F(u_n)(x,y) - F(u)(x,y)\|$$

$$\leq \sum_{k=1}^{m}(\|I_k(u_n(x_k^-,y)) - I_k(u(x_k^-,y))\| + \|I_k(u_n(x_k^-,0)) - I_k(u(x_k^-,0))\|)$$

$$+ \frac{1}{\Gamma(r_1)\Gamma(r_2)}\sum_{k=1}^{m}\int_{x_{k-1}}^{x_k}\int_0^y (x_k-s)^{r_1-1}(y-t)^{r_2-1}$$

$$\times \|f(s,t,u_n(s,t)) - f(s,t,u(s,t))\|\mathrm{d}t\mathrm{d}s$$

$$+\frac{1}{\Gamma(r_1)\Gamma(r_2)}\int_{x_k}^{x}\int_{0}^{y}(x-s)^{r_1-1}(y-t)^{r_2-1}$$

$$\times \|f(s,t,u_n(s,t)) - f(s,t,u(s,t))\|\mathrm{d}t\mathrm{d}s.$$

Since f and I_k; $k = 1,\dots,m$ are continuous functions, we have

$$\|F(u_n) - F(u)\|_\infty \to 0 \quad \text{as } n \to \infty.$$

Step 2: *F maps bounded sets into bounded sets in $PC(J,\mathbb{R}^n)$.* Indeed, it is enough to show that for any $\eta^* > 0$, there exists a positive constant ℓ such that for each $u \in B_{\eta^*} = \{u \in PC(J,\mathbb{R}^n) : \|u\|_\infty \le \eta^*\}$, we have $\|F(u)\|_\infty \le \ell$. (H_{124}) and (H_{125}) imply that for each $(x,y) \in J$,

$$\begin{aligned}\|F(u)(x,y)\| &\le \|\mu(x,y)\| + \sum_{k=1}^{m}(\|I_k(u(x_k^-,y))\| + \|I_k(u(x_k^-,0))\|)\\ &\quad+\frac{1}{\Gamma(r_1)\Gamma(r_2)}\sum_{k=1}^{m}\int_{x_{k-1}}^{x_k}\int_{0}^{y}(x_k-s)^{r_1-1}(y-t)^{r_2-1}\\ &\quad\times\|f(s,t,u(s,t))\|\mathrm{d}t\mathrm{d}s\\ &\quad+\frac{1}{\Gamma(r_1)\Gamma(r_2)}\int_{x_k}^{x}\int_{0}^{y}(x-s)^{r_1-1}(y-t)^{r_2-1}\|f(s,t,u(s,t))\|\mathrm{d}t\mathrm{d}s\\ &\le \|\mu\|_\infty + 2m\psi^*(\eta^*) + \frac{2a^{r_1}b^{r_2}\phi_f^0\psi_*(\eta^*)}{\Gamma(r_1+1)\Gamma(r_2+1)} := \ell.\end{aligned}$$

Step 3: *F maps bounded sets into equicontinuous sets of $PC(J,\mathbb{R}^n)$.* Let (τ_1,y_1), $(\tau_2,y_2) \in [0,a]\times[0,b]$, $\tau_1 < \tau_2$ and $y_1 < y_2$, B_{η^*} be a bounded set of $PC(J,\mathbb{R}^n)$ as in Step 2, and let $u \in B_{\eta^*}$. Then for each $(x,y) \in J$, we have

$$\begin{aligned}&\|F(u)(\tau_2,y_2) - F(u)(\tau_1,y_1)\|\\ &\le \|\mu(\tau_1,y_1) - \mu(\tau_2,y_2)\| + \sum_{k=1}^{m}(\|I_k(u(x_k^-,y_1)) - I_k(u(x_k^-,y_2))\|)\\ &+\frac{1}{\Gamma(r_1)\Gamma(r_2)}\sum_{k=1}^{m}\int_{x_{k-1}}^{x_k}\int_{0}^{y_1}(x_k-s)^{r_1-1}[(y_2-t)^{r_2-1}-(y_1-t)^{r_2-1}]\end{aligned}$$

$$\times f(s,t,u(s,t))dtds$$
$$+\frac{1}{\Gamma(r_1)\Gamma(r_2)}\sum_{k=1}^{m}\int_{x_{k-1}}^{x_k}\int_{y_1}^{y_2}(x_k-s)^{r_1-1}(y_2-t)^{r_2-1}\|f(s,t,u(s,t))\|dtds$$
$$+\frac{1}{\Gamma(r_1)\Gamma(r_2)}\int_0^{\tau_1}\int_0^{y_1}[(\tau_2-s)^{r_1-1}(y_2-t)^{r_2-1}-(\tau_1-s)^{r_1-1}(y_1-t)^{r_2-1}]$$
$$\times f(s,t,u(s,t))dtds$$
$$+\frac{1}{\Gamma(r_1)\Gamma(r_2)}\int_{\tau_1}^{\tau_2}\int_{y_1}^{y_2}(\tau_2-s)^{r_1-1}(y_2-t)^{r_2-1}\|f(s,t,u(s,t))dtds\|$$
$$+\frac{1}{\Gamma(r_1)\Gamma(r_2)}\int_0^{\tau_1}\int_{y_1}^{y_2}(\tau_2-s)^{r_1-1}(y_2-t)^{r_2-1}\|f(s,t,u(s,t))dtds\|$$
$$+\frac{1}{\Gamma(r_1)\Gamma(r_2)}\int_{\tau_1}^{\tau_2}\int_0^{y_1}(\tau_2-s)^{r_1-1}(y_2-t)^{r_2-1}\|f(s,t,u(s,t))dtds\|$$
$$\leq\|\mu(\tau_1,y_1)-\mu(\tau_2,y_2)\|+\sum_{k=1}^{m}(\|I_k(u(x_k^-,y_1))-I_k(u(x_k^-,y_2))\|)$$
$$+\frac{\phi_f^0\psi_*(\eta^*)}{\Gamma(r_1)\Gamma(r_2)}\sum_{k=1}^{m}\int_{x_{k-1}}^{x_k}\int_0^{y_1}(x_k-s)^{r_1-1}[(y_2-t)^{r_2-1}-(y_1-t)^{r_2-1}]dtds$$
$$+\frac{\phi_f^0\psi_*(\eta^*)}{\Gamma(r_1)\Gamma(r_2)}\sum_{k=1}^{m}\int_{x_{k-1}}^{x_k}\int_{y_1}^{y_2}(x_k-s)^{r_1-1}(y_2-t)^{r_2-1}dtds$$
$$+\frac{\phi_f^0\psi_*(\eta^*)}{\Gamma(r_1+1)\Gamma(r_2+1)}[2y_2^{r_2}(\tau_2-\tau_1)^{r_1}+2\tau_2^{r_1}(y_2-y_1)^{r_2}$$
$$+\tau_1^{r_1}y_1^{r_2}-\tau_2^{r_1}y_2^{r_2}-2(\tau_2-\tau_1)^{r_1}(y_2-y_1)^{r_2}].$$

As $\tau_1\longrightarrow\tau_2$ and $y_1\longrightarrow y_2$, the right-hand side of the above inequality tends to zero. As a consequence of Steps 1–3 together with the Arzelá-Ascoli theorem, we can conclude that $F: PC(J,\mathbb{R}^n)\to PC(J,\mathbb{R}^n)$ is completely continuous.

Step 4: *A priori bound.* For $\lambda\in[0,1]$, let u be such that for each $(x,y)\in J$ we have $u(x,y)=\lambda(Fu)(x,y)$. For each $(x,y)\in J$, then from (5.3.1) and (5.3.2) we have

$$\frac{\|u\|_\infty}{\|\mu\|_\infty+2m\psi^*(\|u\|)+\frac{2a^{r_1}b^{r_2}\phi_f^0\psi_*(\|u\|)}{\Gamma(r_1+1)\Gamma(r_2+1)}}\leq 1.$$

By condition (5.3.3), there exists $\overline{M}$ such that $\|u\|_\infty\neq\overline{M}$. Let

$$U=\{u\in PC(J,\mathbb{R}^n):\|u\|_\infty<\overline{M}\}.$$

The operator $F : \overline{U} \to PC(J, \mathbb{R}^n)$ is continuous and completely continuous. From the choice of U, there is no $u \in \partial U$ such that $u = \lambda F(u)$ for some $\lambda \in (0, 1)$. As a consequence of the nonlinear alternative of Leray–Schauder type [136], we deduce that F has a fixed point u in $\overline{U}$ which is a solution of the problems (5.1)–(5.3). □

Now we present two existence results for the nonlocal problems (5.4)–(5.6). Their proofs are similar to those for problems (5.1)–(5.3).

Definition 5.4. A function $u \in PC(J_k, \mathbb{R}^n)$; $k = 0, \dots, m$ whose r-derivative exists on J_k; $k = 0, \dots, m$, is said to be a solution of (5.4)–(5.6) if u satisfies $({}^c D^r_{x_k} u)(x, y) = f(x, y, u(x, y))$ on J_k; $k = 0, \dots, m$ and conditions (5.5), (5.6) are satisfied.

Theorem 5.5. *Assume that (5.2.1), (5.2.2), and the following conditions:*

(5.5.1) There exists $\tilde{l} > 0$ such that

$$\|Q(u) - Q(v)\| \le \tilde{l}\|u - v\|, \textit{ for any } u, v \in PC(J, \mathbb{R}^n),$$

(5.5.2) There exists $\tilde{l}^* > 0$ such that

$$\|K(u) - K(v)\| \le \tilde{l}^*\|u - v\|, \textit{ for any } u, v \in PC(J, \mathbb{R}^n),$$

hold. If

$$\tilde{l} + \tilde{l}^* + 2ml^* + \frac{2la^{r_1}b^{r_2}}{\Gamma(r_1 + 1)\Gamma(r_2 + 1)} < 1,$$

then there exists a unique solution for IVP (5.4)–(5.6) on J.

Theorem 5.6. *Let $f(\cdot, \cdot, u) \in PC(J, \mathbb{R}^n)$ for each $u \in \mathbb{R}^n$. Assume that (5.3.1), (5.3.2), and the following conditions:*

(5.6.1) There exists $\tilde{d} > 0$ such that

$$\|Q(u)\| \le \tilde{d}(1 + \|u\|), \textit{ for any } u \in PC(J, \mathbb{R}^n),$$

(5.6.2) There exists $d^* > 0$ such that

$$\|K(u)\| \le d^*(1 + \|u\|), \textit{ for any } u \in PC(J, \mathbb{R}^n),$$

(5.6.3) There exists a number $\overline{M}_* > 0$ such that

$$\frac{\overline{M}_*}{(\tilde{d} + d^*)(1 + \overline{M}_*) + \|\mu\|_\infty + 2m\psi^*(\overline{M}_*) + \frac{2a^{r_1}b^{r_2}\phi^0_f\psi_*(\overline{M}_*)}{\Gamma(r_1+1)\Gamma(r_2+1)}} > 1,$$

hold. Then there exists at least one solution for IVP (5.4)–(5.6) on J.

5.2.3 An Example

As an application of our results we consider the following impulsive partial hyperbolic differential equations of the form:

$$({}^cD^r_{x_k}u)(x,y) = \frac{1}{(10e^{x+y+2})(1+|u(x,y)|)},$$

$$\text{if } (x,y) \in J_k;\ k = 0,\ldots,m, \tag{5.8}$$

$$u(x_k^+,y) = u(x_k^-,y) + \frac{1}{(6e^{x+y+4})(1+|u(x_k^-,y)|)};\ \text{ if } y \in [0,1],\ k = 1,\ldots,m, \tag{5.9}$$

$$u(x,0) = x,\ u(0,y) = y^2;\ \text{if } x \in [0,1] \text{ and } y \in [0,1]. \tag{5.10}$$

Set

$$f(x,y,u) = \frac{1}{(10e^{x+y+2})(1+|u|)};\ (x,y) \in [0,1]\times[0,1],$$

$$I_k(u(x_k^-,y)) = \frac{1}{(6e^{x+y+4})(1+|u(x_k^-,y)|)},\ y \in [0,1].$$

For each $u,\ \overline{u} \in \mathbb{R}$ and $(x,y) \in [0,1]\times[0,1]$ we have

$$|f(x,y,u) - f(x,y,\overline{u})| \le \frac{1}{10e^2}|u-\overline{u}|,$$

and

$$|I_k(u) - I_k(\overline{u})| \le \frac{1}{6e^4}|u-\overline{u}|.$$

Hence conditions (5.2.1) and (5.2.2) are satisfied with $l = \frac{1}{10e^2}$ and $l^* = \frac{1}{6e^4}$. We shall show that condition (5.7) holds with $a = b = 1$. Indeed, if we assume, for instance, that the number of impulses $m = 3$, then we have

$$2ml^* + \frac{2la^{r_1}b^{r_2}}{\Gamma(r_1+1)\Gamma(r_2+1)} = \frac{1}{e^4} + \frac{1}{5e^2\Gamma(r_1+1)\Gamma(r_2+1)} < 1,$$

which is satisfied for each $(r_1,r_2) \in (0,1]\times(0,1]$. Consequently, Theorem 5.2 implies that problem (5.8)–(5.10) has a unique solution defined on $[0,1]\times[0,1]$.

5.3 Impulsive Partial Hyperbolic Differential Equations at Variable Times

5.3.1 Introduction

In this section, we shall be concerned with the existence and uniqueness of solutions for the following impulsive partial hyperbolic fractional order differential equations at variable times:

$$({}^cD^r_{x_k}u)(x,y) = f(x,y,u(x,y)); \text{ if } (x,y) \in J_k;\ k = 0,\dots,m, \quad (5.11)$$

$$u(x^+,y) = I_k(u(x,y)); \text{ if } (x,y) \in J,\ x = x_k(u(x,y)),\ k = 1,\dots,m, \quad (5.12)$$

$$u(x,0) = \varphi(x),\ u(0,y) = \psi(y);\ x \in [0,a],\ y \in [0,b], \quad (5.13)$$

where $J_0 = [0,x_1] \times [0,b]$, $J_k = (x_k, x_{k+1}] \times [0,b]$; $k = 1,\dots,m$, $x_k = x_k(u(x,y))$; $k = 1,\dots,m$, $a,b > 0$, $0 = x_0 < x_1 < \cdots < x_m < x_{m+1} = a$, $f : J \times \mathbb{R}^n \to \mathbb{R}^n$, $J = [0,a] \times [0,b]$, $I_k : \mathbb{R}^n \to \mathbb{R}^n$; $k = 1,\dots,m$ are given functions and $\varphi : [0,a] \to \mathbb{R}^n$, $\psi : [0,b] \to \mathbb{R}^n$ are absolutely continuous functions with $\varphi(0) = \psi(0)$.

5.3.2 Existence of Solutions

Let us define what we mean by a solution of problem (5.11)–(5.13).

Definition 5.7. A function $u \in \Omega \cap \cup_{k=1}^m AC(J_k, \mathbb{R}^n)$ whose r-derivative exists on J_k; $k = 0,\dots,m$ is said to be a solution of (5.11)–(5.13) if u satisfies $({}^cD^r_{x_k}u)(x,y) = f(x,y,u(x,y))$ on J_k; $k = 0,\dots,m$ and conditions (5.12) and (3.3) are satisfied.

We are now in a position to state and prove our existence result for our problem based on Schaefer's fixed point .

Theorem 5.8. *Assume that*

(5.8.1) *The function* $f : J \times \mathbb{R}^n \to \mathbb{R}^n$ *is continuous*

(5.8.2) *There exists a constant* $M > 0$ *such that* $\|f(x,y,u)\| \le M(1 + \|u\|)$, *for each* $(x,y) \in J$, *and each* $u \in \mathbb{R}^n$

(5.8.3) *The function* $x_k \in C^1(\mathbb{R}^n, \mathbb{R})$ *for* $k = 1,\dots,m$. *Moreover,*

$$0 = x_0(u) < x_1(u) < \cdots < x_m(u) < x_{m+1}(u) = a, \quad \text{for all } u \in \mathbb{R}^n$$

(5.8.4) *There exists a constant* $M^* > 0$ *such that* $\|I_k(u)\| \le M^*(1 + \|u\|)$, *for each* $u \in \mathbb{R}^n$ *and* $k = 1,\dots,m$,

(5.8.5) *For all* $u \in \mathbb{R}^n$, $x_k(I_k(u)) \le x_k(u) < x_{k+1}(I_k(u))$, *for* $k = 1,\dots,m$

(5.8.6) For all $(s,t,u) \in J \times \mathbb{R}^n$, we have

$$x_k'(u)\left[\varphi'(s) + \frac{r_1 - 1}{\Gamma(r_1)\Gamma(r_2)}\int_{x_k}^{s}\right.$$

$$\left.\times \int_0^t (s-\theta)^{r_1-2}(t-\eta)^{r_2-1} f(\theta,\eta,u(\theta,\eta))\mathrm{d}\eta\mathrm{d}\theta\right] \neq 1$$

$k = 1,\ldots,m$. *Then* (5.11)–(5.13) *has at least one solution on* J.

Proof. The proof will be given in several steps.

Step 1: Consider the following problem:

$$({}^cD_0^r u)(x,y) = f(x,y,u(x,y)); \text{ if } (x,y) \in J, \tag{5.14}$$

$$u(x,0) = \varphi(x),\ u(0,y) = \psi(y);\ x \in [0,a],\ y \in [0,b]. \tag{5.15}$$

Transform problem (5.14) and (5.15) into a fixed-point problem. Consider the operator $N : C(J,\mathbb{R}^n) \to C(J,\mathbb{R}^n)$ defined by

$$N(u)(x,y) = \mu(x,y) + \frac{1}{\Gamma(r_1)\Gamma(r_2)}\int_0^x\int_0^y (x-s)^{r_1-1}(y-t)^{r_2-1}$$

$$\times f(s,t,u(s,t))\mathrm{d}t\mathrm{d}s.$$

Lemma 2.14 implies that the fixed points of operator N are solutions of problem (5.14) and (5.15). We shall show that the operator N is continuous and completely continuous.

Claim 1. N is continuous. Let $\{u_n\}$ be a sequence such that $u_n \to u$ in $C(J,\mathbb{R}^n)$. Let $\eta > 0$ be such that $\|u_n\| \leq \eta$. Then for each $(x,y) \in J$, we have

$$\|N(u_n)(x,y) - N(u)(x,y)\|$$

$$\leq \frac{1}{\Gamma(r_1)\Gamma(r_2)}\int_0^x\int_0^y (x-s)^{r_1-1}(y-t)^{r_2-1}$$

$$\times \|f(s,t,u_n(s,t)) - f(s,t,u(s,t))\|\mathrm{d}t\mathrm{d}s$$

$$\leq \frac{\|f(.,.,u_{n(.,.)}) - f(.,.,u)\|_\infty}{\Gamma(r_1)\Gamma(r_2)}\int_0^a\int_0^b (x-s)^{r_1-1}(y-t)^{r_2-1}\mathrm{d}t\mathrm{d}s$$

$$\leq \frac{a^{r_1}b^{r_2}\|f(.,.,u_{n(.,.)}) - f(.,.,u_{(.,.)})\|_\infty}{\Gamma(r_1+1)\Gamma(r_2+1)}.$$

Since f is a continuous function, we have

$$\|N(u_n) - N(u)\|_\infty \to 0 \quad \text{as } n \to \infty.$$

Claim 2. N maps bounded sets into bounded sets in $C(J, \mathbb{R}^n)$. Indeed, it is enough to show that for any $\eta^* > 0$, there exists a positive constant ℓ such that for each $u \in B_{\eta^*} = \{u \in C(J, \mathbb{R}^n) : \|u\|_\infty \leq \eta^*\}$, we have $\|N(u)\|_\infty \leq \ell$. By (5.8.2) we have for each $(x, y) \in J$, we have

$$\begin{aligned}\|N(u)(x,y)\| &\leq \|\mu(x,u)\| + \frac{1}{\Gamma(r_1)\Gamma(r_2)} \int_0^x \int_0^y (x-s)^{r_1-1}(y-t)^{r_2-1} \\ &\quad \times \|f(s,t,u(s,t))\| dt ds \\ &\leq \|\mu(x,u)\| + \frac{M(1+\eta^*)a^{r_1}b^{r_2}}{\Gamma(r_1+1)\Gamma(r_2+1)}.\end{aligned}$$

Thus $\|N(u)\|_\infty \leq \|\mu\|_\infty + \frac{M(1+\eta^*)a^{r_1}b^{r_2}}{\Gamma(r_1+1)\Gamma(r_2+1)} := \ell.$

Claim 3. N maps bounded sets into equicontinuous sets of $C(J, \mathbb{R}^n)$. Let (τ_1, y_1), $(\tau_2, y_2) \in J$, $\tau_1 < \tau_2$ and $y_1 < y_2$, B_{η^*} be a bounded set of $C(J, \mathbb{R})$ as in Claim 2, and let $u \in B_{\eta^*}$. Then for each $(x, y) \in J$, we have

$$\begin{aligned}&\|N(u)(\tau_2, y_2) - N(u)(\tau_1, y_1)\| \\ &= \Big\| \mu(\tau_1, y_1) - \mu(\tau_2, y_2) + \frac{1}{\Gamma(r_1)\Gamma(r_2)} \int_0^{\tau_1} \int_0^{y_1} [(\tau_2 - s)^{r_1-1}(y_2-t)^{r_2-1} \\ &\quad -(\tau_1 - s)^{r_1-1}(y_1-t)^{r_2-1}] f(s,t,u(s,t)) dt ds \\ &\quad + \frac{1}{\Gamma(r_1)\Gamma(r_2)} \int_{\tau_1}^{\tau_2} \int_{y_1}^{y_2} (\tau_2 - s)^{r_1-1}(y_2-t)^{r_2-1} f(s,t,u(s,t)) dt ds \\ &\quad + \frac{1}{\Gamma(r_1)\Gamma(r_2)} \int_0^{\tau_1} \int_{y_1}^{y_2} (\tau_2 - s)^{r_1-1}(y_2-t)^{r_2-1} f(s,t,u(s,t)) dt ds \\ &\quad + \frac{1}{\Gamma(r_1)\Gamma(r_2)} \int_{\tau_1}^{\tau_2} \int_0^{y_1} (\tau_2 - s)^{r_1-1}(y_2-t)^{r_2-1} f(s,t,u(s,t)) dt ds \Big\| \\ &\leq \|\mu(\tau_1, y_1) - \mu(\tau_2, y_2)\| + \frac{M(1+\eta^*)}{\Gamma(r_1+1)\Gamma(r_2+1)} [2y_2^{r_2}(\tau_2 - \tau_1)^{r_1} \\ &\quad + 2\tau_2^{r_1}(y_2 - y_1)^{r_2} + \tau_1^{r_1} y_1^{r_2} - \tau_2^{r_1} y_2^{r_2} - 2(\tau_2 - \tau_1)^{r_1}(y_2 - y_1)^{r_2}].\end{aligned}$$

As $\tau_1 \longrightarrow \tau_2$ and $y_1 \longrightarrow y_2$, the right-hand side of the above inequality tends to zero. As a consequence of Claims 1–3 together with the Arzelá-Ascoli theorem, we can conclude that $N : C(J, \mathbb{R}^n) \to C(J, \mathbb{R}^n)$ is completely continuous.

Claim 4. A priori bounds. Now it remains to show that the set $\mathcal{E} = \{u \in C(J, \mathbb{R}^n) : u = \lambda N(u) \text{ for some } 0 < \lambda < 1\}$ is bounded. Let $u \in \mathcal{E}$, then $u = \lambda N(u)$ for some $0 < \lambda < 1$. Thus, for each $(x, y) \in J$, we have

$$\begin{aligned}\|u(x,y)\| &\leq \|\mu(x,y)\| + \frac{1}{\Gamma(r_1)\Gamma(r_2)}\int_0^x\int_0^y (x-s)^{r_1-1}(y-t)^{r_2-1} \\ &\quad\times\|f(s,t,u(s,t))\|dtds \\ &\leq \|\mu\|_\infty + \frac{Ma^{r_1}b^{r_2}}{\Gamma(r_1+1)\Gamma(r_2+1)} \\ &\quad+\frac{M}{\Gamma(r_1)\Gamma(r_2)}\int_0^x\int_0^y (x-s)^{r_1-1}(y-t)^{r_2-1}\|u(s,t)\|dtds.\end{aligned}$$

Set

$$\omega = \|\mu\|_\infty + \frac{Ma^{r_1}b^{r_2}}{\Gamma(r_1+1)\Gamma(r_2+1)}, \quad c = \frac{M}{\Gamma(r_1)\Gamma(r_2)}.$$

Then Lemma 2.44 implies that for each $(x, y) \in J$,

$$\|u(x,y)\| \leq \omega E_{(r_1,r_2)}(c\Gamma(r_1)\Gamma(r_2)a^{r_1}b^{r_2}) := R.$$

This shows that the set $\mathcal{E}$ is bounded. As a consequence of Schaefer's fixed-point theorem, we deduce that N has a fixed point which is a solution of the problem (5.14) and (5.15). Denote this solution by u_1. Define the function

$$r_{k,1}(x,y) = x_k(u_1(x,y)) - x, \qquad \text{for } x \geq 0, y \geq 0.$$

Hypothesis (5.8.3) implies that $r_{k,1}(0,0) \neq 0$ for $k = 1, \ldots, m$. If $r_{k,1}(x,y) \neq 0$ on J for $k = 1, \ldots, m$; i.e.,

$$x \neq x_k(u_1(x,y)), \quad \text{on } J \quad \text{for } k = 1, \ldots, m,$$

then u_1 is a solution of the problem (5.11)–(5.13). It remains to consider the case when $r_{1,1}(x,y) = 0$ for some $(x,y) \in J$. Now since $r_{1,1}(0,0) \neq 0$ and $r_{1,1}$ is continuous, there exists $x_1 > 0, y_1 > 0$ such that $r_{1,1}(x_1,y_1) = 0$, and $r_{1,1}(x,y) \neq 0$, for all $(x,y) \in [0,x_1) \times [0,y_1)$.
Thus by (5.8.6) we have

$$r_{1,1}(x_1,y_1) = 0 \text{ and } r_{1,1}(x,y) \neq 0, \text{ for all } (x,y) \in [0,x_1) \times [0,y_1] \cup (y_1,b].$$

Suppose that there exist $(\bar{x},\bar{y}) \in [0,x_1) \times [0,y_1] \cup (y_1,b]$ such that $r_{1,1}(\bar{x},\bar{y}) = 0$. The function $r_{1,1}$ attains a maximum at some point $(s,t) \in [0,x_1) \times [0,b]$. Since

$$({}^cD^r_{x_1}u_1)(x,y) = f(x,y,u_1(x,y)), \text{ for } (x,y) \in J_1,$$

then

$$\frac{\partial u_1(x,y)}{\partial x} \text{ exists, and } \frac{\partial r_{1,1}(s,t)}{\partial x} = x_1'(u_1(s,t))\frac{\partial u_1(s,t)}{\partial x} - 1 = 0.$$

Since

$$\frac{\partial u_1(x,y)}{\partial x} = \varphi'(x) + \frac{r_1 - 1}{\Gamma(r_1)\Gamma(r_2)}\int_0^x\int_0^y (x-s)^{r_1-2}(y-t)^{r_2-1} f(s,t,u_1(s,t))\mathrm{d}t\mathrm{d}s,$$

then

$$x_1'(u_1(s,t))\Bigg[\varphi'(s) + \frac{r_1-1}{\Gamma(r_1)\Gamma(r_2)}\int_0^s \times \int_0^t (s-\theta)^{r_1-2}(t-\eta)^{r_2-1} f(\theta,\eta,u_1(\theta,\eta))\mathrm{d}\eta\mathrm{d}\theta\Bigg] = 1,$$

which contradicts (5.8.6). From (5.8.3) we have

$$r_{k,1}(x,y) \neq 0 \text{ for all } (x,y) \in [0,x_1) \times [0,b] \text{ and } k = 1,\dots,m.$$

Step 2: In what follows set

$$X_k := [x_k, a] \times [0,b]; \ k = 1,\dots,m.$$

Consider now the problem

$$({}^cD^r_{x_k}u)(x,y) = f(x,y,u(x,y)), \text{ if } (x,y) \in X_1, \tag{5.16}$$

$$u(x_1^+, y) = I_1(u_1(x_1,y)), \ y \in [0,b]. \tag{5.17}$$

Consider the operator $N_1 : C(X_1,\mathbb{R}^n) \to C(X_1,\mathbb{R}^n)$ defined as

$$N_1(u) = \varphi(x) + I_1(u_1(x_1,y)) - I_1(u_1(x_1,0)) + \frac{1}{\Gamma(r_1)\Gamma(r_2)}\int_{x_1}^x\int_0^y (x-s)^{r_1-1}(y-t)^{r_2-1} f(s,t,u(s,t))\mathrm{d}t\mathrm{d}s.$$

As in Step 1 we can show that N_1 is completely continuous. Now it remains to show that the set $\mathcal{E}^* = \{u \in C(X_1,\mathbb{R}^n) : u = \lambda N_1(u) \text{ for some } 0 < \lambda < 1\}$ is bounded. Let $u \in \mathcal{E}^*$, then $u = \lambda N_1(u)$ for some $0 < \lambda < 1$. Thus, from (5.8.2) and (5.8.4) we get for each $(x,y) \in X_1$,

$$\begin{aligned}\|u(x,y)\| &\leq \|\varphi(x)\| + \|I_1(u_1(x_1,y))\| + \|I_1(u_1(x_1,0))\| \\ &\quad + \frac{1}{\Gamma(r_1)\Gamma(r_2)}\int_{x_1}^{x}\int_0^y (x-s)^{r_1-1}(y-t)^{r_2-1}\|f(s,t,u(s,t))\|dtds \\ &\leq \|\varphi\|_\infty + 2M^*(1+\|u_1\|) + \frac{Ma^{r_1}b^{r_2}}{\Gamma(r_1+1)\Gamma(r_2+1)} \\ &\quad + \frac{M}{\Gamma(r_1)\Gamma(r_2)}\int_0^x\int_0^y (x-s)^{r_1-1}(y-t)^{r_2-1}\|u(s,t)\|dtds.\end{aligned}$$

Set

$$\omega^* = \|\varphi\|_\infty + 2M^*(1+\|u_1\|) + \frac{Ma^{r_1}b^{r_2}}{\Gamma(r_1+1)\Gamma(r_2+1)}, \quad c = \frac{M}{\Gamma(r_1)\Gamma(r_2)}.$$

Then Lemma 2.44 implies that for each $(x,y) \in X_1$,

$$\|u(x,y)\| \leq \omega^* E_{(r_1,r_2)}(c\Gamma(r_1)\Gamma(r_2)a^{r_1}b^{r_2}) := R^*.$$

This shows that the set $\mathcal{E}^*$ is bounded. As a consequence of Schaefer's fixed-point theorem, we deduce that N_1 has a fixed point u which is a solution to problems (5.16) and (5.17). Denote this solution by u_2. Define

$$r_{k,2}(x,y) = x_k(u_2(x,y)) - x, \quad \text{for } (x,y) \in X_1.$$

If $r_{k,2}(x,y) \neq 0$ on $(x_1,a] \times [0,b]$ and for all $k = 1,\ldots,m$, then

$$u(x,y) = \begin{cases} u_1(x,y), & \text{if } (x,y) \in [0,x_1) \times [0,b], \\ u_2(x,y), & \text{if } (x,y) \in [x_1,a] \times [0,b], \end{cases}$$

is a solution of the problems (5.11)–(5.13). It remains to consider the case when $r_{2,2}(x,y) = 0$, for some $(x,y) \in (x_1,a] \times [0,b]$. By (5.8.5), we have

$$\begin{aligned} r_{2,2}(x_1^+,y_1) &= x_2(u_2(x_1^+,y_1) - x_1 \\ &= x_2(I_1(u_1(x_1,y_1))) - x_1 \\ &> x_1(u_1(x_1,y_1)) - x_1 \\ &= r_{1,1}(x_1,y_1) = 0. \end{aligned}$$

Since $r_{2,2}$ is continuous, there exists $x_2 > x_1,\ y_2 > y_1$ such that $r_{2,2}(x_2,y_2) = 0$, and $r_{2,2}(x,y) \neq 0$ for all $(x,y) \in (x_1,x_2) \times [0,b]$. It is clear by (5.8.3) that

$$r_{k,2}(x,y) \neq 0 \quad \text{for all } (x,y) \in (x_1,x_2)] \times [0,b],\ k = 2,\ldots,m.$$

Now suppose that there are $(s,t) \in (x_1,x_2) \times [0,b]$ such that $r_{1,2}(s,t) = 0$. From (5.8.5) it follows that

$$\begin{aligned} r_{1,2}(x_1^+, y_1) &= x_1(u_2(x_1^+, y_1) - x_1 \\ &= x_1(I_1(u_1(x_1, y_1))) - x_1 \\ &\leq x_1(u_1(x_1, y_1)) - x_1 \\ &= r_{1,1}(x_1, y_1) = 0. \end{aligned}$$

Thus $r_{1,2}$ attains a nonnegative maximum at some point $(s_1,t_1) \in (x_1,a) \times [0,x_2) \cup (x_2,b]$. Since

$$({}^cD_{x_1}^r u_2)(x,y) = f(x,y,u_2(x,y)), \text{ for } (x,y) \in X_1,$$

then we get

$$\begin{aligned} u_2(x,y) &= \varphi(x) + I_1(u_1(x_1,y)) - I_1(u_1(x_1,0)) \\ &\quad + \frac{1}{\Gamma(r_1)\Gamma(r_2)} \int_{x_1}^{x} \int_{0}^{y} (x-s)^{r_1-1}(y-t)^{r_2-1} f(s,t,u_2(s,t)) \mathrm{d}t \mathrm{d}s, \end{aligned}$$

hence

$$\frac{\partial u_2}{\partial x}(x,y) = \varphi'(x) + \frac{r_1-1}{\Gamma(r_1)\Gamma(r_2)} \int_{x_1}^{x} \int_{0}^{y} (x-s)^{r_1-2}(y-t)^{r_2-1} f(s,t,u_2(s,t)) \mathrm{d}t \mathrm{d}s,$$

then

$$\frac{\partial r_{1,2}(s_1,t_1)}{\partial x} = x_1'(u_2(s_1,t_1)) \frac{\partial u_2}{\partial x}(s_1,t_1) - 1 = 0.$$

Therefore

$$\begin{aligned} x_1'(u_2(s_1,t_1)) \Bigg[\varphi'(s_1) + \frac{r_1-1}{\Gamma(r_1)\Gamma(r_2)} \int_{x_1}^{s_1} \int_{0}^{t_1} (s_1-\theta)^{r_1-2}(t_1-\eta)^{r_2-1} \\ \times f(\theta,\eta,u_2(\theta,\eta)) \mathrm{d}\eta \mathrm{d}\theta \Bigg] = 1, \end{aligned}$$

which contradicts (5.8.6).

Step 3: We continue this process and take into account that $u_{m+1} := u\Big|_{X_m}$ is a solution to the problem

$$\begin{aligned} &({}^cD_{x_m}^r u)(x,y) = f(x,y,u(x,y)), \quad \text{a.e. } (x,y) \in (x_m,a] \times [0,b], \\ &u(x_m^+, y) = I_m(u_{m-1}(x_m,y)), \ y \in [0,b]. \end{aligned}$$

The solution u of the problems (5.11)–(5.13) is then defined by

$$u(x,y)=\begin{cases} u_1(x,y), & \text{if } (x,y)\in[0,x_1]\times[0,b],\\ u_2(x,y), & \text{if } (x,y)\in(x_1,x_2]\times[0,b],\\ \dots & \\ u_{m+1}(x,y), & \text{if } (x,y)\in(x_m,a]\times[0,b]. \end{cases}$$

□

We give now (without proof) a uniqueness result for the problems (5.11)–(5.13) using the Banach contraction principle.

Theorem 5.9. *Assume (5.8.1), (5.8.3), (5.8.5), (5.8.6) and the following conditions:*

(5.9.1) There exists $d > 0$ such that

$$\|f(x,y,u)-f(x,y,\bar{u})\|\le d\,\|u-\bar{u}\|,\ \textit{for each}\ (x,y)\in J,\ u,\bar{u}\in\mathbb{R}^n,$$

(5.9.2) There exists $c_k > 0$; $k = 1,2,\dots,m$ such that

$$\|I_k(x,y,u)-I_k(x,y,\bar{u})\|\le c_k\,\|u-\bar{u}\|,\ \textit{for each}\ (x,y)\in J,\ u,\bar{u}\in\mathbb{R}^n,$$

hold. If

$$2c_k+\frac{da^{r_1}b^{r_2}}{\Gamma(r_1+1)\Gamma(r_2+1)}<1,$$

then the IVP (5.11)–(5.13) have a unique solution.

5.3.3 Nonlocal Impulsive Partial Differential Equations

Now, we shall present existence results for the following nonlocal initial value problem:

$$({}^cD^r_{x_k}u)(x,y)=f(x,y,u(x,y));\ \text{if } (x,y)\in J_k;\ k=0,\dots,m, \tag{5.18}$$

$$u(x^+,y)=I_k(u(x,y));\ \text{if } (x,y)\in J,\ x=x_k(u(x,y)),\ k=0,\dots,m, \tag{5.19}$$

$$u(x,0)+Q(u)=\varphi(x),\ u(0,y)+K(u)=\psi(y);\ x\in[0,a]\text{ and } y\in[0,b], \tag{5.20}$$

where $f,\ \varphi,\ \psi,\ I_k;\ k=1,\dots,m$, are as in problems (5.11)–(5.13) and $Q, K : C(J,\mathbb{R}^n)\to\mathbb{R}^n$ are continuous functions.

Definition 5.10. A function $u \in \Omega \cap \cup_{k=1}^{m} AC(J_k, \mathbb{R}^n)$ whose r-derivative exists on J' is said to be a solution of (5.18)–(5.20) if u satisfies $({}^c D_0^r u)(x, y) = f(x, y, u(x, y))$ on J' and conditions (5.19) and (5.20) are satisfied.

Theorem 5.11. *Assume that (5.8.1)–(5.8.6) and the following conditions:*

(5.11.1) There exists $\tilde{L} > 0$ such that

$$\|Q(u)\| \leq \tilde{L}(1 + \|u\|_\infty), \text{ for any } u \in C(J, \mathbb{R}^n),$$

(5.11.2) There exists $L^* > 0$ such that

$$\|K(u)\| \leq L^*(1 + \|u\|_\infty), \text{ for any } u \in C(J, \mathbb{R}^n),$$

hold. Then there exists at least one solution for IVP (5.18)–(5.20) on J.

Theorem 5.12. *Assume that (5.8.1), (5.8.3), (5.8.5), (5.8.6), (5.9.1), (5.9.2), and the following conditions:*

(5.12.1) There exists $l > 0$ such that

$$\|Q(u) - Q(v)\| \leq l\|u - v\|_\infty, \text{ for any } u, v \in C(J, \mathbb{R}^n),$$

(5.12.2) There exists $l^* > 0$ such that

$$\|K(u) - K(v)\| \leq l^*\|u - v\|_\infty, \text{ for any } u, v \in C(J, \mathbb{R}^n),$$

hold. If

$$l + l^* + 2c_k + \frac{da^{r_1}b^{r_2}}{\Gamma(r_1 + 1)\Gamma(r_2 + 1)} < 1,$$

then there exists a unique solution for IVP (5.18)–(5.20) on J.

5.3.4 An Example

As an application of our results we consider the following impulsive partial hyperbolic functional differential equations of the form:

$$({}^c D_{x_k}^r u)(x, y) = \frac{1 + u(x, y)}{9 + e^{x+y}}; \text{ if } (x, y) \in J_k, x_k = x_k(u(x, y)), \ k = 0, \ldots, m, \tag{5.21}$$

$$u(x_k^+, y) = d_k u(x_k, y); \ y \in [0, 1], \tag{5.22}$$

$$u(x, 0) = x, \ u(0, y) = y^2; \ x, y \in [0, 1], \tag{5.23}$$

where $J = [0,1] \times [0,1]$, $r = (r_1, r_2)$, $0 < r_1, r_2 \leq 1$, $x_k(u) = 1 - \frac{1}{2^k(1+u^2)}$; $k = 1, \ldots, m$ and $\frac{\sqrt{2}}{2} < d_k \leq 1$; $k = 1, \ldots, m$.
Set

$$f(x,y,u) = \frac{1+u}{9+\mathrm{e}^{x+y}}, \quad (x,y) \in [0,1] \times [0,1], \ u \in \mathbb{R},$$

and

$$I_k(u) = d_k u; \ u \in \mathbb{R}, \ k = 1, \ldots, m.$$

Let $u \in \mathbb{R}$ then we have

$$x_{k+1}(u) - x_k(u) = \frac{1}{2^{k+1}(1+u^2)} > 0; \ k = 1, \ldots, m.$$

Hence $0 < x_1(u) < x_2(u) < \cdots < x_m(u) < 1$, for each $u \in \mathbb{R}$. Also, for each $u \in \mathbb{R}$ we have

$$x_{k+1}(I_k(u)) - x_k(u) = \frac{1 + (2d_k^2 - 1)u^2}{2^{k+1}(1+u^2)(1+d_k^2)} > 0.$$

Finally, for all $(x,y) \in J$ and each $u \in \mathbb{R}$ we get

$$|I_k(u)| = |d_k u| \leq |u| \leq 3(1+|u|); \ k = 1, \ldots, m,$$

and

$$|f(x,y,u)| = \frac{|1+u|}{9+\mathrm{e}^{x+y}} \leq \frac{1}{10}(1+|u|).$$

Since all conditions of Theorem 5.8 are satisfied, problem (5.21)–(5.23) has at least one solution on $[0,1] \times [0,1]$.

5.4 Impulsive Discontinuous Partial Hyperbolic Differential Equations on Banach Algebras

5.4.1 Introduction

This section deals with the existence of solutions to fractional order IVP, for the system

$$^cD^r_{x_k}\left(\frac{u(x,y)}{f(x,y,u(x,y))}\right) = g(x,y,u(x,y)), \text{ if } (x,y) \in J_k; \ k = 0, \ldots, m, \tag{5.24}$$

$$u(x_k^+, y) = u(x_k^-, y) + I_k(u(x_k^-, y)); \ \text{ if } \ y \in [0,b]; \ k = 1, \ldots, m, \tag{5.25}$$

$$u(x,0) = \varphi(x), \ u(0,y) = \psi(y), \ x \in [0,a], \ y \in [0,b], \tag{5.26}$$

where $a, b > 0$, $0 = x_0 < x_1 < \cdots < x_m < x_{m+1} = a$, $f : J \times \mathbb{R} \to \mathbb{R}^*$, $g : J \times \mathbb{R} \to \mathbb{R}$ and $I_k : \mathbb{R} \to \mathbb{R}$, $k = 1, \ldots, m$ are given functions satisfying suitable conditions and $\varphi : [0, a] \to \mathbb{R}$, $\psi : [0, b] \to \mathbb{R}$ are given absolutely continuous functions with $\varphi(0) = \psi(0)$.

5.4.2 Existence of Solutions

Let us start by defining what we mean by a solution of the problems (5.24)–(5.26). Consider the space

$$PC(J, \mathbb{R}) = \big\{u : J \to \mathbb{R} : u \in C(J_k, \mathbb{R});\ k = 1, \ldots, m, \text{ and there exist } u(x_k^-, y) \text{ and } u(x_k^+, y);\ k = 1, \ldots, m, \text{ with } u(x_k^-, y) = u(x_k, y),\ y \in [0, b]\big\}.$$

This set is a Banach space with the norm $\|u\|_{PC} = \sup_{(x,y)\in J} |u(x, y)|$.

Definition 5.13. A function $u \in PC(J, \mathbb{R})$ whose r-derivative exists on J_k; $k = 0, \ldots, m$ is said to be a solution of (5.24)–(5.26) if

(i) The function $(x, y) \mapsto \frac{u(x,y)}{f(x,y,u(x,y))}$ is absolutely continuous
(ii) u satisfies ${}^cD^r_{x_k}\left(\frac{u(x,y)}{f(x,y,u(x,y))}\right) = g(x, y, u(x, y))$ on J_k; $k = 0, \ldots, m$ and conditions (5.25), (5.26) are satisfied

Theorem 5.14. *Assume that the following hypotheses:*

(5.14.1) The function f is continuous on $J \times \mathbb{R}$
(5.14.2) There exists a function $\alpha \in C(J, \mathbb{R}_+)$ such that

$$|f(x, y, u) - f(x, y, \overline{u})| \leq \alpha(x, y)|u - \overline{u}|, \textit{ for all } (x, y) \in J \textit{ and } u, \overline{u} \in \mathbb{R},$$

(5.14.3) The function g is Carathéodory, and there exists $h \in L^\infty(J, \mathbb{R}_+)$ such that

$$|g(x, y, u)| \leq h(x, y); \quad \textit{a.e. } (x, y) \in J, \textit{ for all } u \in \mathbb{R},$$

(5.14.4) There exists a function $\beta \in C(J, \mathbb{R}_+)$ such that

$$\left|\frac{I_k(u)}{f(x, y, u)}\right| \leq \beta(x, y), \quad \textit{for all } (x, y) \in J \textit{ and } u \in \mathbb{R},$$

hold. If

$$\|\alpha\|_\infty \left[\|\mu\|_\infty + 2m\|\beta\|_\infty + \frac{2a^{r_1} b^{r_2} \|h\|_{L^\infty}}{\Gamma(r_1 + 1)\Gamma(r_2 + 1)}\right] < 1, \tag{5.27}$$

Then the IVP (5.24)–(5.26) have at least one solution on J.

Proof. Let $X := PC(J, \mathbb{R})$. Define two operators A and B on X by

$$Au(x,y) = f(x,y,u(x,y)); \quad (x,y) \in J, \tag{5.28}$$

$$\begin{aligned} Bu(x,y) &= \mu(x,y) + \sum_{i=1}^{m} \left(\frac{I_i(u(x_i^-,y))}{f(x_i^+,y,u(x_i^+,y))} - \frac{I_i(u(x_i^-,0))}{f(x_i^+,0,u(x_i^+,0))} \right) \\ &+ \frac{1}{\Gamma(r_1)\Gamma(r_2)} \sum_{i=1}^{m} \int_{x_{i-1}}^{x_i} \int_0^y (x_i - s)^{r_1-1}(y-t)^{r_2-1} g(s,t,u(s,t))\mathrm{d}t\mathrm{d}s \\ &+ \frac{1}{\Gamma(r_1)\Gamma(r_2)} \int_{x_m}^{x} \int_0^y (x-s)^{r_1-1}(y-t)^{r_2-1} \\ &\times g(s,t,u(s,t))\mathrm{d}t\mathrm{d}s; \; (x,y) \in J. \end{aligned} \tag{5.29}$$

Solving (5.24)–(5.26) is equivalent to solving (3.46), which is further equivalent to solving the operator equation

$$Au(x,y)\, Bu(x,y) = u(x,y), \quad (x,y) \in J. \tag{5.30}$$

We show that operators A and B satisfy all the assumptions of Theorem 2.35. First we shall show that A is a Lipschitz. Let $u_1, u_2 \in X$. Then by (5.14.2),

$$\begin{aligned} |Au_1(x,y) - Au_2(x,y)| &= |f(x,y,u_1(x,y)) - f(x,y,u_2(x,y))| \\ &\le \alpha(x,y)|u_1(x,y) - u_2(x,y)| \\ &\le \|\alpha\|_\infty \|u_1 - u_2\|_{PC}. \end{aligned}$$

Taking the maximum over (x,y) in the above inequality yields

$$\|Au_1 - Au_2\|_{PC} \le \|\alpha\|_\infty \|u_1 - u_2\|_{PC},$$

and so A is a Lipschitz with a Lipschitz constant $\|\alpha\|_\infty$. Next, we show that B is compact operator on X. Let $\{u_n\}$ be a sequence in X. From (5.14.3) and (5.14.4) it follows that

$$\|Bu_n\|_{PC} \le \|\mu\|_\infty + 2m\|\beta\|_\infty + \frac{2a^{r_1}b^{r_2}\|h\|_{L^\infty}}{\Gamma(r_1+1)\Gamma(r_2+1)}.$$

As a result $\{Bu_n : n \in \mathbb{N}\}$ is a uniformly bounded set in X. Let $(\tau_1, y_1), (\tau_2, y_2) \in J$, $\tau_1 < \tau_2$ and $y_1 < y_2$, then for each $(x,y) \in J$,

$$
\begin{aligned}
&|B(u_n)(\tau_2, y_2) - B(u_n)(\tau_1, y_1)| \\
&\quad \leq |\mu(\tau_1, y_1) - \mu(\tau_2, y_2)| + \sum_{k=1}^{m} \left| \frac{I_k(u(x_k^-, y_1))}{f(x_k^+, y_1, u(x_i^+, y_1))} - \frac{I_k(u(x_k^-, y_2))}{f(x_k^+, y_2, u(x_i^+, y_2))} \right| \\
&\quad + \frac{1}{\Gamma(r_1)\Gamma(r_2)} \sum_{k=1}^{m} \int\limits_{x_{k-1}}^{x_k} \int\limits_{0}^{y_1} (x_k - s)^{r_1-1}[(y_2 - t)^{r_2-1} - (y_1 - t)^{r_2-1}] \\
&\quad \times g(s, t, u(s, t)) \mathrm{d}t \mathrm{d}s \\
&\quad + \frac{1}{\Gamma(r_1)\Gamma(r_2)} \sum_{k=1}^{m} \int\limits_{x_{k-1}}^{x_k} \int\limits_{y_1}^{y_2} (x_k - s)^{r_1-1}(y_2 - t)^{r_2-1} |g(s, t, u(s, t))| \mathrm{d}t \mathrm{d}s \\
&\quad + \frac{1}{\Gamma(r_1)\Gamma(r_2)} \int_0^{\tau_1} \int_0^{y_1} [(\tau_2 - s)^{r_1-1}(y_2 - t)^{r_2-1} - (\tau_1 - s)^{r_1-1}(y_1 - t)^{r_2-1}] \\
&\quad \times g(s, t, u(s, t)) \mathrm{d}t \mathrm{d}s \\
&\quad + \frac{1}{\Gamma(r_1)\Gamma(r_2)} \int_{\tau_1}^{\tau_2} \int_{y_1}^{y_2} (\tau_2 - s)^{r_1-1}(y_2 - t)^{r_2-1} |g(s, t, u(s, t))| \mathrm{d}t \mathrm{d}s \\
&\quad + \frac{1}{\Gamma(r_1)\Gamma(r_2)} \int_0^{\tau_1} \int_{y_1}^{y_2} (\tau_2 - s)^{r_1-1}(y_2 - t)^{r_2-1} |g(s, t, u(s, t))| \mathrm{d}t \mathrm{d}s \\
&\quad + \frac{1}{\Gamma(r_1)\Gamma(r_2)} \int_{\tau_1}^{\tau_2} \int_0^{y_1} (\tau_2 - s)^{r_1-1}(y_2 - t)^{r_2-1} |g(s, t, u(s, t))| \mathrm{d}t \mathrm{d}s \\
&\quad \leq |\mu(\tau_1, y_1) - \mu(\tau_2, y_2)| + \sum_{k=1}^{m} \left| \frac{I_k(u(x_k^-, y_1))}{f(x_k^+, y_1, u(x_i^+, y_1))} - \frac{I_k(u(x_k^-, y_2))}{f(x_k^+, y_2, u(x_i^+, y_2))} \right| \\
&\quad + \frac{\|h\|_{L^\infty}}{\Gamma(r_1)\Gamma(r_2)} \sum_{k=1}^{m} \int\limits_{x_{k-1}}^{x_k} \int\limits_{0}^{y_1} (x_k - s)^{r_1-1}[(y_2 - t)^{r_2-1} - (y_1 - t)^{r_2-1}] \mathrm{d}t \mathrm{d}s \\
&\quad + \frac{\|h\|_{L^\infty}}{\Gamma(r_1)\Gamma(r_2)} \sum_{k=1}^{m} \int\limits_{x_{k-1}}^{x_k} \int\limits_{y_1}^{y_2} (x_k - s)^{r_1-1}(y_2 - t)^{r_2-1} \mathrm{d}t \mathrm{d}s \\
&\quad + \frac{\|h\|_{L^\infty}}{\Gamma(r_1)\Gamma(r_2)} \int_0^{\tau_1} \int_0^{y_1} [(\tau_2 - s)^{r_1-1}(y_2 - t)^{r_2-1} - (\tau_1 - s)^{r_1-1}(y_1 - t)^{r_2-1}] \mathrm{d}t \mathrm{d}s \\
&\quad + \frac{\|h\|_{L^\infty}}{\Gamma(r_1)\Gamma(r_2)} \int_{\tau_1}^{\tau_2} \int_{y_1}^{y_2} (\tau_2 - s)^{r_1-1}(y_2 - t)^{r_2-1} \mathrm{d}t \mathrm{d}s \\
&\quad + \frac{\|h\|_{L^\infty}}{\Gamma(r_1)\Gamma(r_2)} \int_0^{\tau_1} \int_{y_1}^{y_2} (\tau_2 - s)^{r_1-1}(y_2 - t)^{r_2-1} \mathrm{d}t \mathrm{d}s
\end{aligned}
$$

$$+\frac{\|h\|_{L^\infty}}{\Gamma(r_1)\Gamma(r_2)}\int_{\tau_1}^{\tau_2}\int_0^{y_1}(\tau_2-s)^{r_1-1}(y_2-t)^{r_2-1}dtds$$

$$\leq|\mu(\tau_1,y_1)-\mu(\tau_2,y_2)|+\sum_{k=1}^{m}\left|\frac{I_k(u(x_k^-,y_1))}{f(x_k^+,y_1,u(x_i^+,y_1))}-\frac{I_k(u(x_k^-,y_2))}{f(x_k^+,y_2,u(x_i^+,y_2))}\right|$$

$$+\frac{\|h\|_{L^\infty}}{\Gamma(r_1)\Gamma(r_2)}\sum_{k=1}^{m}\int_{x_{k-1}}^{x_k}\int_0^{y_1}(x_k-s)^{r_1-1}[(y_2-t)^{r_2-1}-(y_1-t)^{r_2-1}]dtds$$

$$+\frac{\|h\|_{L^\infty}}{\Gamma(r_1)\Gamma(r_2)}\sum_{k=1}^{m}\int_{x_{k-1}}^{x_k}\int_{y_1}^{y_2}(x_k-s)^{r_1-1}(y_2-t)^{r_2-1}dtds$$

$$+\frac{\|h\|_{L^\infty}}{\Gamma(r_1+1)\Gamma(r_2+1)}\left[2y_2^{r_2}(\tau_2-\tau_1)^{r_1}+2\tau_2^{r_1}(y_2-y_1)^{r_2}\right.$$
$$\left.+\tau_1^{r_1}y_1^{r_2}-\tau_2^{r_1}y_2^{r_2}-2(\tau_2-\tau_1)^{r_1}(y_2-y_1)^{r_2}\right].$$

As $\tau_1 \longrightarrow \tau_2$ and $y_1 \longrightarrow y_2$, the right-hand side of the above inequality tends to zero. From this we conclude that $\{Bu_n : n \in \mathbb{N}\}$ is an equicontinuous set in X. Hence $B : X \to X$ is compact by Arzelà-Ascoli theorem. Moreover,

$$M = \|B(X)\| \leq \|\mu\|_\infty + 2m\|\beta\|_\infty + \frac{2a^{r_1}b^{r_2}\|h\|_{L^\infty}}{\Gamma(r_1+1)\Gamma(r_2+1)},$$

and so,

$$\alpha M \leq \|\alpha\|_\infty\left(\|\mu\|_\infty + 2m\|\beta\|_\infty + \frac{2a^{r_1}b^{r_2}\|h\|_{L^\infty}}{\Gamma(r_1+1)\Gamma(r_2+1)}\right) < 1,$$

by assumption (5.27). To finish, it remains to show that either the conclusion (i) or the conclusion (ii) of Theorem 2.35 holds. We now will show that the conclusion (ii) is not possible. Let $u \in X$ be any solution to (5.24)–(5.26), then for any $\lambda \in (0,1)$ we have

$$u(x,y)=\lambda f(x,y,u(x,y))\Big[\mu(x,y)+\sum_{0<x_k<x}\left(\frac{I_k(u(x_k^-,y))}{f(x_k^+,y,u(x_i^+,y))}-\frac{I_k(u(x_k^-,0))}{f(x_k^+,0,u(x_i^+,0))}\right)$$

$$+\frac{1}{\Gamma(r_1)\Gamma(r_2)}\sum_{0<x_k<x}\int_{x_{k-1}}^{x_k}\int_0^{y}(x_k-s)^{r_1-1}(y-t)^{r_2-1}g(s,t,u(s,t))dtds$$

$$+\frac{1}{\Gamma(r_1)\Gamma(r_2)}\int_{x_k}^{x}\int_0^{y}(x-s)^{r_1-1}(y-t)^{r_2-1}g(s,t,u(s,t))dtds\Big].$$

for $(x, y) \in J$. Therefore,

$$
\begin{aligned}
|u(x,y)| &\le |f(x,y,u(x,y))| \left(|\mu(x,y)| + 2m\|\beta\|_\infty + \frac{2a^{r_1}b^{r_2}\|h\|_{L^\infty}}{\Gamma(r_1+1)\Gamma(r_2+1)} \right) \\
&\le \left[|f(x,y,u(x,y)) - f(x,y,0)| + |f(x,y,0)| \right] \\
&\quad \times \left(|\mu(x,y)| + 2m\|\beta\|_\infty + \frac{2a^{r_1}b^{r_2}\|h\|_{L^\infty}}{\Gamma(r_1+1)\Gamma(r_2+1)} \right) \\
&\le \left[\|\alpha\|_\infty |u(x,y)| + f^* \right] \left(\|\mu(x,y)\| + 2m\|\beta\|_\infty + \frac{2a^{r_1}b^{r_2}\|h\|_{L^\infty}}{\Gamma(r_1+1)\Gamma(r_2+1)} \right) \\
&\le \left[\|\alpha\|_\infty \|u\|_{PC} + f^* \right] \left(\|\mu\|_\infty + 2m\|\beta\|_\infty + \frac{2a^{r_1}b^{r_2}\|h\|_{L^\infty}}{\Gamma(r_1+1)\Gamma(r_2+1)} \right),
\end{aligned}
$$

where $f^* = \sup\{|f(x,y,0)| : (x,y) \in J\}$, and consequently

$$
\|u\|_{PC} \le \frac{f^* \left[\|\mu\|_\infty + 2m\|\beta\|_\infty + \frac{2a^{r_1}b^{r_2}\|h\|_{L^\infty}}{\Gamma(r_1+1)\Gamma(r_2+1)} \right]}{1 - \|\alpha\|_\infty \left[\|\mu\|_\infty + 2m\|\beta\|_\infty + \frac{2a^{r_1}b^{r_2}\|h\|_{L^\infty}}{\Gamma(r_1+1)\Gamma(r_2+1)} \right]} := M.
$$

Thus the conclusion (ii) of Theorem 2.35 does not hold. Therefore the IVP (5.24)–(5.26) have a solution on J. □

5.4.3 Existence of Extremal Solutions

We equip the space $PC(J, \mathbb{R})$ with the order relation $\le$ with the help of the cone defined by

$$
K = \{u \in PC(J, \mathbb{R}) : u(x,y) \ge 0, \quad \forall (x,y) \in J\}.
$$

Thus $u \le \bar{u}$ if and only if $u(x,y) \le \bar{u}(x,y)$ for each $(x,y) \in J$. It is well known that the cone K is positive and normal in $PC(J, \mathbb{R})$.

Definition 5.15. A function $\underline{u}(\cdot,\cdot) \in PC(J, \mathbb{R})$ is said to be a lower solution of (5.24)–(5.26) if we have

$$
\begin{aligned}
{}^cD^r_{x_k}\left[\frac{\underline{u}(x,y)}{f(x,y,\underline{u}(x,y))} \right] &\le g(x,y,\underline{u}(x,y)), \quad (x,y) \in J_k;\ k = 0,\dots,m, \\
\underline{u}(x_k^+, y) &\le \underline{u}(x_k^-, y) + I_k(\underline{u}(x_k^-, y)), \ \text{if } (x,y) \in J_k;\ k = 1,\dots,m, \\
\underline{u}(x,0) &\le \varphi(x), \quad \underline{u}(0,y) \le \psi(y), \ (x,y) \in J.
\end{aligned}
$$

Similarly a function $\bar{u}(\cdot,\cdot) \in PC(J,\mathbb{R})$ is said to be an upper solution of (5.24)–(5.26) if we have

$$
\begin{aligned}
{}^{c}D^{r}_{x_k}\left[\frac{\bar{u}(x,y)}{f(x,y,\bar{u}(x,y))}\right] &\geq g(x,y,\bar{u}(x,y)), \quad (x,y)\in J_k;\ k=0,\dots,m,\\
\bar{u}(x_k^+,y) \geq \bar{u}(x_k^-,y) &+ I_k(\bar{u}(x_k^-,y)), \ \text{if } (x,y)\in J_k;\ k=1,\dots,m,\\
\bar{u}(x,0)\geq \varphi(x), &\quad \bar{u}(0,y)\geq \psi(y),\ (x,y)\in J.
\end{aligned}
$$

Definition 5.16. A solution u_M of the problems (5.24–5.26) is said to be maximal if for any other solution u to the problems (5.24)–(5.26) one has $u(x,y)\leq u_M(x,y)$, for all $(x,y)\in J$. Again a solution u_m of the problems (5.24)–(5.26) is said to be minimal if $u_m(x,y)\leq u(x,y)$, for all $(x,y)\in J$ where u is any solution of the problems (5.24)–(5.26) on J.

Theorem 5.17. *Assume that hypotheses (5.14.2) and*

(5.17.1) $f : J\times\mathbb{R}_+ \to \mathbb{R}_+^*,\ g : J\times\mathbb{R}_+\to\mathbb{R}_+,\ \psi(y)\geq 0$ *on* $[0,b]$ *and*

$$\frac{\varphi(x)}{f(x,0,\varphi(x))}\geq\frac{\varphi(0)}{f(0,0,\varphi(0))},\ \text{for all } x\in[0,a],$$

(5.17.2) The functions f and g are Chandrabhan

(5.17.3) There exists a function $\tilde{h}\in L^\infty(J,\mathbb{R}_+)$ such that

$$|g(x,y,u)|\leq\tilde{h}(x,y),\quad \text{a.e. } (x,y)\in J,\ \text{for all } u\in\mathbb{R},$$

(5.17.4) There exists a function $\tilde{\beta}\in C(J,\mathbb{R}_+)$ such that

$$\left|\frac{I_k(u)}{f(x,y,u)}\right|\leq\tilde{\beta}(x,y),\quad \text{for all } (x,y)\in J,\ \text{for all } u\in\mathbb{R},$$

(5.17.5) The problem (5.24)–(5.26) have a lower solution $\underline{u}$ and an upper solution $\bar{u}$ with $\underline{u}\leq\bar{u}$,

hold. If

$$\|\alpha\|_\infty\left[\|\mu\|_\infty+2m\|\tilde{\beta}\|+\frac{2a^{r_1}b^{r_2}\|\tilde{h}\|_{L^\infty}}{\Gamma(r_1+1)\Gamma(r_2+1)}\right]<1,$$

then the problems (5.24)–(5.26) have a minimal and a maximal positive solution on J.

Proof. Let $X = PC(J_k,\mathbb{R})$; $k=0,\dots,m$ and consider a closed interval $[\underline{u},\bar{u}]$ in X which is well defined in view of hypothesis (5.17.5). Define two operators $A,B : [\underline{u},\bar{u}]\to X$ by (5.28) and (5.29), respectively. Clearly A and B define the operators $A,B : [\underline{u},\bar{u}]\to K$. Now solving (5.24)–(5.26) is equivalent to solving (3.46), which is further equivalent to solving the operator equation

$$Au(x,y)\,Bu(x,y)=u(x,y),\quad (x,y)\in J_k;\ k=0,\dots,m. \tag{5.31}$$

We show that operators A and B satisfy all the assumptions of Theorem 2.41. As in Theorem 5.14 we can prove that A is Lipschitz with a Lipschitz constant $\|\alpha\|_\infty$ and B is completely continuous operator on $[\underline{u}, \overline{u}]$. Now hypothesis (5.17.2) implies that A and B are nondecreasing on $[\underline{u}, \overline{u}]$. To see this, let $u_1, u_2 \in [\underline{u}, \overline{u}]$ be such that $u_1 \le u_2$. Then by (5.17.2), we get

$$\begin{aligned} Au_1(x,y) &= f(x,y,u_1(x,y)) \le f(x,y,u_2(x,y)) = Au_2(x,y), \\ &\forall (x,y) \in J_k;\ k = 0,\dots,m, \end{aligned}$$

and

$$\begin{aligned} &Bu_1(x,y) \\ &= \mu(x,y) + \sum_{0<x_k<x} \left(\frac{I_k(u_1(x_k^-,y))}{f(x_k^+,y,u(x_i^+,y))} - \frac{I_k(u_1(x_k^-,0))}{f(x_k^+,0,u(x_i^+,0))} \right) \\ &+ \frac{1}{\Gamma(r_1)\Gamma(r_2)} \sum_{0<x_k<x} \int_{x_{k-1}}^{x_k} \int_0^y (x_k-s)^{r_1-1}(y-t)^{r_2-1} g(s,t,u_1(s,t))\mathrm{d}t\mathrm{d}s \\ &+ \frac{1}{\Gamma(r_1)\Gamma(r_2)} \int_{x_k}^{x} \int_0^y (x-s)^{r_1-1}(y-t)^{r_2-1} g(s,t,u_1(s,t))\mathrm{d}t\mathrm{d}s \\ &\le \mu(x,y) + \sum_{0<x_k<x} \left(\frac{I_k(u_2(x_k^-,y))}{f(x_k^+,y,u(x_i^+,y))} - \frac{I_k(u_2(x_k^-,0))}{f(x_k^+,0,u(x_i^+,0))} \right) \\ &+ \frac{1}{\Gamma(r_1)\Gamma(r_2)} \sum_{0<x_k<x} \int_{x_{k-1}}^{x_k} \int_0^y (x_k-s)^{r_1-1}(y-t)^{r_2-1} g(s,t,u_2(s,t))\mathrm{d}t\mathrm{d}s \\ &+ \frac{1}{\Gamma(r_1)\Gamma(r_2)} \int_{x_k}^{x} \int_0^y (x-s)^{r_1-1}(y-t)^{r_2-1} g(s,t,u_2(s,t))\mathrm{d}t\mathrm{d}s \\ &= Bu_2(x,y), \quad \forall (x,y) \in J. \end{aligned}$$

So A and B are nondecreasing operators on $[\underline{u}, \overline{u}]$. Again hypothesis (5.17.5) implies

$$\begin{aligned} \underline{u}(x,y) = [f(x,y,\underline{u}(x,y))] &\left(\mu(x,y) + \sum_{0<x_k<x} \left(\frac{I_k(\underline{u}(x_k^-,y))}{f(x_k^+,y,u(x_i^+,y))} - \frac{I_k(\underline{u}(x_k^-,0))}{f(x_k^+,0,u(x_i^+,0))} \right) \right. \\ &+ \frac{1}{\Gamma(r_1)\Gamma(r_2)} \sum_{0<x_k<x} \int_{x_{k-1}}^{x_k} \int_0^y (x_k-s)^{r_1-1}(y-t)^{r_2-1} g(s,t,\underline{u}(s,t))\mathrm{d}t\mathrm{d}s \end{aligned}$$

$$+\frac{1}{\Gamma(r_1)\Gamma(r_2)}\int_{x_k}^{x}\int_{0}^{y}(x-s)^{r_1-1}(y-t)^{r_2-1}g(s,t,\underline{u}(s,t))\mathrm{d}t\,\mathrm{d}s\Bigg)$$

$$\leq [f(x,y,u(x,y))]\Bigg(\mu(x,y)+\sum_{0<x_k<x}\left(\frac{I_k(u(x_k^-,y))}{f(x_k^+,y,u(x_i^+,y))}-\frac{I_k(u(x_k^-,0))}{f(x_k^+,0,u(x_i^+,0))}\right)$$

$$+\frac{1}{\Gamma(r_1)\Gamma(r_2)}\sum_{0<x_k<x}\int_{x_{k-1}}^{x_k}\int_{0}^{y}(x_k-s)^{r_1-1}(y-t)^{r_2-1}g(s,t,u(s,t))\mathrm{d}t\,\mathrm{d}s$$

$$+\frac{1}{\Gamma(r_1)\Gamma(r_2)}\int_{x_k}^{x}\int_{0}^{y}(x-s)^{r_1-1}(y-t)^{r_2-1}g(s,t,u(s,t))\mathrm{d}t\,\mathrm{d}s\Bigg)$$

$$\leq [f(x,y,\overline{u}(x,y))]\Bigg(\mu(x,y)+\sum_{0<x_k<x}\left(\frac{I_k(\overline{u}(x_k^-,y))}{f(x_k^+,y,u(x_i^+,y))}-\frac{I_k(\overline{u}(x_k^-,0))}{f(x_k^+,0,u(x_i^+,0))}\right)$$

$$+\frac{1}{\Gamma(r_1)\Gamma(r_2)}\sum_{0<x_k<x}\int_{x_{k-1}}^{x_k}\int_{0}^{y}(x_k-s)^{r_1-1}(y-t)^{r_2-1}g(s,t,\overline{u}(s,t))\mathrm{d}t\,\mathrm{d}s$$

$$+\frac{1}{\Gamma(r_1)\Gamma(r_2)}\int_{x_k}^{x}\int_{0}^{y}(x-s)^{r_1-1}(y-t)^{r_2-1}g(s,t,\overline{u}(s,t))\mathrm{d}t\,\mathrm{d}s\Bigg)$$

$$\leq \overline{u}(x,y),$$

for all $(x,y)\in J$ and $u\in[\underline{u},\overline{u}]$. As a result

$$\underline{u}(x,y)\leq Au(x,y)Bu(x,y)\leq\overline{u}(x,y),\quad \forall (x,y)\in J_k;\ k=0,\ldots,m,\ \text{and } u\in[\underline{u},\overline{u}].$$

Hence $Au\,Bu\in[\underline{u},\overline{u}]$, for all $u\in[\underline{u},\overline{u}]$.

Notice for any $u\in[\underline{u},\overline{u}]$,

$$M=\|B([\underline{u},\overline{u}])\|$$

$$\leq|\mu(x,y)|+\Bigg|\sum_{0<x_k<x}\left(\frac{I_k(u(x_k^-,y))}{f(x_k^+,y,u(x_i^+,y))}-\frac{I_k(u(x_k^-,0))}{f(x_k^+,0,u(x_i^+,0))}\right)$$

$$+\frac{1}{\Gamma(r_1)\Gamma(r_2)}\sum_{0<x_k<x}\int_{x_{k-1}}^{x_k}\int_{0}^{y}(x_k-s)^{r_1-1}(y-t)^{r_2-1}g(s,t,u(s,t))\mathrm{d}t\,\mathrm{d}s$$

$$+\frac{1}{\Gamma(r_1)\Gamma(r_2)}\int_{x_k}^{x}\int_{0}^{y}(x-s)^{r_1-1}(y-t)^{r_2-1}g(s,t,u(s,t))\mathrm{d}t\,\mathrm{d}s\Bigg|$$

$$\leq\|\mu\|_\infty+2m\|\tilde{\beta}\|+\frac{2a^{r_1}b^{r_2}\|\tilde{h}\|_{L^\infty}}{\Gamma(r_1+1)\Gamma(r_2+1)}.$$

and so,

$$\alpha M \leq \|\alpha\|_\infty \left(\|\mu\|_\infty + 2m\|\tilde{\beta}\| + \frac{2a^{r_1}b^{r_2}\|\tilde{h}\|_{L^\infty}}{\Gamma(r_1+1)\Gamma(r_2+1)} \right) < 1.$$

Thus the operators A and B satisfy all the conditions of Theorem 2.41 and so the operator equation (5.29) has a least a greatest solution in $[\underline{u}, \overline{u}]$. This further implies that the problems (5.24)–(5.26) have a minimal and a maximal positive solution on J. □

Theorem 5.18. *Assume that hypotheses (5.14.1), (5.17.1)–(5.17.5) hold. Then the problems (5.24)–(5.26) have a minimal and a maximal positive solution on J.*

Proof. Let $X = PC(J, \mathbb{R})$. Consider the order interval $[\underline{u}, \overline{u}]$ in X and define two operators A and B on $[\underline{u}, \overline{u}]$ by (5.28) and (5.29), respectively. Then the problems (5.24)–(5.26) are transformed into an operator equation $Au(x,y)\,Bu(x,y) = u(x,y)$, $(x,y) \in J$ in a Banach algebra X. Notice that (H_{141}) implies $A, B : [\underline{u}, \overline{u}] \to K$. Since the cone K in X is normal, $[\underline{u}, \overline{u}]$ is a norm-bounded set in X. Next we show that A is completely continuous on $[\underline{u}, \overline{u}]$. Now the cone K in X is normal, so the order interval $[\underline{u}, \overline{u}]$ is norm-bounded. Hence there exists a constant $\sigma > 0$ such that $\|u\| \leq \sigma$ for all $u \in [\underline{u}, \overline{u}]$. As f is continuous on compact set $J \times [-\sigma, \sigma]$, it attains its maximum , say M. Therefore, for any subset S of $[\underline{u}, \overline{u}]$ we have

$$\begin{aligned} \|A(S)\| &= \sup\{|Au| : u \in S\} \\ &= \sup\left\{ \sup_{(x,y)\in J} |f(x,y,u(x,y))| : u \in S \right\} \\ &\leq \sup\left\{ \sup_{(x,y)\in J} |f(x,y,u)| : u \in [-\sigma, \sigma] \right\} \\ &\leq M. \end{aligned}$$

This shows that $A(S)$ is a uniformly bounded subset of X. We note that the function $f(x,y,u)$ is uniformly continuous on $J \times [-\sigma, \sigma]$. Therefore, for any $(\tau_1, y_1), (\tau_2, y_2) \in J$ we have

$$|f(\tau_1, y_1, u) - f(\tau_2, y_2, u)| \to 0 \quad \text{as } (\tau_1, y_1) \to (\tau_2, y_2),$$

for all $u \in [-\sigma, \sigma]$. Similarly for any $u_1, u_2 \in [-\sigma, \sigma]$

$$|f(x,y,u_1) - f(x,y,u_2)| \to 0 \quad \text{as } u_1 \to u_2,$$

for all $(x,y) \in J$. Hence for any $(\tau_1, y_1), (\tau_2, y_2) \in J$ and for any $u \in S$ one has

$$\begin{aligned} |Au(\tau_1, y_1) - Au(\tau_2, y_2)| &= |f(\tau_1, y_1, u(\tau_1, y_1)) - f(\tau_2, y_2, u(\tau_2, y_2))| \\ &\leq |f(\tau_1, y_1, u(\tau_1, y_1)) - f(\tau_2, y_2, u(\tau_1, y_1)| \\ &\quad + |f(\tau_2, y_2, u(\tau_1, y_1)) - f(\tau_2, y_2, u(\tau_2, y_2))| \\ &\to 0 \quad \text{as } (\tau_1, y_1) \to (\tau_2, y_2). \end{aligned}$$

This shows that $A(S)$ is an equicontinuous set in K. Now an application of Arzelà-Ascoli theorem yields that A is a completely continuous operator on $[\underline{u}, \overline{u}]$. Next it can be shown as in the proof of Theorem 5.17 that B is a compact operator on $[\underline{u}, \overline{u}]$. Now an application of Theorem 2.40 yields that the problems (5.24)–(5.26) have a minimal and maximal positive solution on J. □

5.4.4 An Example

As an application of our results we consider the following partial hyperbolic functional differential equations of the form:

$$ {}^cD^r_{x_k}\left(\frac{u(x,y)}{f(x,y,u(x,y))}\right) = g(x,y,u(x,y)), \text{ if } (x,y) \in J_k;\ k = 0,1, \quad (5.32)$$

$$ u\left(\frac{1}{2}^+, y\right) = u\left(\frac{1}{2}^-, y\right) + I_1\left(u\left(\frac{1}{2}^-, y\right)\right), \text{ if } y \in [0,1], \quad (5.33)$$

$$ u(x,0) = \varphi(x),\ x \in [0,1],\ u(0,y) = \psi(y),\ y \in [0,1], \quad (5.34)$$

where $J_0 = [0, \frac{1}{2}] \times [0,1]$, $J_1 = (\frac{1}{2}, 1] \times [0,1]$, $f, g : [0,1] \times [0,1] \times \mathbb{R} \to \mathbb{R}$ and $I_1 : \mathbb{R} \to \mathbb{R}$ are defined by

$$ f(x,y,u) = \frac{1}{e^{x+y+10}(1+|u|)},$$

$$ g(x,y,u) = \frac{1}{e^{x+y+8}(1+u^2)},$$

and

$$ I_1(u) = \frac{(8+e^{-10})^2}{512e^{10}(1+|u|)^2}.$$

The functions $\varphi, \psi : [0,1] \to \mathbb{R}$ are defined by

$$ \varphi(x) = \begin{cases} \frac{x^2}{2}e^{-10}; \text{ if } x \in [0, \frac{1}{2}], \\ x^2e^{-10}; \text{ if } x \in (\frac{1}{2}, 1], \end{cases}$$

and

$$ \psi(y) = ye^{-10}, \text{ for all } y \in [0,1].$$

We show that the functions φ, ψ, f, g, and I_1 satisfy all the hypotheses of Theorem 5.14. Clearly, the function f satisfies (5.14.1) and (5.14.2) with $\alpha(x,y) = \frac{1}{e^{x+y+10}}$ and

$$ \|\alpha\|_\infty = \frac{1}{e^{10}}.$$

Also, the function g satisfies (5.14.3) with $h(x,y) = \frac{1}{e^{x+y+8}}$ and

$$\|h\|_{L^\infty} = \frac{1}{e^8}.$$

Finally, condition (5.14.4) holds with $\beta(x,y) = \frac{81e^{x+y}}{512}$ and $\|\beta\|_\infty = \frac{81e^2}{512}$. A simple computation gives $\|\mu\|_\infty < 4e$. Condition (5.27) holds. Indeed

$$\begin{aligned}
&\|\alpha\|_\infty \left[\|\mu\|_\infty + 2m\|\beta\|_\infty + \frac{2a^{r_1}b^{r_2}\|h\|_{L^\infty}}{\Gamma(r_1+1)\Gamma(r_2+1)} \right] \\
&< \frac{1}{e^{10}} \left[4e + \frac{81e^2}{256} + \frac{2}{e^8\Gamma(r_1+1)\Gamma(r_2+1)} \right] \\
&< 1,
\end{aligned}$$

for each $(r_1, r_2) \in (0,1] \times (0,1]$. Hence by Theorem 5.14, the problems (5.32)–(5.34) have a solution defined on $[0,1] \times [0,1]$.

5.5 Impulsive Partial Hyperbolic Differential Equations with Variable Times and Infinite Delay

5.5.1 *Introduction*

In this section, we shall be concerned with the existence of solutions for the following impulsive partial hyperbolic differential equations:

$$({}^cD^r_{x_k}u)(x,y) = f(x,y,u_{(x,y)}); \text{ if } (x,y) \in J_k;\ k = 0,\dots,m,$$

$$x_k = x_k(u(x,y));\ k = 1,\dots,m, \tag{5.35}$$

$$u(x^+,y) = I_k(u(x,y)); \text{ if } y \in [0,b],\ x = x_k(u(x,y)),\ k = 1,\dots,m, \tag{5.36}$$

$$u(x,y) = \phi(x,y); \text{ if } (x,y) \in \tilde{J}', \tag{5.37}$$

$$u(x,0) = \varphi(x),\ x \in [0,a],\ u(0,y) = \psi(y);\ y \in [0,b], \tag{5.38}$$

where $a, b > 0$, $\tilde{J}' = (-\infty,a] \times (-\infty,b] \backslash (0,a] \times (0,b]$, $\phi \in C(\tilde{J}, \mathbb{R}^n)$, φ, ψ are as in problems (3.1)–(3.3), $0 = x_0 < x_1 < \cdots < x_m < x_{m+1} = a$, $f : J \times \mathcal{B} \to \mathbb{R}^n$, $I_k : \mathbb{R}^n \to \mathbb{R}^n$, $k = 1,\dots,m$ are given functions and $\mathcal{B}$ is a phase space.

5.5.2 Main Result

To define the solutions of problems (5.35)–(5.38), we shall consider the space

$$\begin{aligned}\Omega = \big\{u : (-\infty, a] \times (-\infty, b] \to \mathbb{R}^n : u_{(x,y)} \in \mathcal{B} \text{ for } (x,y) \in E \text{ and there exist} \\ 0 = x_0 < x_1 < x_2 < \cdots < x_m < x_{m+1} = a \text{ such that } x_k = x_k(u(x_k,.)), \\ \text{and } u(x_k^-,.),\ u(x_k^+,.) \text{ exist with } u(x_k^-,.) = u(x_k,.);\ k = 1,\ldots,m, \\ \text{and } u \in C(J_k, \mathbb{R}^n); k = 0,\ldots,m\big\},\end{aligned}$$

where $J_k := (x_k, x_{k+1}] \times (0, b]$. Let $\|u\|_\Omega$ be the seminorm in Ω defined by

$$\|u\|_\Omega = \|\phi\|_{\mathcal{B}} + \sup\{\|u_k\|,\ k = 0,\ldots,m\},$$

where u_k is the restriction of u to J_k; $k = 0,\ldots,m$. Let us define what we mean by a solution of problems (5.35)–(5.38).

Definition 5.19. A function $u \in \Omega$ whose r-derivative exists on J_k; $k = 0,\ldots,m$ is said to be a solution of (5.35)–(5.38) if u satisfies $({}^cD^r_{x_k}u)(x,y) = f(x,y,u(x,y))$ on J_k; $k = 0,\ldots,m$ and conditions (5.36)–(5.38) are satisfied.

Theorem 5.20. *Let $f : J \times \mathcal{B} \to \mathbb{R}^n$ be a Carathéodory function. Assume that*

(5.20.1) The function $x_k \in C^1(\mathbb{R}^n, \mathbb{R})$ for $k = 1,\ldots,m$. Moreover,

$$0 = x_0(u) < x_1(u) < \cdots < x_m(u) < x_{m+1}(u) = a, \quad \textit{for all } u \in \mathbb{R}^n,$$

(5.20.2) There exists a constant $M > 0$ such that

$$\|f(x,y,u)\| \le M(1 + \|u\|_{\mathcal{B}}),\ \textit{for each } (x,y) \in J,\ \textit{and each } u \in \mathcal{B},$$

(5.20.3) For all $(s,t,u) \in J \times \mathbb{R}^n$ and $u_{(.,.)} \in \mathcal{B}$, we have

$$x_k'(u)\left[\varphi'(s) + \frac{r_1-1}{\Gamma(r_1)\Gamma(r_2)}\int_{x_k}^{s}\int_0^t (s-\theta)^{r_1-2}(t-\eta)^{r_2-1} f(\theta,\eta,u_{(\theta,\eta)})d\eta d\theta\right] \ne 1;$$

$k=1,\ldots,m$,
(5.20.4) For all $u \in \mathbb{R}^n$, $x_k(I_k(u)) \le x_k(u) < x_{k+1}(I_k(u))$ for $k = 1,\ldots,m$,
(5.20.5) There exists a constant $M^* > 0$ such that

$$\|I_k(u)\| \le M^*(1 + \|u\|_{\mathcal{B}}),\ \textit{for each } u \in \mathcal{B};\ k = 1,\ldots,m.$$

Then the IVP (5.35)–(5.38) has at least one solution on $(-\infty, a] \times (-\infty, b]$.

Proof. The proof will be given in several steps.

Step 1: Set

$$\Omega_0=\{u : (-\infty, a] \times (-\infty, b] \to \mathbb{R}^n : u_{(x,y)} \in \mathcal{B} \text{ for } (x, y) \in E \text{ and } u \in C(J, \mathbb{R}^n)\}.$$

Consider the following problem:

$$({}^c D_0^r u)(x, y) = f(x, y, u_{(x,y)}), \text{ if } (x, y) \in J, \tag{5.39}$$

$$u(x, y) = \phi(x, y); \text{ if } (x, y) \in \tilde{J}', \tag{5.40}$$

$$u(x, 0) = \varphi(x), \ x \in [0, a], \ u(0, y) = \psi(y), \ y \in [0, b]. \tag{5.41}$$

Transform problems (5.39)–(5.41) into a fixed-point problem. Consider the operator $N : \Omega_0 \to \Omega_0$ defined by

$$N(u)(x, y) = \begin{cases} \phi(x, y); & (x, y) \in \tilde{J}, \\ \mu(x, y) \\ +\frac{1}{\Gamma(r_1)\Gamma(r_2)} \int\limits_0^x \int\limits_0^y (x-s)^{r_1-1}(y-t)^{r_2-1} f(s, t, u_{(s,t)}) \mathrm{d}t \mathrm{d}s; & (x, y) \in J. \end{cases} \tag{5.42}$$

Let $v(.,.) : (-\infty, a] \times (-\infty, b] \to \mathbb{R}^n$ be a function defined by

$$v(x, y) = \begin{cases} \phi(x, y), \ (x, y) \in \tilde{J}', \\ \mu(x, y), \ (x, y) \in J. \end{cases}$$

Then $v_{(x,y)} = \phi$ for all $(x, y) \in E$. For each $w \in C(J, \mathbb{R}^n)$ with $w(0, 0) = 0$, we denote by $\overline{w}$ the function defined by

$$\overline{w}(x, y) = \begin{cases} 0, & (x, y) \in \tilde{J}', \\ w(x, y) & (x, y) \in J. \end{cases}$$

If $u(.,.)$ satisfies the integral equation

$$u(x, y) = \mu(x, y) + \frac{1}{\Gamma(r_1)\Gamma(r_2)} \int\limits_0^x \int\limits_0^y (x-s)^{r_1-1}(y-t)^{r_2-1} f(s, t, u_{(s,t)}) \mathrm{d}t \mathrm{d}s,$$

we can decompose $u(.,.)$ as $u(x, y) = \overline{w}(x, y) + v(x, y); \ (x, y) \in J$, which implies $u_{(x,y)} = \overline{w}_{(x,y)} + v_{(x,y)}$, for every $(x, y) \in J$, and the function $w(.,.)$ satisfies

$$w(x, y) = \frac{1}{\Gamma(r_1)\Gamma(r_2)} \int\limits_0^x \int\limits_0^y (x-s)^{r_1-1}(y-t)^{r_2-1} f(s, t, \overline{w}_{(s,t)} + v_{(s,t)}) \mathrm{d}t \mathrm{d}s.$$

Set

$$C_0 = \{w \in \Omega_0 : w(x,y) = 0 \text{ for } (x,y) \in E\},$$

and let $\|.\|_{(a,b)}$ be the norm in C_0 defined by

$$\|w\|_{(a,b)} = \sup_{(x,y)\in E} \|w_{(x,y)}\|_{\mathcal{B}} + \sup_{(x,y)\in J} \|w(x,y)\| = \sup_{(x,y)\in J} \|w(x,y)\|, \; w \in C_0.$$

C_0 is a Banach space with norm $\|.\|_{(a,b)}$. Let the operator $P: C_0 \to C_0$ be defined by

$$P(w)(x,y) = \frac{1}{\Gamma(r_1)\Gamma(r_2)} \int_0^x \int_0^y (x-s)^{r_1-1}(y-t)^{r_2-1} \times f(s,t,\overline{w}_{(s,t)} + v_{(s,t)})\mathrm{d}t\mathrm{d}s, \quad (x,y) \in J. \tag{5.43}$$

The operator N has a fixed point is equivalent to P has a fixed point, and so we turn to proving that P has a fixed point. We shall use the Leray–Schauder alternative to prove that P has fixed point. We shall show that the operator P is continuous and completely continuous.

Claim 1. P is continuous. Let$\{w_n\}$ be a sequence such that $w_n \to w$ in C_0. Then

$$\begin{aligned}
&\|P(w_n)(x,y) - P(w)(x,y)\| \\
&\le \frac{1}{\Gamma(r_1)\Gamma(r_2)} \int_0^a \int_0^b (x-s)^{r_1-1}(y-t)^{r_2-1} \\
&\quad \times \|f(s,t,\overline{w_{n(s,t)}} + v_{n(s,t)}) - f(s,t,\overline{w}_{(s,t)} + v_{(s,t)})\|\mathrm{d}t\mathrm{d}s.
\end{aligned}$$

Since f is a Carathéodory function, then we have

$$\begin{aligned}
\|P(w_n) - P(w)\|_\infty &\le \frac{a^{r_1}b^{r_2}\|f(.,.,\overline{w_n}_{(.,.)} + v_{n(.,.)}) - f(.,.,\overline{w}_{(.,.)} + v_{(.,.)})\|_\infty}{\Gamma(r_1+1)\Gamma(r_2+1)} \\
&\to 0 \text{ as } n \to \infty.
\end{aligned}$$

Claim 2. P maps bounded sets into bounded sets in C_0. Indeed, it is enough to show that, for any $\eta > 0$, there exists a positive constant $\widetilde{\ell}$ such that, for each $w \in B_\eta = \{w \in C_0 : \|w\|_{(a,b)} \le \eta\}$, we have $\|P(w)\|_\infty \le \widetilde{\ell}$. Let $w \in B_\eta$. By (H_{162}) we have for each $(x,y) \in J$,

$$\begin{aligned}\|P(w)(x,y)\| &\leq \frac{1}{\Gamma(r_1)\Gamma(r_2)}\int_0^x\int_0^y (x-s)^{r_1-1}(y-t)^{r_2-1}\\ &\quad\times\|f(s,t,\overline{w}_{(s,t)}+v_{(s,t)})\|\mathrm{d}t\mathrm{d}s\\ &\leq \frac{1}{\Gamma(r_1)\Gamma(r_2)}\int_0^x\int_0^y (x-s)^{r_1-1}(y-t)^{r_2-1}\\ &\quad\times M(1+\|\overline{w}_{(s,t)}+v_{(s,t)}\|_{\mathcal{B}})\mathrm{d}t\mathrm{d}s\\ &\leq \frac{M(1+\eta^*)}{\Gamma(r_1)\Gamma(r_2)}\int_0^a\int_0^b (x-s)^{r_1-1}(y-t)^{r_2-1}\mathrm{d}t\mathrm{d}s\\ &\leq \frac{Ma^{r_1}b^{r_2}(1+\eta^*)}{\Gamma(r_1+1)\Gamma(r_2+1)} := \ell^*,\end{aligned}$$

where

$$\begin{aligned}\|\overline{w}_{(s,t)}+v_{(s,t)}\|_{\mathcal{B}} &\leq \|\overline{w}_{(s,t)}\|_{\mathcal{B}}+\|v_{(s,t)}\|_{\mathcal{B}}\\ &\leq K\eta+K\|\phi(0,0)\|+M\|\phi\|_{\mathcal{B}} := \eta^*.\end{aligned}$$

Hence $\|P(w)\|_\infty \leq \ell^*$.

Claim 3. P maps bounded sets into equicontinuous sets in C_0. Let $(x_1,y_1),(x_2,y_2)\in(0,a]\times(0,b]$, $x_1<x_2$, $y_1<y_2$, B_η be a bounded set as in Claim 2, and let $w\in B_\eta$. Then

$$\begin{aligned}&\|P(w)(x_2,y_2)-P(w)(x_1,y_1)\|\\ &\leq \frac{1}{\Gamma(r_1)\Gamma(r_2)}\left\|\int_0^{x_1}\int_0^{y_1}\left[(x_2-s)^{r_1-1}(y_2-t)^{r_2-1}\right.\right.\\ &\quad\left.-(x_1-s)^{r_1-1}(y_1-t)^{r_2-1}\right]f(s,t,u_{(s,t)})\mathrm{d}t\mathrm{d}s\\ &\quad\left.+\frac{1}{\Gamma(r_1)\Gamma(r_2)}\int_{x_1}^{x_2}\int_{y_1}^{y_2}(x_2-s)^{r_1-1}(y_2-t)^{r_2-1}f(s,t,\overline{w}_{(s,t)}+v_{(s,t)})\mathrm{d}t\mathrm{d}s\right\|\\ &\quad+\frac{1}{\Gamma(r_1)\Gamma(r_2)}\int_0^{x_1}\int_{y_1}^{y_2}(x_2-s)^{r_1-1}(y_2-t)^{r_2-1}\|f(s,t,\overline{w}_{(s,t)}+v_{(s,t)})\|\mathrm{d}t\mathrm{d}s\\ &\quad+\frac{1}{\Gamma(r_1)\Gamma(r_2)}\int_{x_1}^{x_2}\int_0^{y_1}(x_2-s)^{r_1-1}(y_2-t)^{r_2-1}\|f(s,t,\overline{w}_{(s,t)}+v_{(s,t)})\|\mathrm{d}t\mathrm{d}s\end{aligned}$$

$$\leq \frac{M(1+\eta)}{\Gamma(r_1)\Gamma(r_2)}\int_0^{x_1}\int_0^{y_1}[(x_1-s)^{r_1-1}(y_1-t)^{r_2-1}-(x_2-s)^{r_1-1}(y_2-t)^{r_2-1}]dtds$$

$$+\frac{M(1+\eta)}{\Gamma(r_1)\Gamma(r_2)}\int_{x_1}^{x_2}\int_{y_1}^{y_2}(x_2-s)^{r_1-1}(y_2-t)^{r_2-1}dtds$$

$$+\frac{M(1+\eta)}{\Gamma(r_1)\Gamma(r_2)}\int_0^{x_1}\int_{y_1}^{y_2}(x_2-s)^{r_1-1}(y_2-t)^{r_2-1}dtds$$

$$+\frac{M(1+\eta)}{\Gamma(r_1)\Gamma(r_2)}\int_{x_1}^{x_2}\int_0^{y_1}(x_2-s)^{r_1-1}(y_2-t)^{r_2-1}dtds$$

$$\leq \frac{M(1+\eta)}{\Gamma(r_1+1)\Gamma(r_2+1)}[2y_2^{r_2}(x_2-x_1)^{r_1}+2x_2^{r_1}(y_2-y_1)^{r_2}$$
$$+x_1^{r_1}y_1^{r_2}-x_2^{r_1}y_2^{r_2}-2(x_2-x_1)^{r_1}(y_2-y_1)^{r_2}].$$

As $x_1 \to x_2$, $y_1 \to y_2$ the right-hand side of the above inequality tends to zero. The equicontinuity for the cases $x_1 < x_2 < 0$, $y_1 < y_2 < 0$ and $x_1 \leq 0 \leq x_2$, $y_1 \leq 0 \leq y_2$ is obvious. As a consequence of Claims 1–3, together with the Arzela–Ascoli theorem , we can conclude that $P : C_0 \to C_0$ is continuous and completely continuous.

Claim 4. (A priori bounds): We now show there exists an open set $U \subseteq C_0$ with $w \neq \lambda P(w)$, for $\lambda \in (0,1)$ and $w \in \partial U$. Let $w \in C_0$ and $w = \lambda P(w)$ for some $0 < \lambda < 1$. Thus for each $(x,y) \in J$,

$$w(x,y) = \frac{\lambda}{\Gamma(r_1)\Gamma(r_2)}\int_0^x\int_0^y(x-s)^{r_1-1}(y-t)^{r_2-1}f(s,t,u_{(s,t)})dtds.$$

This implies by (5.20.2) that, for each $(x,y) \in J$, we have

$$\|w(x,y)\| \leq \frac{1}{\Gamma(r_1)\Gamma(r_2)}\int_0^x\int_0^y(x-s)^{r_1-1}(y-t)^{r_2-1}M[1+\|\overline{w}_{(s,t)}+v_{(s,t)}\|_{\mathcal{B}}]dtds.$$

But

$$\|\overline{w}_{(s,t)}+v_{(s,t)}\|_{\mathcal{B}} \leq \|\overline{w}_{(s,t)}\|_{\mathcal{B}}+\|v_{(s,t)}\|_{\mathcal{B}}$$
$$\leq K\sup\{w(\tilde{s},\tilde{t}) : (\tilde{s},\tilde{t}) \in [0,s]\times[0,t]\}$$
$$+M\|\phi\|_{\mathcal{B}}+K\|\phi(0,0)\|. \quad (5.44)$$

If we name z(s,t) the right-hand side of (5.44), then we have

$$\|\overline{w}_{(s,t)} + v_{(s,t)}\|_{\mathcal{B}} \le z(x, y),$$

and therefore, for each $(x, y) \in J$ we obtain

$$\|w(x, y)\| \le \frac{M}{\Gamma(r_1)\Gamma(r_2)} \int_0^x \int_0^y (x-s)^{r_1-1}(y-t)^{r_2-1}(1+z(s,t))dtds. \quad (5.45)$$

Using the above inequality and the definition of z we have that

$$\begin{aligned} z(x, y) &\le M\|\phi\|_{\mathcal{B}} + K\|\phi(0,0)\| \\ &+ \frac{KM}{\Gamma(r_1)\Gamma(r_2)} \int_0^x \int_0^y (x-s)^{r_1-1}(y-t)^{r_2-1}(1+z(s,t))dtds \\ &\le M\|\phi\|_{\mathcal{B}} + K\|\phi(0,0)\| + \frac{KMa^{r_1}b^{r_2}}{\Gamma(r_1+1)\Gamma(r_2+1)} \\ &+ \frac{KM}{\Gamma(r_1)\Gamma(r_2)} \int_0^x \int_0^y (x-s)^{r_1-1}(y-t)^{r_2-1}z(s,t)dtds, \end{aligned}$$

for each $(x, y) \in J$. Set

$$R = M\|\phi\|_{\mathcal{B}} + K\|\phi(0,0)\| + \frac{KMa^{r_1}b^{r_2}}{\Gamma(r_1+1)\Gamma(r_2+1)}.$$

Then for Lemma 2.43, there exists $\delta = \delta(r_1, r_2)$ such that

$$\|z\|_\infty \le R\left[1 + \frac{\delta KMa^{r_1}b^{r_2}}{\Gamma(r_1+1)\Gamma(r_2+1)}\right] := \widetilde{M}.$$

Then, (5.45) implies that

$$\|w\|_\infty \le \frac{M(1+\widetilde{M})a^{r_1}b^{r_2}}{\Gamma(r_1+1)\Gamma(r_2+1)} := M^*.$$

Set

$$U = \{w \in C_0 : \|w\|_{(a,b)} < M^* + 1\}.$$

$P : \overline{U} \to C_0$ is continuous and completely continuous. By our choice of U, there is no $w \in \partial U$ such that $w = \lambda P(w)$, for $\lambda \in (0, 1)$. As a consequence of the nonlinear alternative of Leray–Schauder type [136], we deduce that N has a fixed point which is a solution to problems (5.39)–(5.41). Denote this solution by u_1. Define the functions

$$r_{k,1}(x,y) = x_k(u_1(x,y)) - x, \qquad \text{for } x \geq 0, y \geq 0.$$

Hypothesis (5.20.1) implies that $r_{k,1}(0,0) \neq 0$ for $k = 1, \ldots, m$.

If $r_{k,1}(x,y) \neq 0$ on J for $k = 1, \ldots, m$; i.e.,

$$x \neq x_k(u_1(x,y)), \quad \text{on } J \quad \text{for } k = 1, \ldots, m,$$

then u_1 is a solution of the problems (5.35)–(5.38). It remains to consider the case when $r_{1,1}(x,y) = 0$ for some $(x,y) \in J$. Now since $r_{1,1}(0,0) \neq 0$ and $r_{1,1}$ is continuous, there exists $x_1 > 0, y_1 > 0$ such that $r_{1,1}(x_1,y_1) = 0$, and $r_{1,1}(x,y) \neq 0$, for all $(x,y) \in [0,x_1) \times [0,y_1)$.

Thus, we have

$$r_{1,1}(x_1,y_1) = 0 \text{ and } r_{1,1}(x,y) \neq 0, \text{ for all } (x,y) \in [0,x_1) \times [0,y_1] \cup (y_1,b].$$

Suppose that there exist $(\bar{x},\bar{y}) \in [0,x_1) \times [0,y_1] \cup (y_1,b]$ such that $r_{1,1}(\bar{x},\bar{y}) = 0$. The function $r_{1,1}$ attains a maximum at some point $(s,t) \in [0,x_1) \times [0,b]$. Since

$$({}^cD_0^r u_1)(x,y) = f(x,y,u_{1(x,y)}), \text{ for } (x,y) \in J,$$

then

$$\frac{\partial u_1(x,y)}{\partial x} \text{ exists, and } \frac{\partial r_{1,1}(s,t)}{\partial x} = x_1'(u_1(s,t))\frac{\partial u_1(s,t)}{\partial x} - 1 = 0.$$

Since

$$\frac{\partial u_1(x,y)}{\partial x} = \varphi'(x) + \frac{r_1 - 1}{\Gamma(r_1)\Gamma(r_2)} \int_0^x \int_0^y (x-s)^{r_1-2}(y-t)^{r_2-1} f(s,t,u_{1(s,t)})\mathrm{d}t\mathrm{d}s,$$

then

$$x_1'(u_1(s,t)) \left[\varphi'(s) + \frac{r_1-1}{\Gamma(r_1)\Gamma(r_2)} \int_0^s \int_0^t (s-\theta)^{r_1-2}(t-\eta)^{r_2-1} f(\theta,\eta,u_{1(\theta,\eta)})\mathrm{d}\theta\mathrm{d}\eta \right] = 1,$$

which contradicts (5.20.3). From (5.20.1) we have

$$r_{k,1}(x,y) \neq 0 \text{ for all } (x,y) \in [0,x_1) \times [0,b] \text{ and } k = 1, \ldots m.$$

Step 2: In what follows set

$$\begin{aligned}\Omega_k =\{u : (-\infty, a] \times (-\infty, b] \to \mathbb{R}^n : u_{(x,y)} \in \mathcal{B} \text{ for } (x, y) \in E \text{ and there exist} \\ 0 = x_0 < x_1 < x_2 < \cdots < x_m < x_{m+1} = a \text{ such that } x_k = x_k(u(x_k, .)), \\ \text{and } u(x_k^-, .),\ u(x_k^+, .) \text{ exist with } u(x_k^-, .) = u(x_k, .);\ k = 1, \dots, m, \\ \text{and } u \in C(X_k, \mathbb{R}^n); k = 0, \dots, m\},\end{aligned}$$

where

$$X_k := [x_k, a] \times [0, b];\ k = 1, \dots, m.$$

Consider now the problem

$$({}^c D_{x_1}^r u)(x, y) = f(x, y, u_{(x,y)}),\ \text{if } (x, y) \in X_1, \tag{5.46}$$

$$u(x_1^+, y) = I_1(u_1(x_1, y)) \tag{5.47}$$

$$u(x, y) = u_1(x, y),\ \text{if } (x, y) \in \tilde{J} \cup [0, x_1) \times [0, b]. \tag{5.48}$$

Consider the operator $N_1 : \Omega_1 \to \Omega_1$ defined as

$$N_1(u)(x, y)= \begin{cases} u_1(x, y), & (x, y) \in \tilde{J} \cup [0, x_1) \times [0, b], \\ \varphi(x) + I_1(u_1(x_1, y)) - I_1(u_1(x_1, 0)) & \\ +\frac{1}{\Gamma(r_1)\Gamma(r_2)} \int_{x_1}^{x} \int_0^y (x-s)^{r_1-1}(y-t)^{r_2-1} & \\ \times f(s, t, u_{(s,t)})dt\,ds, & (x, y) \in X_1. \end{cases}$$

As in Step 1 we can show that N_1 is completely continuous. We now show there exists an open set $U' \subseteq \Omega_1$ with $w \neq \lambda N_1(w)$, for $\lambda \in (0, 1)$ and $w \in \partial U'$. Let $w \in \Omega_1$ and $w = \lambda N_1(w)$ for some $0 < \lambda < 1$. Thus, from (5.20.2) and (5.20.5) we get for each $(x, y) \in X_1$,

$$\begin{aligned}\|w(x, y)\| &\leq \|\varphi(x)\| + \|I_1(u_1(x_1, y))\| + \|I_1(u_1(x_1, 0))\| \\ &\quad + \frac{1}{\Gamma(r_1)\Gamma(r_2)} \int_{x_1}^{x} \int_0^y (x-s)^{r_1-1}(y-t)^{r_2-1} \| f(s, t, u_{(s,t)})\| dt\,ds \\ &\leq \|\varphi\|_\infty + 2M^*(1 + \|u_1\|) \\ &\quad + \frac{M}{\Gamma(r_1)\Gamma(r_2)} \int_0^x \int_0^y (x-s)^{r_1-1}(y-t)^{r_2-1}(1 + \|z(s, t)\|)dt\,ds \\ &\leq \|\varphi\|_\infty + 2M^*(1 + \|u_1\|) + \frac{M(1 + \widetilde{M})a^{r_1}b^{r_2}}{\Gamma(r_1 + 1)\Gamma(r_2 + 1)} := R^*.\end{aligned}$$

Set

$$U' = \{w \in \Omega_1 : \|w\| < R^* + 1\}.$$

$N_1 : \overline{U'} \to \Omega_1$ is continuous and completely continuous . By our choice of U', there is no $w \in \partial U'$ such that $w = \lambda N_1(w)$, for $\lambda \in (0, 1)$. As a consequence of the nonlinear alternative of Leray–Schauder type [136], we deduce that N_1 has a fixed point u which is a solution to problem (5.46)–(5.48). Denote this solution by u_2. Define

$$r_{k,2}(x, y) = x_k(u_2(x, y)) - x, \quad \text{for } (x, y) \in X_1.$$

If $r_{k,2}(x, y) \neq 0$ on $(x_1, a] \times [0, b]$ and for all $k = 1, \dots, m$, then

$$u(x, y) = \begin{cases} u_1(x, y), & \text{if } (x, y) \in \tilde{J} \cup [0, x_1) \times [0, b], \\ u_2(x, y), & \text{if } (x, y) \in [x_1, a] \times [0, b], \end{cases}$$

is a solution of the problems (5.35)–(5.38). It remains to consider the case when $r_{2,2}(x, y) = 0$, for some $(x, y) \in (x_1, a] \times [0, b]$. By (5.20.4), we have

$$\begin{aligned} r_{2,2}(x_1^+, y_1) &= x_2(u_2(x_1^+, y_1) - x_1 \\ &= x_2(I_1(u_1(x_1, y_1))) - x_1 \\ &> x_1(u_1(x_1, y_1)) - x_1 \\ &= r_{1,1}(x_1, y_1) = 0. \end{aligned}$$

Since $r_{2,2}$ is continuous, there exists $x_2 > x_1,\ y_2 > y_1$ such that $r_{2,2}(x_2, y_2) = 0$, and $r_{2,2}(x, y) \neq 0$ for all $(x, y) \in (x_1, x_2) \times [0, b]$. It is clear by (5.20.1) that

$$r_{k,2}(x, y) \neq 0 \quad \text{for all } (x, y) \in (x_1, x_2) \times [0, b],\ k = 2, \dots, m.$$

Now suppose that there are $(s, t) \in (x_1, x_2) \times [0, b]$ such that $r_{1,2}(s, t) = 0$. From (5.20.4) it follows that

$$\begin{aligned} r_{1,2}(x_1^+, y_1) &= x_1(u_2(x_1^+, y_1) - x_1 \\ &= x_1(I_1(u_1(x_1, y_1))) - x_1 \\ &\leq x_1(u_1(x_1, y_1)) - x_1 \\ &= r_{1,1}(x_1, y_1) = 0. \end{aligned}$$

Thus $r_{1,2}$ attains a nonnegative maximum at some point $(s_1, t_1) \in (x_1, a) \times [0, x_2) \cup (x_2, b]$. Since

$$({}^cD^r_{x_1} u_2)(x, y) = f(x, y, u_{2(x,y)}), \text{ for } (x, y) \in X_1,$$

then we get

$$\begin{aligned} u_2(x, y) &= \varphi(x) + I_1(u_1(x_1, y)) - I_1(u_1(x_1, 0)) \\ &+ \frac{1}{\Gamma(r_1)\Gamma(r_2)} \int_{x_1}^{x} \int_{0}^{y} (x - s)^{r_1 - 1}(y - t)^{r_2 - 1} f(s, t, u_{2(s,t)}) dt ds, \end{aligned}$$

hence

$$\frac{\partial u_2}{\partial x}(x,y) = \varphi'(x) + \frac{r_1 - 1}{\Gamma(r_1)\Gamma(r_2)} \int_{x_1}^{x} \int_{0}^{y} (x-s)^{r_1-2}(y-t)^{r_2-1} f(s,t,u_{2(s,t)}) \mathrm{d}t \mathrm{d}s,$$

then

$$\frac{\partial r_{1,2}(s_1,t_1)}{\partial x} = x_1'(u_2(s_1,t_1)) \frac{\partial u_2}{\partial x}(s_1,t_1) - 1 = 0.$$

Therefore

$$x_1'(u_2(s_1,t_1)) \left[\varphi'(s_1) + \frac{r_1-1}{\Gamma(r_1)\Gamma(r_2)} \int_{x_1}^{s_1} \int_{0}^{t_1} (s_1-\theta)^{r_1-2}(t_1-\eta)^{r_2-1} f(\theta,\eta,u_{2(\theta,\eta)}) \mathrm{d}\eta \mathrm{d}\theta \right] = 1,$$

which contradicts (5.20.3).

Step 3: We continue this process and take into account that $u_{m+1} := u\Big|_{X_m}$ is a solution to the problem

$$\begin{cases} ({}^c D_{x_m}^r u)(x,y) = f(x,y,u_{(x,y)}), & \text{a.e. } (x,y) \in (x_m,a] \times [0,b], \\ u(x_m^+,y) = I_m(u_{m-1}(x_m,y)), \\ u(x,y) = u_1(x,y), \text{ if } (x,y) \in \tilde{J} \cup [0,x_1) \times [0,b], \\ u(x,y) = u_2(x,y), \text{ if } (x,y) \in [x_1,x_2) \times [0,b], \\ \dots \\ u(x,y) = u_m(x,y), \text{ if } (x,y) \in [x_{m-1},x_m) \times [0,b]. \end{cases}$$

The solution u of the problems (5.35)–(5.38) is then defined by

$$u(x,y) = \begin{cases} u_1(x,y), & \text{if } (x,y) \in \tilde{J} \cup [0,x_1] \times [0,b], \\ u_2(x,y), & \text{if } (x,y) \in (x_1,x_2] \times [0,b], \\ \dots \\ u_{m+1}(x,y), & \text{if } (x,y) \in (x_m,a] \times [0,b]. \end{cases}$$

□

5.5.3 An Example

As an application of our results we consider the following impulsive partial hyperbolic differential equations of the form:

$$({}^c D^r_{x_k} u)(x,y) = \frac{1 + \|u_{(x,y)}\|}{9 + e^{x+y}}; \text{ if } (x,y) \in J_k; \ k = 0, \dots, m,$$

$$x_k = x_k(u(x,y)); \ k = 1, \dots, m, \tag{5.49}$$

$$u(x_k^+, y) = d_k u(x_k, y); \ y \in [0,1], \ k = 1, \dots, m, \tag{5.50}$$

$$u(x,y) = x + y^2, \ (x,y) \in (-\infty, 1] \times (-\infty, 1] \backslash (0,1] \times (0,1], \tag{5.51}$$

$$u(x,0) = x, \ u(0,y) = y^2; \ x, y \in [0,1], \tag{5.52}$$

where $J = [0,1] \times [0,1]$, $r = (r_1, r_2)$, $0 < r_1, r_2 \leq 1$, $x_k(u) = 1 - \frac{1}{2^k(1+u^2)}$; $k = 1, \dots, m$ and $\frac{\sqrt{2}}{2} < d_k \leq 1$; $k = 1, \dots, m$. Let $\mathcal{B}_\gamma$ be the phase space defined in the Example of Sect. 3.7. Set

$$f(x,y,u_{(x,y)}) = \frac{1 + \|u_{(x,y)}\|}{9 + e^{x+y}}, \ (x,y) \in [0,1] \times [0,1], \ u_{(x,y)} \in \mathcal{B}_\gamma,$$

and

$$I_k(u) = d_k u; \ u \in \mathbb{R}, \ k = 1, \dots, , m.$$

Let $u \in \mathbb{R}$ then we have

$$x_{k+1}(u) - x_k(u) = \frac{1}{2^{k+1}(1+u^2)} > 0; \ k = 1, \dots, m.$$

Hence $0 < x_1(u) < x_2(u) < \cdots < x_m(u) < 1$, for each $u \in \mathbb{R}$. Also, for each $u \in \mathbb{R}$ we have

$$x_{k+1}(I_k(u)) - x_k(u) = \frac{1 + (2d_k^2 - 1)u^2}{2^{k+1}(1+u^2)(1+d_k^2)} > 0.$$

Finally, for all $(x,y) \in J$ and each $u \in \mathbb{R}$ we get

$$|I_k(u)| = |d_k u| \leq |u| \leq 3(1 + |u|); \ k = 1, \dots, m,$$

and

$$|f(x,y,u)| = \frac{|1+u|}{9 + e^{x+y}} \leq \frac{1}{10}(1 + |u|).$$

Since all conditions of Theorem 5.20 are satisfied, problem (5.49)–(5.52) have at least one solution on $(-\infty, 1] \times (-\infty, 1]$.

5.6 Impulsive Partial Hyperbolic Functional Differential Equations of Fractional Order with State-Dependent Delay

5.6.1 Introduction

In this section, we start by studying the existence result to fractional order IVP , for the system

$$({}^cD^r_{x_k}u)(x,y) = f(x,y,u_{(\rho_1(x,y,u_{(x,y)}),\rho_2(x,y,u_{(x,y)}))});\ \text{if } (x,y)\in J_k;\ k=0,\dots,m, \tag{5.53}$$

$$u(x_k^+,y) = u(x_k^-,y) + I_k(u(x_k^-,y));\ \text{if } y\in[0,b];\ k=1,\dots,m, \tag{5.54}$$

$$u(x,y) = \phi(x,y),\ \text{if } (x,y)\in\tilde{J} := [-\alpha,a]\times[-\beta,b]\backslash(0,a]\times(0,b], \tag{5.55}$$

$$u(x,0) = \varphi(x),\ x\in[0,a],\ u(0,y) = \psi(y),\ y\in[0,b], \tag{5.56}$$

where $a,b,\alpha,\beta>0$, $0=x_0<x_1<\cdots<x_m<x_{m+1}=a$, $\phi:\tilde{J}\to\mathbb{R}^n$, $J=[0,a]\times[0,b]$, is a given continuous function, φ,ψ are as in problems (3.1)–(3.3), $f:J\times C\to\mathbb{R}^n$, $\rho_1:J\times C\to[-\alpha,a]$, $\rho_2:J\times C\to[-\beta,b]$, $I_k:\mathbb{R}^n\to\mathbb{R}^n$; $k=1,\dots,m$ are given functions and C is the Banach space defined by

$$C = C_{(\alpha,\beta)} = \{u:[-\alpha,0]\times[-\beta,0]\to\mathbb{R}^n:\ \text{continuous and there exist}$$
$$\tau_k\in(-\alpha,0)\ \text{with } u(\tau_k^-,\tilde{y})\ \text{and } u(x_k^+,\tilde{y}),\ k=1,\dots,m,\ \text{exist for any}$$
$$\tilde{y}\in[-\beta,0]\ \text{with } u(\tau_k^-,\tilde{y}) = u(\tau_k,\tilde{y})\},$$

with norm

$$\|u\|_{PC} = \sup_{(x,y)\in[-\alpha,0]\times[-\beta,0]}\|u(x,y)\|.$$

Next we consider the following system of partial hyperbolic differential equations of fractional order with infinite delay

$$({}^cD^r_{x_k}u)(x,y) = f(x,y,u_{(\rho_1(x,y,u_{(x,y)}),\rho_2(x,y,u_{(x,y)}))});\ \text{if } (x,y)\in J_k;\ k=0,\dots,m, \tag{5.57}$$

$$u(x_k^+,y) = u(x_k^-,y) + I_k(u(x_k^-,y));\ \text{if } (x,y)\in J_k;\ k=1,\dots,m, \tag{5.58}$$

$$u(x,y) = \phi(x,y),\ \text{if } (x,y)\in\tilde{J}' := (-\infty,a]\times(-\infty,b]\backslash(0,a]\times(0,b], \tag{5.59}$$

$$u(x,0) = \varphi(x),\ u(0,y) = \psi(y);\ x,y\in[0,b], \tag{5.60}$$

where φ, ψ, I_k are as in problems (5.53)–(5.56), $f:J\times\mathcal{B}\to\mathbb{R}^n$, $\rho_1:J\times\mathcal{B}\to(-\infty,a]$, $\rho_2:J\times\mathcal{B}\to(-\infty,b]$, $\phi:\tilde{J}'\to\mathbb{R}^n$ and $\mathcal{B}$ is a phase space.

5.6.2 Impulsive Partial Differential Equations with Finite Delay

Consider the Banach space

$$\begin{aligned} PC &:= PC(J,\mathbb{R}^n) \\ &= \big\{u : J \to \mathbb{R}^n : u \in C(J_k,\mathbb{R}^n);\ k = 1,\ldots,m, \text{ and there exist } u(x_k^-,y) \\ &\quad \text{and } u(x_k^+,y);\, k = 1,\ldots,m, \text{ with } u(x_k^-,y) = u(x_k,y),\ y \in [0,b]\big\}, \end{aligned}$$

with the norm

$$\|u\|_{PC} = \sup_{(x,y)\in J} \|u(x,y)\|.$$

Set $\widetilde{PC} := PC([-\alpha,a]\times[-\beta,b],\mathbb{R}^n)$, which is a Banach space with the norm

$$\|u\|_{\widetilde{PC}} = \sup\{\|u(x,y)\| : (x,y) \in [-\alpha,a]\times[-\beta,b]\}.$$

Definition 5.21. A function $u \in \widetilde{PC}$ such that its mixed derivative D^2_{xy} exists on $J_k;\ k = 0,\ldots,m$ is said to be a solution of (5.53)–(5.56) if u satisfies the condition (5.55) on $\tilde{J}$, (5.53) on $J_k;\ k = 0,\ldots,m$ and conditions (5.54) and (5.56) are satisfied.

Set $\mathcal{R} := \mathcal{R}_{(\rho_1^-,\rho_2^-)}$

$$= \{(\rho_1(s,t,u),\rho_2(s,t,u)) : (s,t,u) \in J\times C,\ \rho_i(s,t,u) \le 0;\ i = 1,2\}.$$

We always assume that $\rho_i : J\times C \to \mathbb{R};\ i = 1,2$ are continuous and the function $(s,t) \longmapsto u_{(s,t)}$ is continuous from $\mathcal{R}$ into C.

The first result is based on Banach fixed-point theorem.

Theorem 5.22. *Let $f : J\times C \longrightarrow \mathbb{R}^n$ be continuous. Assume that*

(5.22.1) There exists a constant $l > 0$ such that

$$\|f(x,y,u)-f(x,y,\overline{u})\| \le l\,\|u-\overline{u}\|_C, \text{ for each } (x,y)\in J \text{ and each } u,\overline{u}\in C,$$

(5.22.2) There exists a constant $l^* > 0$ such that

$$\|I_k(u) - I_k(\overline{u})\| \le l^*\|u-\overline{u}\|, \text{ for each } u,\overline{u} \in \mathbb{R}^n;\ k = 1,\ldots,m.$$

If

$$2ml^* + \frac{2la^{r_1}b^{r_2}}{\Gamma(r_1+1)\Gamma(r_2+1)} < 1, \tag{5.61}$$

then (5.53)–(5.56) have a unique solution on $[-\alpha,a]\times[-\beta,b]$.

Proof. We transform the problems (5.53)–(5.56) into a fixed-point problem. Consider the operator $F : \widetilde{PC} \to \widetilde{PC}$ defined by

$$F(u)(x,y) = \begin{cases} \phi(x,y), & (x,y) \in \tilde{J}, \\ \mu(x,y) + \sum\limits_{0<x_k<x} (I_k(u(x_k^-,y)) - I_k(u(x_k^-,0))) & \\ + \frac{1}{\Gamma(r_1)\Gamma(r_2)} \sum\limits_{0<x_k<x} \int\limits_{x_{k-1}}^{x_k} \int\limits_{0}^{y} (x_k - s)^{r_1-1}(y-t)^{r_2-1} & \\ \times f(s,t,u_{(\rho_1(s,t,u_{(s,t)}),\rho_2(s,t,u_{(s,t)}))})\mathrm{d}t\mathrm{d}s & \\ + \frac{1}{\Gamma(r_1)\Gamma(r_2)} \int\limits_{x_k}^{x} \int\limits_{0}^{y} (x-s)^{r_1-1}(y-t)^{r_2-1} & \\ \times f(s,t,u_{(\rho_1(s,t,u_{(s,t)}),\rho_2(s,t,u_{(s,t)}))})\mathrm{d}t\mathrm{d}s, & (x,y) \in J. \end{cases}$$

Clearly, from Lemma 2.15 the fixed points of the operator F are solutions of the problems (5.53)–(5.56). We shall use the Banach contraction principle to prove that F has a fixed point. For this, we show that F *is a contraction*. Let $u, v \in \widetilde{PC}$, then for each $(x,y) \in J$, we have

$$\begin{aligned} &\|F(u)(x,y) - F(v)(x,y)\| \\ &\le \sum_{k=1}^{m} (\|I_k(u(x_k^-,y)) - I_k(v(x_k^-,y))\| + \|I_k(u(x_k^-,0)) - I_k(v(x_k^-,0))\|) \\ &+ \frac{1}{\Gamma(r_1)\Gamma(r_2)} \sum_{k=1}^{m} \int\limits_{x_{k-1}}^{x_k} \int\limits_{0}^{y} (x_k-s)^{r_1-1}(y-t)^{r_2-1} \| f(s,t,u_{(\rho_1(s,t,u_{(s,t)}),\rho_2(s,t,u_{(s,t)}))}) \\ &- f(s,t,v_{(\rho_1(s,t,u_{(s,t)}),\rho_2(s,t,u_{(s,t)}))})\|\mathrm{d}t\mathrm{d}s \\ &+ \frac{1}{\Gamma(r_1)\Gamma(r_2)} \int\limits_{x_k}^{x} \int\limits_{0}^{y} (x-s)^{r_1-1}(y-t)^{r_2-1} \| f(s,t,u_{(\rho_1(s,t,u_{(s,t)}),\rho_2(s,t,u_{(s,t)}))}) \\ &- f(s,t,v_{(\rho_1(s,t,u_{(s,t)}),\rho_2(s,t,u_{(s,t)}))})\|\mathrm{d}t\mathrm{d}s \\ &\le \sum_{k=1}^{m} l^*(\|u(x_k^-,y) - v(x_k^-,y)\| + \|u(x_k^-,0) - v(x_k^-,0)\|) \\ &+ \frac{l}{\Gamma(r_1)\Gamma(r_2)} \sum_{k=1}^{m} \int\limits_{x_{k-1}}^{x_k} \int\limits_{0}^{y} (x_k - s)^{r_1-1}(y-t)^{r_2-1} \\ &\times \|u_{(\rho_1(s,t,u_{(s,t)}),\rho_2(s,t,u_{(s,t)}))} - v_{(\rho_1(s,t,u_{(s,t)}),\rho_2(s,t,u_{(s,t)}))}\|_C \mathrm{d}t\mathrm{d}s \end{aligned}$$

$$+\frac{l}{\Gamma(r_1)\Gamma(r_2)}\int_{x_k}^{x}\int_{0}^{y}(x-s)^{r_1-1}(y-t)^{r_2-1}$$

$$\times \|u_{(\rho_1(s,t,u_{(s,t)}),\rho_2(s,t,u_{(s,t)}))}-v_{(\rho_1(s,t,u_{(s,t)}),\rho_2(s,t,u_{(s,t)}))}\|_C dtds$$

$$\leq \left[2ml^* + \frac{2la^{r_1}b^{r_2}}{\Gamma(r_1+1)\Gamma(r_2+1)}\right]\|u-v\|_C.$$

By the condition (5.61), we conclude that F is a contraction . As a consequence of Banach's fixed-point theorem, we deduce that F has a unique fixed point which is a solution of the problems (5.53)–(5.56). □

In the following theorem we give an existence result for the problems (5.53)–(5.56) by applying the nonlinear alternative of Leray–Schauder type [136].

Theorem 5.23. *Let $f : J \times C \longrightarrow \mathbb{R}^n$ be continuous. Assume that the following conditions hold:*

(5.23.1) There exists $\phi_f \in C(J, \mathbb{R}_+)$ and $\psi : [0,\infty) \to (0,\infty)$ continuous and nondecreasing such that

$$\|f(x,y,u)\| \leq \phi_f(x,y)\psi(\|u\|_C), \quad \textit{for all } (x,y) \in J,\ u \in C,$$

(5.23.2) There exists $\psi^* : [0,\infty) \to (0,\infty)$ continuous and nondecreasing such that

$$\|I_k(u)\| \leq \psi^*(\|u\|), \quad \textit{for all } u \in \mathbb{R}^n,$$

(5.23.3) There exists a number $\overline{M} > 0$ such that

$$\frac{\overline{M}}{\|\mu\|_\infty + 2m\psi^*(\overline{M}) + \frac{2a^{r_1}b^{r_2}\phi_f^0\psi(\overline{M})}{\Gamma(r_1+1)\Gamma(r_2+1)}} > 1,$$

where $\phi_f^0 = \sup\{\phi_f(x,y) : (x,y) \in J\}$.

Then (5.53)–(5.56) have at least one solution on $[-\alpha, a] \times [-\beta, b]$.

Proof. Consider the operator F defined in Theorem 5.22. We shall show that the operator F is continuous and completely continuous.

A priori estimate. For $\lambda \in [0,1]$, let u be such that for each $(x,y) \in J$ we have $u(x,y) = \lambda(Fu)(x,y)$. For each $(x,y) \in J$, then from (5.23.1) and (5.23.2) we have

$$\|u(x,y)\| \leq \|\mu(x,y)\| + \sum_{k=1}^{m}(\|I_k(u(x_k^-,y))\| + \|I_k(u(x_k^-,0))\|)$$

$$+\frac{1}{\Gamma(r_1)\Gamma(r_2)}\sum_{k=1}^{m}\int_{x_{k-1}}^{x_k}\int_{0}^{y}(x_k-s)^{r_1-1}(y-t)^{r_2-1}$$

$$\times\|f(s,t,u_{(\rho_1(s,t,u_{(s,t)}),\rho_2(s,t,u_{(s,t)}))})\|dtds$$

$$+\frac{1}{\Gamma(r_1)\Gamma(r_2)}\int_{x_k}^{x}\int_{0}^{y}(x-s)^{r_1-1}(y-t)^{r_2-1}$$

$$\times\|f(s,t,u_{(\rho_1(s,t,u_{(s,t)}),\rho_2(s,t,u_{(s,t)}))})\|dtds$$

$$\leq \|\mu\|_\infty + 2m\psi^*(\|u\|) + \frac{2a^{r_1}b^{r_2}\phi_f^0\psi(\|u\|)}{\Gamma(r_1+1)\Gamma(r_2+1)}.$$

Thus

$$\frac{\|u\|_{PC}}{\|\mu\|_\infty + 2m\psi^*(\|u\|_{PC}) + \frac{2a^{r_1}b^{r_2}\phi_f^0\psi(\|u\|_{PC})}{\Gamma(r_1+1)\Gamma(r_2+1)}} \leq 1.$$

By condition (5.23.3), there exists $\overline{M}$ such that $\|u\|_\infty \neq \overline{M}$.

Let

$$U = \{u \in \widetilde{PC} : \|u\|_{\widetilde{PC}} < \overline{M}\}.$$

The operator $F : \overline{U} \to \widetilde{PC}$ is continuous and completely continuous. From the choice of U, there is no $u \in \partial U$ such that $u = \lambda F(u)$ for some $\lambda \in (0,1)$. As a consequence of the nonlinear alternative of Leray–Schauder type [136], we deduce that F has a fixed point u in $\overline{U}$ which is a solution of the problems (5.53)–(5.56). □

5.6.3 *Impulsive Partial Differential Equations with Infinite Delay*

Now we present two existence results for the problems (5.57)–(5.60). Let us start in this section by defining what we mean by a solution of the problems (5.57)–(5.60). Let the space

$$\Omega := \{u : (-\infty,a]\times(-\infty,b] \to \mathbb{R}^n : u_{(x,y)} \in \mathcal{B} \text{ for } (x,y) \in E \text{ and } u|_J \in PC\}.$$

Definition 5.24. A function $u \in \Omega$ such that its mixed derivative D^2_{xy} exists on J_k; $k = 0,\ldots,m$ is said to be a solution of (5.57)–(5.60) if u satisfies the condition (5.59) on $\tilde{J}'$, (5.57) on J_k, and conditions (5.58) and (5.60) are satisfied on J.

Set $\mathcal{R}' := \mathcal{R}'_{(\rho_1^-,\rho_2^-)}$

$$= \{(\rho_1(s,t,u),\rho_2(s,t,u)) : (s,t,u) \in J\times\mathcal{B},\ \rho_i(s,t,u) \leq 0;\ i = 1,2\}.$$

We always assume that $\rho_1 : J\times\mathcal{B} \to (-\infty,a]$, $\rho_2 : J\times\mathcal{B} \to (-\infty,b]$ are continuous and the function $(s,t) \longmapsto u_{(s,t)}$ is continuous from $\mathcal{R}'$ into $\mathcal{B}$.

We will need to introduce the following hypothesis:

(H_ϕ) There exists a continuous bounded function $L : \mathcal{R}'_{(\rho_1^-,\rho_2^-)} \to (0,\infty)$ such that

$$\|\phi_{(s,t)}\|_{\mathcal{B}} \leq L(s,t)\|\phi\|_{\mathcal{B}}, \text{ for any}(s,t) \in \mathcal{R}'.$$

In the sequel we will make use of the following generalization of a consequence of the phase space axioms ([148]).

Lemma 5.25. *If $u \in \Omega$, then*

$$\|u_{(s,t)}\|_{\mathcal{B}} = (M + L')\|\phi\|_{\mathcal{B}} + K \sup_{(\theta,\eta)\in[0,\max\{0,s\}]\times[0,\max\{0,t\}]} \|u(\theta,\eta)\|,$$

where

$$L' = \sup_{(s,t)\in\mathcal{R}'} L(s,t).$$

Our first existence result for the IVP (5.57)–(5.60) is based on the Banach contraction principle.

Theorem 5.26. *Assume that the following hypotheses hold:*

(5.26.1) There exists $\ell' > 0$ such that

$$\|f(x,y,u) - f(x,y,v)\| \leq \ell'\|u-v\|_{\mathcal{B}}, \textit{ for any } u,\ v \in \mathcal{B} \textit{ and } (x,y) \in J,$$

(5.26.2) There exists a constant $l^* > 0$ such that

$$\|I_k(u) - I_k(\overline{u})\| \leq l^*\|u - \overline{u}\|, \textit{ for each } u, \overline{u} \in \mathbb{R}^n;\ k = 1,\ldots,m.$$

If

$$2ml^* + \frac{2K\ell' a^{r_1} b^{r_2}}{\Gamma(r_1+1)\Gamma(r_2+1)} < 1, \tag{5.62}$$

then there exists a unique solution for IVP (5.57)–(5.60) on $(-\infty,a]\times(-\infty,b]$.

Proof. Transform the problems (5.57)–(5.60) into a fixed-point problem. Consider the operator $N : \Omega \to \Omega$ defined by

$$N(u)(x,y) = \begin{cases} \phi(x,y), & (x,y) \in \tilde{J}', \\ \mu(x,y) + \sum\limits_{0<x_k<x} (I_k(u(x_k^-,y)) - I_k(u(x_k^-,0))) & \\ + \frac{1}{\Gamma(r_1)\Gamma(r_2)} \sum\limits_{0<x_k<x} \int\limits_{x_{k-1}}^{x_k}\int\limits_0^y (x_k-s)^{r_1-1}(y-t)^{r_2-1} & \\ \times f(s,t,u_{(\rho_1(s,t,u_{(s,t)}),\rho_2(s,t,u_{(s,t)}))})dtds & \\ + \frac{1}{\Gamma(r_1)\Gamma(r_2)} \int\limits_{x_k}^{x}\int\limits_0^y (x-s)^{r_1-1}(y-t)^{r_2-1} & \\ \times f(s,t,u_{(\rho_1(s,t,u_{(s,t)}),\rho_2(s,t,u_{(s,t)}))})dtds, & (x,y) \in J. \end{cases} \tag{5.63}$$

Let $v(.,.) : (-\infty, a] \times (-\infty, b] \to \mathbb{R}^n$ be a function defined by

$$v(x,y) = \begin{cases} \phi(x,y), & (x,y) \in \tilde{J}', \\ \mu(x,y), & (x,y) \in J. \end{cases}$$

Then $v_{(x,y)} = \phi$ for all $(x,y) \in E$. For each $w \in C(J, \mathbb{R}^n)$ with $w(x,y) = 0$ for each $(x,y) \in E$ we denote by $\overline{w}$ the function defined by

$$\overline{w}(x,y) = \begin{cases} 0, & (x,y) \in \tilde{J}', \\ w(x,y) & (x,y) \in J. \end{cases}$$

If $u(.,.)$ satisfies the integral equation

$$u(x,y) = \mu(x,y) + \frac{1}{\Gamma(r_1)\Gamma(r_2)} \int_0^x \int_0^y (x-s)^{r_1-1}(y-t)^{r_2-1} \times f(s,t,u_{(\rho_1(s,t,u_{(s,t)}),\rho_2(s,t,u_{(s,t)}))})dtds,$$

we can decompose $u(.,.)$ as $u(x,y) = \overline{w}(x,y) + v(x,y)$; $(x,y) \in J$, which implies $u_{(x,y)} = \overline{w}_{(x,y)} + v_{(x,y)}$, for every $(x,y) \in J$, and the function $w(.,.)$ satisfies

$$\begin{aligned} w(x,y) = & \sum_{0<x_k<x} (I_k(u(x_k^-, y)) - I_k(u(x_k^-, 0))) \\ & + \frac{1}{\Gamma(r_1)\Gamma(r_2)} \sum_{0<x_k<x} \int_{x_{k-1}}^{x_k} \int_0^y (x_k-s)^{r_1-1}(y-t)^{r_2-1} \\ & \times f(s,t,u_{(\rho_1(s,t,u_{(s,t)}),\rho_2(s,t,u_{(s,t)}))})dtds \\ & + \frac{1}{\Gamma(r_1)\Gamma(r_2)} \int_{x_k}^x \int_0^y (x-s)^{r_1-1}(y-t)^{r_2-1} \\ & \times f(s,t,u_{(\rho_1(s,t,u_{(s,t)}),\rho_2(s,t,u_{(s,t)}))})dtds. \end{aligned}$$

Set

$$C_0 = \{w \in \Omega : \ w(x,y) = 0 \ \text{for} \ (x,y) \in E\},$$

and let $\|.\|_{(a,b)}$ be the seminorm in C_0 defined by

$$\|w\|_{(a,b)} = \sup_{(x,y)\in E} \|w_{(x,y)}\|_{\mathcal{B}} + \sup_{(x,y)\in J} \|w(x,y)\| = \sup_{(x,y)\in J} \|w(x,y)\|, \ w \in C_0.$$

C_0 is a Banach space with norm $\|.\|_{(a,b)}$. Let the operator $P : C_0 \to C_0$ be defined by

$$P(x,y) = \sum_{0<x_k<x} (I_k(u(x_k^-,y)) - I_k(u(x_k^-,0)))$$

$$+\frac{1}{\Gamma(r_1)\Gamma(r_2)} \sum_{0<x_k<x} \int_{x_{k-1}}^{x_k} \int_0^y (x_k-s)^{r_1-1}(y-t)^{r_2-1}$$

$$\times f(s,t,\overline{w}_{(\rho_1(s,t,u_{(s,t)}),\rho_2(s,t,u_{(s,t)}))} + v_{(\rho_1(s,t,u_{(s,t)}),\rho_2(s,t,u_{(s,t)}))})\mathrm{d}t\mathrm{d}s$$

$$+\frac{1}{\Gamma(r_1)\Gamma(r_2)} \int_{x_k}^{x} \int_0^y (x-s)^{r_1-1}(y-t)^{r_2-1}$$

$$\times f(s,t,\overline{w}_{(\rho_1(s,t,u_{(s,t)}),\rho_2(s,t,u_{(s,t)}))} + v_{(\rho_1(s,t,u_{(s,t)}),\rho_2(s,t,u_{(s,t)}))})\mathrm{d}t\mathrm{d}s, \tag{5.64}$$

for each $(x,y) \in J$. The operator N has a fixed point is equivalent to P has a fixed point, and so we turn to proving that P has a fixed point. We can easily show that $P : C_0 \to C_0$ is a contraction map, and hence it has a unique fixed point by Banach's contraction principle. □

Now we give an existence result based on the nonlinear alternative of Leray–Schauder type [136].

Theorem 5.27. *Assume* (H_ϕ) *and*

(5.27.1) There exist $p,q \in C(J,\mathbb{R}_+)$ *such that*

$$\|f(x,y,u)\| \le p(x,y) + q(x,y)\|u\|_{\mathcal{B}}, \text{ for } (x,y) \in J \text{ and each } u \in \mathcal{B},$$

(5.27.2) There exist $c_k > 0$; $k = 1,\dots,m$ *such that*

$$\|I_k(u)\| \le c_k \quad \text{for all } u \in \mathbb{R}^n.$$

Then the IVP (5.57)–(5.60) have at least one solution on $(-\infty,a] \times (-\infty,b]$.

Proof. Let $P : C_0 \to C_0$ defined as in (5.64). As in Theorem 5.22, we can show that the operator P is continuous and completely continuous. We now show that there exists an open set $U \subseteq C_0$ with $w \ne \lambda P(w)$, for $\lambda \in (0,1)$ and $w \in \partial U$. Let $w \in C_0$ and $w = \lambda P(w)$ for some $0 < \lambda < 1$. By (5.27.1) and (5.27.2) for each $(x,y) \in J$, we have

$$\|w(x,y)\| \le \sum_{k=1}^m 2c_k + \frac{2\|p\|_\infty a^{r_1} b^{r_2}}{\Gamma(r_1+1)\Gamma(r_2+1)}$$

$$+\frac{2}{\Gamma(r_1)\Gamma(r_2)} \int_0^x \int_0^y (x-s)^{r_1-1}(y-t)^{r_2-1} q(s,t)$$

$$\times \|\overline{w}_{(s,t)} + v_{(s,t)}\|_{\mathcal{B}}\mathrm{d}t\mathrm{d}s.$$

But Lemma 5.25 implies that

$$\begin{aligned}\|\overline{w}_{(s,t)} + v_{(s,t)}\|_{\mathcal{B}} &\leq \|\overline{w}_{(s,t)}\|_{\mathcal{B}} + \|v_{(s,t)}\|_{\mathcal{B}} \\ &\leq K \sup\{w(\tilde{s},\tilde{t}) : (\tilde{s},\tilde{t}) \in [0,s] \times [0,t]\} \\ &\quad +(M + L')\|\phi\|_{\mathcal{B}} + K\|\phi(0,0)\|. \end{aligned} \tag{5.65}$$

If we name $z(s,t)$ the right-hand side of (5.65), then we have

$$\|\overline{w}_{(s,t)} + v_{(s,t)}\|_{\mathcal{B}} \leq z(x,y),$$

and therefore, for each $(x,y) \in J$ we obtain

$$\begin{aligned}\|w(x,y)\| \leq 2\sum_{k=1}^{m} c_k + \frac{2\|p\|_\infty a^{r_1} b^{r_2}}{\Gamma(r_1+1)\Gamma(r_2+1)} \\ +\frac{2}{\Gamma(r_1)\Gamma(r_2)} \int_0^x \int_0^y (x-s)^{r_1-1}(y-t)^{r_2-1} q(s,t) z(s,t) \mathrm{d}t \mathrm{d}s.\end{aligned} \tag{5.66}$$

Using the above inequality and the definition of z for each $(x,y) \in J$ we have

$$\begin{aligned}z(x,y) \leq (M+L')\|\phi\|_{\mathcal{B}} + K\|\phi(0,0)\| + 2\sum_{k=1}^{m} c_k + \frac{2\|p\|_\infty a^{r_1} b^{r_2}}{\Gamma(r_1+1)\Gamma(r_2+1)} \\ +\frac{2K\|q\|_\infty}{\Gamma(r_1)\Gamma(r_2)} \int_0^x \int_0^y (x-s)^{r_1-1}(y-t)^{r_2-1} z(s,t) \mathrm{d}t \mathrm{d}s.\end{aligned}$$

Then by Lemma 2.43, there exists $\delta = \delta(r_1, r_2)$ such that we have

$$\|z(x,y)\| \leq R + \delta \frac{2K\|q\|_\infty}{\Gamma(r_1)\Gamma(r_2)} \int_0^x \int_0^y (x-s)^{r_1-1}(y-t)^{r_2-1} R \mathrm{d}t \mathrm{d}s,$$

where

$$R = (M+L')\|\phi\|_{\mathcal{B}} + K\|\phi(0,0)\| + 2\sum_{k=1}^{m} c_k + \frac{2\|p\|_\infty a^{r_1} b^{r_2}}{\Gamma(r_1+1)\Gamma(r_2+1)}.$$

Hence

$$\|z\|_\infty \leq R + \frac{2R\delta K\|q\|_\infty a^{r_1} b^{r_2}}{\Gamma(r_1+1)\Gamma(r_2+1)} := \widetilde{M}.$$

Then, (5.66) implies that

$$\|w\|_\infty \leq 2\sum_{k=1}^{m} c_k + \frac{2a^{r_1}b^{r_2}}{\Gamma(r_1+1)\Gamma(r_2+1)}(\|p\|_\infty + \widetilde{M}\|q\|_\infty) := M^*.$$

Set

$$U = \{w \in C_0 : \|w\|_{(a,b)} < M^* + 1\}.$$

$P : \overline{U} \to C_0$ is continuous and completely continuous. By our choice of U, there is no $w \in \partial U$ such that $w = \lambda P(w)$, for $\lambda \in (0,1)$. As a consequence of the nonlinear alternative of Leray–Schauder type [136], we deduce that N has a fixed point which is a solution to problems (5.57)–(5.60). □

5.6.4 *Examples*

5.6.4.1 Example 1

As an application of our results we consider the following impulsive partial hyperbolic functional differential equations of the form:

$$({}^cD^r_{x_k}u)(x,y) = \frac{e^{-x-y}}{9+e^{x+y}} \times \frac{|u(x-\sigma_1(u(x,y)), y-\sigma_2(u(x,y)))|}{1+|u(x-\sigma_1(u(x,y)), y-\sigma_2(u(x,y)))|},$$
$$\text{if } (x,y) \in J_k;\ k=0,1, \tag{5.67}$$

$$u\left(\left(\frac{1}{2}\right)^+, y\right) = u\left(\left(\frac{1}{2}\right)^-, y\right) + \frac{|u((\frac{1}{2})^-, y)|}{3+|u((\frac{1}{2})^-, y)|} :\ y \in [0,1], \tag{5.68}$$

$$u(x,y) = x + y^2,\ (x,y) \in [-1,1]\times[-2,1]\backslash(0,1]\times(0,1], \tag{5.69}$$

$$u(x,0) = x,\ u(0,y) = y^2,\ \ x \in [0,1],\ y \in [0,1], \tag{5.70}$$

where $J_0 = [0,\frac{1}{2}]\times[0,1],\ J_1 = (\frac{1}{2},1]\times[0,1],\ \sigma_1 \in C(\mathbb{R},[0,1]),\ \sigma_2 \in C(\mathbb{R},[0,2])$. Set

$$\rho_1(x,y,\varphi) = x - \sigma_1(\varphi(0,0)),\ \ (x,y,\varphi) \in J \times C,$$

$$\rho_2(x,y,\varphi) = y - \sigma_2(\varphi(0,0)),\ \ (x,y,\varphi) \in J \times C,$$

where $C := C_{(1,2)}$. Set

$$f(x,y,\varphi) = \frac{e^{-x-y}|\varphi|}{(9+e^{x+y})(1+|\varphi|)},\ \ (x,y) \in [0,1]\times[0,1],\ \varphi \in C,$$

and

$$I_k(u) = \frac{u}{3+u},\ \ u \in \mathbb{R}_+.$$

A simple computation shows that conditions of Theorem 5.22 are satisfied which implies that problems (5.67)–(5.70) have a unique solution defined on $[-1,1] \times [-2,1]$.

5.6.4.2 Example 2

We consider now the following impulsive fractional order partial hyperbolic differential equations with infinite delay of the form:

$$({}^c D^r_{x_k} u)(x,y) = \frac{c e^{x+y-\gamma(x+y)} |u(x-\sigma_1(u(x,y)), y-\sigma_2(u(x,y)))|}{(e^{x+y}+e^{-x-y})(1+|u(x-\sigma_1(u(x,y)), y-\sigma_2(u(x,y)))|)};$$
$$\text{if } (x,y) \in J_k;\ k=0,\dots,m, \tag{5.71}$$

$$u\left(\left(\frac{k}{k+1}\right)^+, y\right) = u\left(\left(\frac{k}{k+1}\right)^-, y\right) + \frac{\left|u\left(\left(\frac{k}{k+1}\right)^-, y\right)\right|}{3mk + \left|u\left(\left(\frac{k}{k+1}\right)^-, y\right)\right|};$$
$$y \in [0,1],\ k=1,\dots,m, \tag{5.72}$$

$$u(x,0) = x,\ u(0,y) = y^2,\quad x \in [0,1],\ y \in [0,1], \tag{5.73}$$

$$u(x,y) = x + y^2,\ (x,y) \in \tilde{J}' := (-\infty,1] \times (-\infty,1] \backslash (0,1] \times (0,1], \tag{5.74}$$

where $c = \frac{10}{\Gamma(r_1+1)\Gamma(r_2+1)}$, γ a positive real constant, and $\sigma_1, \sigma_2 \in C(\mathbb{R},[0,\infty))$. Let $\mathcal{B}_\gamma$ be the phase space defined in the Example of Sect. 3.7. Set

$$\rho_1(x,y,\varphi) = x - \sigma_1(\varphi(0,0)),\quad (x,y,\varphi) \in J \times \mathcal{B}_\gamma,$$

$$\rho_2(x,y,\varphi) = y - \sigma_2(\varphi(0,0)),\quad (x,y,\varphi) \in J \times \mathcal{B}_\gamma,$$

$$f(x,y,\varphi) = \frac{c e^{x+y-\gamma(x+y)}|\varphi|}{(e^{x+y}+e^{-x-y})(1+|\varphi|)},\quad (x,y) \in [0,1]\times[0,1],\ \varphi \in \mathcal{B}_\gamma$$

and

$$I_k(u) = \frac{u}{3mk+u};\ u \in \mathbb{R}_+,\ k=1,\dots,m.$$

We can easily show that conditions of Theorem 5.23 are satisfied, and hence problem (5.71)–(5.74) has a unique solution defined on $(-\infty,1] \times (-\infty,1]$.

5.7 Impulsive Partial Hyperbolic Functional Differential Equations with Variable Times and State-Dependent Delay

5.7.1 Introduction

In this section, we start by studying the existence and uniqueness of solutions for the following impulsive partial hyperbolic differential equations with variable times:

$$({}^cD^r_{x_k}u)(x,y) = f(x,y,u_{(\rho_1(x,y,u_{(x,y)}),\rho_2(x,y,u_{(x,y)}))});$$

$$\text{if } (x,y)\in J_k;\ k=0,\ldots,m,\ \ x_k = x_k(u(x,y)),\ k=1,\ldots,m, \tag{5.75}$$

$$u(x^+,y) = I_k(u(x,y));\ \text{ if } y\in[0,b],\ x = x_k(u(x,y)),\ k=1,\ldots,m, \tag{5.76}$$

$$u(x,y) = \phi(x,y);\ \text{if } (x,y)\in\tilde{J} := [-\alpha,a]\times[-\beta,b]\backslash(0,a]\times(0,b], \tag{5.77}$$

$$u(x,0)=\varphi(x);\ x\in[0,a],\ u(0,y)=\psi(y);\ y\in[0,b], \tag{5.78}$$

where $a,b,\alpha,\beta>0,\ \ 0=x_0<x_1<\cdots<x_m<x_{m+1}=a,\ \phi\in C(\tilde{J},\mathbb{R}^n),\ \varphi,\psi$ are as in problem (3.1)–(3.3), $f:J\times C\to\mathbb{R}^n,\ \rho_1:J\times C\to[-\alpha,a],\ \rho_2: J\times C\to[-\beta,b],\ I_k:\mathbb{R}^n\to\mathbb{R}^n,\ k=1,\ldots,m$ are given functions, $J:=[0,a]\times[0,b]$, and C is the space defined by

$$\begin{aligned} C = C_{(\alpha,\beta)} = \big\{&u:[-\alpha,0]\times[-\beta,0]\to\mathbb{R}^n:\ \text{continuous and there exist}\\ &\tau_k\in(-\alpha,0)\ \text{ such that } \tau_k=\tau_k(u(\tau_k,.)),\ \text{with } u(\tau_k^-,\tilde{y})\ \text{ and } u(\tau_k^+,\tilde{y}),\\ &k=1,\ldots,m,\ \text{exist for any } \tilde{y}\in[-\beta,0] \text{ with } u(\tau_k^-,\tilde{y})=u(\tau_k,\tilde{y})\big\}. \end{aligned}$$

C is a Banach space with norm

$$\|u\|_C = \sup_{(x,y)\in[-\alpha,0]\times[-\beta,0]}\|u(x,y)\|.$$

Next we consider the following system of partial hyperbolic differential equations of fractional order with infinite delay:

$$({}^cD^r_{x_k}u)(x,y) = f(x,y,u_{(\rho_1(x,y,u_{(x,y)}),\rho_2(x,y,u_{(x,y)}))});$$

$$\text{if } (x,y)\in J_k;\ k=0,\ldots,m,\ \ x_k = x_k(u(x,y)),\ k=1,\ldots,m, \tag{5.79}$$

$$u(x^+,y) = I_k(u(x,y));\ \text{ if } y\in[0,b],\ x = x_k(u(x,y)),\ k=1,\ldots,m, \tag{5.80}$$

$$u(x,y) = \phi(x,y);\ \text{if } (x,y)\in\tilde{J}' := (-\infty,a]\times(-\infty,b]\backslash(0,a]\times(0,b], \tag{5.81}$$

$$u(x,0)=\varphi(x);\ x\in[0,a],\ u(0,y)=\psi(y);\ y\in[0,b], \tag{5.82}$$

where $\varphi,\ \psi,\ I_k$ are as in problems (5.75)–(5.78), $\phi \in C(\tilde{J}',\mathbb{R}^n),\ f : J \times \mathcal{B} \to \mathbb{R}^n,\ \rho_1 : J \times \mathcal{B} \to (-\infty, a],\ \rho_2 : J \times \mathcal{B} \to (-\infty, b]$ are given functions and $\mathcal{B}$ is a phase space.

5.7.2 *Impulsive Partial Differential Equations with Finite Delay*

Let us start in this section by defining what we mean by a solution of the problem (5.75)–(5.78). Let $J_k,\ PC$, and $\widetilde{PC}$ be defined as in Sect. 4.7.

Definition 5.28. A function $u \in \widetilde{PC}$ such that its mixed derivative D^2_{xy} exists on $J_k;\ k = 0,\dots,m$ is said to be a solution of (5.75)–(5.78) if u satisfies the condition (5.77) on $\tilde{J}$, (5.75) on J_k and conditions (5.76) and (5.78) are satisfied on J.

Set $\mathcal{R} := \mathcal{R}_{(\rho_1^-,\rho_2^-)}$

$$= \{(\rho_1(s,t,u), \rho_2(s,t,u)) : (s,t,u) \in J \times C,\ \rho_i(s,t,u) \le 0;\ i = 1,2\}.$$

We always assume that $\rho_1 : J \times C \to [-\alpha, a],\ \rho_2 : J \times C \to [-\beta, b]$ are continuous and the function $(s,t) \longmapsto u_{(s,t)}$ is continuous from $\mathcal{R}$ into C.

Theorem 5.29. *Assume that*

(5.29.1) The function $f : J \times C \to \mathbb{R}^n$ is continuous
(5.29.2) There exists a constant $M > 0$ such that

$$\|f(x,y,u)\| \le M(1 + \|u\|),\ \textit{for each}\ (x,y) \in J,\ u \in C,$$

(5.29.3) The function $x_k \in C^1(\mathbb{R}^n,\mathbb{R})$ for $k = 1,\dots,m$. Moreover,

$$0 = x_0(u) < x_1(u) < \cdots < x_m(u) < x_{m+1}(u) = a, \quad \textit{for all}\ u \in \mathbb{R}^n,$$

(5.29.4) There exists a constant $M^* > 0$ such that

$$\|I_k(u)\| \le M^*(1 + \|u\|);\ k = 1,\dots,m,\ \textit{for each}\ u \in C,$$

(5.29.5) For all $u \in C$, $x_k(I_k(u)) \le x_k(u) < x_{k+1}(I_k(u))$, for $k = 1,\dots,m$,
(5.29.6) For all $(s,t,u) \in J \times C$, we have

$$x_k'(u)[\varphi'(s) + \frac{r_1 - 1}{\Gamma(r_1)\Gamma(r_2)} \int_{x_k}^{s} \int_{0}^{t} (s-\theta)^{r_1-2}(t-\eta)^{r_2-1}$$

$$\times f(\theta,\eta,u_{(\rho_1(\theta,\eta,u_{(\theta,\eta)}),\rho_2(\theta,\eta,u_{(\theta,\eta)}))})\mathrm{d}\eta\mathrm{d}\theta] \ne 1,$$

$k = 1,\dots,m$. Then (5.75)–(5.78) has at least one solution on $[-\alpha,a] \times [-\beta,b]$.

Proof. The proof will be given in several steps.

Step 1: Set

$$PC_0=\{u:[-\alpha,a]\times[-\beta,b]\to\mathbb{R}^n : u_{(x,y)}\in C \text{ for } (x,y)\in J \text{ and } u\in PC(J,\mathbb{R}^n)\}.$$

Consider the following problem:

$$({}^cD_0^r u)(x,y) = f(x,y,u_{(\rho_1(x,y,u_{(x,y)}),\rho_2(x,y,u_{(x,y)}))});\ \text{if } (x,y)\in J, \tag{5.83}$$

$$u(x,y)=\phi(x,y);\ \text{if } (x,y)\in\tilde{J}, \tag{5.84}$$

$$u(x,0)=\varphi(x),\ u(0,y)=\psi(y);\ x\in[0,a] \text{ and } y\in[0,b]. \tag{5.85}$$

Transform problems (5.83)–(5.85) into a fixed-point problem. Consider the operator $N:PC_0\to PC_0$ defined by

$$N(u)(x,y)=\begin{cases}\phi(x,y); & (x,y)\in\tilde{J},\\ \mu(x,y)+\frac{1}{\Gamma(r_1)\Gamma(r_2)}\int_0^x\int_0^y(x-s)^{r_1-1}(y-t)^{r_2-1} & \\ \times f(s,t,u_{(\rho_1(s,t,u_{(s,t)}),\rho_2(s,t,u_{(s,t)}))})\mathrm{d}t\,\mathrm{d}s; & (x,y)\in J.\end{cases}$$

Lemma 2.12 implies that the fixed points of operator N are solutions of problems (5.83)–(5.85). We shall show that the operator N is continuous and completely continuous.

Claim 1. N is continuous. Let $\{u_n\}$ be a sequence such that $u_n\to u$ in PC_0. Let $\eta>0$ be such that $\|u_n\|\le\eta$. Then for each $(x,y)\in J$, we have

$$\|N(u_n)(x,y)-N(u)(x,y)\|\le\frac{1}{\Gamma(r_1)\Gamma(r_2)}\int_0^x\int_0^y(x-s)^{r_1-1}(y-t)^{r_2-1}$$

$$\|f(s,t,u_{n(\rho_1(s,t,u_{(s,t)}),\rho_2(s,t,u_{(s,t)}))})-f(s,t,u_{(\rho_1(s,t,u_{(s,t)}),\rho_2(s,t,u_{(s,t)}))})\|\mathrm{d}t\,\mathrm{d}s$$

$$\le\frac{\|f(.,.,u_{n(.,.)})-f(.,.,u)\|_\infty}{\Gamma(r_1)\Gamma(r_2)}\int_0^a\int_0^b(x-s)^{r_1-1}(y-t)^{r_2-1}\mathrm{d}s\mathrm{d}t$$

$$\le\frac{a^{r_1}b^{r_2}\|f(.,.,u_{n(.,.)})-f(.,.,u_{(.,.)})\|_\infty}{\Gamma(r_1+1)\Gamma(r_2+1)}.$$

Since f is a continuous function, we have

$$\|N(u_n)-N(u)\|_\infty\to 0\quad\text{as } n\to\infty.$$

Claim 2. N maps bounded sets into bounded sets in PC_0. Indeed, it is enough to show that for any $\eta^* > 0$, there exists a positive constant ℓ such that for each $u \in B_{\eta^*} = \{u \in PC_0 : \|u\|_\infty \leq \eta^*\}$, we have $\|N(u)\|_\infty \leq \ell$. By (5.29.2) for each $(x, y) \in J$, we have

$$\|N(u)(x,y)\| \leq \|\mu(x,u)\| + \frac{1}{\Gamma(r_1)\Gamma(r_2)} \int_0^x \int_0^y (x-s)^{r_1-1}(y-t)^{r_2-1}$$

$$\times \|f(s,t,u_{(\rho_1(s,t,u_{(s,t)}),\rho_2(s,t,u_{(s,t)}))})\| dt ds$$

$$\leq \|\mu(x,u)\| + \frac{M(1+\eta^*)a^{r_1}b^{r_2}}{\Gamma(r_1+1)\Gamma(r_2+1)}.$$

Thus $\|N(u)\|_\infty \leq \|\mu\|_\infty + \frac{M(1+\eta^*)a^{r_1}b^{r_2}}{\Gamma(r_1+1)\Gamma(r_2+1)} := \ell$.

Claim 3. N maps bounded sets into equicontinuous sets of PC_0.
Let $(\tau_1, y_1), (\tau_2, y_2) \in J$, $\tau_1 < \tau_2$ and $y_1 < y_2$, B_{η^*} be a bounded set of PC_0 as in Claim 2, and let $u \in B_{\eta^*}$. Then for each $(x, y) \in J$, we have

$$\|N(u)(\tau_2, y_2) - N(u)(\tau_1, y_1)\|$$

$$= \Big\| \mu(\tau_1, y_1) - \mu(\tau_2, y_2) + \frac{1}{\Gamma(r_1)\Gamma(r_2)} \int_0^{\tau_1} \int_0^{y_1} [(\tau_2-s)^{r_1-1}(y_2-t)^{r_2-1}$$

$$-(\tau_1-s)^{r_1-1}(y_1-t)^{r_2-1}] f(s,t,u_{(\rho_1(s,t,u_{(s,t)}),\rho_2(s,t,u_{(s,t)}))}) dt ds$$

$$+ \frac{1}{\Gamma(r_1)\Gamma(r_2)} \int_{\tau_1}^{\tau_2} \int_{y_1}^{y_2} (\tau_2-s)^{r_1-1}(y_2-t)^{r_2-1}$$

$$\times f(s,t,u_{(\rho_1(s,t,u_{(s,t)}),\rho_2(s,t,u_{(s,t)}))}) dt ds$$

$$+ \frac{1}{\Gamma(r_1)\Gamma(r_2)} \int_0^{\tau_1} \int_{y_1}^{y_2} (\tau_2-s)^{r_1-1}(y_2-t)^{r_2-1}$$

$$\times f(s,t,u_{(\rho_1(s,t,u_{(s,t)}),\rho_2(s,t,u_{(s,t)}))}) dt ds$$

$$+ \frac{1}{\Gamma(r_1)\Gamma(r_2)} \int_{\tau_1}^{\tau_2} \int_0^{y_1} (\tau_2-s)^{r_1-1}(y_2-t)^{r_2-1}$$

$$\times f(s,t,u_{(\rho_1(s,t,u_{(s,t)}),\rho_2(s,t,u_{(s,t)}))}) dt ds \Big\|$$

$$\leq \|\mu(\tau_1, y_1) - \mu(\tau_2, y_2)\|$$

$$+ \frac{M(1+\eta^*)}{\Gamma(r_1)\Gamma(r_2)} \int_0^{\tau_1} \int_0^{y_1} [(\tau_1-s)^{r_1-1}(y_1-t)^{r_2-1}$$

$$-(\tau_2-s)^{r_1-1}(y_2-t)^{r_2-1}] dt ds$$

$$+\frac{M(1+\eta^*)}{\Gamma(r_1)\Gamma(r_2)}\int_{\tau_1}^{\tau_2}\int_{y_1}^{y_2}(\tau_2-s)^{r_1-1}(y_2-t)^{r_2-1}\mathrm{d}t\mathrm{d}s$$

$$+\frac{M(1+\eta^*)}{\Gamma(r_1)\Gamma(r_2)}\int_{0}^{\tau_1}\int_{y_1}^{y_2}(\tau_2-s)^{r_1-1}(y_2-t)^{r_2-1}\mathrm{d}t\mathrm{d}s$$

$$+\frac{M(1+\eta^*)}{\Gamma(r_1)\Gamma(r_2)}\int_{\tau_1}^{\tau_2}\int_{0}^{y_1}(\tau_2-s)^{r_1-1}(y_2-t)^{r_2-1}\mathrm{d}t\mathrm{d}s$$

$$\leq \|\mu(\tau_1,y_1)-\mu(\tau_2,y_2)\|$$
$$+\frac{M(1+\eta^*)}{\Gamma(r_1+1)\Gamma(r_2+1)}[2y_2^{r_2}(\tau_2-\tau_1)^{r_1}+2\tau_2^{r_1}(y_2-y_1)^{r_2}$$
$$+\tau_1^{r_1}y_1^{r_2}-\tau_2^{r_1}y_2^{r_2}-2(\tau_2-\tau_1)^{r_1}(y_2-y_1)^{r_2}].$$

As $\tau_1 \longrightarrow \tau_2$ and $y_1 \longrightarrow y_2$, the right-hand side of the above inequality tends to zero. As a consequence of Claims 1–3 together with the Arzelá-Ascoli theorem, we can conclude that N is completely continuous.

Claim 4. A priori bounds. Now it remains to show that the set $\mathcal{E} = \{u \in PC_0 : u = \lambda N(u)$ for some $0 < \lambda < 1\}$ is bounded. Let $u \in \mathcal{E}$, then $u = \lambda N(u)$ for some $0 < \lambda < 1$. Thus, for each $(x,y) \in J$, we have

$$\|u(x,y)\| \leq \|\mu(x,y)\| + \frac{1}{\Gamma(r_1)\Gamma(r_2)}\int_0^x\int_0^y (x-s)^{r_1-1}(y-t)^{r_2-1}$$
$$\times\|f(s,t,u_{(\rho_1(s,t,u_{(s,t)}),\rho_2(s,t,u_{(s,t)}))})\|\mathrm{d}t\mathrm{d}s$$
$$\leq \|\mu\|_\infty + \frac{Ma^{r_1}b^{r_2}}{\Gamma(r_1+1)\Gamma(r_2+1)}$$
$$+\frac{M}{\Gamma(r_1)\Gamma(r_2)}\int_0^x\int_0^y (x-s)^{r_1-1}(y-t)^{r_2-1}\|u_{(s,t)}\|\mathrm{d}t\mathrm{d}s.$$

Set

$$\omega = \|\mu\|_\infty + \frac{Ma^{r_1}b^{r_2}}{\Gamma(r_1+1)\Gamma(r_2+1)}.$$

Then Lemma 2.43 implies that for each $(x,y) \in J$, there exists $\delta = \delta(r_1,r_2)$ such that

$$\|u(x,y)\| \leq \omega\left[1+\frac{M\delta}{\Gamma(r_1)\Gamma(r_2)}\int_0^x\int_0^y (x-s)^{r_1-1}(y-t)^{r_2-1}\mathrm{d}t\mathrm{d}s\right]$$
$$\leq \omega\left[1+\frac{M\delta a^{r_1}b^{r_2}}{\Gamma(r_1+1)\Gamma(r_2+1)}\right] := R.$$

This shows that the set $\mathcal{E}$ is bounded. As a consequence of Schaefer's fixed-point theorem (Theorem 2.34), we deduce that N has a fixed point which is a solution of the problems (5.83)–(5.85). Denote this solution by u_1. Define the function

$$r_{k,1}(x,y) = x_k(u_1(x,y)) - x, \qquad \text{for } x \geq 0,\, y \geq 0.$$

Hypothesis (5.29.3) implies that $r_{k,1}(0,0) \neq 0$ for $k = 1,\dots,m$. If $r_{k,1}(x,y) \neq 0$ on J for $k = 1,\dots,m$; i.e.,

$$x \neq x_k(u_1(x,y)); \quad \text{on } J \quad \text{for } k = 1,\dots,m,$$

then u_1 is a solution of the problems (5.75)–(5.78).

It remains to consider the case when $r_{1,1}(x,y) = 0$ for some $(x,y) \in J$. Now since $r_{1,1}(0,0) \neq 0$ and $r_{1,1}$ is continuous, there exists $x_1 > 0$, $y_1 > 0$ such that $r_{1,1}(x_1,y_1) = 0$, and $r_{1,1}(x,y) \neq 0$, for all $(x,y) \in [0,x_1) \times [0,y_1)$.

Thus by (5.29.6) we have

$$r_{1,1}(x_1,y_1) = 0 \text{ and } r_{1,1}(x,y) \neq 0, \text{ for all } (x,y) \in [0,x_1) \times [0,y_1] \cup (y_1,b].$$

Suppose that there exist $(\bar{x},\bar{y}) \in [0,x_1) \times [0,y_1] \cup (y_1,b]$ such that $r_{1,1}(\bar{x},\bar{y}) = 0$. The function $r_{1,1}$ attains a maximum at some point $(s,t) \in [0,x_1) \times [0,b]$. Since

$$({}^cD_0^r u_1)(x,y) = f(x,y,u_{1(\rho_1(x,y,u_{(x,y)}),\rho_2(x,y,u_{(x,y)}))}), \text{ for } (x,y) \in J,$$

then

$$\frac{\partial u_1(x,y)}{\partial x} \text{ exists, and } \frac{\partial r_{1,1}(s,t)}{\partial x} = x_1'(u_1(s,t))\frac{\partial u_1(s,t)}{\partial x} - 1 = 0.$$

Since

$$\frac{\partial u_1(x,y)}{\partial x} = \varphi'(x) + \frac{r_1 - 1}{\Gamma(r_1)\Gamma(r_2)}\int_0^x\int_0^y (x-s)^{r_1-2}(y-t)^{r_2-1} \times f(s,t,u_{1(\rho_1(s,t,u_{(s,t)}),\rho_2(s,t,u_{(s,t)}))})dtds,$$

then

$$x_1'(u_1(s,t))\left[\varphi'(s) + \frac{r_1-1}{\Gamma(r_1)\Gamma(r_2)}\int_0^s\int_0^t (s-\theta)^{r_1-2}(t-\eta)^{r_2-1} \times f(\theta,\eta,u_{(\rho_1(\theta,\eta,u_{(\theta,\eta)}),\rho_2(\theta,\eta,u_{(\theta,\eta)}))})d\theta d\eta\right] = 1,$$

which contradicts (5.29.6). From (5.29.1) we have

$$r_{k,1}(x,y) \neq 0 \text{ for all } (x,y) \in [0,x_1) \times [0,b] \text{ and } k = 1,\dots m.$$

Step 2: In what follows set

$$\begin{aligned} PC_k = &\{u : [-\alpha, a] \times [-\beta, b] \to \mathbb{R}^n : u_{(x,y)} \in C \text{ for } (x,y) \in J, \text{ and there exist} \\ &0 = x_0 < x_1 < x_2 < \cdots < x_m < x_{m+1} = a \text{ such that } x_k = x_k(u(x_k, .)), \\ &\text{and } u(x_k^-, .),\ u(x_k^+, .) \text{ exist with } u(x_k^-, .) = u(x_k, .);\ k = 1, \ldots, m, \\ &\text{and } u \in C(X_k, \mathbb{R}^n); k = 0, \ldots, m\}, \end{aligned}$$

where

$$X_k := [x_k, a] \times [0, b];\ k = 1, \ldots, m.$$

Consider now the problem

$$({}^cD_{x_1}^r u)(x,y) = f(x, y, u_{(\rho_1(x,y,u_{(x,y)}),\rho_2(x,y,u_{(x,y)}))}), \text{ if } (x,y) \in X_1, \tag{5.86}$$

$$u(x_1^+, y) = I_1(u_1(x_1, y)), \tag{5.87}$$

$$u(x,y) = u_1(x,y), \text{ if } (x,y) \in \tilde{J} \cup [0, x_1) \times [0, b]. \tag{5.88}$$

Consider the operator $N_1 : PC_1 \to PC_1$ defined as

$$N_1(u)(x,y) = \begin{cases} u_1(x,y), & (x,y) \in \tilde{J} \cup [0,x_1) \times [0,b], \\ \varphi(x) + I_1(u_1(x_1,y)) - I_1(u_1(x_1,0)) & \\ +\frac{1}{\Gamma(r_1)\Gamma(r_2)} \int\limits_{x_1}^{x} \int\limits_{0}^{y} (x-s)^{r_1-1}(y-t)^{r_2-1} & \\ \times f(s,t,u_{(\rho_1(s,t,u_{(s,t)}),\rho_2(s,t,u_{(s,t)}))})dtds, & (x,y) \in X_1. \end{cases}$$

As in Step 1 we can show that N_1 is completely continuous. Now it remains to show that the set $\mathcal{E}^* = \{u \in PC_1 : u = \lambda N_1(u) \text{ for some } 0 < \lambda < 1\}$ is bounded. Let $u \in \mathcal{E}^*$, then $u = \lambda N_1(u)$ for some $0 < \lambda < 1$. Thus, from (5.29.2) and (5.29.4) we get for each $(x,y) \in X_1$,

$$\begin{aligned} \|u(x,y)\| \le\ &\|\varphi(x)\| + \|I_1(u_1(x_1,y))\| + \|I_1(u_1(x_1,0))\| \\ &+\frac{1}{\Gamma(r_1)\Gamma(r_2)} \int_{x_1}^{x} \int_0^y (x-s)^{r_1-1}(y-t)^{r_2-1} \\ &\times \|f(s,t,u_{(\rho_1(s,t,u_{(s,t)}),\rho_2(s,t,u_{(s,t)}))})\|dtds \\ \le\ &\|\varphi\|_\infty + 2M^*(1 + \|u_1\|) + \frac{Ma^{r_1}b^{r_2}}{\Gamma(r_1+1)\Gamma(r_2+1)} \\ &+\frac{M}{\Gamma(r_1)\Gamma(r_2)} \int_0^x \int_0^y (x-s)^{r_1-1}(y-t)^{r_2-1}\|u_{(s,t)}\|dtds. \end{aligned}$$

Set

$$\omega^* = \|\varphi\|_\infty + 2M^*(1 + \|u_1\|) + \frac{Ma^{r_1}b^{r_2}}{\Gamma(r_1+1)\Gamma(r_2+1)}.$$

Then Lemma 2.43 implies that for each $(x, y) \in X_1$, there exists $\delta = \delta(r_1, r_2)$ such that

$$\begin{aligned}\|u(x,y)\| &\leq \omega^* \left[1 + \frac{M\delta}{\Gamma(r_1)\Gamma(r_2)} \int_0^x \int_0^y (x-s)^{r_1-1}(y-t)^{r_2-1} \mathrm{d}t\mathrm{d}s\right] \\ &\leq \omega^* \left[1 + \frac{M\delta a^{r_1} b^{r_2}}{\Gamma(r_1+1)\Gamma(r_2+1)}\right] := R^*.\end{aligned}$$

This shows that the set $\mathcal{E}^*$ is bounded. As a consequence of Schaefer's fixed-point theorem (Theorem 2.34), we deduce that N_1 has a fixed point u which is a solution to problems (5.86)–(5.88). Denote this solution by u_2. Define

$$r_{k,2}(x,y) = x_k(u_2(x,y)) - x, \quad \text{for } (x,y) \in X_1.$$

If $r_{k,2}(x,y) \neq 0$ on $(x_1, a] \times [0,b]$ and for all $k = 1, \ldots, m$, then

$$u(x,y) = \begin{cases} u_1(x,y), & \text{if } (x,y) \in \tilde{J} \cup [0,x_1) \times [0,b], \\ u_2(x,y), & \text{if } (x,y) \in [x_1,a] \times [0,b], \end{cases}$$

is a solution of the problems (5.75)–(5.77). It remains to consider the case when $r_{2,2}(x,y) = 0$, for some $(x,y) \in (x_1,a] \times [0,b]$. By (5.29.5), we have

$$\begin{aligned} r_{2,2}(x_1^+, y_1) &= x_2(u_2(x_1^+, y_1) - x_1 \\ &= x_2(I_1(u_1(x_1,y_1))) - x_1 \\ &> x_1(u_1(x_1,y_1)) - x_1 \\ &= r_{1,1}(x_1,y_1) = 0. \end{aligned}$$

Since $r_{2,2}$ is continuous, there exists $x_2 > x_1,\ y_2 > y_1$ such that $r_{2,2}(x_2,y_2) = 0$, and $r_{2,2}(x,y) \neq 0$ for all $(x,y) \in (x_1,x_2) \times [0,b]$. It is clear by (5.29.3) that

$$r_{k,2}(x,y) \neq 0 \quad \text{for all } (x,y) \in (x_1,x_2)] \times [0,b];\ k = 2, \ldots, m.$$

Now suppose that there are $(s,t) \in (x_1,x_2) \times [0,b]$ such that $r_{1,2}(s,t) = 0$. From (5.29.5) it follows that

$$\begin{aligned} r_{1,2}(x_1^+, y_1) &= x_1(u_2(x_1^+, y_1) - x_1 \\ &= x_1(I_1(u_1(x_1,y_1))) - x_1 \\ &\leq x_1(u_1(x_1,y_1)) - x_1 \\ &= r_{1,1}(x_1,y_1) = 0. \end{aligned}$$

Thus $r_{1,2}$ attains a nonnegative maximum at some point $(s_1,t_1)\in(x_1,a)\times[0,x_2)\cup(x_2,b]$. Since

$$({}^cD^r_{x_1}u_2)(x,y)=f(x,y,u_2(x,y)),\text{ for }(x,y)\in X_1,$$

then we get

$$\begin{aligned}u_2(x,y)&=\varphi(x)+I_1(u_1(x_1,y))-I_1(u_1(x_1,0))\\&\quad+\frac{1}{\Gamma(r_1)\Gamma(r_2)}\int_{x_1}^{x}\int_0^y(x-s)^{r_1-1}(y-t)^{r_2-1}\\&\quad\times f(s,t,u_{2(\rho_1(s,t,u_{(s,t)}),\rho_2(s,t,u_{(s,t)}))})\,dt\,ds,\end{aligned}$$

hence

$$\begin{aligned}\frac{\partial u_2}{\partial x}(x,y)&=\varphi'(x)+\frac{r_1-1}{\Gamma(r_1)\Gamma(r_2)}\int_{x_1}^{x}\int_0^y(x-s)^{r_1-2}(y-t)^{r_2-1}\\&\quad\times f(s,t,u_{2(\rho_1(s,t,u_{(s,t)}),\rho_2(s,t,u_{(s,t)}))})\,dt\,ds,\end{aligned}$$

then

$$\frac{\partial r_{1,2}(s_1,t_1)}{\partial x}=x_1'(u_2(s_1,t_1))\frac{\partial u_2}{\partial x}(s_1,t_1)-1=0.$$

Therefore

$$\begin{aligned}x_1'(u_2(s_1,t_1))&\left[\varphi'(s_1)+\frac{r_1-1}{\Gamma(r_1)\Gamma(r_2)}\int_{x_1}^{s_1}\int_0^{t_1}(s_1-\theta)^{r_1-2}(t_1-\eta)^{r_2-1}\right.\\&\left.\times f(\theta,\eta,u_{2(\rho_1(\theta,\eta,u_{(\theta,\eta)}),\rho_2(\theta,\eta,u_{(\theta,\eta)}))})\,d\eta\,d\theta\right]=1,\end{aligned}$$

which contradicts (5.29.6).

Step 3: We continue this process and take into account that $u_{m+1}:=u\big|_{X_m}$ is a solution to the problem

$$\begin{cases}({}^cD^r_{x_m}u)(x,y)=f(x,y,u_{(\rho_1(x,y,u_{(x,y)}),\rho_2(x,y,u_{(x,y)}))}),\text{ a.e. }(x,y)\in(x_m,a]\times[0,b],\\u(x_m^+,y)=I_m(u_{m-1}(x_m,y)),\\u(x,y)=u_1(x,y),\text{ if }(x,y)\in\tilde{J}\cup[0,x_1)\times[0,b],\\u(x,y)=u_2(x,y),\text{ if }(x,y)\in[x_1,x_2)\times[0,b],\\\dots\\u(x,y)=u_m(x,y),\text{ if }(x,y)\in[x_{m-1},x_m)\times[0,b].\end{cases}$$

The solution u of the problems (5.75)–(5.77) is then defined by

$$u(x,y)=\begin{cases} u_1(x,y), & \text{if } (x,y)\in[0,x_1]\times[0,b],\\ u_2(x,y), & \text{if } (x,y)\in(x_1,x_2]\times[0,b],\\ \dots & \\ u_{m+1}(x,y), & \text{if } (x,y)\in(x_m,a]\times[0,b]. \end{cases}$$

□

5.7.3 Impulsive Partial Differential Equations with Infinite Delay

Now we present an existence result for the problems (5.79)–(5.82). Consider the space

$$\Omega=\big\{u:(-\infty,a]\times(-\infty,b]\to\mathbb{R}^n: u_{(x,y)}\in\mathcal{B} \text{ for } (x,y)\in E \text{ and } u|_J\in PC\big\}.$$

Let $\|u\|_\Omega$ be the seminorm in Ω defined by

$$\|u\|_\Omega=\|\phi\|_{\mathcal{B}}+\sup\{\|u_k\|,\ k=0,\dots,m\},$$

where u_k is the restriction of u to J_k, $k=0,\dots,m$.

Definition 5.30. A function $u\in\Omega$ such that its mixed derivative D^2_{xy} exists on J_k; $k=0,\dots,m$ is said to be a solution of (5.79)–(5.82) if u satisfies $({}^cD^r_{x_k}u)(x,y)=f(x,y,u(x,y))$ on J_k; $k=0,\dots,m$ and conditions (5.80)—–(5.82) are satisfied.

Set $\mathcal{R}':=\mathcal{R}'_{(\rho_1^-,\rho_2^-)}$

$$=\{(\rho_1(s,t,u),\rho_2(s,t,u)):(s,t,u)\in J\times\mathcal{B},\ \rho_i(s,t,u)\le 0;\ i=1,2\}.$$

We always assume that $\rho_1:J\times\mathcal{B}\to(-\infty,a]$, $\rho_2:J\times\mathcal{B}\to(-\infty,b]$ are continuous and the function $(s,t)\longmapsto u_{(s,t)}$ is continuous from $\mathcal{R}'$ into $\mathcal{B}$.

We will need to introduce the following hypothesis:

(H_ϕ) There exists a continuous bounded function $L:\mathcal{R}'_{(\rho_1^-,\rho_2^-)}\to(0,\infty)$ such that

$$\|\phi_{(s,t)}\|_B\le L(s,t)\|\phi\|_B,\ \text{for any}(s,t)\in\mathcal{R}'.$$

In the sequel we will make use of Lemma 5.25.

Theorem 5.31. *Let $f:J\times\mathcal{B}\to\mathbb{R}^n$ be a Carathéodory function. Assume that*

(5.31.1) The function $x_k\in C^1(\mathbb{R}^n,\mathbb{R})$ for $k=1,\dots,m$. Moreover,

$$0=x_0(u)<x_1(u)<\dots<x_m(u)<x_{m+1}(u)=a,\quad \textit{for all } u\in\mathbb{R}^n,$$

(5.31.2) There exists a constant $M' > 0$ *such that*

$$\|f(x,y,u)\| \leq M'(1+\|u\|_{\mathcal{B}}),\ \textit{for each}\ (x,y)\in J,\ \textit{and each}\ u\in\mathcal{B},$$

(5.31.3) For all $(s,t,u)\in J\times\mathbb{R}^n$ *and* $u_{(\ldots)}\in\mathcal{B}$, *we have*

$$x'_k(u)\left[\varphi'(s)+\frac{r_1-1}{\Gamma(r_1)\Gamma(r_2)}\int_{x_k}^{s}\int_{0}^{t}(s-\theta)^{r_1-2}(t-\eta)^{r_2-1}\right.$$

$$\left. f(\theta,\eta,u_{(\rho_1(\theta,\eta,u_{(\theta,\eta)}),\rho_2(\theta,\eta,u_{(\theta,\eta)}))})\mathrm{d}\eta\mathrm{d}\theta\right]\neq 1,$$

$k=1,\ldots,m,$

(5.31.4) For all $u\in\mathbb{R}^n$, $x_k(I_k(u))\leq x_k(u)<x_{k+1}(I_k(u))$ *for* $k=1,\ldots,m$,

(5.31.5) There exists a constant $M^*>0$ *such that*

$$\|I_k(u)\|\leq M^*(1+\|u\|_{\mathcal{B}}),\ \textit{for each}\ u\in\mathcal{B},\ k=1,\ldots,m.$$

Then the IVP (5.79)–(5.82) have at least one solution on $(-\infty,a]\times(-\infty,b]$.

Proof. The proof will be given in several steps.

Step 1: Set

$$\Omega_0=\{u:(-\infty,a]\times(-\infty,b]\to\mathbb{R}^n : u_{(x,y)}\in\mathcal{B}\text{ for }(x,y)\in E\text{ and }u\in C(J,\mathbb{R}^n)\}.$$

Consider the following problem:

$$({}^cD_0^r u)(x,y)=f(x,y,u_{(\rho_1(x,y,u_{(x,y)}),\rho_2(x,y,u_{(x,y)}))});\ \text{if }(x,y)\in J,\tag{5.89}$$

$$u(x,y)=\phi(x,y);\ \text{if }(x,y)\in\tilde{J}',\tag{5.90}$$

$$u(x,0)=\varphi(x),\ u(0,y)=\psi(y);\ x\in[0,a],\ y\in[0,b].\tag{5.91}$$

Transform problems (5.89)–(5.91) into a fixed-point problem. Consider the operator $N:\Omega_0\to\Omega_0$ defined by

$$N(u)(x,y)=\begin{cases}\phi(x,y); & (x,y)\in\tilde{J}',\\ \mu(x,y) & \\ +\frac{1}{\Gamma(r_1)\Gamma(r_2)}\int_0^x\int_0^y(x-s)^{r_1-1}(y-t)^{r_2-1} & \\ \times f(s,t,u_{(\rho_1(s,t,u_{(s,t)}),\rho_2(s,t,u_{(s,t)}))})\mathrm{d}t\mathrm{d}s; & (x,y)\in J.\end{cases}\tag{5.92}$$

Let $v(.,.):(-\infty,a]\times(-\infty,b]\to\mathbb{R}^n$. be a function defined by

$$v(x,y)=\begin{cases}\phi(x,y),\ (x,y)\in\tilde{J}',\\ \mu(x,y),\ (x,y)\in J.\end{cases}$$

Then $v_{(x,y)} = \phi$ for all $(x, y) \in E$. For each $w \in C(J, \mathbb{R}^n)$ with $w(0,0) = 0$, we denote by $\overline{w}$ the function defined by

$$\overline{w}(x,y) = \begin{cases} 0, & (x,y) \in \tilde{J}', \\ w(x,y) & (x,y) \in J. \end{cases}$$

If $u(.,.)$ satisfies the integral equation

$$u(x,y) = \mu(x,y) + \frac{1}{\Gamma(r_1)\Gamma(r_2)} \int_0^x \int_0^y (x-s)^{r_1-1}(y-t)^{r_2-1} \times f(s,t,u_{(\rho_1(s,t,u_{(s,t)}),\rho_2(s,t,u_{(s,t)}))})dtds,$$

we can decompose $u(.,.)$ as $u(x,y) = \overline{w}(x,y) + v(x,y)$; $(x,y) \in J$, which implies $u_{(x,y)} = \overline{w}_{(x,y)} + v_{(x,y)}$, for every $(x,y) \in J$, and the function $w(.,.)$ satisfies

$$w(x,y) = \frac{1}{\Gamma(r_1)\Gamma(r_2)} \int_0^x \int_0^y (x-s)^{r_1-1}(y-t)^{r_2-1} \times f(s,t,\overline{w}_{(\rho_1(s,t,u_{(s,t)}),\rho_2(s,t,u_{(s,t)}))} + v_{(\rho_1(s,t,u_{(s,t)}),\rho_2(s,t,u_{(s,t)}))})dtds.$$

Set

$$C_0 = \{w \in \Omega_0 : \ w(x,y) = 0 \text{ for } (x,y) \in E\},$$

and let $\|.\|_{(a,b)}$ be the seminorm in C_0 defined by

$$\|w\|_{(a,b)} = \sup_{(x,y)\in E} \|w_{(x,y)}\|_{\mathcal{B}} + \sup_{(x,y)\in J} \|w(x,y)\| = \sup_{(x,y)\in J} \|w(x,y)\|, \ w \in C_0.$$

C_0 is a Banach space with norm $\|.\|_{(a,b)}$. Let the operator $P : C_0 \to C_0$ be defined by

$$P(w)(x,y) = \frac{1}{\Gamma(r_1)\Gamma(r_2)} \int_0^x \int_0^y (x-s)^{r_1-1}(y-t)^{r_2-1} \times f(s,t,\overline{w}_{(\rho_1(s,t,u_{(s,t)}),\rho_2(s,t,u_{(s,t)}))} + v_{(\rho_1(s,t,u_{(s,t)}),\rho_2(s,t,u_{(s,t)}))})dtds, \quad (x,y) \in J. \tag{5.93}$$

The operator N has a fixed point is equivalent to P has a fixed point, and so we turn to proving that P has a fixed point. We shall use Schaefer's fixed point theorem to prove that P has fixed point. We shall show that the operator P is continuous and completely continuous.

Claim 1. P is continuous. Let$\{w_n\}$ be a sequence such that $w_n \to w$ in C_0. Then

$$\|P(w_n)(x,y) - P(w)(x,y)\|$$
$$\leq \frac{1}{\Gamma(r_1)\Gamma(r_2)} \int_0^a \int_0^b (x-s)^{r_1-1}(y-t)^{r_2-1}$$
$$\times \|f(s,t,\overline{w}_{n(\rho_1(s,t,u_{(s,t)}),\rho_2(s,t,u_{(s,t)}))} + v_{n(\rho_1(s,t,u_{(s,t)}),\rho_2(s,t,u_{(s,t)}))})$$
$$-f(s,t,\overline{w}_{(\rho_1(s,t,u_{(s,t)}),\rho_2(s,t,u_{(s,t)}))} + v_{(\rho_1(s,t,u_{(s,t)}),\rho_2(s,t,u_{(s,t)}))})\|dtds.$$

Since f is a Carathéodory function, then we have

$$\|P(w_n) - P(w)\|_\infty \leq \frac{a^{r_1}b^{r_2}\|f(.,.,\overline{w_n}_{(...)} + v_{n(...)}) - f(.,.,\overline{w}_{(...)} + v_{(...)})\|_\infty}{\Gamma(r_1+1)\Gamma(r_2+1)}$$
$$\to 0 \text{ as } n \to \infty.$$

Claim 2. P maps bounded sets into bounded sets in C_0. Indeed, it is enough to show that, for any $\eta > 0$, there exists a positive constant $\widetilde{\ell}$ such that, for each $w \in B_\eta = \{w \in C_0 : \|w\|_{(a,b)} \leq \eta\}$, we have $\|P(w)\|_\infty \leq \widetilde{\ell}$.

Lemma 5.25 implies that

$$\|\overline{w}_{(\rho_1(s,t,u_{(s,t)}),\rho_2(s,t,u_{(s,t)}))} + v_{(\rho_1(s,t,u_{(s,t)}),\rho_2(s,t,u_{(s,t)}))}\|_B$$
$$\leq \|\overline{w}_{(\rho_1(s,t,u_{(s,t)}),\rho_2(s,t,u_{(s,t)}))}\|_B + \|v_{(\rho_1(s,t,u_{(s,t)}),\rho_2(s,t,u_{(s,t)}))}\|_B$$
$$\leq K\eta + K\|\phi(0,0)\| + (M+L')\|\phi\|_{\mathcal{B}}.$$

Set

$$\eta^* := K\eta + K\|\phi(0,0)\| + (M+L')\|\phi\|_{\mathcal{B}}$$

Let $w \in B_\eta$. By (5.31.2) we have for each $(x,y) \in J$,

$$\|P(w)(x,y)\| \leq \frac{1}{\Gamma(r_1)\Gamma(r_2)} \int_0^x \int_0^y (x-s)^{r_1-1}(y-t)^{r_2-1}$$
$$\times \|f(s,t,\overline{w}_{(\rho_1(s,t,u_{(s,t)}),\rho_2(s,t,u_{(s,t)}))} + v_{(\rho_1(s,t,u_{(s,t)}),\rho_2(s,t,u_{(s,t)}))})\|dtds$$
$$\leq \frac{1}{\Gamma(r_1)\Gamma(r_2)} \int_0^x \int_0^y (x-s)^{r_1-1}(y-t)^{r_2-1}$$
$$\times M'(1 + \|\overline{w}_{(\rho_1(s,t,u_{(s,t)}),\rho_2(s,t,u_{(s,t)}))} + v_{(\rho_1(s,t,u_{(s,t)}),\rho_2(s,t,u_{(s,t)}))}\|_{\mathcal{B}})dtds$$
$$\leq \frac{M'(1+\eta^*)}{\Gamma(r_1)\Gamma(r_2)} \int_0^a \int_0^b (x-s)^{r_1-1}(y-t)^{r_2-1}dtds$$
$$\leq \frac{M'a^{r_1}b^{r_2}(1+\eta^*)}{\Gamma(r_1+1)\Gamma(r_2+1)} := \ell^*.$$

Hence $\|P(w)\|_\infty \leq \ell^*$.

Claim 3. P maps bounded sets into equicontinuous sets in C_0. Let $(\tau_1, y_1), (\tau_2, y_2) \in (0,a] \times (0,b]$, $\tau_1 < \tau_2$, $y_1 < y_2$, B_η be a bounded set as in Claim 2, and let $w \in B_\eta$. Then

$$
\begin{aligned}
&\|P(w)(\tau_2, y_2) - P(w)(\tau_1, y_1)\| \\
&\leq \frac{1}{\Gamma(r_1)\Gamma(r_2)}\|\int_0^{\tau_1}\int_0^{y_1}[(\tau_2 - s)^{r_1-1}(y_2 - t)^{r_2-1} \\
&\quad -(\tau_1 - s)^{r_1-1}(y_1 - t)^{r_2-1}]f(s,t,u_{(\rho_1(s,t,u_{(s,t)}),\rho_2(s,t,u_{(s,t)}))})dt\,ds \\
&\quad +\frac{1}{\Gamma(r_1)\Gamma(r_2)}\int_{\tau_1}^{\tau_2}\int_{y_1}^{y_2}(\tau_2 - s)^{r_1-1}(y_2 - t)^{r_2-1} \\
&\quad \times f(s,t,\overline{w}_{(\rho_1(s,t,u_{(s,t)}),\rho_2(s,t,u_{(s,t)}))} + v_{(\rho_1(s,t,u_{(s,t)}),\rho_2(s,t,u_{(s,t)}))})dt\,ds \\
&\quad +\frac{1}{\Gamma(r_1)\Gamma(r_2)}\int_0^{\tau_1}\int_{y_1}^{y_2}(\tau_2 - s)^{r_1-1}(y_2 - t)^{r_2-1} \\
&\quad \times f(s,t,\overline{w}_{(\rho_1(s,t,u_{(s,t)}),\rho_2(s,t,u_{(s,t)}))} + v_{(\rho_1(s,t,u_{(s,t)}),\rho_2(s,t,u_{(s,t)}))})dt\,ds \\
&\quad +\frac{1}{\Gamma(r_1)\Gamma(r_2)}\int_{\tau_1}^{\tau_2}\int_0^{y_1}(\tau_2 - s)^{r_1-1}(y_2 - t)^{r_2-1} \\
&\quad \times f(s,t,\overline{w}_{(\rho_1(s,t,u_{(s,t)}),\rho_2(s,t,u_{(s,t)}))} + v_{(\rho_1(s,t,u_{(s,t)}),\rho_2(s,t,u_{(s,t)}))})dt\,ds\| \\
&\leq \frac{M'(1+\eta)}{\Gamma(r_1)\Gamma(r_2)}\int_0^{\tau_1}\int_0^{y_1}[(\tau_1 - s)^{r_1-1}(y_1 - t)^{r_2-1} - (\tau_2 - s)^{r_1-1}(y_2 - t)^{r_2-1}]dt\,ds \\
&\quad +\frac{M'(1+\eta)}{\Gamma(r_1)\Gamma(r_2)}\int_{\tau_1}^{\tau_2}\int_{y_1}^{y_2}(\tau_2 - s)^{r_1-1}(y_2 - t)^{r_2-1}dt\,ds \\
&\quad +\frac{M'(1+\eta)}{\Gamma(r_1)\Gamma(r_2)}\int_0^{\tau_1}\int_{y_1}^{y_2}(\tau_2 - s)^{r_1-1}(y_2 - t)^{r_2-1}dt\,ds \\
&\quad +\frac{M'(1+\eta)}{\Gamma(r_1)\Gamma(r_2)}\int_{\tau_1}^{\tau_2}\int_0^{y_1}(\tau_2 - s)^{r_1-1}(y_2 - t)^{r_2-1}dt\,ds \\
&\leq \frac{M'(1+\eta)}{\Gamma(r_1+1)\Gamma(r_2+1)}[2y_2^{r_2}(\tau_2 - x_1)^{r_1} + 2\tau_2^{r_1}(y_2 - y_1)^{r_2} \\
&\quad +\tau_1^{r_1}y_1^{r_2} - \tau_2^{r_1}y_2^{r_2} - 2(\tau_2 - \tau_1)^{r_1}(y_2 - y_1)^{r_2}].
\end{aligned}
$$

As $\tau_1 \to \tau_2$, $y_1 \to y_2$ the right-hand side of the above inequality tends to zero. The equicontinuity for the cases $x_1 < x_2 < 0$, $y_1 < y_2 < 0$ and $x_1 \leq 0 \leq x_2$, $y_1 \leq 0 \leq y_2$ is obvious. As a consequence of Claims 1–3, together with the Arzela–Ascoli theorem, we can conclude that $P : C_0 \to C_0$ is continuous and completely continuous.

Claim 4. (A priori bounds): Now it remains to show that the set $\mathcal{F} = \{u \in C_0 : u = \lambda P(u)$ for some $0 < \lambda < 1\}$ is bounded. Let $u \in \mathcal{F}$, then $u = \lambda P(u)$ for some $0 < \lambda < 1$. Thus, for each $(x, y) \in J$, we have

$$w(x,y) = \frac{\lambda}{\Gamma(r_1)\Gamma(r_2)} \int_0^x \int_0^y (x-s)^{r_1-1}(y-t)^{r_2-1} \times f(s,t,u_{(\rho_1(s,t,u_{(s,t)}),\rho_2(s,t,u_{(s,t)}))})dtds.$$

This implies by (5.31.2) that, for each $(x, y) \in J$, we have

$$\|w(x,y)\| \leq \frac{1}{\Gamma(r_1)\Gamma(r_2)} \int_0^x \int_0^y (x-s)^{r_1-1}(y-t)^{r_2-1} \times M'[1 + \|\overline{w}_{(\rho_1(s,t,u_{(s,t)}),\rho_2(s,t,u_{(s,t)}))} + v_{(\rho_1(s,t,u_{(s,t)}),\rho_2(s,t,u_{(s,t)}))}\|_{\mathcal{B}}]dtds.$$

But

$$\begin{aligned} &\|\overline{w}_{(\rho_1(s,t,u_{(s,t)}),\rho_2(s,t,u_{(s,t)}))} + v_{(\rho_1(s,t,u_{(s,t)}),\rho_2(s,t,u_{(s,t)}))}\|_{\mathcal{B}} \\ &\leq \|\overline{w}_{(\rho_1(s,t,u_{(s,t)}),\rho_2(s,t,u_{(s,t)}))}\|_{\mathcal{B}} + \|v_{(\rho_1(s,t,u_{(s,t)}),\rho_2(s,t,u_{(s,t)}))}\|_{\mathcal{B}} \\ &\leq K \sup\{w(\tilde{s},\tilde{t}) : (\tilde{s},\tilde{t}) \in [0,s] \times [0,t]\} \\ &\quad +(M + L')\|\phi\|_{\mathcal{B}} + K\|\phi(0,0)\|. \end{aligned} \tag{5.94}$$

If we name z(s,t) the right-hand side of (5.94), then we have

$$\|\overline{w}_{(\rho_1(s,t,u_{(s,t)}),\rho_2(s,t,u_{(s,t)}))} + v_{(\rho_1(s,t,u_{(s,t)}),\rho_2(s,t,u_{(s,t)}))}\|_{\mathcal{B}} \leq z(s,t),$$

and therefore, for each $(x, y) \in J$ we obtain

$$\|w(x,y)\| \leq \frac{M'}{\Gamma(r_1)\Gamma(r_2)} \int_0^x \int_0^y (x-s)^{r_1-1}(y-t)^{r_2-1}(1 + z(s,t))dtds. \tag{5.95}$$

Using the above inequality and the definition of z we have that

$$\begin{aligned} z(x,y) &\leq (M + L')\|\phi\|_{\mathcal{B}} + K\|\phi(0,0)\| \\ &\quad + \frac{KM'}{\Gamma(r_1)\Gamma(r_2)} \int_0^x \int_0^y (x-s)^{r_1-1}(y-t)^{r_2-1}(1 + z(s,t))dtds, \end{aligned}$$

for each $(x, y) \in J$. Then for Lemma 2.43, there exists $\delta = \delta(r_1, r_2)$ such that we have

$$\|z\|_\infty \le R\left[1 + \frac{K\delta M' a^{r_1} b^{r_2}}{\Gamma(r_1+1)\Gamma(r_2+1)}\right] := M^*,$$

where

$$R = (M + L')\|\phi\|_{\mathcal{B}} + K\|\phi(0,0)\| + \frac{KM' a^{r_1} b^{r_2}}{\Gamma(r_1+1)\Gamma(r_2+1)}.$$

Then, (5.95) implies that

$$\|w\|_\infty \le \frac{KM' a^{r_1} b^{r_2}(1+\widetilde{M})}{\Gamma(r_1+1)\Gamma(r_2+1)} := M^*.$$

This shows that the set $\mathcal{F}$ is bounded. As a consequence of Schaefer's fixed-point theorem (Theorem 2.34), we deduce that P has a fixed point u which is a solution to problems (5.89)–(5.91). Denote this solution by u_1. Define the functions

$$r_{k,1}(x, y) = x_k(u_1(x, y)) - x, \qquad \text{for } x \ge 0,\ y \ge 0.$$

Hypothesis (5.31.1) implies that $r_{k,1}(0,0) \ne 0$ for $k = 1, \dots, m$. If $r_{k,1}(x, y) \ne 0$ on J for $k = 1, \dots, m$; i.e.,

$$x \ne x_k(u_1(x, y)), \quad \text{on } J \quad \text{for } k = 1, \dots, m,$$

then u_1 is a solution of the problems (5.79)–(5.82). It remains to consider the case when $r_{1,1}(x, y) = 0$ for some $(x, y) \in J$. Now since $r_{1,1}(0,0) \ne 0$ and $r_{1,1}$ is continuous, there exists $x_1 > 0, y_1 > 0$ such that $r_{1,1}(x_1, y_1) = 0$, and $r_{1,1}(x, y) \ne 0$, for all $(x, y) \in [0, x_1) \times [0, y_1)$.

Thus, we have

$$r_{1,1}(x_1, y_1) = 0 \text{ and } r_{1,1}(x, y) \ne 0, \text{ for all } (x, y) \in [0, x_1) \times [0, y_1] \cup (y_1, b].$$

Suppose that there exist $(\bar{x}, \bar{y}) \in [0, x_1) \times [0, y_1] \cup (y_1, b]$ such that $r_{1,1}(\bar{x}, \bar{y}) = 0$. The function $r_{1,1}$ attains a maximum at some point $(s, t) \in [0, x_1) \times [0, b]$. Since

$$({}^cD_0^r u_1)(x, y) = f(x, y, u_{1(\rho_1(x,y,u_{(x,y)}),\rho_2(x,y,u_{(x,y)}))}), \text{ for } (x, y) \in J,$$

then

$$\frac{\partial u_1(x, y)}{\partial x} \text{ exists, and } \frac{\partial r_{1,1}(s, t)}{\partial x} = x_1'(u_1(s, t))\frac{\partial u_1(s, t)}{\partial x} - 1 = 0.$$

Since

$$\frac{\partial u_1(x,y)}{\partial x} = \varphi'(x) + \frac{r_1 - 1}{\Gamma(r_1)\Gamma(r_2)} \int_0^x \int_0^y (x-s)^{r_1-2}(y-t)^{r_2-1}$$
$$\times f(s,t,u_{1(\rho_1(s,t,u_{(s,t)}),\rho_2(s,t,u_{(s,t)}))})\mathrm{d}t\,\mathrm{d}s,$$

then

$$x_1'(u_1(s,t))\left[\varphi'(s) + \frac{r_1-1}{\Gamma(r_1)\Gamma(r_2)}\int_0^s\int_0^t (s-\theta)^{r_1-2}(t-\eta)^{r_2-1}\right.$$
$$\left.\times f(\theta,\eta,u_{1(\rho_1(\theta,\eta,u_{(\theta,\eta)}),\rho_2(\theta,\eta,u_{(\theta,\eta)}))})\mathrm{d}\eta\mathrm{d}\theta\right] = 1,$$

which contradicts (5.31.3). From (5.31.1) we have

$$r_{k,1}(x,y) \neq 0 \text{ for all } (x,y) \in [0,x_1) \times [0,b] \text{ and } k = 1,\dots m.$$

Step 2: In what follows set

$$\begin{aligned}\Omega_k = \big\{u : (-\infty,a] \times (-\infty,b] \to \mathbb{R}^n : u_{(x,y)} \in \mathcal{B} \text{ for } (x,y) \in E \text{ and there exist}\\ 0 = x_0 < x_1 < x_2 < \cdots < x_m < x_{m+1} = a \text{ such that } x_k = x_k(u(x_k,.)),\\ \text{and } u(x_k^-,.),\ u(x_k^+,.) \text{ exist with } u(x_k^-,.) = u(x_k,.);\ k = 1,\dots,m,\\ \text{and } u \in C(X_k,\mathbb{R}^n); k = 0,\dots,m\big\},\end{aligned}$$

where

$$X_k := [x_k,a] \times [0,b];\ k = 1,\dots,m.$$

Consider now the problem

$$({}^cD_{x_1}^r u)(x,y) = f(x,y,u_{(\rho_1(x,y,u_{(x,y)}),\rho_2(x,y,u_{(x,y)}))});\ \text{if } (x,y) \in X_1, \quad (5.96)$$

$$u(x_1^+,y) = I_1(u_1(x_1,y)) \quad (5.97)$$

$$u(x,y) = u_1(x,y),\ \text{if } (x,y) \in \tilde{J}' \cup [0,x_1) \times [0,b]. \quad (5.98)$$

Consider the operator $N_1 : \Omega_1 \to \Omega_1$ defined as

$$N_1(u)(x,y) = \begin{cases} u_1(x,y), & (x,y) \in \tilde{J}' \cup [0,x_1) \times [0,b], \\ \varphi(x) + I_1(u_1(x_1,y)) - I_1(u_1(x_1,0)) & \\ + \frac{1}{\Gamma(r_1)\Gamma(r_2)} \int_{x_1}^x \int_0^y (x-s)^{r_1-1}(y-t)^{r_2-1} & \\ \times f(s,t,u_{(\rho_1(s,t,u_{(s,t)}),\rho_2(s,t,u_{(s,t)}))})\mathrm{d}t\,\mathrm{d}s, & (x,y) \in X_1. \end{cases}$$

As in Step 1 we can show that N_1 is completely continuous. Now it remains to show that the set $\mathcal{F}^* = \{u \in C(X_1, \mathbb{R}^n) : u = \lambda N_1(u) \text{ for some } 0 < \lambda < 1\}$ is bounded. Let $u \in \mathcal{F}^*$, then $u = \lambda N_1(u)$ for some $0 < \lambda < 1$. Thus, from (5.31.2) and (5.31.5) we get for each $(x, y) \in X_1$,

$$\begin{aligned}\|w(x,y)\| &\leq \|\varphi(x)\| + \|I_1(u_1(x_1,y))\| + \|I_1(u_1(x_1,0))\| \\ &\quad + \frac{1}{\Gamma(r_1)\Gamma(r_2)} \int_{x_1}^{x} \int_0^y (x-s)^{r_1-1}(y-t)^{r_2-1} \| \\ &\quad \times f(s,t,u_{(\rho_1(s,t,u_{(s,t)}),\rho_2(s,t,u_{(s,t)}))}) \| dt ds \\ &\leq \|\varphi\|_\infty + 2M^*(1+\|u_1\|) \\ &\quad + \frac{M}{\Gamma(r_1)\Gamma(r_2)} \int_0^x \int_0^y (x-s)^{r_1-1}(y-t)^{r_2-1}(1+\|z(s,t)\|) dt ds.\end{aligned}$$

Set

$$C^* = \|\varphi\|_\infty + 2M^*(1+\|u_1\|) + \frac{KM' a^{r_1} b^{r_2}}{\Gamma(r_1+1)\Gamma(r_2+1)}.$$

Then Lemma 2.43 implies that there exists $\delta = \delta(r_1, r_2) > 0$ such that for each $(x, y) \in X_1$,

$$\|w(x,y)\| \leq C^* \left[1 + \delta \frac{KM'(1+C^*) a^{r_1} b^{r_2}}{\Gamma(r_1+1)\Gamma(r_2+1)}\right] := R^*.$$

This shows that the set $\mathcal{F}^*$ is bounded. As a consequence of Schaefer's fixed-point theorem (Theorem 2.34), we deduce that N_1 has a fixed point u which is a solution to problems (5.96)–(5.98). Denote this solution by u_2. Define

$$r_{k,2}(x,y) = x_k(u_2(x,y)) - x, \quad \text{for } (x,y) \in X_1.$$

If $r_{k,2}(x, y) \neq 0$ on $(x_1, a] \times [0, b]$ and for all $k = 1, \ldots, m$, then

$$u(x,y) = \begin{cases} u_1(x,y); & \text{if } (x,y) \in \tilde{J}' \cup [0,x_1) \times [0,b], \\ u_2(x,y), & \text{if } (x,y) \in [x_1,a] \times [0,b], \end{cases}$$

is a solution of the problems (5.79)–(5.82). It remains to consider the case when $r_{2,2}(x, y) = 0$, for some $(x, y) \in (x_1, a] \times [0, b]$. By (5.31.4), we have

$$\begin{aligned} r_{2,2}(x_1^+, y_1) &= x_2(u_2(x_1^+, y_1) - x_1 \\ &= x_2(I_1(u_1(x_1,y_1))) - x_1 \\ &> x_1(u_1(x_1,y_1)) - x_1 \\ &= r_{1,1}(x_1,y_1) = 0.\end{aligned}$$

Since $r_{2,2}$ is continuous, there exists $x_2 > x_1,\ y_2 > y_1$ such that $r_{2,2}(x_2, y_2) = 0$, and $r_{2,2}(x, y) \neq 0$ for all $(x, y) \in (x_1, x_2) \times [0, b]$.

It is clear by (5.31.1) that

$$r_{k,2}(x, y) \neq 0 \quad \text{for all } (x, y) \in (x_1, x_2) \times [0, b],\ k = 2, \ldots, m.$$

Now suppose that there are $(s, t) \in (x_1, x_2) \times [0, b]$ such that $r_{1,2}(s, t) = 0$. From (5.31.4) it follows that

$$\begin{aligned} r_{1,2}(x_1^+, y_1) &= x_1(u_2(x_1^+, y_1) - x_1 \\ &= x_1(I_1(u_1(x_1, y_1))) - x_1 \\ &\leq x_1(u_1(x_1, y_1)) - x_1 \\ &= r_{1,1}(x_1, y_1) = 0. \end{aligned}$$

Thus $r_{1,2}$ attains a nonnegative maximum at some point $(s_1, t_1) \in (x_1, a) \times [0, x_2) \cup (x_2, b]$. Since

$$({}^c D^r_{x_1} u_2)(x, y) = f(x, y, u_{2(\rho_1(x,y,u_{(x,y)}),\rho_2(x,y,u_{(x,y)}))}); \quad \text{for } (x, y) \in X_1,$$

then we get

$$\begin{aligned} u_2(x, y) = \varphi(x) &+ I_1(u_1(x_1, y)) - I_1(u_1(x_1, 0)) \\ &+ \frac{1}{\Gamma(r_1)\Gamma(r_2)} \int_{x_1}^{x} \int_0^{y} (x - s)^{r_1 - 1}(y - t)^{r_2 - 1} \\ &\times f(s, t, u_{2(\rho_1(s,t,u_{(s,t)}),\rho_2(s,t,u_{(s,t)}))}) \mathrm{d}t\, \mathrm{d}s, \end{aligned}$$

hence

$$\begin{aligned} \frac{\partial u_2}{\partial x}(x, y) = \varphi'(x) &+ \frac{r_1 - 1}{\Gamma(r_1)\Gamma(r_2)} \int_{x_1}^{x} \int_0^{y} (x - s)^{r_1 - 2}(y - t)^{r_2 - 1} \\ &\times f(s, t, u_{2(\rho_1(s,t,u_{(s,t)}),\rho_2(s,t,u_{(s,t)}))}) \mathrm{d}t\, \mathrm{d}s, \end{aligned}$$

then

$$\frac{\partial r_{1,2}(s_1, t_1)}{\partial x} = x_1'(u_2(s_1, t_1)) \frac{\partial u_2}{\partial x}(s_1, t_1) - 1 = 0.$$

Therefore

$$\begin{aligned} x_1'(u_2(s_1, t_1)) \Bigg[\varphi'(s_1) &+ \frac{r_1 - 1}{\Gamma(r_1)\Gamma(r_2)} \int_{x_1}^{s_1} \int_0^{t_1} (s_1 - \theta)^{r_1 - 2}(t_1 - \eta)^{r_2 - 1} \\ &\times f(\theta, \eta, u_{2(\rho_1(\theta,\eta,u_{(\theta,\eta)}),\rho_2(\theta,\eta,u_{(\theta,\eta)}))}) \mathrm{d}\eta \mathrm{d}\theta \Bigg] = 1, \end{aligned}$$

which contradicts (5.31.3).

Step 3: We continue this process and take into account that $u_{m+1} := u\Big|_{X_m}$ is a solution to the problem

$$\begin{cases} ({}^cD^r_{x_m}u)(x,y)=f(x,y,u_{(\rho_1(x,y,u_{(x,y)}),\rho_2(x,y,u_{(x,y)}))}); \quad \text{a.e. } (x,y)\in(x_m,a]\times[0,b], \\ u(x_m^+,y) = I_m(u_{m-1}(x_m,y)), \\ u(x,y) = u_1(x,y), \text{ if } (x,y)\in \tilde{J}'\cup[0,x_1)\times[0,b], \\ u(x,y) = u_2(x,y), \text{ if } (x,y)\in[x_1,x_2)\times[0,b], \\ \dots \\ u(x,y) = u_m(x,y), \text{ if } (x,y)\in[x_{m-1},x_m)\times[0,b]. \end{cases}$$

The solution u of the problems (5.79)–(5.82) is then defined by

$$u(x,y) = \begin{cases} u_1(x,y), & \text{if } (x,y)\in\tilde{J}'\cup[0,x_1]\times[0,b], \\ u_2(x,y), & \text{if } (x,y)\in(x_1,x_2]\times[0,b], \\ \dots \\ u_{m+1}(x,y), & \text{if } (x,y)\in(x_m,a]\times[0,b]. \end{cases}$$

□

5.7.4 Examples

5.7.4.1 Example 1

As an application of our results we consider the following impulsive partial hyperbolic functional differential equations of the form:

$$({}^cD^r_{x_k}u)(x,y) = \frac{1+|u(x-\sigma_1(u(x,y)),y-\sigma_2(u(x,y)))|}{9+e^{x+y}};$$

$$\text{if } (x,y)\in J_k;\ k=0,\dots,m,\ x_k = x_k(u(x,y));\ k=1,\dots,m, \tag{5.99}$$

$$u(x_k^+,y) = d_k u(x_k,y);\ y\in[0,1], \tag{5.100}$$

$$u(x,0) = x,\ u(0,y) = y^2;\ x,y\in[0,1], \tag{5.101}$$

$$u(x,y) = x+y^2,\ (x,y)\in[-1,1]\times[-2,1]\backslash(0,1]\times(0,1], \tag{5.102}$$

where $r=(r_1,r_2)$, $0<r_1,r_2\le 1$, $x_k(u) = 1-\frac{1}{2^k(1+u^2)}$; $k=1,\dots,m$ and $\frac{\sqrt{2}}{2}<d_k\le 1$; $k=1,\dots m$, $\sigma_1\in C(\mathbb{R},[0,1])$, $\sigma_2\in C(\mathbb{R},[0,2])$. Set

$$\rho_1(x,y,\varphi) = x-\sigma_1(\varphi(0,0)),\ (x,y,\varphi)\in J\times C,$$

$$\rho_2(x,y,\varphi) = y-\sigma_2(\varphi(0,0)),\ (x,y,\varphi)\in J\times C,$$

where $C := C_{(1,2)}$. Set

$$f(x,y,\varphi) = \frac{1+|\varphi|}{9+e^{x+y}}, \ (x,y) \in [0,1]\times[0,1], \ \varphi \in C.$$

and

$$I_k(u) = d_k u; \ u \in \mathbb{R}, \ k = 1,\ldots,m.$$

Let $u \in \mathbb{R}$ then we have

$$x_{k+1}(u) - x_k(u) = \frac{1}{2^{k+1}(1+u^2)} > 0; \ k = 1,\ldots,m.$$

Hence $0 < x_1(u) < x_2(u) < \cdots < x_m(u) < 1$, for each $u \in \mathbb{R}$.

Also, for each $u \in \mathbb{R}$ we have

$$x_{k+1}(I_k(u)) - x_k(u) = \frac{1+(2d_k^2-1)u^2}{2^{k+1}(1+u^2)(1+d_k^2)} > 0.$$

Finally, for all $(x,y) \in J$ and each $u \in \mathbb{R}$ we get

$$|I_k(u)| = |d_k u| \leq |u| \leq 3(1+|u|); \ k = 1,\ldots,m,$$

and

$$|f(x,y,u)| = \frac{|1+u|}{9+e^{x+y}} \leq \frac{1}{10}(1+|u|).$$

Since all conditions of Theorem 5.29 are satisfied, problems (5.99)–(5.102) have at least one solution on $[-1,1]\times[-2,1]$.

5.7.4.2 Example 2

As an application of our results we consider the following impulsive partial hyperbolic differential equations of the form:

$$({}^cD^r_{x_k}u)(x,y) = \frac{1+|u(x-\sigma_1(u(x,y)), y-\sigma_2(u(x,y)))|}{9+e^{x+y}};$$

$$\text{if } (x,y) \in J_k; \ k = 0,\ldots,m, \ x_k = x_k(u(x,y)); \ k = 1,\ldots,m, \tag{5.103}$$

$$u(x_k^+, y) = d_k u(x_k, y), \ y \in [0,1]; \ k = 1,\ldots,m, \tag{5.104}$$

$$u(x,y) = x + y^2, \ (x,y) \in (-\infty,1]\times(-\infty,1]\backslash(0,1]\times(0,1], \tag{5.105}$$

$$u(x,0) = x, \ x \in [0,1], \ u(0,y) = y^2, \ y \in [0,1], \tag{5.106}$$

where $r = (r_1, r_2)$, $0 < r_1, r_2 \leq 1$, $x_k(u) = 1 - \frac{1}{2^k(1+u^2)}$; $k = 1, \ldots, m$ and $\frac{\sqrt{2}}{2} < d_k \leq 1$; $k = 1, \ldots, m$ and $\sigma_1, \sigma_2 \in C(\mathbb{R}, [0, \infty))$. Let $\mathcal{B}_\gamma$ be the phase space defined in the Example of Sect. 3.7. Set

$$\rho_1(x, y, \varphi) = x - \sigma_1(\varphi(0,0)), \quad (x, y, \varphi) \in J \times \mathcal{B}_\gamma,$$

$$\rho_2(x, y, \varphi) = y - \sigma_2(\varphi(0,0)), \quad (x, y, \varphi) \in J \times \mathcal{B}_\gamma,$$

$$f(x, y, \varphi) = \frac{1 + |\varphi|}{9 + e^{x+y}}, \quad (x, y) \in [0,1] \times [0,1], \ u_{(x,y)} \in \mathcal{B}_\gamma,$$

and

$$I_k(u) = d_k u; \ u \in \mathbb{R}, \ k = 1, \ldots, m.$$

Let $u \in \mathbb{R}$ then we have

$$x_{k+1}(u) - x_k(u) = \frac{1}{2^{k+1}(1 + u^2)} > 0; \ k = 1, \ldots, m.$$

Hence $0 < x_1(u) < x_2(u) < \cdots < x_m(u) < 1$, for each $u \in \mathbb{R}$. Also, for each $u \in \mathbb{R}$ we have

$$x_{k+1}(I_k(u)) - x_k(u) = \frac{1 + (2d_k^2 - 1)u^2}{2^{k+1}(1 + u^2)(1 + d_k^2)} > 0.$$

Finally, for all $(x, y) \in J$ and each $u \in \mathbb{R}$ we get

$$|I_k(u)| = |d_k u| \leq |u| \leq 3(1 + |u|); \ k = 1, \ldots, m,$$

and

$$|f(x, y, u)| = \frac{|1 + u|}{9 + e^{x+y}} \leq \frac{1}{10}(1 + |u|).$$

Since all conditions of Theorem 5.31 are satisfied, problem (5.103)–(5.106) has at least one solution on $(-\infty, 1] \times (-\infty, 1]$.

5.8 Upper and Lower Solutions Method for Impulsive Partial Hyperbolic Differential Equations

5.8.1 Introduction

This section deals with the existence of solutions to impulsive fractional order IVP, for the system

$$({}^c D_{x_k}^r u)(x, y) = f(x, y, u(x, y)), \text{ if } (x, y) \in J_k; \ k = 0, \ldots, m, \tag{5.107}$$

$$u(x_k^+, y) = u(x_k^-, y) + I_k(u(x_k^-, y)); \text{ if } y \in [0, b],\ k = 1, \dots, m, \tag{5.108}$$

$$u(x, 0) = \varphi(x),\ x \in [0, a],\ u(0, y) = \psi(y),\ y \in [0, b], \tag{5.109}$$

where $a, b > 0$, $0 = x_0 < x_1 < \cdots < x_m < x_{m+1} = a$, $f : J \times \mathbb{R}^n \to \mathbb{R}^n$ and $I_k : \mathbb{R}^n \to \mathbb{R}^n$, $k = 0, 1, \dots, m$, $\varphi : [0, a] \to \mathbb{R}^n$, $\psi : [0, b] \to \mathbb{R}^n$ are given absolutely continuous functions with $\varphi(0) = \psi(0)$. Here $u(x_k^+, y)$, $J = [0, a] \times [0, b]$ and $u(x_k^-, y)$ denote the right and left limits of $u(x, y)$ at $x = x_k$, respectively.

5.8.2 Main Result

In what follows set

$$J_k := (x_k, x_{k+1}] \times (0, b].$$

To define the solutions of problem (5.107)–(5.109), we shall consider the Banach space

$$\begin{aligned} PC(J, \mathbb{R}^n) = \{ & u : J \to \mathbb{R}^n : u \in C(J_k, \mathbb{R}^n);\ k = 1, \dots, m, \text{ and there} \\ & \text{exist } u(x_k^-, y) \text{ and } u(x_k^+, y);\ y \in [0, b],\ k = 1, \dots, m, \\ & \text{with } u(x_k^-, y) = u(x_k, y)\}, \end{aligned}$$

with the norm

$$\|u\|_{PC} = \sup_{(x,y) \in J} \|u(x, y)\|.$$

Definition 5.32. A function $u \in PC(J, \mathbb{R}^n) \bigcap \bigcup_{k=0}^{m} C^1((x_k, x_{k+1}) \times [0, b], \mathbb{R}^n)$ such that its mixed derivative D_{xy}^2 exists on J_k; $k = 0, \dots, m$ is said to be a solution of (5.107)–(5.109) if u satisfies $({}^c D_{x_k}^r u)(x, y) = f(x, y, u(x, y))$ on J_k; $k = 0, \dots, m$ and conditions (5.108), (5.109) are satisfied.

Definition 5.33. A function $z \in PC(J, \mathbb{R}^n) \bigcap \bigcup_{k=0}^{m} C^1((x_k, x_{k+1}) \times [0, b], \mathbb{R}^n)$ is said to be a lower solution of (5.107)–(5.109) if z satisfies

$$({}^c D_{x_k}^r z)(x, y) \le f(x, y, z(x, y)),\ z(x, 0) \le \varphi(x),\ z(0, y) \le \psi(y) \text{ on } J_k;\ k = 0, \dots, m,$$

$$z(x_k^+, y) \le z(x_k^-, y) + I_k(z(x_k^-, y)), \text{ if } y \in [0, b];\ k = 1, \dots, m,$$

$$z(x, 0) \le \varphi(x),\ z(0, y) \le \psi(y) \text{ on } J,$$

$$\text{and } z(0, 0) \le \varphi(0).$$

The function z is said to be an upper solution of (5.107)–(5.109) if the reversed inequalities hold.

Let $z,\ \bar{z} \in C(J, \mathbb{R}^n)$ be such that

$$z(x,y) = (z_1(x,y), z_2(x,y), \ldots, z_n(x,y)),\ (x,y) \in J,$$

and

$$\bar{z}(x,y) = (\bar{z}_1(x,y), \bar{z}_2(x,y), \ldots, \bar{z}_n(x,y)),\ (x,y) \in J.$$

The notation $z \leq \bar{z}$ means that

$$z_i(x,y) \leq \bar{z}_i(x,y),\ i = 1, \ldots, n.$$

Further, we present conditions for the existence of a solution of our problem.

Theorem 5.34. *Assume that the following hypotheses hold:*

(5.34.1) The function $f : J \times \mathbb{R}^n \to \mathbb{R}^n$ is jointly continuous.
(5.34.2) There exist v and $w \in PC \bigcap C^1((x_k, x_{k+1}) \times [0,b], \mathbb{R}^n),\ k = 0, \ldots, m$ lower and upper solutions for the problems (5.107)–(5.109) such that $v \leq w$.
(5.34.3) For each $y \in [0,b]$, we have

$$v(x_k^+, y) \leq \min_{u \in [v(x_k^-, y), w(x_k^-, y)]} I_k(u) \leq \max_{u \in [v(x_k^-, y), w(x_k^-, y)]} I_k(u)$$
$$\leq w(x_k^+, y),\ k = 1, \ldots, m.$$

Then the problems (5.107)–(5.109) have at least one solution u such that

$$v(x,y) \leq u(x,y) \leq w(x,y) \text{ for all } (x,y) \in J.$$

Proof. Transform the problems (5.107)–(5.109) into a fixed-point problem. Consider the following modified problem:

$$({}^cD_0^r u)(x,y) = g(x,y,u(x,y));\ \text{if } (x,y) \in J,\ x \neq x_k;\ k = 1, \ldots, m, \quad (5.110)$$

$$u(x_k^+, y) = u(x_k^-, y) + I_k(h(x_k^-, y, u(x_k^-, y)));\ \text{ if } y \in [0,b];\ k = 1, \ldots, m, \quad (5.111)$$

$$u(x,0) = \varphi(x),\ u(0,y) = \psi(y);\ x \in [0,a] \text{ and } y \in [0,b], \quad (5.112)$$

where

$$g(x,y,u(x,y)) = f(x,y,h(x,y,u(x,y))),$$
$$h(x,y,u(x,y)) = \max\{v(x,y), \min\{u(x,y), w(x,y)\}\},$$

for each $(x,y) \in J$. A solution to (5.110)–(5.112) is a fixed point of the operator $N : PC(J, \mathbb{R}^n) \to PC(J, \mathbb{R}^n)$ defined by

$$N(u)(x,y)=\mu(x,y)+\sum_{0<x_k<x}(I_k(h(x_k^-,y,u(x_k^-,y)))-I_k(h(x_k^-,0,u(x_k^-,0))))$$
$$+\frac{1}{\Gamma(r_1)\Gamma(r_2)}\sum_{0<x_k<x}\int_{x_{k-1}}^{x_k}\int_0^y(x_k-s)^{r_1-1}(y-t)^{r_2-1}$$
$$\times g(s,t,u(s,t))\mathrm{d}t\mathrm{d}s$$
$$+\frac{1}{\Gamma(r_1)\Gamma(r_2)}\int_{x_k}^{x}\int_0^y(x-s)^{r_1-1}(y-t)^{r_2-1}g(s,t,u(s,t))\mathrm{d}t\mathrm{d}s.$$

Notice that g is a continuous function, and from (5.34.2) there exists $M>0$ such that

$$\|g(x,y,u)\|\le M;\ \text{for each } (x,y)\in J,\ \text{and } u\in\mathbb{R}^n. \tag{5.113}$$

Also, by the definition of h and from (5.34.3) we have

$$v(x_k^+,y)\le I_k(h(x_k,y,u(x_k,y)))\le w(x_k^+,y);\ y\in[0,b];\ k=1,\ldots,m, \tag{5.114}$$

Set

$$\eta=\|\mu\|_\infty+2\sum_{k=1}^{m}\max_{y\in[0,b]}(\|v(x_k^+,y)\|,\|w(x_k^+,y)\|)+\frac{2Ma^{r_1}b^{r_2}}{\Gamma(r_1+1)\Gamma(r_2+1)},$$

and

$$D=\{u\in PC(J,\mathbb{R}^n):\|u\|_{PC}\le\eta\}.$$

Clearly D is a closed convex subset of $PC(J,\mathbb{R}^n)$ and that N maps D into D. We shall show that N satisfies the assumptions of Schauder's fixed-point theorem. The proof will be given in several steps.

Step 1: *N is continuous.* Let $\{u_n\}$ be a sequence such that $u_n\to u$ in D. Then

$$\|N(u_n)(x,y)-N(u)(x,y)\|$$
$$\le\sum_{k=1}^{m}(\|I_k(h(x_k^-,y,u_n(x_k^-,y)))-I_k(h(x_k^-,y,u(x_k^-,y)))\|)$$
$$+\sum_{k=1}^{m}(\|I_k(h(x_k^-,0,u_n(x_k^-,0)))-I_k(h(x_k^-,0,u(x_k^-,0)))\|)$$
$$+\frac{1}{\Gamma(r_1)\Gamma(r_2)}\sum_{k=1}^{m}\int_{x_{k-1}}^{x_k}\int_0^y(x_k-s)^{r_1-1}(y-t)^{r_2-1}\|g(s,t,u_n(s,t))-g(s,t,u(s,t))\|\mathrm{d}t\mathrm{d}s$$
$$+\frac{1}{\Gamma(r_1)\Gamma(r_2)}\int_{x_k}^{x}\int_0^y(x-s)^{r_1-1}(y-t)^{r_2-1}\|g(s,t,u_n(s,t))-g(s,t,u(s,t))\|\mathrm{d}t\mathrm{d}s.$$

Since g and $I_k, k = 1, \dots, m$ are continuous functions, we have

$$\|N(u_n) - N(u)\|_{PC} \to 0 \quad \text{as } n \to \infty.$$

Step 2: *$N(D)$ is bounded.* This is clear since $N(D) \subset D$ and D is bounded.

Step 3: *$N(D)$ is equicontinuous.* Let $(\tau_1, y_1), (\tau_2, y_2) \in [0, a] \times [0, b]$, $\tau_1 < \tau_2$ and $y_1 < y_2$, and $u \in D$. Then

$$\begin{aligned}
&\|N(u)(\tau_2, y_2) - N(u)(\tau_1, y_1)\| \\
&\quad \le \|\mu(\tau_1, y_1) - \mu(\tau_2, y_2)\| \\
&\qquad + \sum_{k=1}^{m} (\|I_k(h(x_k^-, y_1, u(x_k^-, y_1))) - I_k(h(x_k^-, y_2, u(x_k^-, y_2)))\|) \\
&\qquad + \frac{1}{\Gamma(r_1)\Gamma(r_2)} \sum_{k=1}^{m} \int\limits_{x_{k-1}}^{x_k} \int\limits_{0}^{y_1} (x_k - s)^{r_1 - 1}[(y_2 - t)^{r_2 - 1} - (y_1 - t)^{r_2 - 1}] \\
&\qquad \times g(s, t, u(s, t))\mathrm{d}t\mathrm{d}s \\
&\qquad + \frac{1}{\Gamma(r_1)\Gamma(r_2)} \sum_{k=1}^{m} \int\limits_{x_{k-1}}^{x_k} \int\limits_{y_1}^{y_2} (x_k - s)^{r_1 - 1}(y_2 - t)^{r_2 - 1}\|g(s, t, u(s, t))\|\mathrm{d}t\mathrm{d}s \\
&\qquad + \frac{1}{\Gamma(r_1)\Gamma(r_2)} \int_0^{\tau_1} \int_0^{y_1} [(\tau_2 - s)^{r_1 - 1}(y_2 - t)^{r_2 - 1} - (\tau_1 - s)^{r_1 - 1}(y_1 - t)^{r_2 - 1}] \\
&\qquad \times g(s, t, u(s, t))\mathrm{d}t\mathrm{d}s \\
&\qquad + \frac{1}{\Gamma(r_1)\Gamma(r_2)} \int_{\tau_1}^{\tau_2} \int_{y_1}^{y_2} (\tau_2 - s)^{r_1 - 1}(y_2 - t)^{r_2 - 1}\|g(s, t, u(s, t))\|\mathrm{d}t\mathrm{d}s \\
&\qquad + \frac{1}{\Gamma(r_1)\Gamma(r_2)} \int_0^{\tau_1} \int_{y_1}^{y_2} (\tau_2 - s)^{r_1 - 1}(y_2 - t)^{r_2 - 1}\|g(s, t, u(s, t))\|\mathrm{d}t\mathrm{d}s \\
&\qquad + \frac{1}{\Gamma(r_1)\Gamma(r_2)} \int_{\tau_1}^{\tau_2} \int_0^{y_1} (\tau_2 - s)^{r_1 - 1}(y_2 - t)^{r_2 - 1}\|g(s, t, u(s, t))\|\mathrm{d}t\mathrm{d}s \\
&\quad \le \|\mu(\tau_1, y_1) - \mu(\tau_2, y_2)\| \\
&\qquad + \sum_{k=1}^{m} (\|I_k(h(x_k^-, y_1, u(x_k^-, y_1))) - I_k(h(x_k^-, y_2, u(x_k^-, y_2)))\|) \\
&\qquad + \frac{M}{\Gamma(r_1)\Gamma(r_2)} \sum_{k=1}^{m} \int\limits_{x_{k-1}}^{x_k} \int\limits_{0}^{y_1} (x_k - s)^{r_1 - 1}[(y_2 - t)^{r_2 - 1} - (y_1 - t)^{r_2 - 1}]\mathrm{d}t\mathrm{d}s
\end{aligned}$$

$$+\frac{M}{\Gamma(r_1)\Gamma(r_2)}\sum_{k=1}^{m}\int_{x_{k-1}}^{x_k}\int_{y_1}^{y_2}(x_k-s)^{r_1-1}(y_2-t)^{r_2-1}dtds$$

$$+\frac{M}{\Gamma(r_1)\Gamma(r_2)}\int_0^{\tau_1}\int_0^{y_1}[(\tau_2-s)^{r_1-1}(y_2-t)^{r_2-1}-(\tau_1-s)^{r_1-1}(y_1-t)^{r_2-1}]dtds$$

$$+\frac{M}{\Gamma(r_1)\Gamma(r_2)}\int_{\tau_1}^{\tau_2}\int_{y_1}^{y_2}(\tau_2-s)^{r_1-1}(y_2-t)^{r_2-1}dtds$$

$$+\frac{M}{\Gamma(r_1)\Gamma(r_2)}\int_0^{\tau_1}\int_{y_1}^{y_2}(\tau_2-s)^{r_1-1}(y_2-t)^{r_2-1}dtds$$

$$+\frac{M}{\Gamma(r_1)\Gamma(r_2)}\int_{\tau_1}^{\tau_2}\int_0^{y_1}(\tau_2-s)^{r_1-1}(y_2-t)^{r_2-1}dtds.$$

As $\tau_1 \longrightarrow \tau_2$ and $y_1 \longrightarrow y_2$, the right-hand side of the above inequality tends to zero. As a consequence of Steps 1–3 together with the Arzelá-Ascoli theorem, we can conclude that $N : D \to D$ is continuous and compact . From an application of Schauder's theorem, we deduce that N has a fixed point u which is a solution of the problems (5.110)–(5.112).

Step 4: *The solution u of (5.110)–(5.112) satisfies*

$$v(x,y) \le u(x,y) \le w(x,y) \ \text{ for all } \ (x,y) \in J.$$

Let u be the above solution to (5.110)–(5.112). We prove that

$$u(x,y) \le w(x,y) \ \text{ for all } \ (x,y) \in J.$$

Assume that $u - w$ attains a positive maximum on $[x_k^+, x_{k+1}^-] \times [0,b]$ at $(\overline{x}_k, \overline{y}) \in [x_k^+, x_{k+1}^-] \times [0,b]$ for some $k = 0, \ldots, m$; i.e.,

$$(u-w)(\overline{x}_k, \overline{y}) = \max\{u(x,y) - w(x,y) : (x,y) \in [x_k^+, x_{k+1}^-] \times [0,b]\} > 0;$$

for some $k = 0, \ldots, m$. We distinguish the following cases.

Case 1. If $(\overline{x}_k, \overline{y}) \in (x_k^+, x_{k+1}^-) \times [0,b]$ there exists $(x_k^*, \overline{y}^*) \in (x_k^+, x_{k+1}^-) \times [0,b]$ such that

$$[u(x,y^*) - w(x,y^*)] + [u(x_k^*,y) - w(x_k^*,y)] - [u(x_k^*,y^*) - w(x_k^*,y^*)] \le 0;$$
$$\text{for all } (x,y) \in ([x_k^*, \overline{x}_k] \times \{y^*\}) \cup (\{x_k^*\} \times [y^*, b]) \tag{5.115}$$

and

$$u(x,y) - w(x,y) > 0, \text{ for all } (x,y) \in (x_k^*, \overline{x}_k] \times [\overline{y}^*, b]. \tag{5.116}$$

By the definition of h one has

$${}^cD^r_{x_k^*}u(x,y) = f(x,y,w(x,y)) \ \text{for all } \ (x,y) \in [x_k^*, \overline{x}_k] \times [\overline{y}^*, b].$$

An integration on $[x_k^*, x] \times [\overline{y}^*, y]$ for each $(x, y) \in [x_k^*, \overline{x}_k] \times [\overline{y}^*, b]$ yields

$$\begin{aligned} &u(x, y) + u(x_k^*, y^*) - u(x, y^*) - u(x_k^*, y) \\ &= \frac{1}{\Gamma(r_1)\Gamma(r_2)} \int_{x_k^*}^{x} \int_{\overline{y}^*}^{y} (x-s)^{r_1-1}(y-t)^{r_2-1} f(s, t, w(s, t)) \mathrm{d}t \mathrm{d}s. \end{aligned} \tag{5.117}$$

From (5.117) and using the fact that w is an upper solution to (5.107)–(5.109) we get

$$\begin{aligned} u(x, y) + u(x_k^*, y^*) - u(x, y^*) - u(x_k^*, y) &\le w(x, y) \\ + w(x_k^*, y^*) - w(x, y^*) - w(x_k^*, y), \end{aligned}$$

which gives,

$$\begin{aligned} [u(x, y) - w(x, y)] \le [u(x, y^*) - w(x, y^*)] &+ [u(x_k^*, y) - w(x_k^*, y)] \\ -[u(x_k^*, y^*) - w(x_k^*, y^*)]. \end{aligned} \tag{5.118}$$

Thus from (5.115), (5.116), and (5.118) we obtain the contradiction

$$\begin{aligned} 0 < [u(x, y) - w(x, y)] &\le [u(x, y^*) - w(x, y^*)] + [u(x_k^*, y) - w(x_k^*, y)] \\ -[u(x_k^*, y^*) - w(x_k^*, y^*)] &\le 0; \text{ for all } (x, y) \in [x_k^*, \overline{x}_k] \times [y^*, b]. \end{aligned}$$

Case 2. If $\overline{x}_k = x_k^+$, $k = 1, \ldots, m$. Then

$$w(x_k^+, \overline{y}) < I_k(h(x_k^-, u(x_k^-, \overline{y}))) \le w(x_k^+, \overline{y})$$

which is a contradiction. Thus

$$u(x, y) \le w(x, y) \text{ for all } (x, y) \in J.$$

Analogously, we can prove that

$$u(x, y) \ge v(x, y); \text{ for all } (x, y) \in J.$$

This shows that the problems (5.110)–(5.112) have a solution u satisfying $v \le u \le w$ which is solution of (5.107)–(5.109). □

5.9 Notes and Remarks

The results of Chap. 5 are taken from Abbas and Benchohra [11, 13], and Abbas et al. [2, 3, 25, 27]. Other results may be found in [58, 77, 78, 110].

Chapter 6
Impulsive Partial Hyperbolic Functional Differential Inclusions

6.1 Introduction

In this chapter, we shall present existence results for some classes of initial value problems for impulsive partial hyperbolic differential inclusions with fractional order.

6.2 Impulsive Partial Hyperbolic Differential Inclusions

6.2.1 Introduction

This section concerns the existence results to impulsive fractional order IVP for the system

$$({}^cD^r_{x_k}u)(x,y) \in F(x,y,u(x,y)), \text{ if } (x,y) \in J_k;\ k = 0,\dots,m, \tag{6.1}$$

$$u(x_k^+,y) = u(x_k^-,y) + I_k(u(x_k^-,y)), \text{ if } y \in [0,b];\ k = 1,\dots,m, \tag{6.2}$$

$$u(x,0) = \varphi(x);\ x \in [0,a],\ u(0,y) = \psi(y);\ y \in [0,b], \tag{6.3}$$

where $J_0 = [0,x_1] \times [0,b]$, $J_k = (x_k,x_{k+1}]$; $k = 1,\dots,m$, $a,b > 0$, $0 = x_0 < x_1 < \dots < x_m < x_{m+1} = a$, $F : J \times \mathbb{R}^n \to \mathcal{P}(\mathbb{R}^n)$ is a compact-valued multivalued map, $J = [0,a] \times [0,b]$, $\mathcal{P}(\mathbb{R}^n)$ is the family of all subsets of $\mathbb{R}^n$, $I_k : \mathbb{R}^n \to \mathbb{R}^n$, $k = 0,1,\dots,m$ are given functions and $\varphi : [0,a] \to \mathbb{R}^n$, $\psi : [0,b] \to \mathbb{R}^n$ are given absolutely continuous functions with $\varphi(0) = \psi(0)$.

To define the solutions of (6.1)–(6.3), we shall consider the Banach space

$$PC(J,\mathbb{R}^n)=\{u : J \to \mathbb{R}^n : \text{ there exist } 0 = x_0 < x_1 < x_2 < \dots < x_m < x_{m+1}=a$$

S. Abbas et al., *Topics in Fractional Differential Equations*, Developments in Mathematics 27, DOI 10.1007/978-1-4614-4036-9_6,

such that $u(x_k^-, y)$ and $u(x_k^+, y)$ exist with $u(x_k^-, y) = u(x_k, y)$;

$k = 0, \ldots, m$, and $u \in C(J_k, \mathbb{R}^n); k = 0, \ldots, m\}$.

Definition 6.1. A function $u \in PC(J, \mathbb{R}^n) \cap \cup_{k=0}^{m} AC((x_k, x_{k+1}) \times (0, b], \mathbb{R}^n)$ such that its mixed derivative D_{xy}^2 exists on J_k; $k = 0, \ldots, m$ is said to be a solution of (6.1)–(6.3) if there exists a function $f \in L^1(J, \mathbb{R}^n)$ with $f(x, y) \in F(x, y, u(x, y))$ such that $({}^c D_{x_k}^r u)(x, y) = f(x, y)$ on J_k; $k = 0, \ldots, m$ and u satisfies conditions (6.2) and (6.3).

6.2.2 The Convex Case

Now we are concerned with the existence of solutions for the problems (6.1)–(6.3) when the right-hand side is compact and convex valued.

Theorem 6.2. *Assume the following hypotheses hold:*

(6.2.1) $F : J_k \times \mathbb{R}^n \longrightarrow \mathcal{P}_{cp,cv}(\mathbb{R}^n)$, $k = 0, \ldots, m$, *is a Carathéodory multivalued map.*

(6.2.2) There exist $p \in L^\infty(J, \mathbb{R}_+)$ *and* $\psi_* : [0, \infty) \to (0, \infty)$ *continuous and nondecreasing such that* $\|F(x, y, u)\|_{\mathcal{P}} \leq p(x, y)\psi_*(\|u\|)$ *for* $(x, y) \in J, x \neq x_k$, $k = 0, \ldots, m$, *and each* $u \in \mathbb{R}^n$.

(6.2.3) There exists $l \in L^\infty(J, \mathbb{R}_+)$ *such that*

$$H_d(F(x, y, u), F(x, y, \overline{u})) \leq l(x, y)\|u - \overline{u}\| \text{ for every } u, \overline{u} \in \mathbb{R}^n,$$

and

$$d(0, F(x, y, 0)) \leq l(x, y), \text{ a.e. } (x, y) \in J_k,\ k = 0, \ldots, m.$$

(6.2.4) There exist constants c_k, *such that* $\|I_k(u)\| \leq c_k$, $k = 1, \ldots, m$ *for each* $u \in \mathbb{R}^n$.

(6.2.5) There exist constants c_k^*, *such that*

$$\|I_k(u) - I_k(\overline{u})\| \leq c_k^* \|u - \overline{u}\|, \text{ for each } u, \overline{u} \in \mathbb{R}^n,\ k = 1, \ldots, m.$$

(6.2.6) There exists a number $M > 0$ *such that*

$$\frac{M}{\|\mu\|_\infty + 2\sum_{k=1}^{m} c_k + \dfrac{2p^* a^{r_1} b^{r_2} \psi_*(M)}{\Gamma(r_1 + 1)\Gamma(r_2 + 1)}} > 1, \tag{6.4}$$

where $p^* = \|p\|_{L^\infty}$. *Then the IVP (6.1)–(6.3) have at least one solution on* J.

Proof. Transform the problem (6.1)–(6.3) into a fixed-point problem. Consider the multivalued operator $N : PC(J,\mathbb{R}^n) \to \mathcal{P}(PC(J,\mathbb{R}^n))$ defined by $N(u) = \{h \in PC(J,\mathbb{R}^n)\}$ where for $f \in S_{F,u}$,

$$h(u)(x,y) = \mu(x,y) + \sum_{0<x_k<x} (I_k(u(x_k^-,y)) - I_k(u(x_k^-,0)))$$

$$+ \frac{1}{\Gamma(r_1)\Gamma(r_2)} \sum_{0<x_k<x} \int_{x_{k-1}}^{x_k} \int_0^y (x_k-s)^{r_1-1}(y-t)^{r_2-1} f(s,t)\mathrm{d}t\mathrm{d}s$$

$$+ \frac{1}{\Gamma(r_1)\Gamma(r_2)} \int_{x_k}^{x} \int_0^y (x-s)^{r_1-1}(y-t)^{r_2-1} f(s,t)\mathrm{d}t\mathrm{d}s.$$

Remark 6.3. Clearly, from Lemma 2.15, the fixed points of N are solutions to (6.1)–(6.3).

We shall show that N satisfies the assumptions of the nonlinear alternative of Leray–Schauder type. The proof of this theorem will be given in several steps.

Step 1: *$N(u)$ is convex for each $u \in PC(J,\mathbb{R}^n)$.* Indeed, if $h_1,\ h_2$ belong to $N(u)$, then there exist $f_1,\ f_2 \in S_{F,u}$ such that for each $(x,y) \in J$ we have

$$h_i(u)(x,y) = \mu(x,y) + \sum_{0<x_k<x} (I_k(u(x_k^-,y)) - I_k(u(x_k^-,0)))$$

$$+ \frac{1}{\Gamma(r_1)\Gamma(r_2)} \sum_{0<x_k<x} \int_{x_{k-1}}^{x_k} \int_0^y (x_k-s)^{r_1-1}(y-t)^{r_2-1} f_i(s,t)\mathrm{d}t\mathrm{d}s$$

$$+ \frac{1}{\Gamma(r_1)\Gamma(r_2)} \int_{x_k}^{x} \int_0^y (x-s)^{r_1-1}(y-t)^{r_2-1} f_i(s,t)\mathrm{d}t\mathrm{d}s,$$

where $f_i \in S_{F,u}\ i = 1,2$. Let $0 \le \xi \le 1$. Then, for each $(x,y) \in J$, we have

$$(\xi h_1 + (1-\xi)h_2)(x,y) = \mu(x,y) + \sum_{0<x_k<x} (I_k(u(x_k^-,y)) - I_k(u(x_k^-,0)))$$

$$+ \frac{1}{\Gamma(r_1)\Gamma(r_2)} \sum_{0<x_k<x} \int_{x_{k-1}}^{x_k} \int_0^y (x_k-s)^{r_1-1}(y-t)^{r_2-1}$$

$$\times [\xi f_1(s,t) + (1-\xi) f_2(s,t)]\mathrm{d}t\mathrm{d}s$$

$$+ \frac{1}{\Gamma(r_1)\Gamma(r_2)} \int_{x_k}^{x} \int_0^y (x-s)^{r_1-1}(y-t)^{r_2-1}$$

$$\times [\xi f_1(s,t) + (1-\xi) f_2(s,t)]\mathrm{d}t\mathrm{d}s.$$

Since $S_{F,u}$ is convex (because F has convex values), we have

$$\xi h_1 + (1-\xi)h_2 \in N(u).$$

Step 2: *N maps bounded sets into bounded sets in $PC(J,\mathbb{R}^n)$.* Let $B_{\eta^*} = \{u \in PC(J,\mathbb{R}^n) : \|u\|_\infty \leq \eta^*\}$ be bounded set in $PC(J,\mathbb{R}^n)$ and $u \in B_{\eta^*}$. Then for each $h \in N(u)$, there exists $f \in S_{F,u}$ such that

$$\begin{aligned} h(u)(x,y) =& \mu(x,y) + \sum_{0<x_k<x} (I_k(u(x_k^-,y)) - I_k(u(x_k^-,0))) \\ &+ \frac{1}{\Gamma(r_1)\Gamma(r_2)} \sum_{0<x_k<x} \int_{x_{k-1}}^{x_k} \int_0^y (x_k-s)^{r_1-1}(y-t)^{r_2-1} f(s,t)\mathrm{d}t\mathrm{d}s \\ &+ \frac{1}{\Gamma(r_1)\Gamma(r_2)} \int_{x_k}^{x} \int_0^y (x-s)^{r_1-1}(y-t)^{r_2-1} f(s,t)\mathrm{d}t\mathrm{d}s. \end{aligned}$$

By (6.2.2) and (6.2.4) we have for each $(x,y) \in J$,

$$\begin{aligned} \|h(x,y)\| \leq& \|\mu(x,y)\| + \sum_{x_1<x_k<x} (\|I_k(u(x_k^-,y))\| + \|I_k(u(x_k^-,0))\|) \\ &+ \frac{1}{\Gamma(r_1)\Gamma(r_2)} \sum_{x_1<x_k<x} \int_{x_{k-1}}^{x_k} \int_0^y (x_k-s)^{r_1-1}(y-t)^{r_2-1} \|f(s,t)\|\mathrm{d}t\mathrm{d}s \\ &+ \frac{1}{\Gamma(r_1)\Gamma(r_2)} \int_{x_k}^{x} \int_0^y (x-s)^{r_1-1}(y-t)^{r_2-1} \|f(s,t)\|\mathrm{d}t\mathrm{d}s \\ \leq& \|\mu(x,y)\| + \sum_{k=1}^{m} (\|I_k(u(x_k^-,y))\| + \|I_k(u(x_k^-,0))\|) \\ &+ \frac{1}{\Gamma(r_1)\Gamma(r_2)} \sum_{k=1}^{m} \int_{x_{k-1}}^{x_k} \int_0^y (x_k-s)^{r_1-1}(y-t)^{r_2-1} p(s,t)\psi_*(\|u\|)\mathrm{d}t\mathrm{d}s \\ &+ \frac{1}{\Gamma(r_1)\Gamma(r_2)} \int_{x_k}^{x} \int_0^y (x-s)^{r_1-1}(y-t)^{r_2-1} p(s,t)\psi_*(\|u\|)\mathrm{d}t\mathrm{d}s. \end{aligned}$$

Thus

$$\|h\|_\infty \leq \|\mu\|_\infty + 2\sum_{k=1}^{m} c_k + \frac{2a^{r_1}b^{r_2}p^*\psi_*(\eta^*)}{\Gamma(r_1+1)\Gamma(r_2+1)} := \ell.$$

Step 3: *N maps bounded sets into equicontinuous sets of $PC(J,\mathbb{R}^n)$*. Let (τ_1, y_1), $(\tau_2, y_2) \in J$, $\tau_1 < \tau_2$ and $y_1 < y_2$, B_{η^*} be a bounded set of $PC(J,\mathbb{R}^n)$ as in Step 2, let $u \in B_{\eta^*}$ and $h \in N(u)$, then for each $(x,y) \in J$,

$$
\begin{aligned}
&\|h(u)(\tau_2, y_2) - h(u)(\tau_1, y_1)\| \\
&\leq \|\mu(\tau_1, y_1) - \mu(\tau_2, y_2)\| + \sum_{k=1}^{m} \left(\|I_k(u(x_k^-, y_1)) - I_k(u(x_k^-, y_2))\|\right) \\
&\quad + \frac{1}{\Gamma(r_1)\Gamma(r_2)} \sum_{k=1}^{m} \int_{x_{k-1}}^{x_k} \int_0^{y_1} (x_k - s)^{r_1-1} \left[(y_2 - t)^{r_2-1} - (y_1 - t)^{r_2-1}\right] \\
&\quad \times f(s,t,u(s,t))\mathrm{d}t\mathrm{d}s \\
&\quad + \frac{1}{\Gamma(r_1)\Gamma(r_2)} \sum_{k=1}^{m} \int_{x_{k-1}}^{x_k} \int_{y_1}^{y_2} (x_k - s)^{r_1-1} (y_2 - t)^{r_2-1} \|f(s,t,u(s,t))\|\mathrm{d}t\mathrm{d}s \\
&\quad + \frac{1}{\Gamma(r_1)\Gamma(r_2)} \int_0^{\tau_1} \int_0^{y_1} \left[(\tau_2 - s)^{r_1-1}(y_2 - t)^{r_2-1} - (\tau_1 - s)^{r_1-1}(y_1 - t)^{r_2-1}\right] \\
&\quad \times f(s,t,u(s,t))\mathrm{d}t\mathrm{d}s \\
&\quad + \frac{1}{\Gamma(r_1)\Gamma(r_2)} \int_{\tau_1}^{\tau_2} \int_{y_1}^{y_2} (\tau_2 - s)^{r_1-1}(y_2 - t)^{r_2-1} \|f(s,t,u(s,t))\|\mathrm{d}t\mathrm{d}s \\
&\quad + \frac{1}{\Gamma(r_1)\Gamma(r_2)} \int_0^{\tau_1} \int_{y_1}^{y_2} (\tau_2 - s)^{r_1-1}(y_2 - t)^{r_2-1} \|f(s,t,u(s,t))\|\mathrm{d}t\mathrm{d}s \\
&\quad + \frac{1}{\Gamma(r_1)\Gamma(r_2)} \int_{\tau_1}^{\tau_2} \int_0^{y_1} (\tau_2 - s)^{r_1-1}(y_2 - t)^{r_2-1} \|f(s,t,u(s,t))\|\mathrm{d}t\mathrm{d}s \\
&\leq \|\mu(\tau_1, y_1) - \mu(\tau_2, y_2)\| + \sum_{k=1}^{m} \left(\|I_k(u(x_k^-, y_1)) - I_k(u(x_k^-, y_2))\|\right) \\
&\quad + \frac{\phi_f^0 \psi_*(\eta^*)}{\Gamma(r_1)\Gamma(r_2)} \sum_{k=1}^{m} \int_{x_{k-1}}^{x_k} \int_0^{y_1} (x_k - s)^{r_1-1} \left[(y_2 - t)^{r_2-1} - (y_1 - t)^{r_2-1}\right] \mathrm{d}t\mathrm{d}s \\
&\quad + \frac{\phi_f^0 \psi_*(\eta^*)}{\Gamma(r_1)\Gamma(r_2)} \sum_{k=1}^{m} \int_{x_{k-1}}^{x_k} \int_{y_1}^{y_2} (x_k - s)^{r_1-1} (y_2 - t)^{r_2-1} \mathrm{d}t\mathrm{d}s \\
&\quad + \frac{\phi_f^0 \psi_*(\eta^*)}{\Gamma(r_1)\Gamma(r_2)} \int_0^{\tau_1} \int_0^{y_1} [(\tau_2 - s)^{r_1-1}(y_2 - t)^{r_2-1} - (\tau_1 - s)^{r_1-1}(y_1 - t)^{r_2-1}]\mathrm{d}t\mathrm{d}s \\
&\quad + \frac{\phi_f^0 \psi_*(\eta^*)}{\Gamma(r_1)\Gamma(r_2)} \int_{\tau_1}^{\tau_2} \int_{y_1}^{y_2} (\tau_2 - s)^{r_1-1}(y_2 - t)^{r_2-1} \mathrm{d}t\mathrm{d}s
\end{aligned}
$$

$$+\frac{\phi_f^0\psi_*(\eta^*)}{\Gamma(r_1)\Gamma(r_2)}\int_0^{\tau_1}\int_{y_1}^{y_2}(\tau_2-s)^{r_1-1}(y_2-t)^{r_2-1}dtds$$

$$+\frac{\phi_f^0\psi_*(\eta^*)}{\Gamma(r_1)\Gamma(r_2)}\int_{\tau_1}^{\tau_2}\int_0^{y_1}(\tau_2-s)^{r_1-1}(y_2-t)^{r_2-1}dtds.$$

As $\tau_1 \longrightarrow \tau_2$ and $y_1 \longrightarrow y_2$, the right-hand side of the above inequality tends to zero. As a consequence of Steps 1–3 together with the Arzelá-Ascoli theorem, we can conclude that N is completely continuous.

Step 4: *N has a closed graph.* Let $u_n \to u_*$, $h_n \in N(u_n)$ and $h_n \to h_*$. We need to show that $h_* \in N(u_*)$.

$h_n \in N(u_n)$ means that there exists $f_n \in S_{F,u_n}$ such that, for each $(x,y) \in J$,

$$h_n(x,y) = \mu(x,y) + \sum_{0<x_k<x}(I_k(u_n(x_k^-,y)) - I_k(u_n(x_k^-,0)))$$

$$+\frac{1}{\Gamma(r_1)\Gamma(r_2)}\sum_{0<x_k<x}\int_{x_{k-1}}^{x_k}\int_0^y(x_k-s)^{r_1-1}(y-t)^{r_2-1}f_n(s,t)dtds$$

$$+\frac{1}{\Gamma(r_1)\Gamma(r_2)}\int_{x_k}^{x}\int_0^y(x-s)^{r_1-1}(y-t)^{r_2-1}f_n(s,t)dtds.$$

We must show that there exists $f_* \in S_{F,u*}$ such that, for each $(x,y) \in J$,

$$h_*(x,y) = \mu(x,y) + \sum_{0<x_k<x}(I_k(u_*(x_k^-,y)) - I_k(u_*(x_k^-,0)))$$

$$+\frac{1}{\Gamma(r_1)\Gamma(r_2)}\sum_{0<x_k<x}\int_{x_{k-1}}^{x_k}\int_0^y(x_k-s)^{r_1-1}(y-t)^{r_2-1}$$

$$f_*(s,t)dtds$$

$$+\frac{1}{\Gamma(r_1)\Gamma(r_2)}\int_{x_k}^{x}\int_0^y(x-s)^{r_1-1}(y-t)^{r_2-1}f_*(s,t)dtds.$$

Since $F(x,y,\cdot)$ is upper semicontinuous, then for every $\varepsilon > 0$, there exist $n_0(\epsilon) \geq 0$ such that for every $n \geq n_0$, we have

$$f_n(x,y) \in F(x,y,u_n(x,y)) \subset F(x,y,u_*(x,y)) + \varepsilon B(0,1), \text{ a.e. } (x,y) \in J.$$

Since $F(.,.,.)$ has compact values, then there exists a subsequence f_{n_m} such that

$$f_{n_m}(\cdot,\cdot) \to f_*(\cdot,\cdot) \text{ as } m \to \infty$$

and

$$f_*(x,y) \in F(x,y,u_*(x,y)), \text{ a.e. } (x,y) \in J.$$

For every $w(x,y) \in F(x,y,u_*(x,y))$, we have

$$\|f_{n_m}(x,y) - u_*(x,y)\| \leq \|f_{n_m}(x,y) - w(x,y)\| + \|w(x,y) - f_*(x,y)\|.$$

Then

$$\|f_{n_m}(x,y) - f_*(x,y)\| \leq d\Big(f_{n_m}(x,y), F(x,y,u_*(x,y))\Big).$$

By an analogous relation, obtained by interchanging the roles of f_{n_m} and f_*, it follows that

$$\begin{aligned}\|f_{n_m}(x,y) - u_*(x,y)\| &\leq H_d\Big(F(x,y,u_n(x,y)), F(x,y,u_*(x,y))\Big)\\ &\leq l(x,y)\|u_n - u_*\|_\infty.\end{aligned}$$

Let $l^* := \|l\|_{L^\infty}$, then by (6.2.3) and (6.2.5) we obtain for each $(x,y) \in J$,

$$\begin{aligned}&\|h_n(x,y) - h_*(x,y)\|\\ &\leq \sum_{k=1}^{m} \|I_k(u_n(x_k^-,y)) - I_k(u_*(x_k^-,y))\|\\ &\quad+\sum_{k=1}^{m} \|I_k(u_n(x_k^-,0)) - I_k(u_*(x_k^-,0))\|\\ &\quad+\frac{1}{\Gamma(r_1)\Gamma(r_2)}\sum_{x_1<x_k<x}\int_{x_{k-1}}^{x_k}\int_0^y (x_k-s)^{r_1-1}(y-t)^{r_2-1}\\ &\quad\times\|f_{n_m}(s,t) - f_*(s,t)\|\mathrm{d}t\mathrm{d}s\\ &\quad+\frac{1}{\Gamma(r_1)\Gamma(r_2)}\int_{x_k}^{x}\int_0^y (x-s)^{r_1-1}(y-t)^{r_2-1}\\ &\quad\times\|f_{n_m}(s,t) - f_*(s,t)\|\mathrm{d}t\mathrm{d}s\\ &\leq \|u_{n_m} - u_*\|_\infty\sum_{k=1}^{m} 2c_k^*\\ &\quad+\frac{2\|u_{n_m} - u_*\|_\infty}{\Gamma(r_1)\Gamma(r_2)}\int_0^x\int_0^y (x-s)^{r_1-1}(y-t)^{r_2-1}l(s,t)\mathrm{d}t\mathrm{d}s.\end{aligned}$$

Hence

$$\|h_{n_m} - h_*\|_\infty \leq \left[2\sum_{k=1}^{m} c_k^* + \frac{2l^* a^{r_1} b^{r_2}}{\Gamma(r_1+1)\Gamma(r_2+1)}\right]\|u_{n_m} - u_*\|_\infty, \to 0 \text{ as } m \to \infty.$$

Step 5: *A priori bounds on solutions.* Let u be a possible solution of the problems (6.1)–(6.3). Then, there exists $f \in S_{F,u}$ such that, for each $(x,y) \in J$,

$$\|u(x,y)\| \leq \|\mu(x,y)\| + \sum_{x_1<x_k<x} (\|I_k(u(x_k^-,y))\| + \|I_k(u(x_k^-,0))\|)$$

$$+\frac{\psi_*(\|u\|)}{\Gamma(r_1)\Gamma(r_2)} \sum_{x_1<x_k<x} \int_{x_{k-1}}^{x_k} \int_0^y (x_k-s)^{r_1-1}(y-t)^{r_2-1} p(s,t)\mathrm{d}t\mathrm{d}s$$

$$+\frac{\psi_*(\|u\|)}{\Gamma(r_1)\Gamma(r_2)} \int_{x_k}^{x} \int_0^y (x-s)^{r_1-1}(y-t)^{r_2-1} p(s,t)\mathrm{d}t\mathrm{d}s$$

$$\leq \|\mu\|_\infty + 2\sum_{k=1}^m c_k + \frac{2p^* a^{r_1} b^{r_2} \psi_*(\|u\|_\infty)}{\Gamma(r_1+1)\Gamma(r_2+1)},$$

then

$$\frac{\|u\|_\infty}{\|\mu\|_\infty + 2\sum_{k=1}^m c_k + \dfrac{2p^* a^{r_1} b^{r_2} \psi_*(\|u\|_\infty)}{\Gamma(r_1+1)\Gamma(r_2+1)}} \leq 1.$$

Thus by condition (6.4), there exists M such that $\|u\|_\infty \neq M$. Let

$$U = \{u \in PC(J,\mathbb{R}^n) : \|u\|_\infty < M\}.$$

The operator $N : \overline{U} \to \mathcal{P}(PC(J,\mathbb{R}^n))$ is upper semicontinuous and completely continuous. From the choice of U, there is no $u \in \partial U$ such that $u \in \lambda N(u)$ for some $\lambda \in (0,1)$. As a consequence of the nonlinear alternative of Leray–Schauder type, we deduce that N has a fixed point u in $\overline{U}$ which is a solution of the problems (6.1)–(6.3). □

Theorem 6.4. *Assume that (6.2.1), (6.2.4), (6.2.6), and the following condition:*

(6.4.1) there exists $p \in L^\infty(J,\mathbb{R}_+)$ *such that*

$$\|F(x,y,u)\|_{\mathcal{P}} \leq p(x,y)(1+\|u\|), \text{ for } (x,y) \in J \text{ and each } u \in \mathbb{R}^n,$$

hold. If

$$\Gamma(r_1+1)\Gamma(r_2+1) > 2p^* a^{r_1} b^{r_2}, \tag{6.5}$$

$p^* = \|p\|_{L^\infty}$, *then problem (6.1)–(6.3) has at least one solution.*

Proof. We shall show that the operator N defined in Theorem 6.2 satisfies fixed-point theorem. Let $\rho > 0$ be such that

$$\rho > \frac{(\|\mu\|_\infty + 2\sum_{k=1}^m c_k)\Gamma(r_1+1)\Gamma(r_2+1) + 2p^* a^{r_1} b^{r_2}}{\Gamma(r_1+1)\Gamma(r_2+1) - 2p^* a^{r_1} b^{r_2}},$$

and consider the subset

$$D_\rho = \{u \in PC(J,\mathbb{R}^n) : \|u\|_\infty \le \rho\}.$$

Clearly, the subset D_ρ is closed, bounded, and convex. From (6.5) we have $N(D_\rho) \subseteq D_\rho$. As before the multivalued operator $N : D_\rho \to \mathcal{P}(D_\rho)$ is upper semicontinuous and completely continuous. Hence Lemma 2.38 implies that N has a fixed point which is a solution to problems (6.1)–(6.3). □

6.2.3 The Nonconvex Case

We present now a result for the problems (6.1)–(6.3) with a nonconvex valued right-hand side. Our considerations are based on the fixed-point theorem for contraction multivalued maps given by Covitz and Nadler [96].

Theorem 6.5. *Assume (6.2.3), (6.2.5) and the following hypothesis holds:*

(6.5.1) $F : J \times \mathbb{R}^n \longrightarrow \mathcal{P}_{cp}(\mathbb{R}^n)$ *has the property that*

$$F(\cdot,\cdot,u) : J \to \mathcal{P}_{cp}(\mathbb{R}^n) \text{ is measurable for each } u \in \mathbb{R}^n.$$

If

$$2\sum_{k=1}^m c_k^* + \frac{2l^* a^{r_1} b^{r_2}}{\Gamma(r_1+1)\Gamma(r_2+1)} < 1, \tag{6.6}$$

then the IVP (6.1)–(6.3) have at least one solution on J.

Remark 6.6. For each $u \in PC(J,\mathbb{R}^n)$, the set $S_{F,u}$ is nonempty since by (6.4.1), F has a measurable selection (see [93], Theorem III.6).

Proof. We shall show that N defined in Theorem 6.2 satisfies the assumptions of Lemma 2.39. The proof will be given in two steps.

Step 6: $N(u) \in \mathcal{P}_{cl}(PC(J,\mathbb{R}^n))$ *for each* $u \in PC(J,\mathbb{R}^n)$. Indeed, let $(u_n)_{n\ge 0} \in N(u)$ such that $u_n \longrightarrow \tilde{u}$ in $PC(J,\mathbb{R}^n)$. Then, $\tilde{u} \in PC(J,\mathbb{R}^n)$ and there exists $f_n \in S_{F,u}$ such that, for each $(x,y) \in J$,

$$u_n(x,y)=\mu(x,y)+\sum_{0<x_k<x}(I_k(u(x_k^-,y))-I_k(u(x_k^-,0)))$$

$$+\frac{1}{\Gamma(r_1)\Gamma(r_2)}\sum_{0<x_k<x}\int_{x_{k-1}}^{x_k}\int_0^y(x_k-s)^{r_1-1}(y-t)^{r_2-1}f_n(s,t)dtds$$

$$+\frac{1}{\Gamma(r_1)\Gamma(r_2)}\int_{x_k}^{x}\int_0^y(x-s)^{r_1-1}(y-t)^{r_2-1}f_n(s,t)dtds.$$

Using the fact that F has compact values and from (6.5.1), we may pass to a subsequence if necessary to get that $f_n(.,.)$ converges weakly to f in $L^1_w(J,\mathbb{R}^n)$ (the space endowed with the weak topology). A standard argument shows that $f_n(.,.)$ converges strongly to f and hence $f\in S_{F,u}$. Then, for each $(x,y)\in J$, $u_n(x,y)\longrightarrow\tilde{u}(x,y)$,
where

$$\tilde{u}(x,y)=\mu(x,y)+\sum_{0<x_k<x}(I_k(u(x_k^-,y))-I_k(u(x_k^-,0)))$$

$$+\frac{1}{\Gamma(r_1)\Gamma(r_2)}\sum_{0<x_k<x}\int_{x_{k-1}}^{x_k}\int_0^y(x_k-s)^{r_1-1}(y-t)^{r_2-1}f(s,t)dtds$$

$$+\frac{1}{\Gamma(r_1)\Gamma(r_2)}\int_{x_k}^{x}\int_0^y(x-s)^{r_1-1}(y-t)^{r_2-1}f(s,t)dtds.$$

So, $\tilde{u}\in N(u)$.

Step 7: *There exists $\gamma<1$ such that*

$$H_d(N(u),N(\overline{u}))\le\gamma\|u-\overline{u}\|_\infty \textit{ for each } u,\overline{u}\in PC(J,\mathbb{R}^n).$$

Let $u,\overline{u}\in PC(J,\mathbb{R}^n)$ and $h\in N(u)$. Then, there exists $f(x,y)\in F(x,y,u(x,y))$ such that for each $(x,y)\in J$

$$h(x,y)=\mu(x,y)+\sum_{0<x_k<x}(I_k(u(x_k^-,y))-I_k(u(x_k^-,0)))$$

$$+\frac{1}{\Gamma(r_1)\Gamma(r_2)}\sum_{0<x_k<x}\int_{x_{k-1}}^{x_k}\int_0^y(x_k-s)^{r_1-1}(y-t)^{r_2-1}f(s,t)dtds$$

$$+\frac{1}{\Gamma(r_1)\Gamma(r_2)}\int_{x_k}^{x}\int_0^y(x-s)^{r_1-1}(y-t)^{r_2-1}f(s,t)dtds.$$

From (6.2.3) it follows that

$$H_d(F(x,y,u(x,y)), F(x,y,\overline{u}(x,y))) \le l(x,y)\|u(x,y)-\overline{u}(x,y)\|.$$

Hence, there exists $w(x,y) \in F(x,y,\overline{u}(x,y))$ such that

$$\|f(x,y)-w(x,y)\| \le l(x,y)\|u(x,y)-\overline{u}(x,y)\|,\ (x,y) \in J.$$

Consider $U : J \to \mathcal{P}(\mathbb{R}^n)$ given by

$$U(x,y) = \{w \in PC(J,\mathbb{R}^n) : \|f(x,y)-w(x,y)\| \le l(x,y)\|u(x,y)-\overline{u}(x,y)\|\}.$$

Since the multivalued operator $u(x,y) = U(x,y) \cap F(x,y,\overline{u}(x,y))$ is measurable (see Proposition III.4 in [93]), there exists a function $u_2(x,y)$ which is a measurable selection for u. So, $\overline{f}(x,y) \in F(x,y,\overline{u}(x,y))$, and for each $(x,y) \in J$,

$$\|f(x,y)-\overline{f}(x,y)\| \le l(x,y)\|u(x,y)-\overline{u}(x,y)\|.$$

Let us define for each $(x,y) \in J$

$$\begin{aligned}\overline{h}(x,y) = {} & \mu(x,y) + \sum_{0<x_k<x} (I_k(\overline{u}(x_k^-,y)) - I_k(\overline{u}(x_k^-,0))) \\ & + \frac{1}{\Gamma(r_1)\Gamma(r_2)} \sum_{0<x_k<x} \int_{x_{k-1}}^{x_k} \int_0^y (x_k-s)^{r_1-1}(y-t)^{r_2-1}\overline{f}(s,t)\,dt\,ds \\ & + \frac{1}{\Gamma(r_1)\Gamma(r_2)} \int_{x_k}^{x} \int_0^y (x-s)^{r_1-1}(y-t)^{r_2-1}\overline{f}(s,t)\,dt\,ds,\end{aligned}$$

then we get

$$\begin{aligned}\|h(x,y)-\overline{h}(x,y)\| \le {} & \sum_{k=1}^{m} \|I_k(u(x_k^-,y)) - I_k(\overline{u}(x_k^-,y))\| \\ & + \sum_{k=1}^{m} \|I_k(u(x_k^-,0)) - I_k(\overline{u}(x_k^-,0))\| \\ & + \frac{1}{\Gamma(r_1)\Gamma(r_2)} \sum_{0<x_k<x} \int_{x_{k-1}}^{x_k} \int_0^y (x_k-s)^{r_1-1}(y-t)^{r_2-1} \\ & \times \|f(s,t)-\overline{f}(s,t)\|\,dt\,ds\end{aligned}$$

$$+\frac{1}{\Gamma(r_1)\Gamma(r_2)}\int_{x_k}^{x}\int_{0}^{y}(x-s)^{r_1-1}(y-t)^{r_2-1}\|f(s,t))$$

$$-\overline{f}(s,t)\|\mathrm{d}t\mathrm{d}s$$

$$\leq \|u-\overline{u}\|_\infty 2\sum_{k=1}^{m}c_k^*$$

$$+\frac{2\|u-\overline{u}\|_\infty}{\Gamma(r_1)\Gamma(r_2)}\int_{0}^{x}\int_{0}^{y}(x-s)^{r_1-1}(y-t)^{r_2-1}l(s,t)\mathrm{d}t\mathrm{d}s$$

$$\leq \left[2\sum_{k=1}^{m}c_k^* + \frac{2l^*a^{r_1}b^{r_2}}{\Gamma(r_1+1)\Gamma(r_2+1)}\right]\|u-\overline{u}\|_\infty.$$

Thus, for each $(x,y)\in J$

$$\|h_1-h_2\|_\infty \leq \left[2\sum_{k=1}^{m}c_k^* + \frac{2l^*a^{r_1}b^{r_2}}{\Gamma(r_1+1)\Gamma(r_2+1)}\right]\|u-\overline{u}\|_\infty.$$

By an analogous relation, obtained by interchanging the roles of u and $\overline{u}$, it follows that

$$H_d(N(u),N(\overline{u})) \leq \left[2\sum_{k=1}^{m}c_k^* + \frac{2l^*a^{r_1}b^{r_2}}{\Gamma(r_1+1)\Gamma(r_2+1)}\right]\|u-\overline{u}\|_\infty.$$

So by (6.6), N is a contraction and thus, by Lemma 2.39, N has a fixed point u which is solution to (6.1)–(6.3). □

Now we present a result for the problems (6.1)–(6.3) in the spirit of the nonlinear alternative of Leray–Schauder type for single-valued maps combined with a selection theorem due to Bressan-Colombo.

Theorem 6.7. *Assume (6.2.1), (6.2.4) and the following assumption holds:*

(6.7.1) $F: J\times\mathbb{R}^n \longrightarrow \mathcal{P}_{cp}(\mathbb{R}^n)$ *is a nonempty compact-valued multivalued map such that*

- *(a)* $(x,y,u)\mapsto F(x,y,u)$ *is* $\mathcal{L}\otimes\mathcal{B}$ *measurable for each* $u\in\mathbb{R}^n$
- *(b)* $u\mapsto F(x,y,u)$ *is lower semicontinuous for a.e.* $(x,y)\in J$

Then problems (6.1)–(6.3) have at least one solution.

Proof. (6.2.2) and (6.7.1) imply by Lemma 2.2 in Frigon [126] that F is of lower semicontinuous type. The from Lemma 2.26 there exists a continuous function

$$g: PC(J,\mathbb{R}^n)\to PC(J,\mathbb{R}^n),$$

such that $g(u) \in \mathcal{F}(u)$ for all $u \in PC(J, \mathbb{R}^n)$. Consider the following problem:

$$({}^cD^r_{x_k}u)(x,y) = g(u(x,y)); \text{ if } (x,y) \in J_k\ k = 0,\ldots,m, \tag{6.7}$$

$$u(x_k^+, y) = u(x_k^-, y) + I_k(u(x_k^-, y)); \text{ if } y \in [0,b];\ k = 1,\ldots,m, \tag{6.8}$$

$$u(x,0) = \varphi(x);\ x \in [0,a];\ y \in [0,b]. \tag{6.9}$$

Clearly, if $u \in PC(J, \mathbb{R}^n)$ is a solution of the problems (6.7)–(6.9), then u is a solution to the problems (6.1)–(6.3). Transform the problems (6.7)–(6.9) into a fixed-point problem. Consider the operator $N_1 : PC(J, \mathbb{R}^n) \to PC(J, \mathbb{R}^n)$ defined by

$$\begin{aligned} N_1(u)(x,y) =& \mu(x,y) + \sum_{0<x_k<x} (I_k(u(x_k^-, y)) - I_k(u(x_k^-, 0))) \\ &+ \frac{1}{\Gamma(r_1)\Gamma(r_2)} \sum_{0<x_k<x} \int_{x_{k-1}}^{x_k} \int_0^y (x_k - s)^{r_1-1}(y-t)^{r_2-1} g(u(s,t))\mathrm{d}t\mathrm{d}s \\ &+ \frac{1}{\Gamma(r_1)\Gamma(r_2)} \int_{x_k}^{x} \int_0^y (x-s)^{r_1-1}(y-t)^{r_2-1} g(u(s,t))\mathrm{d}t\mathrm{d}s. \end{aligned}$$

We can easily show as Theorem 6.2 that N_1 is continuous and completely continuous and there is no $u \in \partial U$ such that $u = \lambda N_1(u)$ for some $\lambda \in (0,1)$. We omit the details. As a consequence of the nonlinear alternative of Leray–Schauder type [136], we deduce that N_1 has a fixed point u in U which is a solution of the problems (6.7)–(6.9). Hence, u is a solution to the problems (6.1)–(6.3). □

6.2.4 An Example

As an application of the main results, we consider the following impulsive fractional differential inclusion:

$$\begin{aligned} &{}^cD^r_{x_k}u(x,y) \in F(x,y,u(x,y));\ \text{a.e. } (x,y) \in J_k;\ k = 0,\ldots,m, \\ &x_k = \frac{k}{k+1},\ k = 1,\ldots,m, \end{aligned} \tag{6.10}$$

$$u\left(\left(\frac{k}{k+1}\right)^{+},y\right)=u\left(\left(\frac{k}{k+1}\right)^{-},y\right)+\frac{|u((\frac{k}{k+1})^{-},y)|}{3k+|u((\frac{k}{k+1})^{-},y)|};\ y\in[0,1],\ k=1,\ldots,m, \tag{6.11}$$

$$u(x,0)=x,\ u(0,y)=y^2;\ x,y\in[0,1], \tag{6.12}$$

where $J=[0,1]\times[0,1]$, $r=(r_1,r_2)$ and $0<r_1,r_2\leq 1$. Set

$$I_k(u)=\frac{u}{3k+u};\ u\in\mathbb{R}_+;\ u\in\mathbb{R}_+,\ k=1,\ldots,m.$$

Then for each $u\in\mathbb{R}_+$ and $k=1,\ldots,m$, we have $|I_k(u)|\leq k$. Hence condition (6.2.4) is satisfied with $c_k=k,\ k=1,\ldots,m$. Let $u,\ \overline{u}\in\mathbb{R}_+$ then for each $(x,y)\in[0,1]\times[0,1]$, we have

$$|I_k(u)-I_k(\overline{u})|\leq\left|\frac{u}{3k+u}-\frac{\overline{u}}{3k+\overline{u}}\right|\leq\frac{1}{3}|u-\overline{u}|.$$

Hence condition (6.2.5) is satisfied with $c_k^*=\frac{1}{3},\ k=1,\ldots,m$. Set

$$F(x,y,u(x,y))=\{u\in\mathbb{R}:f_1(x,y,u(x,y))\leq u\leq f_2(x,y,u(x,y))\},$$

where $f_1,f_2:[0,1]\times[0,1]\times\mathbb{R}\to\mathbb{R}$. We assume that for each $(x,y)\in J,\ f_1(x,y,.)$ is lower semicontinuous (i.e., the set $\{z\in\mathbb{R}:f_1(x,y,z)>\nu\}$ is open for each $\nu\in\mathbb{R}$), and assume that for each $(x,y)\in J,\ f_2(x,y,.)$ is upper semicontinuous (i.e., the set $\{z\in\mathbb{R}:f_2(x,y,z)<\nu\}$ is open for each $\nu\in\mathbb{R}$). Assume that there are $P\in C(J,\mathbb{R}_+)$ and $\Psi:[0,\infty)\to(0,\infty)$ continuous and nondecreasing such that

$$\max(|f_1(x,y,z)|,\ |f_2(x,y,z)|)\leq P(x,y)\Psi(|z|),$$

for each $(x,y)\in J$ and all $z\in\mathbb{R}$. It is clear that F is compact and convex valued, and it is upper semicontinuous (see [104]). Since all the conditions of Theorem 6.2 are satisfied, problems (6.10)–(6.12) have at least one solution u on $[0,1]\times[0,1]$.

6.3 Impulsive Partial Hyperbolic Differential Inclusions with Variable Times

6.3.1 Introduction

This section concerns with the existence of solutions for the following impulsive partial fractional order IVP, for the system:

$$\begin{aligned}&({}^cD^r_{x_k}u)(x,y) \in F(x,y,u(x,y)),\\ &\quad \text{where } (x,y) \in J_k;\ k=0,\dots,m,\ x_k = x_k(u(x,y));\ k=1,\dots,m, \end{aligned} \tag{6.13}$$

$$u(x^+,y) = I_k(u(x,y)), \text{ where } y \in [0,b],\ x = x_k(u(x,y));\ k=1,\dots,m, \tag{6.14}$$

$$u(x,0)=\varphi(x),\ u(0,y)=\psi(y),\ x\in[0,a],\ y\in[0,b], \tag{6.15}$$

where $a,b>0$, $0=x_0<x_1<\cdots<x_m<x_{m+1}=a$, $F: J\times\mathbb{R}^n \to \mathcal{P}(\mathbb{R}^n)$ is a compact-valued multivalued map, $\mathcal{P}(\mathbb{R}^n)$ is the family of all subsets of $\mathbb{R}^n$, $x_k: \mathbb{R}^n \to \mathbb{R}$, $I_k: \mathbb{R}^n \to \mathbb{R}^n$, $k=1,2,\dots,m$ are given functions and φ, ψ are as in problems (6.1)–(6.3). Next we consider the following IVP for impulsive partial neutral functional differential inclusions:

$$\begin{aligned}&{}^cD^r_{x_k}[u(x,y)-g(x,y,u(x,y))] \in F(x,y,u(x,y));\\ &(x,y)\in J_k;\ k=0,\dots,m,\ x_k = x_k(u(x,y));\ k=1,\dots,m,\end{aligned} \tag{6.16}$$

$$u(x^+,y) = I_k(u(x,y)), \text{ where } y \in [0,b],\ x = x_k(u(x,y));\ k=1,\dots,m, \tag{6.17}$$

$$u(x,0)=\varphi(x),\ u(0,y)=\psi(y),\ x\in[0,a],\ y\in[0,b], \tag{6.18}$$

where $F,\varphi,\psi,x_k,I_k; k=1,2,\dots,m$ are as in problems (6.13)–(6.15) and $g: J\times\mathbb{R}^n\to\mathbb{R}^n$ is a given function.

6.3.2 Existence of Solutions

To define the solutions of problems (6.13)–(6.15), we shall consider the Banach space

$$\begin{aligned}\Omega = \{u : J \to \mathbb{R}^n :\ &\text{there exist } 0=x_0<x_1<x_2<\cdots<x_m<x_{m+1}=a\\ &\text{such that } x_k = x_k(u(x_k,.)), \text{ and } u(x_k^-,.),\ u(x_k^+,.)\\ &\text{exist with } u(x_k^-,.) = u(x_k,.);\\ &k=0,\dots,m, \text{ and } u\in C(J_k,\mathbb{R}^n); k=0,\dots,m\},\end{aligned}$$

where $J_k := (x_k, x_{k+1}] \times [0, b]$, with the norm

$$\|u\|_{\Omega} = \max\{\|u_k\|,\ k = 0, \ldots, m\},$$

where u_k is the restriction of u to J_k, $k = 0, \ldots, m$.

In what follows, we will assume that F is a Carathéodory multivalued map. Let us start by defining what we mean by a solution of the problems (3.1)–(3.3).

Definition 6.8. A function $u \in \Omega \cap \cup_{k=1}^{m} AC(J_k, \mathbb{R}^n)$ such that its mixed derivative D^2_{xy} exists on J_k, $k = 0, \ldots, m$ is said to be a solution of (6.13)–(6.15) if there exists a function $f(x, y) \in F(x, y, u(x, y))$ such that u satisfies $({}^c D^r_{x_k} u)(x, y) = f(x, y)$ on J_k; $k = 0, \ldots, m$ and conditions (6.14) and (6.15) are satisfied.

Theorem 6.9. *Assume that the hypotheses*

(6.9.1) The function $x_k \in C^1(\mathbb{R}^n, \mathbb{R})$ *for* $k = 1, \ldots, m$. *Moreover,*

$$0 = x_0(u) < x_1(u) < \cdots < x_m(u) < x_{m+1}(u) = a, \quad \textit{for all } u \in \mathbb{R}^n,$$

(6.9.2) For each $u \in \mathbb{R}^n$, *there exist constants* $c_k, d_k > 0$ *such that*

$$\|I_k(u)\| \le c_k \|u\| + d_k, \textit{ for each } u \in \mathbb{R}^n,\ k = 1, \ldots, m,$$

(6.9.3) There exists a continuous nondecreasing function $\delta : [0, \infty) \to (0, \infty)$, *and* $p \in L^\infty(J, \mathbb{R}_+)$ *such that*

$$\|F(x, y, u)\|_{\mathcal{P}} \le p(x, y)\delta(\|u\|) \quad \textit{a.e. } (x, y) \in J, \textit{ and each } u \in \mathbb{R}^n,$$

(6.9.4) There exists $l \in L^\infty(J, \mathbb{R}_+)$; $k = 1, \ldots, m$, *such that*

$$H_d(F(x, y, u), F(x, y, \overline{u})) \le l(x, y)\|u - \overline{u}\| \textit{ for every } u, \overline{u} \in \mathbb{R}^n,$$

and

$$d(0, F(x, y, 0)) \le l(x, y), \textit{ a.e. } (x, y) \in J_k,\ k = 0, \ldots, m,$$

(6.9.5) For all $(s, t, u) \in J \times \mathbb{R}^n$, *there exists* $f \in S_{F,u}$ *such that*

$$x'_k(u)\left[\varphi'(s) + \frac{r_1 - 1}{\Gamma(r_1)\Gamma(r_2)} \int_{x_k}^{s} \right.$$

$$\left. \times \int_0^t (s-\theta)^{r_1-2}(t-\eta)^{r_2-1} f(\theta, \eta) d\eta d\theta \right] \ne 1;\ k = 1, \ldots, m,$$

(6.9.6) For all $u \in \mathbb{R}^n$, $x_k(I_k(u)) \leq x_k(u) < x_{k+1}(I_k(u))$ for $k = 1, \dots, m$,
(6.9.7) There exists $M > 0$ such that

$$\min\left\{\frac{M}{\|\mu\|_\infty + \frac{p^* a^{r_1} b^{r_2} \delta(M)}{\Gamma(r_1+1)\Gamma(r_2+1)}},\right.$$

$$\left.\frac{M}{\|\varphi\|_\infty + 2c_k M + 2d_k + \frac{p^* a^{r_1} b^{r_2} \delta(M)}{\Gamma(r_1+1)\Gamma(r_2+1)}};\ k = 1, \dots, m\right\} > 1, \tag{6.19}$$

where $p^* = \|p\|_{L^\infty}$, hold. Then the initial-value problems (6.13)–(6.15) have at least one solution on J.

Proof. The proof of this theorem will be given in several steps.

Step 1: Consider the problem

$$({}^c D_0^r u)(x,y) \in F(x,y,u(x,y)), \text{ where } (x,y) \in J, \tag{6.20}$$

$$u(x,0) = \varphi(x),\ u(0,y) = \psi(y),\ x \in [0,a],\ y \in [0,b]. \tag{6.21}$$

Transform the problem into a fixed-point problem. Consider the operator $N : C(J,\mathbb{R}^n) \to \mathcal{P}(C(J,\mathbb{R}^n))$ defined as

$$N(u) = \left\{h \in C(J,\mathbb{R}^n) : \left\{\begin{array}{l} h(x,y) = \mu(x,y) + \frac{1}{\Gamma(r_1)\Gamma(r_2)} \int\limits_0^x \int\limits_0^y (x-s)^{r_1-1} \\ \times (y-t)^{r_2-1} f(s,t)\mathrm{d}t\mathrm{d}s;\ (x,y) \in J,\ f \in S_{F,u} \end{array}\right\}\right.$$

Clearly, the fixed points of N are solutions to (6.20) and (6.21). We shall show that the operator N is completely continuous. The proof will be given in several claims.

Claim 1. $N(u)$ is convex for each $u \in C(J,\mathbb{R}^n)$). Indeed, if h_1, h_2 belong to $N(u)$, then there exist $f_1, f_2 \in S_{F,u}$ such that for each $(x,y) \in J$, we have

$$h_i(x,y) = \mu(x,y) + \frac{1}{\Gamma(r_1)\Gamma(r_2)} \int\limits_0^x \int\limits_0^y (x-s)^{r_1-1}(y-t)^{r_2-1} f_i(s,t)\mathrm{d}t\mathrm{d}s,\ i = 1,2.$$

Let $0 \leq d \leq 1$. Then for each $(x, y) \in J$ we have

$$(dh_1 + (1-d)h_2)(x, y) = \mu(x, y) + \frac{1}{\Gamma(r_1)\Gamma(r_2)} \int_0^x \int_0^y (x-s)^{r_1-1}(y-t)^{r_2-1} \times [f_1(s,t) + (1-d) f_2(s,t)] \mathrm{d}t \mathrm{d}s.$$

Since $S_{F,u}$ is convex (because F has convex values) then for each $(x, y) \in J$,

$$dh_1 + (1-d)h_2 \in N(u).$$

Claim 2. N maps bounded sets into bounded sets in $C(J, \mathbb{R}^n)$. Indeed, it is sufficient to show that for any $q > 0$ there exists a positive constant ℓ such that for each $u \in B_q = \{u \in C(J, \mathbb{R}^n) : \|u\|_\infty \leq q\}$ we have $\|N(u)\| \leq \ell$. Let $u \in B_q$ and $h \in N(u)$ then there exists $f \in S_{F,u}$ such that for each $(x, y) \in J$ we have

$$h(x, y) = \mu(x, y) + \frac{1}{\Gamma(r_1)\Gamma(r_2)} \int_0^x \int_0^y (x-s)^{r_1-1}(y-t)^{r_2-1} f(s,t) \mathrm{d}t \mathrm{d}s.$$

Thus,

$$\begin{aligned} \|h(x, y)\| &\leq \|\mu(x, y)\| \\ &\quad + \frac{1}{\Gamma(r_1)\Gamma(r_2)} \int_0^x \int_0^y (x-s)^{r_1-1}(y-t)^{r_2-1} p(s,t)\delta(\|u\|) \mathrm{d}t \mathrm{d}s \\ &\leq \|\mu(x, y)\| + \frac{p^*\delta(q)}{\Gamma(r_1)\Gamma(r_2)} \int_0^a \int_0^b (x-s)^{r_1-1}(y-t)^{r_2-1} \mathrm{d}t \mathrm{d}s. \end{aligned}$$

Then we obtain that

$$\|h\| \leq \|\mu\|_\infty + \frac{a^{r_1} b^{r_2} p^* \delta(q)}{\Gamma(r_1+1)\Gamma(r_2+1)} := \ell.$$

Claim 3. N maps bounded sets into equicontinuous sets of $C(J, \mathbb{R}^n)$. Let (τ_1, y_1), $(\tau_2, y_2) \in J$, $\tau_1 < \tau_2$ and $y_1 < y_2$, B_q the bounded set of $C(J, \mathbb{R}^n)$ as in Claim 2. Let $u \in B_q$ and $h \in N(u)$. Then there exists $f \in S_{F,u}$ such that for each $(x, y) \in J$, we have

$$h(x, y) = \mu(x, y) + \frac{1}{\Gamma(r_1)\Gamma(r_2)} \int_0^x \int_0^y (x-s)^{r_1-1}(y-t)^{r_2-1} f(s,t) \mathrm{d}t \mathrm{d}s.$$

Then, for each $(x, y) \in J$, we have

$$\|h(\tau_2, y_2) - h(\tau_1, y_1)\| = \Big\|\mu(\tau_1, y_1) - \mu(\tau_2, y_2)$$
$$+\frac{1}{\Gamma(r_1)\Gamma(r_2)}\int_0^{\tau_1}\int_0^{y_1}[(\tau_2 - s)^{r_1-1}(y_2 - t)^{r_2-1}$$
$$-(\tau_1 - s)^{r_1-1}$$
$$\times(y_1 - t)^{r_2-1}]f(s,t)\mathrm{d}t\,\mathrm{d}s$$

$$+\frac{1}{\Gamma(r_1)\Gamma(r_2)}\int_{\tau_1}^{\tau_2}\int_{y_1}^{y_2}(\tau_2 - s)^{r_1-1}(y_2 - t)^{r_2-1}f(s,t)\mathrm{d}t\,\mathrm{d}s$$
$$+\frac{1}{\Gamma(r_1)\Gamma(r_2)}\int_0^{\tau_1}\int_{y_1}^{y_2}(\tau_2 - s)^{r_1-1}(y_2 - t)^{r_2-1}f(s,t)\mathrm{d}t\,\mathrm{d}s$$
$$+\frac{1}{\Gamma(r_1)\Gamma(r_2)}\int_{\tau_1}^{\tau_2}\int_0^{y_1}(\tau_2 - s)^{r_1-1}(y_2 - t)^{r_2-1}f(s,t)\mathrm{d}t\,\mathrm{d}s\Big\|$$
$$\leq \|\mu(\tau_1, y_1) - \mu(\tau_2, y_2)\|$$
$$+\frac{p^*\delta(q)}{\Gamma(r_1+1)\Gamma(r_2+1)}[2y_2^{r_2}(\tau_2 - \tau_1)^{r_1} + 2\tau_2^{r_1}(y_2 - y_1)^{r_2}$$
$$+\tau_1^{r_1}y_1^{r_2} - \tau_2^{r_1}y_2^{r_2} - 2(\tau_2 - \tau_1)^{r_1}(y_2 - y_1)^{r_2}].$$

As $\tau_1 \longrightarrow \tau_2$ and $y_1 \longrightarrow y_2$, the right-hand side of the above inequality tends to zero. As a consequence of Claims 1–3 together and the Arzela-Ascoli theorem we can conclude that $N : C(J, \mathbb{R}^n) \to \mathcal{P}(C(J, \mathbb{R}^n))$ is a completely continuous multivalued operator.

Claim 4. N has a closed graph. Let $u_n \to u_*$, $h_n \in N(u_n)$ and $h_n \to h_*$. We need to show that $h_* \in N(u_*)$.

$h_n \in N(u_n)$ means that there exists $f_n \in S_{F,u_n}$ such that, for each $(x, y) \in J$,

$$h_n(x,y) = \mu(x,y) + \frac{1}{\Gamma(r_1)\Gamma(r_2)}\int_0^x\int_0^y(x - s)^{r_1-1}(y - t)^{r_2-1}f_n(s,t)\mathrm{d}t\mathrm{d}s.$$

We must show that there exists $f_* \in S_{F,u*}$ such that, for each $(x, y) \in J$,

$$h_*(x,y) = \mu(x,y) + \frac{1}{\Gamma(r_1)\Gamma(r_2)}\int_0^x\int_0^y(x - s)^{r_1-1}(y - t)^{r_2-1}f_*(s,t)\mathrm{d}t\mathrm{d}s.$$

Since $F(x, y, \cdot)$ is upper semicontinuous, then for every $\varepsilon > 0$, there exist $n_0(\epsilon) \geq 0$ such that for every $n \geq n_0$, we have

$$f_n(x,y) \in F(x, y, u_{n(x,y)}) \subset F(x, y, u_{*(x,y)}) + \varepsilon B(0,1), \text{ a.e. } (x,y) \in J.$$

Since $F(.,.,.)$ has compact values, then there exists a subsequence f_{n_m} such that

$$f_{n_m}(\cdot,\cdot) \to f_*(\cdot,\cdot) \text{ as } m \to \infty$$

and

$$f_*(x,y) \in F(x,y,u_{*(x,y)}), \text{ a.e. } (x,y) \in J.$$

For every $w \in F(x,y,u_{*(x,y)})$, we have

$$\|f_{n_m}(x,y) - u_*(x,y)\| \le \|f_{n_m}(x,y) - w\| + \|w - f_*(x,y)\|.$$

Then

$$\|f_{n_m}(x,y) - f_*(x,y)\| \le d\Big(f_{n_m}(x,y), F(x,y,u_{*(x,y)})\Big).$$

By an analogous relation, obtained by interchanging the roles of f_{n_m} and f_*, it follows that

$$\begin{aligned}\|f_{n_m}(x,y) - u_*(x,y)\| &\le H_d\Big(F(x,y,u_{n(x,y)}), F(x,y,u_{*(x,y)})\Big)\\ &\le l(x,y)\|u_n - u_*\|_\infty.\end{aligned}$$

Let $l^* := \|l\|_{L^\infty}$, then by (6.9.4) we obtain for each $(x,y) \in J$,

$$\begin{aligned}&\|h_{n(x,y)} - h_{*(x,y)}\|\\ &\le \frac{1}{\Gamma(r_1)\Gamma(r_2)}\int_0^x\int_0^y (x-s)^{r_1-1}(y-t)^{r_2-1}\|f_m(s,t) - f_*(s,t)\|dtds\\ &\le \frac{\|u_{n_m} - u_*\|_\infty}{\Gamma(r_1)\Gamma(r_2)}\int_0^x\int_0^y (x-s)^{r_1-1}(y-t)^{r_2-1}l(s,t)dtds.\end{aligned}$$

Hence

$$\|h_{n_m} - h_*\|_\infty \le \frac{a^{r_1}b^{r_2}l^*\|u_{n_m} - u_*\|_\infty}{\Gamma(r_1+1)\Gamma(r_2+1)} \to 0 \text{ as } m \to \infty.$$

Claim 5. A priori bounds on solutions. Let u be a possible solution of the problems (6.13)–(6.15). Then, there exists $f \in S_{F,u}$ such that, for each $(x,y) \in J$,

$$\begin{aligned}\|u(x,y)\| &\le \|\mu(x,y)\| + \frac{1}{\Gamma(r_1)\Gamma(r_2)}\int_0^x\int_0^y (x-s)^{r_1-1}(y-t)^{r_2-1}\|f(s,t)\|dtds\\ &\le \|\mu(x,y)\| + \frac{1}{\Gamma(r_1)\Gamma(r_2)}\int_0^x\\ &\quad\times\int_0^y (x-s)^{r_1-1}(y-t)^{r_2-1}p(s,t)\delta(\|u(s,t)\|)dtds\\ &\le \|\mu\|_\infty + \frac{p^*a^{r_1}b^{r_2}\delta(\|u\|_\infty)}{\Gamma(r_1+1)\Gamma(r_2+1)}.\end{aligned}$$

This implies that for each $(x, y) \in J$, we have

$$\frac{\|u\|_\infty}{\|\mu\|_\infty + \frac{p^* a^{r_1} b^{r_2} \delta(\|u\|_\infty)}{\Gamma(r_1+1)\Gamma(r_2+1)}} < 1$$

Then by condition (6.19), there exists M such that $\|u\|_\infty \neq M$.
Let

$$U = \{u \in C(J, \mathbb{R}^n) : \|u\|_\infty < M\}.$$

The operator $N : \overline{U} \to \mathcal{P}(C(J, \mathbb{R}^n))$ is upper semicontinuous and completely continuous. From the choice of U, there is no $u \in \partial U$ such that $u \in \lambda N(u)$ for some $\lambda \in (0, 1)$. As a consequence of the nonlinear alternative of Leray–Schauder type (Theorem 2.33), we deduce that N has a fixed point which is a solution of (6.20) and (6.21). Denote this solution by u_1. Define the function

$$r_{k,1}(x, y) = x_k(u_1(x, y)) - x, \qquad \text{for } x \geq 0,\ y \geq 0.$$

Hypothesis (6.9.1) implies that $r_{k,1}(0,0) \neq 0$ for $k = 1, \dots, m$. If $r_{k,1}(x, y) \neq 0$ on J for $k = 1, \dots, m$; i.e.,

$$x \neq x_k(u_1(x, y)), \quad \text{on } J \quad \text{for } k = 1, \dots, m,$$

then u_1 is a solution of the problems (6.13)–(6.15). It remains to consider the case when $r_{1,1}(x, y) = 0$ for some $(x, y) \in J$. Now since $r_{1,1}(0,0) \neq 0$ and $r_{1,1}$ is continuous, there exists $x_1 > 0, y_1 > 0$ such that $r_{1,1}(x_1, y_1) = 0$, and $r_{1,1}(x, y) \neq 0$, for all $(x, y) \in [0, x_1) \times [0, y_1)$. Thus by (6.9.1) we have

$$r_{1,1}(x_1, y_1) = 0 \text{ and } r_{1,1}(x, y) \neq 0, \text{ for all } (x, y) \in [0, x_1) \times [0, y_1] \cup (y_1, b].$$

Suppose that there exist $(\bar{x}, \bar{y}) \in [0, x_1) \times [0, y_1] \cup (y_1, b]$ such that $r_{1,1}(\bar{x}, \bar{y}) = 0$. The function $r_{1,1}$ attains a maximum at some point $(s, t) \in [0, x_1) \times [0, b]$. Since

$$({}^c D_0^r u_1)(x, y) \in F(x, y, u_1(x, y)), \text{ for } (x, y) \in J,$$

then there exists a function $f(x, y) \in F(x, y, u(x, y))$ such that

$$({}^c D_0^r u_1)(x, y) = f(x, y), \text{ for } (x, y) \in J.$$

Hence

$$\frac{\partial u_1(x, y)}{\partial x} \text{ exists, and } \frac{\partial r_{1,1}(s, t)}{\partial x} = x_1'(u_1(s, t))\frac{\partial u_1(s, t)}{\partial x} - 1 = 0.$$

Since

$$\frac{\partial u_1(x,y)}{\partial x} = \varphi'(x) + \frac{r_1 - 1}{\Gamma(r_1)\Gamma(r_2)} \int\limits_0^x \int\limits_0^y (x-s)^{r_1-2}(y-t)^{r_2-1} f(s,t)\mathrm{d}t\mathrm{d}s,$$

then

$$x_1'(u_1(s,t)) \left[\varphi'(s) + \frac{r_1 - 1}{\Gamma(r_1)\Gamma(r_2)} \int\limits_0^s \int\limits_0^t (s-\theta)^{r_1-2}(t-\eta)^{r_2-1} f(\theta,\eta)\mathrm{d}\theta\mathrm{d}\eta \right] = 1,$$

which contradicts (6.9.5). From (6.9.1) we have

$$r_{k,1}(x,y) \neq 0 \text{ for all } (x,y) \in [0,x_1) \times [0,b] \text{ and } k = 1,\ldots,m.$$

Step 2: In what follows set

$$X_k := [x_k, a] \times [0,b];\ k = 1,\ldots,m.$$

Consider now the problem

$$({}^cD^r_{x_1}u)(x,y) \in F(x,y,u(x,y));\ \text{where } (x,y) \in X_1, \tag{6.22}$$

$$u(x_1^+, y) = I_1(u_1(x_1,y)). \tag{6.23}$$

Consider the operator $N_1 : C(X_1,\mathbb{R}^n) \to \mathcal{P}(C(X_1,\mathbb{R}^n))$ defined as

$$N_1(u) = \left\{ h \in C(X_1,\mathbb{R}^n) : \left\{ \begin{array}{l} h(x,y) = \varphi(x) + I_1(u_1(x_1,y)) - I_1(u_1(x_1,0)) \\ +\frac{1}{\Gamma(r_1)\Gamma(r_2)} \int\limits_{x_1}^x \int\limits_0^y (x-s)^{r_1-1}(y-t)^{r_2-1} \\ \times f(s,t)\mathrm{d}t\mathrm{d}s;\ (x,y) \in X_1,\ f \in S_{F,u}. \end{array} \right. \right\}$$

As in Step 1 we can show that N_1 is upper semicontinuous and completely continuous. Let u be a possible solution of the problems (6.22) and (6.23). Then, there exists $f \in S_{F,u}$ such that, for each $(x,y) \in X_1$,

$$\begin{aligned} \|u(x,y)\| &\leq \|\varphi(x)\| + \|I_1(u_1(x_1,y))\| + \|I_1(u_1(x_1,0))\| \\ &\quad + \frac{1}{\Gamma(r_1)\Gamma(r_2)} \int_{x_1}^x \int_0^y (x-s)^{r_1-1}(y-t)^{r_2-1}\|f(s,t)\|\mathrm{d}t\mathrm{d}s \\ &\leq \|\varphi\|_\infty + 2c_1\|u\| + 2d_1 \\ &\quad + \frac{1}{\Gamma(r_1)\Gamma(r_2)} \int\limits_0^x \int\limits_0^y (x-s)^{r_1-1}(y-t)^{r_2-1} p(s,t)\delta(\|u\|)\mathrm{d}t\mathrm{d}s. \end{aligned}$$

Then

$$\frac{\|u\|_\infty}{\|\varphi\|_\infty + 2c_1\|u\|_\infty + 2d_1 + \dfrac{p^* a^{r_1} b^{r_2} \delta(\|u\|_\infty)}{\Gamma(r_1+1)\Gamma(r_2+1)}} < 1$$

Then by condition (6.19), there exists M' such that $\|u\|_\infty \neq M'$. Let

$$U' = \{u \in C(X_1, \mathbb{R}^n) : \|u\|_\infty < M'\}.$$

The operator $N_1 : \overline{U}' \to \mathcal{P}(C(X_1, \mathbb{R}^n))$ is upper semicontinuous and completely continuous. From the choice of U', there is no $u \in \partial U'$ such that $u \in \lambda N_1(u)$ for some $\lambda \in (0, 1)$. As a consequence of the nonlinear alternative of Leray–Schauder type (Theorem 2.33), we deduce that N_1 has a fixed point which is a solution of (6.22) and (6.23). Denote this solution by u_2. Define

$$r_{k,2}(x, y) = x_k(u_2(x, y)) - x, \quad \text{for } (x, y) \in X_1.$$

If $r_{k,2}(x, y) \neq 0$ on $(x_1, a] \times [0, b]$ and for all $k = 1, \dots, m$, then

$$u(x, y) = \begin{cases} u_1(x, y); & \text{if } (x, y) \in [0, x_1) \times [0, b], \\ u_2(x, y); & \text{if } (x, y) \in [x_1, a] \times [0, b], \end{cases}$$

is a solution of the problems (6.13)–(6.15). It remains to consider the case when $r_{2,2}(x, y) = 0$, for some $(x, y) \in (x_1, a] \times [0, b]$. By (6.9.6), we have

$$\begin{aligned} r_{2,2}(x_1^+, y_1) &= x_2(u_2(x_1^+, y_1) - x_1 \\ &= x_2(I_1(u_1(x_1, y_1))) - x_1 \\ &> x_1(u_1(x_1, y_1)) - x_1 \\ &= r_{1,1}(x_1, y_1) = 0. \end{aligned}$$

Since $r_{2,2}$ is continuous, there exists $x_2 > x_1,\ y_2 > y_1$ such that $r_{2,2}(x_2, y_2) = 0$, and $r_{2,2}(x, y) \neq 0$ for all $(x, y) \in (x_1, x_2) \times [0, b]$. It is clear by (6.9.1) that

$$r_{k,2}(x, y) \neq 0 \quad \text{for all } (x, y) \in (x_1, x_2) \times [0, b];\ k = 2, \dots, m.$$

Now suppose that there are $(s, t) \in (x_1, x_2) \times [0, b]$ such that $r_{1,2}(s, t) = 0$. From (6.9.6) it follows that

$$\begin{aligned} r_{1,2}(x_1^+, y_1) &= x_1(u_2(x_1^+, y_1) - x_1 \\ &= x_1(I_1(u_1(x_1, y_1))) - x_1 \\ &\le x_1(u_1(x_1, y_1)) - x_1 \\ &= r_{1,1}(x_1, y_1) = 0. \end{aligned}$$

Thus $r_{1,2}$ attains a nonnegative maximum at some point $(s_1, t_1) \in (x_1, a) \times [0, x_2) \cup (x_2, b]$. Since

$$({}^c D^r_{x_1} u_2)(x, y) \in F(x, y, u_2(x, y)), \text{ for } (x, y) \in X_1,$$

then there exists a function $f(x, y) \in F(x, y, u(x, y))$ such that

$$({}^c D^r_{x_1} u_2)(x, y) = f(x, y), \text{ for } (x, y) \in X_1.$$

Then we get

$$u_2(x, y) = \varphi(x) + I_1(u_1(x_1, y)) - I_1(u_1(x_1, 0)) + \frac{1}{\Gamma(r_1)\Gamma(r_2)} \int_{x_1}^{x} \int_{0}^{y} (x-s)^{r_1-1}(y-t)^{r_2-1} f(s,t)\mathrm{d}t\mathrm{d}s,$$

hence

$$\frac{\partial u_2}{\partial x}(x, y) = \varphi'(x) + \frac{r_1 - 1}{\Gamma(r_1)\Gamma(r_2)} \int_{x_1}^{x} \int_{0}^{y} (x-s)^{r_1-2}(y-t)^{r_2-1} f(s,t)\mathrm{d}t\mathrm{d}s,$$

then

$$\frac{\partial r_{1,2}(s_1, t_1)}{\partial x} = x_1'(u_2(s_1, t_1)) \frac{\partial u_2}{\partial x}(s_1, t_1) - 1 = 0.$$

Therefore

$$x_1'(u_2(s_1, t_1)) \left[\varphi'(s_1) + \frac{r_1 - 1}{\Gamma(r_1)\Gamma(r_2)} \int_{x_1}^{s_1} \int_{0}^{t_1} (s_1-\theta)^{r_1-2}(t_1-\eta)^{r_2-1} f(\theta,\eta)\mathrm{d}\eta\mathrm{d}\theta \right] = 1,$$

which contradicts (6.9.5).

Step 3: We continue this process and take into account that $u_{m+1} := u\Big|_{X_m}$ is a solution to the problem

$$({}^c D^r_{x_m} u)(x, y) \in F(x, y, u(x, y)); \quad \text{a.e. } (x, y) \in (x_m, a] \times [0, b],$$
$$u(x_m^+, y) = I_m(u_{m-1}(x_m, y)).$$

The solution u of the problems (6.13)–(6.15) is then defined by

$$u(x, y) = \begin{cases} u_1(x, y); & \text{if } (x, y) \in [0, x_1] \times [0, b], \\ u_2(x, y); & \text{if } (x, y) \in (x_1, x_2] \times [0, b], \\ \dots & \\ u_{m+1}(x, y); & \text{if } (x, y) \in (x_m, a] \times [0, b]. \end{cases}$$

□

Now we present (without proof) an existence result as an extension of the result presented in Theorem 6.9 to problems (6.16)–(6.18).

Definition 6.10. A function $u \in \Omega \cap \cup_{k=1}^{m} AC(J_k, \mathbb{R}^n)$ such that its mixed derivative D_{xy}^2 exists on J' is said to be a solution of (6.16)–(6.18) if there exists a function $f(x,y) \in F(x,y,u(x,y))$ such that u satisfies $({}^cD_0^r u)(x,y) = f(x,y)$ on J' and conditions (6.17) and (6.18) are satisfied.

Theorem 6.11. *Assume that (6.9.1)–(6.9.4), (6.9.6) and the following conditions:*

(6.11.1) The function g is nonnegative and completely continuous and there exist constants $0 \le l_1 < 1,\ l_2 \ge 0$ *such that*

$$\|g(x,y,u)\| \le l_1\|u\| + l_2;\ (x,y) \in J, u \in \mathbb{R}^n,$$

(6.11.2) For all $(s,t,u) \in J \times \mathbb{R}^n$, *there exists* $f \in S_{F,u}$ *such that*

$$x_k'(u)\left[\varphi'(s) + \frac{\partial g(s,t,u(s,t))}{\partial x} - \frac{\partial g(s,0,u(s,0))}{\partial x} + \frac{r_1 - 1}{\Gamma(r_1)\Gamma(r_2)}\right.$$
$$\left. \times \int_{x_k}^{s}\int_{0}^{t}(s-\theta)^{r_1-2}(t-\eta)^{r_2-1}f(\theta,\eta)\mathrm{d}\eta\mathrm{d}\theta\right] \ne 1;\ k = 1,\ldots,m,$$

(6.11.3) There exists a number $M' > 0$ *such that*

$$\min\left\{\frac{M'}{\|\mu\|_\infty + 4l_1M' + 4l_2 + \frac{p^*a^{r_1}b^{r_2}\delta(M')}{\Gamma(r_1+1)\Gamma(r_2+1)}},\right.$$
$$\left.\frac{M'}{\|\varphi\|_\infty + (2c_k + 4l_1)M' + 2d_k + 4l_2 + \frac{p^*a^{r_1}b^{r_2}\delta(M')}{\Gamma(r_1+1)\Gamma(r_2+1)}};\ k = 1,\ldots,m\right\} > 1,$$
(6.24)

hold. Then IVP (6.16)–(6.18) has at least one solution on J.

6.3.3 An Example

As an application of our results we consider the following impulsive partial hyperbolic functional differential inclusions of the form:

$$({}^cD_{x_k}^r u)(x,y) \in F(x,y,u(x,y));\ \text{ where } (x,y) \in J_k;\ k = 0,\ldots,m,$$

$$x_k = x_k(u(x,y));\ k = 1,\dots,m, \tag{6.25}$$

$$u(x_k^+, y) = I_k(u(x_k,y));\ y \in [0,1],\ k = 1,\dots,m, \tag{6.26}$$

$$u(x,0) = x,\ u(0,y) = y^2;\ x \in [0,1],\ y \in [0,1], \tag{6.27}$$

where $r = (r_1, r_2)$, $0 < r_1, r_2 \leq 1$, $x_k(u) = 1 - \frac{1}{2^k(1+u^2)}$; $k = 1,\dots,m$, and for each $u \in \mathbb{R}$ we have $x_{k+1}(I_k(u)) > x_k(u)$; $k = 1,\dots,m$, and also there exist constants $c_k, d_k > 0$, such that

$$|I_k(u(x_k,y))| \leq c_k|u(x_k,y)| + d_k,\ k = 1,\dots,m.$$

Let $u \in \mathbb{R}$ then we have

$$x_{k+1}(u) - x_k(u) = \frac{1}{2^{k+1}(1+u^2)} > 0;\ k = 1,\dots,m.$$

Hence $0 < x_1(u) < x_2(u) < \cdots < x_m(u) < 1$, for each $u \in \mathbb{R}$. Set

$$F(x,y,u(x,y)) = \{u \in \mathbb{R} : f_1(x,y,u(x,y)) \leq u \leq f_2(x,y,u(x,y))\},$$

where $f_1, f_2 : [0,1]\times[0,1]\times\mathbb{R} \to \mathbb{R}$. We assume that for each $(x,y) \in J$, $f_1(x,y,.)$ is lower semicontinuous (i.e., the set $\{z \in \mathbb{R} : f_1(x,y,z) > \nu\}$ is open for each $\nu \in \mathbb{R}$), and assume that for each $(x,y) \in J$, $f_2(x,y,.)$ is upper semicontinuous (i.e., the set $\{z \in \mathbb{R} : f_2(x,y,z) < \nu\}$ is open for each $\nu \in \mathbb{R}$). Assume that there are $p \in C([0,1]\times[0,1], \mathbb{R}^+)$ and $\delta : [0,\infty) \to (0,\infty)$ continuous and nondecreasing such that

$$\max(|f_1(x,y,z)|,\ |f_2(x,y,z)|) \leq p(x,y)\delta(|z|),\ \text{for a.e. } (x,y) \in J \text{ and } z \in \mathbb{R}.$$

It is clear that F is compact and convex valued, and it is upper semi-continuous (see [104]). Moreover, assume that conditions (6.9.5) and (6.9.7) are satisfied. Since all conditions of Theorem 6.9 are satisfied, problems (6.25)–(6.27) have at least one solution on $[0,1]\times[0,1]$.

6.4 The Method of Upper and Lower Solutions for Partial Hyperbolic Fractional Order Differential Inclusions with Impulses

6.4.1 Introduction

This section deals with the existence of solutions to impulsive fractional order IVP, for the system

$$({}^cD^r_{x_k}u)(x,y) \in F(x,y,u(x,y)); \text{ if } (x,y) \in J_k;\ k = 0,\dots,m, \quad (6.28)$$

$$u(x_k^+,y) = u(x_k^-,y) + I_k(u(x_k^-,y)); \text{ if } y \in [0,b],\ k = 1,\dots,m, \quad (6.29)$$

$$u(x,0) = \varphi(x);\ x \in [0,a],\ u(0,y) = \psi(y);\ y \in [0,b], \quad (6.30)$$

where $a,b > 0$, $0 = x_0 < x_1 < \cdots < x_m < x_{m+1} = a$, $F : J \times \mathbb{R}^n \to \mathcal{P}(\mathbb{R}^n)$ is a compact-valued multivalued map, $J = [0,a] \times [0,b]$, $\mathcal{P}(\mathbb{R}^n)$ is the family of all subsets of $\mathbb{R}^n$, $I_k : \mathbb{R}^n \to \mathbb{R}^n$, $k = 1,\dots,m$, φ, ψ are as in problem (6.1)–(6.3). Here $u(x_k^+,y)$ and $u(x_k^-,y)$ denote the right and left limits of $u(x,y)$ at $x = x_k$, respectively.

6.4.2 Main Result

To define the solutions of problems (6.28)–(6.30), we shall consider the Banach space $PC(J,\mathbb{R}^n)$ defined in Sect. 6.2.

Definition 6.12. A function $u \in PC(J,\mathbb{R}^n) \bigcap \bigcup_{k=0}^m AC((x_k,x_{k+1}) \times [0,b],\mathbb{R}^n)$ such that its mixed derivative D^2_{xy} exists J_k; $k = 0,\dots,m$ is said to be a solution of (6.28)–(6.30) if there exists a function $f \in L^1(J,\mathbb{R}^n)$ with $f(x,y) \in F(x,y,u(x,y))$ such that u satisfies $({}^cD^r_{0x_k}u)(x,y) = f(x,y)$ on J_k; $k = 0,\dots,m$ and conditions (6.29) and (6.30) are satisfied.

Let $z,\ \bar{z} \in C(J,\mathbb{R}^n)$ be such that

$$z(x,y) = (z_1(x,y), z_2(x,y), \dots, z_n(x,y)),\ (x,y) \in J,$$

and

$$\bar{z}(x,y) = (\bar{z}_1(x,y), \bar{z}_2(x,y), \dots, \bar{z}_n(x,y)),\ (x,y) \in J.$$

The notation $z \le \bar{z}$ means that

$$z_i(x,y) \le \bar{z}_i(x,y),\ i = 1,\dots,n.$$

Definition 6.13. A function $z \in PC(J,\mathbb{R}^n) \bigcap \bigcup_{k=0}^m AC((x_k,x_{k+1}) \times [0,b],\mathbb{R}^n)$ is said to be a lower solution of (6.28)–(6.30) if there exists a function $f \in L^1(J,\mathbf{R}^n)$ with $f(x,y) \in F(x,y,u(x,y))$ such that z satisfies

$$({}^cD^r_{x_k}z)(x,y) \le f(x,y,z(x,y)),\ z(x,0) \le \varphi(x),\ z(0,y) \le \psi(y) \text{ on } J_k;$$
$$k = 0,\dots,m,$$

$$z(x_k^+,y) \le z(x_k^-,y) + I_k(z(x_k^-,y)); \text{ if } y \in [0,b],\ k = 1,\dots,m,$$
$$z(x,0) \le \varphi(x),\ z(0,y) \le \psi(y) \text{ on } J,$$

and $z(0,0) \le \varphi(0)$.

The function z is said to be an upper solution of (6.28)–(6.30) if the reversed inequalities hold.

Theorem 6.14. *Assume that the following hypotheses:*

(6.14.1) $F : J \times \mathbb{R}^n \longrightarrow \mathcal{P}_{cp,cv}(\mathbb{R}^n)$ *is* L^1*-Carathéodory*
(6.14.2) *There exists* $l \in C(J, \mathbb{R}^+)$ *such that*

$$H_d(F(x,y,u), F(x,y,\overline{u})) \leq l(x,y)\|u-\overline{u}\| \text{ for every } u, \overline{u} \in \mathbb{R}^n$$

and

$$d(0, F(x,y,0)) \leq l(x,y), \text{ a.e. } (x,y) \in J$$

(6.14.3) *There exist* v *and* $w \in PC(J,\mathbb{R}^n)\bigcap AC((x_k, x_{k+1}) \times [0,b], \mathbb{R}^n)$, $k = 0, \ldots, m$, *lower and upper solutions for the problem (6.28)–(6.30) such that* $v(x,y) \leq w(x,y)$ *for each* $(x,y) \in J$,
(6.14.4) *For each* $y \in [0,b]$, *we have*

$$\begin{aligned} v(x_k^+, y) &\leq \min_{u \in [v(x_k^-, y), w(x_k^-, y)]} I_k(u) \\ &\leq \max_{u \in [v(x_k^-, y), w(x_k^-, y)]} I_k(u) \\ &\leq w(x_k^+, y), \ k = 1, \ldots, m \end{aligned}$$

hold. Then the problem (6.28)–(6.30) has at least one solution u *such that*

$$v(x,y) \leq u(x,y) \leq w(x,y) \text{ for all } (x,y) \in J.$$

Proof. Transform the problem (6.28)–(6.30) into a fixed-point problem. Consider the following modified problem:

$$({}^cD_0^r u)(x,y) \in F(x,y,g(u(x,y))); \text{ if }; (x,y) \in J_k; \ k = 0, \ldots, m, \tag{6.31}$$

$$u(x_k^+, y) = u(x_k^-, y) + I_k(g(x_k^-, y, u(x_k^-, y))); \text{ if } y \in [0,b]; \ k = 1, \ldots, m, \tag{6.32}$$

$$u(x,0) = \varphi(x); \ x \in [0,a], \ u(0,y) = \psi(y); \ y \in [0,b], \tag{6.33}$$

where $g : PC(J,\mathbb{R}^n) \longrightarrow PC(J,\mathbb{R}^n)$ be the truncation operator defined by

$$g(u)(x,y) = \begin{cases} v(x,y), & u(x,y) < v(x,y) \\ u(x,y), & v(x,y) \leq u(x,y) \leq w(x,y) \\ w(x,y), & u(x,y) > w(x,y). \end{cases}$$

A solution to (6.31)–(6.33) is a fixed point of the operator $G : PC(J,\mathbb{R}^n) \longrightarrow \mathcal{P}(PC(J,\mathbb{R}^n))$ defined by

$$G(u) = \begin{cases} h \in PC(J,\mathbb{R}^n) : \\ h(x,y) = \mu(x,y) \\ + \sum_{0<x_k<x} (I_k(g(x_k^-, y, u(x_k^-, y))) - I_k(g(x_k^-, 0, u(x_k^-, 0)))) \\ + \frac{1}{\Gamma(r_1)\Gamma(r_2)} \sum_{0<x_k<x} \int_{x_{k-1}}^{x_k} \int_0^y (x_k - s)^{r_1-1}(y-t)^{r_2-1} f(s,t)\mathrm{d}t\mathrm{d}s \\ + \frac{1}{\Gamma(r_1)\Gamma(r_2)} \int_{x_k}^{x} \int_0^y (x-s)^{r_1-1}(y-t)^{r_2-1} f(s,t)\mathrm{d}t\mathrm{d}s, \end{cases}$$

where

$$\begin{aligned} f \in \tilde{S}^1_{F,g(u)} &= \{f \in S^1_{F,g(u)} : f(x,y) \geq f_1(x,y) \text{ on } A_1 \\ &\qquad \text{and } f(x,y) \leq f_2(x,y) \text{ on } A_2\}, \\ A_1 &= \{(x,y) \in J : u(x,y) < v(x,y) \leq w(x,y)\}, \\ A_2 &= \{(x,y) \in J : u(x,y) \leq w(x,y) < u(x,y)\}, \end{aligned}$$

and

$$S^1_{F,g(u)} = \{f \in L^1(J,\mathbb{R}^n) : f(x,y) \in F(x,y,g(u(x,y))), \text{ for } (x,y) \in J\}.$$

Remark 6.15. (A) For each $u \in PC(J,\mathbb{R}^n)$, the set $\tilde{S}_{F,g(u)}$ is nonempty. In fact, (6.14.1) implies there exists $f_3 \in S_{F,g(u)}$, so we set

$$f = f_1\chi_{A_1} + f_2\chi_{A_2} + f_3\chi_{A_3},$$

where χ_{A_i} is the characteristic function of A_i; $i = 1,2$, and

$$A_3 = \{(x,y) \in J : v(x,y) \leq u(x,y) \leq w(x,y)\}.$$

Then, by decomposability, $f \in \tilde{S}_{F,g(u)}$,

(B) By the definition of g it is clear that $F(.,.,g(u)(.,.))$ is an L^1-Carathéodory multivalued map with compact convex values and there exists $\phi_1 \in C(J,\mathbb{R}_+)$ such that

$$\|F(x,y,g(u(x,y)))\|_{\mathcal{P}} \leq \phi_1(x,y) \text{ for each } (x,y) \in J \text{and } u \in \mathbb{R}^n,$$

(C) By the definition of g and from (6.14.4) we have

$$u(x_k^+, y) \leq I_k(g(x_k, y, u(x_k, y))) \leq w(x_k^+, y);\ y \in [0,b];\ k = 1,\ldots,m.$$

Set

$$\phi_1^* := \sup\{\phi_1(x,y) : (x,y) \in J\},$$

$$\eta = \|\mu\|_\infty + 2\sum_{k=1}^{m} \max_{y\in[0,b]} (\|v(x_k^+,y)\|, \|w(x_k^+,y)\|) + \frac{2a^{r_1}b^{r_2}\phi_1^*}{\Gamma(r_1+1)\Gamma(r_2+1)},$$

and

$$D = \{u \in PC(J,\mathbb{R}^n) : \|u\|_{PC} \le \eta\}.$$

Clearly D is a closed convex subset of $PC(J,\mathbb{R}^n)$ and that G maps D into D. We shall show that D satisfies the assumptions of Lemma 2.38. The proof will be given in several steps.

Step 1: *$G(u)$ is convex for each $u \in D$.* Indeed, if $h_1,\ h_2$ belong to $G(u)$, then there exist $f_1,\ f_2 \in \tilde{S}^1_{F,g(u)}$ such that for each $(x,y) \in J$ we have

$$\begin{aligned} h_i(u)(x,y) =& \mu(x,y) + \sum_{0<x_k<x} (I_k(g(x_k^-,y,u(x_k^-,y))) - I_k(g(x_k^-,0,u(x_k^-,0)))) \\ &+ \frac{1}{\Gamma(r_1)\Gamma(r_2)} \sum_{0<x_k<x} \int_{x_{k-1}}^{x_k} \int_0^y (x_k-s)^{r_1-1}(y-t)^{r_2-1} f_i(s,t)\mathrm{d}t\mathrm{d}s \\ &+ \frac{1}{\Gamma(r_1)\Gamma(r_2)} \int_{x_k}^{x} \int_0^y (x-s)^{r_1-1}(y-t)^{r_2-1} f_i(s,t)\mathrm{d}t\mathrm{d}s. \end{aligned}$$

Let $0 \le \xi \le 1$. Then, for each $(x,y) \in J$, we have

$$\begin{aligned} (\xi h_1 + (1-\xi)h_2)(x,y) = \mu(x,y) &+ \sum_{0<x_k<x} (I_k(g(x_k^-,y,u(x_k^-,y)))) \\ &- \sum_{0<x_k<x} (I_k(g(x_k^-,0,u(x_k^-,0)))) \\ &+ \frac{1}{\Gamma(r_1)\Gamma(r_2)} \sum_{0<x_k<x} \int_{x_{k-1}}^{x_k} \int_0^y (x_k-s)^{r_1-1}(y-t)^{r_2-1} \\ &\times [\xi f_1(s,t) + (1-\xi) f_2(s,t)]\,\mathrm{d}t\mathrm{d}s \\ &+ \frac{1}{\Gamma(r_1)\Gamma(r_2)} \int_{x_k}^{x} \int_0^y (x-s)^{r_1-1}(y-t)^{r_2-1} \\ &\times [\xi f_1(s,t) + (1-\xi) f_2(s,t)]\mathrm{d}t\mathrm{d}s. \end{aligned}$$

Since $\tilde{S}^1_{F,g(u)}$ is convex (because F has convex values), we have $\xi h_1 + (1-\xi)h_2 \in G(u)$.

Step 2: *$G(D)$ is bounded.* This is clear since $G(D) \subset D$ and D is bounded.

Step 3: $G(D)$ *is equicontinuous.* Let $(\tau_1, y_1), (\tau_2, y_2) \in J$, $\tau_1 < \tau_2$ and $y_1 < y_2$, let $u \in D$ and $h \in G(u)$, then there exists $f \in \tilde{S}^1_{F,g(u)}$ such that for each $(x, y) \in J$ we have

$$\begin{aligned}
&\|h(u)(\tau_2, y_2) - h(u)(\tau_1, y_1)\| \\
&\leq \|\mu(\tau_1, y_1) - \mu(\tau_2, y_2)\| + \sum_{k=1}^{m}(\|I_k(g(x_k^-, y_1, u(x_k^-, y_1))) \\
&\quad -I_k(g(x_k^-, y_2, u(x_k^-, y_2)))\|) \\
&\quad +\frac{1}{\Gamma(r_1)\Gamma(r_2)}\sum_{k=1}^{m}\int_{x_{k-1}}^{x_k}\int_0^{y_1}(x_k - s)^{r_1-1}\left[(y_2 - t)^{r_2-1} - (y_1 - t)^{r_2-1}\right] \\
&\quad \times\|f(s,t)\|dtds \\
&\quad +\frac{1}{\Gamma(r_1)\Gamma(r_2)}\sum_{k=1}^{m}\int_{x_{k-1}}^{x_k}\int_{y_1}^{y_2}(x_k - s)^{r_1-1}(y_2 - t)^{r_2-1}\|f(s,t)\|dtds \\
&\quad +\frac{1}{\Gamma(r_1)\Gamma(r_2)}\int_0^{\tau_1}\int_0^{y_1}\left[(\tau_2 - s)^{r_1-1}(y_2 - t)^{r_2-1} - (\tau_1 - s)^{r_1-1}(y_1 - t)^{r_2-1}\right] \\
&\quad \times\|f(s,t)\|dtds \\
&\quad +\frac{1}{\Gamma(r_1)\Gamma(r_2)}\int_{\tau_1}^{\tau_2}\int_{y_1}^{y_2}(\tau_2 - s)^{r_1-1}(y_2 - t)^{r_2-1}\|f(s,t)\|dtds \\
&\quad +\frac{1}{\Gamma(r_1)\Gamma(r_2)}\int_0^{\tau_1}\int_{y_1}^{y_2}(\tau_2 - s)^{r_1-1}(y_2 - t)^{r_2-1}\|f(s,t)\|dtds \\
&\quad +\frac{1}{\Gamma(r_1)\Gamma(r_2)}\int_{\tau_1}^{\tau_2}\int_0^{y_1}(\tau_2 - s)^{r_1-1}(y_2 - t)^{r_2-1}\|f(s,t)\|dtds \\
&\leq \|\mu(\tau_1, y_1) - \mu(\tau_2, y_2)\| \\
&\quad +\sum_{k=1}^{m}(\|I_k(g(x_k^-, y_1, u(x_k^-, y_1))) - I_k(g(x_k^-, y_2, u(x_k^-, y_2)))\|) \\
&\quad +\frac{\phi_1^*}{\Gamma(r_1)\Gamma(r_2)}\sum_{k=1}^{m}\int_{x_{k-1}}^{x_k}\int_0^{y_1}(x_k - s)^{r_1-1}\left[(y_2 - t)^{r_2-1}\right. \\
&\quad \left.-(y_1 - t)^{r_2-1}\right]dtds \\
&\quad +\frac{\phi_1^*}{\Gamma(r_1)\Gamma(r_2)}\sum_{k=1}^{m}\int_{x_{k-1}}^{x_k}\int_{y_1}^{y_2}(x_k - s)^{r_1-1}(y_2 - t)^{r_2-1}dtds
\end{aligned}$$

$$+\frac{\phi_1^*}{\Gamma(r_1)\Gamma(r_2)}\int_0^{\tau_1}\int_0^{y_1}\left[(\tau_2-s)^{r_1-1}(y_2-t)^{r_2-1}\right.$$
$$\left.-(\tau_1-s)^{r_1-1}(y_1-t)^{r_2-1}\right]\mathrm{d}t\mathrm{d}s$$
$$+\frac{\phi_1^*}{\Gamma(r_1)\Gamma(r_2)}\int_{\tau_1}^{\tau_2}\int_{y_1}^{y_2}(\tau_2-s)^{r_1-1}(y_2-t)^{r_2-1}\mathrm{d}t\mathrm{d}s$$
$$+\frac{\phi_1^*}{\Gamma(r_1)\Gamma(r_2)}\int_0^{\tau_1}\int_{y_1}^{y_2}(\tau_2-s)^{r_1-1}(y_2-t)^{r_2-1}\mathrm{d}t\mathrm{d}s$$
$$+\frac{\phi_1^*}{\Gamma(r_1)\Gamma(r_2)}\int_{\tau_1}^{\tau_2}\int_0^{y_1}(\tau_2-s)^{r_1-1}(y_2-t)^{r_2-1}\mathrm{d}t\mathrm{d}s.$$

As $\tau_1 \longrightarrow \tau_2$ and $y_1 \longrightarrow y_2$, the right-hand side of the above inequality tends to zero. As a consequence of Steps 1–3 together with the Arzelá-Ascoli theorem, we can conclude that $G : D \longrightarrow \mathcal{P}(D)$ is compact.

Step 4: *G has a closed graph.* Let $u_n \to u_*$, $h_n \in G(u_n)$ and $h_n \to h_*$. We need to show that $h_* \in G(u_*)$.

$h_n \in G(u_n)$ means that there exists $f_n \in \tilde{S}^1_{F,u_n}$ such that, for each $(x,y) \in J$,

$$h_n(x,y) = \mu(x,y) + \sum_{0<x_k<x}\left(I_k(g(x_k^-,y,u_n(x_k^-,y))) - I_k(g(x_k^-,0,u_n(x_k^-,0)))\right)$$
$$+\frac{1}{\Gamma(r_1)\Gamma(r_2)}\sum_{0<x_k<x}\int_{x_{k-1}}^{x_k}\int_0^y(x_k-s)^{r_1-1}(y-t)^{r_2-1}f_n(s,t)\mathrm{d}t\mathrm{d}s$$
$$+\frac{1}{\Gamma(r_1)\Gamma(r_2)}\int_{x_k}^{x}\int_0^y(x-s)^{r_1-1}(y-t)^{r_2-1}f_n(s,t)\mathrm{d}t\mathrm{d}s.$$

We must show that there exists $f_* \in \tilde{S}^1_{F,u_*}$ such that, for each $(x,y) \in J$,

$$h_*(x,y) = \mu(x,y) + \sum_{0<x_k<x}\left(I_k(g(x_k^-,y,u_*(x_k^-,y))) - I_k(g(x_k^-,0,u_*(x_k^-,0)))\right)$$
$$+\frac{1}{\Gamma(r_1)\Gamma(r_2)}\sum_{0<x_k<x}\int_{x_{k-1}}^{x_k}\int_0^y(x_k-s)^{r_1-1}(y-t)^{r_2-1}f_*(s,t)\mathrm{d}t\mathrm{d}s$$
$$+\frac{1}{\Gamma(r_1)\Gamma(r_2)}\int_{x_k}^{x}\int_0^y(x-s)^{r_1-1}(y-t)^{r_2-1}f_*(s,t)\mathrm{d}t\mathrm{d}s.$$

Since $F(x,y,\cdot)$ is upper semicontinuous, then for every $\varepsilon > 0$, there exist $n_0(\epsilon) \geq 0$ such that for every $n \geq n_0$, we have

$$f_n(x,y) \in F(x,y,g(u_n(x,y))) \subset F(x,y,g(u_*(x,y)))+\varepsilon B(0,1), \text{ a.e. } (x,y) \in J.$$

Since $F(.,.,.)$ has compact values, then there exists a subsequence f_{n_m} such that

$$f_{n_m}(\cdot,\cdot) \to f_*(\cdot,\cdot) \text{ as } m \to \infty$$

and

$$f_*(x,y) \in F(x,y,g(u_*(x,y))), \text{ a.e. } (x,y) \in J.$$

For every $w(x,y) \in F(x,y,g(u_*(x,y)))$, we have

$$\|f_{n_m}(x,y) - u_*(x,y)\| \le \|f_{n_m}(x,y) - w(x,y)\| + \|w(x,y) - f_*(x,y)\|.$$

Then

$$\|f_{n_m}(x,y) - f_*(x,y)\| \le d\Big(f_{n_m}(x,y), F(x,y,g(u_*(x,y)))\Big).$$

By an analogous relation obtained by interchanging the roles of f_{n_m} and f_*, it follows that

$$\begin{aligned}\|f_{n_m}(x,y) - u_*(x,y)\| &\le H_d\Big(F(x,y,g(u_n(x,y))), F(x,y,g(u_*(x,y)))\Big)\\ &\le l(x,y)\|u_n - u_*\|_\infty.\end{aligned}$$

Let

$$l^* := \sup\{l(x,y) : (x,y) \in J\},$$

then by (6.14.2) we have for each $(x,y) \in J$,

$$\begin{aligned}\|h_n(x,y) - h_*(x,y)\| \le& \sum_{k=1}^{m} \|I_k(g(x_k^-, y, u_{n_m}(x_k^-, y))) - I_k(g(x_k^-, y, u_*(x_k^-, y)))\|\\ &+ \sum_{k=1}^{m} \|I_k(g(x_k^-, 0, u_{n_m}(x_k^-, 0))) - I_k(g(x_k^-, 0, u_*(x_k^-, 0)))\|\\ &+ \frac{1}{\Gamma(r_1)\Gamma(r_2)} \sum_{x_1 < x_k < x} \int_{x_{k-1}}^{x_k} \int_0^y (x_k - s)^{r_1 - 1}(y - t)^{r_2 - 1}\\ &\times \|f_{n_m}(s,t) - f_*(s,t)\| dt ds\\ &+ \frac{1}{\Gamma(r_1)\Gamma(r_2)} \int_{x_k}^{x} \int_0^y (x - s)^{r_1 - 1}(y - t)^{r_2 - 1}\\ &\times \|f_{n_m}(s,t) - f_*(s,t)\| dt ds\\ \le& \sum_{k=1}^{m} \|I_k(g(x_k^-, y, u_{n_m}(x_k^-, y))) - I_k(g(x_k^-, y, u_*(x_k^-, y)))\|\end{aligned}$$

$$+\sum_{k=1}^{m}\|I_k(g(x_k^-,0,u_{n_m}(x_k^-,0)))-I_k(g(x_k^-,0,u_*(x_k^-,0)))\|$$

$$+\frac{2\|u_{n_m}-u_*\|_\infty}{\Gamma(r_1)\Gamma(r_2)}\int_0^x\int_0^y(x-s)^{r_1-1}(y-t)^{r_2-1}l(s,t)\mathrm{d}t\,\mathrm{d}s$$

$$\leq\sum_{k=1}^{m}\|I_k(g(x_k^-,y,u_{n_m}(x_k^-,y)))-I_k(g(x_k^-,y,u_*(x_k^-,y)))\|$$

$$+\sum_{k=1}^{m}\|I_k(g(x_k^-,0,u_{n_m}(x_k^-,0)))-I_k(g(x_k^-,0,u_*(x_k^-,0)))\|$$

$$+\frac{2l^*a^{r_1}b^{r_2}}{\Gamma(r_1+1)\Gamma(r_2+1)}\|u_{n_m}-u_*\|_\infty.$$

Hence

$$\|h_{n_m}-h_*\|_\infty\to 0 \text{ as } m\to\infty.$$

Step 5: *The solution u of (6.31)–(6.33) satisfies*

$$v(x,y)\leq u(x,y)\leq w(x,y) \text{ for all } (x,y)\in J.$$

Let u be the above solution to (6.31)–(6.33). We prove that

$$u(x,y)\leq w(x,y) \text{ for all } (x,y)\in J.$$

Assume that $u-w$ attains a positive maximum on $[x_k^+,x_{k+1}^-]\times[0,b]$ at $(\overline{x}_k,\overline{y})\in[x_k^+,x_{k+1}^-]\times[0,b]$ for some $k=0,\ldots,m$; i.e.,

$$(u-w)(\overline{x}_k,\overline{y})=\max\{u(x,y)-w(x,y):(x,y)\in[x_k^+,x_{k+1}^-]\times[0,b]\}>0;$$

for some $k=0,\ldots,m$. We distinguish the following cases.

Case 1. If $(\overline{x}_k,\overline{y})\in(x_k^+,x_{k+1}^-)\times[0,b]$ there exists $(x_k^*,\overline{y}^*)\in(x_k^+,x_{k+1}^-)\times[0,b]$ such that

$$[u(x,y^*)-w(x,y^*)]+[u(x_k^*,y)-w(x_k^*,y)]$$
$$-[u(x_k^*,y^*)-w(x_k^*,y^*)]\leq 0;\text{ for all }(x,y)\in([x_k^*,\overline{x}_k]\times\{y^*\})\cup(\{x_k^*\}\times[y^*,b]),\tag{6.34}$$

and

$$u(x,y)-w(x,y)>0,\text{ for all }(x,y)\in(x_k^*,\overline{x}_k]\times[\overline{y}^*,b].\tag{6.35}$$

By the definition of h one has

$${}^cD_{x_k^*}^r u(x,y)\in F(x,y,w(x,y)) \text{ for all } (x,y)\in[x_k^*,\overline{x}_k]\times[\overline{y}^*,b].$$

An integration on $[x_k^*, x] \times [\overline{y}^*, y]$ for each $(x, y) \in [x_k^*, \overline{x}_k] \times [\overline{y}^*, b]$ yields

$$u(x, y) + u(x_k^*, y^*) - u(x, y^*) - u(x_k^*, y)$$
$$= \frac{1}{\Gamma(r_1)\Gamma(r_2)} \int_{x_k^*}^{x} \int_{y^*}^{y} (x - s)^{r_1 - 1}(y - t)^{r_2 - 1} f(s, t) \mathrm{d}t \mathrm{d}s. \quad (6.36)$$

where $f(x, y) \in F(x, y, w(x, y))$. From (6.36) and using the fact that w is an upper solution to (6.28)–(6.30) we get

$$u(x, y) + u(x_k^*, y^*) - u(x, y^*) - u(x_k^*, y)$$
$$\leq w(x, y) + w(x_k^*, y^*) - w(x, y^*) - w(x_k^*, y),$$

which gives

$$[u(x, y) - w(x, y)] \leq [u(x, y^*) - w(x, y^*)]$$
$$+[u(x_k^*, y) - w(x_k^*, y)]$$
$$-[u(x_k^*, y^*) - w(x_k^*, y^*)]. \quad (6.37)$$

Thus from (6.34), (6.35), and (6.37) we obtain the contradiction

$$0 < [u(x, y) - w(x, y)]$$
$$\leq [u(x, y^*) - w(x, y^*)]$$
$$+[u(x_k^*, y) - w(x_k^*, y)]$$
$$-[u(x_k^*, y^*) - w(x_k^*, y^*)] \leq 0; \text{ for all } (x, y) \in [x_k^*, \overline{x}_k] \times [y^*, b].$$

Case 2. If $\overline{x}_k = x_k^+$; $k = 1, \ldots, m$. Then

$$w(x_k^+, \overline{y}) < I_k(h(x_k^-, u(x_k^-, \overline{y}))) \leq w(x_k^+, \overline{y})$$

which is a contradiction. Thus

$$u(x, y) \leq w(x, y) \text{ for all } (x, y) \in J.$$

Analogously, we can prove that

$$u(x, y) \geq v(x, y), \text{ for all } (x, y) \in J.$$

Finally the problems (6.31)–(6.33) has a solution u satisfying $v \leq u \leq w$, and hence it is solution of (6.28)–(6.30). □

6.5 Notes and Remarks

The results of Chap. 6 are taken from Abbas and Benchohra [8], and Abbas et al. [26]. Other results may be found in [49, 55, 65, 230].

Chapter 7
Implicit Partial Hyperbolic Functional Differential Equations

7.1 Introduction

In this chapter, we shall present existence results for some classes of initial value problems for partial hyperbolic implicit differential equations with fractional order.

7.2 Darboux Problem for Implicit Differential Equations

7.2.1 Introduction

This section concerns the existence results to fractional order IVP , for the system

$$\overline{D}_0^r u(x,y) = f(x,y,u(x,y),\overline{D}_\theta^r u(x,y)); \text{if}(x,y) \in J := [0,a]\times[0,b], \quad (7.1)$$

$$\begin{cases} u(x,0) = \varphi(x);\ x \in [0,a],\\ u(0,y) = \psi(y);\ y \in [0,b],\\ \varphi(0) = \psi(0), \end{cases} \quad (7.2)$$

where $a, b > 0$, $\overline{D}_0^r$ is the mixed regularized derivative of order $r = (r_1, r_2) \in (0,1]\times(0,1]$, $f : J\times\mathbb{R}^n\times\mathbb{R}^n \to \mathbb{R}^n$ is a given function, $\varphi \in AC([0,a],\mathbb{R}^n)$ and $\psi \in AC([0,b],\mathbb{R}^n)$.

We present two results for the problems (7.1)–(7.2), the first one is based on Banach's contraction principle and the second one on the nonlinear alternative of Leray–Schauder type.

S. Abbas et al., *Topics in Fractional Differential Equations*, Developments in Mathematics 27, DOI 10.1007/978-1-4614-4036-9_7,

7.2.2 Riemann–Liouville and Caputo Partial Fractional Derivatives

For a function $h \in L^1([0,b],\mathbb{R}^n)$; $b > 0$ and $\alpha \in (0,1]$. The connection between D_0^α and ${}^cD_0^\alpha$ is given by

$$ {}^cD_0^\alpha h(t) = D_0^\alpha h(t) - \frac{t^{-\alpha}}{\Gamma(1-\alpha)} h(0^+), \text{ for almost all } t \in [0,b]. \tag{7.3}$$

For more detail see [166].

Corollary 7.1. *For a function $u \in L^1(J,\mathbb{R}^n)$ and $r = (r_1,r_2) \in (0,1] \times (0,1]$. The connection between $D_{0,x}^{r_1}u(x,y)$ and ${}^cD_{0,x}^{r_1}u(x,y)$ with respect to x is given by*

$$\left({}^cD_{0,x}^{r_1}u\right)(x,y) = \left(D_{0,x}^{r_1}u\right)(x,y) - \frac{x^{-r_1}}{\Gamma(1-r_1)}u\left(0^+,y\right). \tag{7.4}$$

Analogously, the connection between $D_{0,y}^{r_2}u(x,y)$ and ${}^cD_{0,y}^{r_2}u(x,y)$ with respect to y is given by

$$\left({}^cD_{0,y}^{r_2}u\right)(x,y) = \left(D_{0,y}^{r_2}u\right)(x,y) - \frac{y^{-r_2}}{\Gamma(1-r_2)}u\left(x,0^+\right). \tag{7.5}$$

Now, let us give the relation between D_0^r and ${}^cD_0^r$, where $r = (r_1,r_2) \in (0,1] \times (0,1]$.

Theorem 7.2. *For $u(x,y) \in AC(J,\mathbb{R}^n)$ and $r = (r_1,r_2) \in (0,1] \times (0,1]$ we have*

$$\begin{aligned}\left({}^cD_0^r u\right)(x,y) = \overline{D}_0^r u(x,y) = \left(D_0^r u\right)(x,y) - \frac{x^{-r_1}}{\Gamma(1-r_1)}\left(D_{0,y}^{r_2}u\right)(0,y)\\ -\frac{y^{-r_2}}{\Gamma(1-r_2)}\left(D_{0,x}^{r_1}u\right)(x,0) + \frac{x^{-r_1}y^{-r_2}}{\Gamma(1-r_1)\Gamma(1-r_2)}u(0,0).\end{aligned}$$

Proof. According to ([246], Lemma 1) $(\overline{D}_0^r u)(x,y) = ({}^cD_0^r u)(x,y)$.
Then

$$\left(\overline{D}_0^r u\right)(x,y) = \left(D_0^r q\right)(x,y) = D_{xy}I_0^{1-r}q(x,y),$$

$$q(x,y) = u(x,y) - \gamma(x,y),\ \gamma(x,y) = u(x,0) + u(0,y) - u(0,0),$$

$$I_0^{1-r}q(x,y) = I_0^{1-r}u(x,y) - I_0^{1-r}\gamma(x,y).$$

As $u(x,y) = \gamma(x,y) + I_0^\sigma v(x,y)$, then $q(x,y) = I_0^\sigma v(x,y)$, where $v(x,y) = D_{xy}u(x,y)$.

Hence

$$I_\theta^{1-r}q(x,y) = I_\theta^{1-r}(I_\theta^\sigma v(x,y))$$
$$= \frac{1}{\Gamma(1-r_1)\Gamma(1-r_2)}\int_0^x\int_0^y\left(\int_0^s\int_0^t (s-z)^{-r_1}(t-\tau)^{-r_2}\right.$$
$$\left.\times v(\tau,z)\mathrm{d}z\mathrm{d}\tau\right)\mathrm{d}t\mathrm{d}s \in AC(J),$$

$$I_0^{1-r}\gamma(x,y) = \frac{y^{1-r_2}}{(1-r_2)\Gamma(1-r_2)}I_{0,x}^{1-r_1}u(x,0)$$
$$+\frac{x^{1-r_1}}{(1-r_1)\Gamma(1-r_1)}I_{0,y}^{1-r_2}u(0,y)$$
$$-\frac{x^{1-r_1}y^{1-r_2}}{(1-r_1)(1-r_2)\Gamma(1-r_1)\Gamma(1-r_2)}u(0,0),$$

besides ([225], Lemma 2.1) $I_0^{1-r}\gamma(x,y) \in AC(J,\mathbb{R}^n)$.

Then

$$I_0^{1-r}u(x,y) = I_0^{1-r}q(x,y) + I_0^{1-r}\gamma(x,y) \in AC(J,\mathbb{R}^n).$$

Finally

$$\overline{D}_0^r u(x,y) = D_{xy}I_0^{1-r}q(x,y) = \left(D_0^r u\right)(x,y) - \frac{y^{-r_2}}{\Gamma(1-r_2)}\left(D_{0,x}^{r_1}u\right)(x,0)$$
$$-\frac{x^{-r_1}}{\Gamma(1-r_1)}\left(D_{0,y}^{r_2}u\right)(0,y) + \frac{x^{-r_1}y^{-r_2}}{\Gamma(1-r_1)\Gamma(1-r_2)}u(0,0).$$

□

7.2.3 *Existence of Solutions*

Let us start by defining what we mean by a solution of the problems (7.1)–(7.2).

Definition 7.3. A function $u \in C(J,\mathbb{R}^n)$ such that $\overline{D}_{0,x}^{r_1}u(x,y)$, $\overline{D}_{0,y}^{r_2}u(x,y)$, $\overline{D}_0^r u(x,y)$ are continuous for $(x,y)\in J$ and $I_0^{1-r}u(x,y) \in AC(J,\mathbb{R}^n)$ is said to be a solution of (7.1)–(7.2) if u satisfies (7.1) and conditions (7.2) on J.

For the existence of solutions for the problems (7.1)–(7.2) we need the following lemma.

Lemma 7.4 ([247]). *Let a function $f(x,y,u,z): J\times\mathbb{R}^n\times\mathbb{R}^n \to \mathbb{R}^n$ be continuous. Then problem (7.1)–(7.2) are equivalent to the problem of the solution of the equation*

$$g(x,y) = f(x,y,\mu(x,y) + I_\theta^r g(x,y), g(x,y)),$$

and if $g(x,y) \in C(J,\mathbb{R}^n)$ is the solution of this equation, then $u(x,y) = \mu(x,y) + I_\theta^r g(x,y)$, where

$$\mu(x,y) = \varphi(x) + \psi(y) - \varphi(0).$$

Further, we present conditions for the existence and uniqueness of a solution of problems (7.1)–(7.2).

Theorem 7.5. *Assume*

(7.5.1) The function $f: J\times\mathbb{R}^n\times\mathbb{R}^n \to \mathbb{R}^n$ is continuous.
(7.5.2) For any $u,v,w,z \in \mathbb{R}^n$ and $(x,y)\in J$, there exist constants $k>0$ and $0<l<1$ such that

$$\|f(x,y,u,z) - f(x,y,v,w)\| \le k\|u-v\| + l\|z-w\|.$$

If

$$\frac{ka^{r_1}b^{r_2}}{(1-l)\Gamma(1+r_1)\Gamma(1+r_2)} < 1, \tag{7.6}$$

then there exists a unique solution for IVP (7.1)–(7.2) on J.

Proof. Transform the problems (7.1)–(7.2) into a fixed-point problem. Consider the operator $N: C(J,\mathbb{R}^n) \to C(J,\mathbb{R}^n)$ defined by

$$N(u)(x,y) = \mu(x,y) + I_\theta^r g(x,y), \tag{7.7}$$

where $g \in C(J,\mathbb{R}^n)$ such that

$$g(x,y) = f(x,y,u(x,y),g(x,y)),$$

By Lemma 7.4, the problem of finding the solutions of the IVP (7.1)–(7.2) is reduced to finding the solutions of the operator equation $N(u) = u$.

Let $v,w \in C(J,\mathbb{R}^n)$. Then, for $(x,y)\in J$, we have

$$\|N(v)(x,y) - N(w)(x,y)\| \le \frac{1}{\Gamma(r_1)\Gamma(r_2)}\int_0^x\int_0^y (x-s)^{r_1-1}(y-t)^{r_2-1} \times\|g(s,t)-h(s,t)\|dtds, \tag{7.8}$$

where $g,h \in C(J,\mathbb{R}^n)$ such that

$$g(x,y) = f(x,y,v(x,y),g(x,y))$$

and

$$h(x,y) = f(x,y,w(x,y),h(x,y)).$$

By (7.5.2), we get

$$\|g(x,y) - h(x,y)\| \leq k\|v(x,y) - w(x,y)\| + l\|g(x,y) - h(x,y)\|.$$

Then

$$\begin{aligned}\|g(x,y) - h(x,y)\| &\leq \frac{k}{1-l}\|v(x,y) - w(x,y)\| \\ &\leq \frac{k}{1-l}\|v - w\|_\infty.\end{aligned}$$

Thus, (7.8) implies that

$$\begin{aligned}\|N(v) - N(w)\|_\infty &\leq \frac{1}{\Gamma(r_1)\Gamma(r_2)}\int_0^x\int_0^y (x-s)^{r_1-1}(y-t)^{r_2-1} \\ &\quad \times \frac{k}{1-l}\|v - w\|_\infty \mathrm{d}t\,\mathrm{d}s \\ &\leq \frac{ka^{r_1}b^{r_2}}{(1-l)\Gamma(1+r_1)\Gamma(1+r_2)}\|v - w\|_\infty.\end{aligned}$$

Hence

$$\|N(v) - N(w)\|_\infty \leq \frac{ka^{r_1}b^{r_2}}{(1-l)\Gamma(1+r_1)\Gamma(1+r_2)}\|v - w\|_\infty.$$

By (7.6), N is a contraction , and hence N has a unique fixed-point by Banach's contraction principle. □

Theorem 7.6. *Assume* (7.5.1) *and the following hypothesis hold:*

(7.6.3) There exist $p, q, d \in C(J, \mathbb{R}_+)$ *such that*

$$\|f(x,y,u,z)\| \leq p(x,y) + q(x,y)\|u\| + d(x,y)\|z\|$$

for $(x,y) \in J$ *and each* $u, z \in \mathbb{R}^n$. *If*

$$d^* + \frac{q^* a^{r_1} b^{r_2}}{\Gamma(1+r_1)\Gamma(1+r_2)} < 1, \tag{7.9}$$

where $d^* = \sup_{(x,y)\in J} d(x,y)$ *and* $q^* = \sup_{(x,y)\in J} q(x,y)$, *then the IVP (7.1)–(7.2) have at least one solution on* J.

Proof. Transform the problem (7.1)–(7.2) into a fixed-point problem. Consider the operator N defined in (7.7). We shall show that the operator N is continuous and compact.

Step 1: *N is continuous*

Let $\{u_n\}_{n\in\mathbb{N}}$ be a sequence such that $u_n \to u$ in $C(J,\mathbb{R}^n)$. Let $\eta > 0$ be such that $\|u_n\| \le \eta$. Then

$$\|N(u_n)(x,y) - N(u)(x,y)\| \le \frac{1}{\Gamma(r_1)\Gamma(r_2)}\int_0^x\int_0^y (x-s)^{r_1-1}(y-t)^{r_2-1} \times\|g_n(s,t)-g(s,t)\|dtds, \qquad (7.10)$$

where $g_n, g \in C(J,\mathbb{R}^n)$ such that

$$g_n(x,y) = f(x,y,u_n(x,y),g_n(x,y))$$

and

$$g(x,y) = f(x,y,u(x,y),g(x,y)).$$

Since $u_n \to u$ as $n\to\infty$ and f is a continuous function, we get

$$g_n(x,y) \to g(x,y) \text{ as } n\to\infty, \text{ for each } (x,y)\in J.$$

Hence, (7.10) gives

$$\|N(u_n)-N(u)\|_\infty \le \frac{a^{r_1}b^{r_2}}{\Gamma(1+r_1)\Gamma(1+r_2)}\|g_n-g\|_\infty \to 0 \text{ as } n\to\infty.$$

Step 2: *N maps bounded sets into bounded sets in* $C(J,\mathbb{R}^n)$ Indeed, it is enough to show that for any $\eta^* > 0$, there exists a positive constant M^* such that, for each

$$u\in B_{\eta^*} = \{u\in C(J) : \|u\|_\infty \le \eta^*\},$$

we have $\|N(u)\|_\infty \le M^*$. For $(x,y)\in J$, we have

$$\|N(u)(x,y)\| \le \|\mu(x,y)\| + \frac{1}{\Gamma(r_1)\Gamma(r_2)}\int_0^x\int_0^y (x-s)^{r_1-1}(y-t)^{r_2-1}\|g(s,t)\|dtds, \qquad (7.11)$$

where $g\in C(J,\mathbb{R}^n)$ such that

$$g(x,y) = f(x,y,u(x,y),g(x,y)).$$

By (7.6.3) we have for each $(x, y) \in J$,

$$\|g(x,y)\| \leq p(x,y) + q(x,y)\|\mu(x,y) + I_\theta^r g(x,y)\| + d(x,y)\|g(x,y)\|$$
$$\leq p^* + q^*\left(\|\mu\|_\infty + \frac{a^{r_1}b^{r_2}\|g(x,y)\|}{\Gamma(1+r_1)\Gamma(1+r_2)}\right) + d^*\|g(x,y)\|,$$

where $p^* = \sup_{(x,y)\in J} p(x,y)$.

Then, by (7.9) we have

$$\|g(x,y)\| \leq \frac{p^* + q^*\|\mu\|_\infty}{1 - d^* - \frac{q^* a^{r_1} b^{r_2}}{\Gamma(1+r_1)\Gamma(1+r_2)}} := M.$$

Thus, (7.11) implies that

$$\|N(u)\|_\infty \leq \|\mu\|_\infty + \frac{M a^{r_1} b^{r_2}}{\Gamma(1+r_1)\Gamma(1+r_2)} := M^*.$$

Step 3: *N maps bounded sets into equicontinuous sets in $C(J, \mathbb{R}^n)$.* Let (x_1, y_1), $(x_2, y_2) \in J$, $x_1 < x_2$, $y_1 < y_2$, B_{η^*} be a bounded set of $C(J, \mathbb{R}^n)$ as in Step 2, and let $u \in B_{\eta^*}$. Then

$$\|N(u)(x_2,y_2) - N(u)(x_1,y_1)\| \leq \|\mu(x_2,y_2) - \mu(x_1,y_1)\|$$
$$+\frac{1}{\Gamma(r_1)\Gamma(r_2)}\int_0^{x_1}\int_0^{y_1}\left[(x_2-s)^{r_1-1}(y_2-t)^{r_2-1}\right.$$
$$\left.-(x_1-s)^{r_1-1}(y_1-t)^{r_2-1}\right]\|g(s,t)\|dtds$$
$$+\frac{1}{\Gamma(r_1)\Gamma(r_2)}\int_{x_1}^{x_2}\int_{y_1}^{y_2}(x_2-s)^{r_1-1}(y_2-t)^{r_2-1}\|g(s,t)\|dtds$$
$$+\frac{1}{\Gamma(r_1)\Gamma(r_2)}\int_0^{x_1}\int_{y_1}^{y_2}(x_2-s)^{r_1-1}(y_2-t)^{r_2-1}\|g(s,t)\|dtds$$
$$+\frac{1}{\Gamma(r_1)\Gamma(r_2)}\int_{x_1}^{x_2}\int_0^{y_1}(x_2-s)^{r_1-1}(y_2-t)^{r_2-1}\|g(s,t)\|dtds,$$

where $g \in C(J, \mathbb{R}^n)$ such that

$$g(x,y) = f(x,y,u(x,y),g(x,y)).$$

But $\|g(x,y)\| \leq M$. Thus

$$\begin{aligned}\|N(u)(x_2,y_2)-N(u)(x_1,y_1)\| &\leq \|\mu(x_2,y_2)-\mu(x_1,y_1)\| \\ &+\frac{M}{\Gamma(1+r_1)\Gamma(1+r_2)}\left[2y_2^{r_2}(x_2-x_1)^{r_1}+2x_2^{r_1}(y_2-y_1)^{r_2}\right. \\ &\left.+x_1^{r_1}y_1^{r_2}-x_2^{r_1}y_2^{r_2}-2(x_2-x_1)^{r_1}(y_2-y_1)^{r_2}\right].\end{aligned}$$

As $x_1 \to x_2$, $y_1 \to y_2$ the right-hand side of the above inequality tends to zero. As a consequence of Steps 1–3, together with the Arzela-Ascoli theorem, we can conclude that N is continuous and completely continuous.

Step 4: *A priori bounds* We now show there exists an open set $U \subseteq C(J,\mathbb{R}^n)$ with $u \neq \lambda N(u)$, for $\lambda \in (0,1)$ and $u \in \partial U$. Let $u \in C(J,\mathbb{R}^n)$ and $u = \lambda N(u)$ for some $0<\lambda<1$. Thus for each $(x,y) \in J$, we have

$$u(x,y)=\lambda\mu(x,y)+\frac{\lambda}{\Gamma(r_1)\Gamma(r_2)}\int_0^x\int_0^y (x-s)^{r_1-1}(y-t)^{r_2-1}g(s,t)\mathrm{d}t\mathrm{d}s.$$

This implies by (7.6.3) and as in step 2 that, for each $(x,y) \in J$, we get $\|u\| \leq M^*$. Set

$$U=\{u \in C(J,\mathbb{R}^n): \|u\|_\infty < M^*+1\}.$$

By our choice of U, there is no $u \in \partial U$ such that $u = \lambda N(u)$, for $\lambda \in (0,1)$. As a consequence of Theorem 2.32, we deduce that N has a fixed-point u in $\overline{U}$ which is a solution to problems (7.1)–(7.2). □

7.2.4 An Example

As an application of our results we consider the following partial hyperbolic functional differential equations of the form:

$$\overline{D}_0^r u(x,y)=\frac{1}{5e^{x+y+2}(1+|u(x,y)|+|\overline{D}_0^r u(x,y)|)};$$

$$\text{if } (x,y) \in [0,1]\times[0,1], \tag{7.12}$$

$$u(x,0)=x,\ u(0,y)=y^2;\ x,y \in [0,1]. \tag{7.13}$$

Set

$$f(x,y,u(x,y),\overline{D}_0^r u(x,y))=\frac{1}{5e^{x+y+2}(1+|u(x,y)|+|\overline{D}_0^r u(x,y)|)};$$

$(x, y) \in [0, 1] \times [0, 1]$. Clearly, the function f is continuous. For each $u, v, \overline{u}, \overline{v} \in \mathbb{R}$ and $(x, y) \in [0, 1] \times [0, 1]$ we have

$$|f(x, y, u(x, y), v(x, y)) - f(x, y, \overline{u}(x, y), \overline{v}(x, y))| \leq \frac{1}{5e^2}(\|u - \overline{u}\| + \|v - \overline{v}\|).$$

Hence condition (7.5.2) is satisfied with $k = l = \dfrac{1}{5e^2}$. We shall show that condition (7.6) holds with $a = b = 1$. Indeed

$$\frac{ka^{r_1}b^{r_2}}{(1-l)\Gamma(1+r_1)\Gamma(1+r_2)} = \frac{1}{(5e^2-1)\Gamma(1+r_1)\Gamma(1+r_2)} < 1,$$

which is satisfied for each $(r_1, r_2) \in (0, 1] \times (0, 1]$. Consequently Theorem 7.5 implies that problems (7.12)–(7.13) have a unique solution defined on $[0, 1] \times [0, 1]$.

7.3 A Global Uniqueness Result for Implicit Differential Equations

7.3.1 Introduction

In the present section we are concerned with the global existence and uniqueness of solutions to fractional order IVP for the system

$$\overline{D}_0^r u(x, y) = f(x, y, u(x, y), \overline{D}_0^r u(x, y)); \text{ if } (x, y) \in J := [0, \infty) \times [0, \infty), \tag{7.14}$$

$$\begin{cases} u(x, 0) = \varphi(x),\ u(0, y) = \psi(y);\ x, y \in [0, \infty), \\ \varphi(0) = \psi(0), \end{cases} \tag{7.15}$$

where $\overline{D}_0^r$ is the mixed regularized derivative of order $r = (r_1, r_2) \in (0, 1] \times (0, 1]$, $f : J \times \mathbb{R}^n \times \mathbb{R}^n \to \mathbb{R}^n$ is a given function and $\varphi, \psi \in AC([0, \infty), \mathbb{R}^n)$. We make use of the nonlinear alternative of Leray-Schauder type for contraction maps on Fréchet spaces.

7.3.2 Existence of Solutions

Let us start by defining what we mean by a solution of the problems (7.14)–(7.15).

Definition 7.7. A function $u \in C(J, \mathbb{R}^n)$ such that $\overline{D}_{0,x}^{r_1} u(x, y), \overline{D}_{0,y}^{r_2} u(x, y)$, $\overline{D}_\theta^r u(x, y)$ are continuous for $(x, y) \in J$ and $I_0^{1-r} u(x, y) \in AC(J, \mathbb{R}^n)$ is said to be a solution of (7.14)–(7.15) if u satisfies (7.14) and conditions (7.15) on J.

Further, we present conditions for the existence and uniqueness of a solution of problems (7.14)–(7.15).

Theorem 7.8. *Assume*

(7.8.1) The function $f : J \times \mathbb{R}^n \times \mathbb{R}^n \to \mathbb{R}^n$ is continuous
(7.8.2) For each $p \in \mathbb{N}$, there exists constants $k_p > 0$ and $0 < l_p < 1$ such that for each $(x, y) \in J_0 := [0, p] \times [0, p]$

$$\|f(x, y, u, z) - f(x, y, v, w)\| \leq k_p\|u - v\| + l_p\|z - w\|,$$

for each $u, v, w, z \in \mathbb{R}^n$.

If

$$\frac{k_p p^{r_1+r_2}}{(1 - l_p)\Gamma(1 + r_1)\Gamma(1 + r_2)} < 1, \tag{7.16}$$

then there exists a unique solution for IVP (7.14)–(7.15) on $[0, \infty) \times [0, \infty)$.

Proof. Transform the problem (7.14)–(7.15) into a fixed-point problem. Consider the operator $N : C(J, \mathbb{R}^n) \to C(J, \mathbb{R}^n)$ defined by

$$N(u)(x, y) = \mu(x, y) + I_0^r g(x, y), \tag{7.17}$$

where $g \in C(J, \mathbb{R}^n)$ such that

$$g(x, y) = f(x, y, u(x, y), g(x, y)),$$

Let u be a possible solution of the problem $u = \lambda N(u)$ for some $0 < \lambda < 1$. This implies that for each $(x, y) \in J_0$, we have

$$u(x, y) = \lambda\mu(x, y) + \frac{\lambda}{\Gamma(r_1)\Gamma(r_2)} \int_0^x \int_0^y (x-s)^{r_1-1}(y-t)^{r_2-1}g(s, t)\mathrm{d}t\mathrm{d}s. \tag{7.18}$$

By (7.8.2) we get

$$\|g(x, y)\| \leq f^* + k_p\|u(x, y)\| + l_p\|g(x, y)\|,$$

where

$$f^* = \sup_{(x,y)\in J_0} \|f(x, y, 0, 0)\|.$$

Then

$$\|g(x, y)\| \leq \frac{f^* + k_p\|u(x, y)\|}{1 - l_p}.$$

Thus, (7.18) implies that

$$\|u(x,y)\| \leq \|\mu(x,y)\| + \frac{1}{(1-l_p)\Gamma(r_1)\Gamma(r_2)} \int_0^x \int_0^y (x-s)^{r_1-1}(y-t)^{r_2-1} \times \left(f^* + k_p\|u(s,t)\|\right) \mathrm{d}t\mathrm{d}s$$
$$\leq \|\mu\|_p + \frac{f^* p^{r_1+r_2}}{(1-l_p)\Gamma(1+r_1)\Gamma(1+r_2)} + \frac{k_p}{(1-l_p)\Gamma(r_1)\Gamma(r_2)} \int_0^x \int_0^y (x-s)^{r_1-1}(y-t)^{r_2-1}\|u(s,t)\|\mathrm{d}t\mathrm{d}s.$$

Set

$$w = \|\mu\|_p + \frac{f^* p^{r_1+r_2}}{(1-l_p)\Gamma(1+r_1)\Gamma(1+r_2)}$$

Lemma 2.43 implies that there exists a constant $\delta = \delta(r_1, r_2)$ such that

$$\|u(x,y)\| \leq w\left(1 + \frac{\delta k_p}{(1-l_p)\Gamma(r_1)\Gamma(r_2)} \int_0^x \int_0^y (x-s)^{r_1-1}(y-t)^{r_2-1}\mathrm{d}t\mathrm{d}s\right)$$
$$\leq w\left(1 + \frac{\delta k_p p^{r_1+r_2}}{(1-l_p)\Gamma(1+r_1)\Gamma(1+r_2)}\right) := M_p.$$

Since for every $(x,y) \in J_0$, $\|u\|_p \leq M_p$. Set

$$U = \{u \in C(J, \mathbb{R}^n) : \|u\|_p \leq M_p + 1 \text{ for all } p \in \mathbb{N}\}.$$

We shall show that $N : U \longrightarrow C(J_0)$ is a contraction map. Indeed, consider $v, w \in C(J_0, \mathbb{R}^n)$. Then, for $(x,y) \in J_0$, we have

$$\|N(v)(x,y) - N(w)(x,y)\| \leq \frac{1}{\Gamma(r_1)\Gamma(r_2)} \int_0^x \int_0^y (x-s)^{r_1-1}(y-t)^{r_2-1} \times \|g(s,t) - h(s,t)\|\mathrm{d}t\mathrm{d}s, \tag{7.19}$$

where $g, h \in C(J_0, \mathbb{R}^n)$ such that

$$g(x,y) = f(x,y,v(x,y),g(x,y))$$

and

$$h(x,y) = f(x,y,w(x,y),h(x,y)).$$

By (7.8.2), we get

$$\|g(x,y)-h(x,y)\| \le k_p\|v(x,y)-w(x,y)\| + l_p\|g(x,y)-h(x,y)\|.$$

Then

$$\|g(x,y)-h(x,y)\| \le \frac{k_p}{1-l_p}\|v-w\|_p.$$

Thus, (7.19) implies that

$$\begin{aligned}\|N(v)-N(w)\|_p &\le \frac{k_p}{(1-l_p)\Gamma(r_1)\Gamma(r_2)}\int_0^x\int_0^y (x-s)^{r_1-1}(y-t)^{r_2-1}\|v-w\|_p dt ds\\ &\le \frac{k_p p^{r_1+r_2}}{(1-l_p)\Gamma(1+r_1)\Gamma(1+r_2)}\|v-w\|_p.\end{aligned}$$

Hence

$$\|N(v)-N(w)\|_p \le \frac{k_p p^{r_1+r_2}}{(1-l_p)\Gamma(1+r_1)\Gamma(1+r_2)}\|v-w\|_p.$$

By (7.16), $N : U \longrightarrow C(J_0, \mathbb{R}^n)$ is a contraction . By our choice of U, there is no $u \in \partial_n U^n$ such that $u = \lambda N(u)$, for $\lambda \in (0,1)$. As a consequence of Theorem 3.48, we deduce that N has a unique fixed-point u in U which is a solution to problem (7.14)–(7.15). □

7.3.3 An Example

As an application of our results we consider the following partial hyperbolic functional differential equations of the form:

$$\overline{D}_0^r u(x,y) = \frac{c_p}{7e^{x+y+2}(1+c_p|u(x,y)|+|\overline{D}_0^r u(x,y)|)}; \text{ if } (x,y)\in[0,\infty)\times[0,\infty), \tag{7.20}$$

$$u(x,0)=x,\ u(0,y)=y^2;\ x,y\in[0,\infty), \tag{7.21}$$

where

$$c_p = \frac{\Gamma(1+r_1)\Gamma(1+r_2)}{p^{r_1+r_2}};\ p\in\mathbb{N}^*.$$

Set

$$f(x,y,u(x,y),\overline{D}_0^r u(x,y)) = \frac{c_p}{7e^{x+y+2}(1+c_p|u(x,y)|+|\overline{D}_0^r u(x,y)|)};$$

$(x, y) \in [0, 1] \times [0, 1]$. Clearly, the function f is continuous. For each $p \in \mathbb{N}^*$ and $(x, y) \in J_0$ we have

$$|f(x, y, u(x, y), v(x, y)) - f(x, y, \overline{u}(x, y), \overline{v}(x, y))| \leq \frac{1}{7e^2}(\|u - \overline{u}\| + c_p\|v - \overline{v}\|),$$

for each $u, v, \overline{u}, \overline{v} \in \mathbb{R}$. Hence condition (7.8.2) is satisfied with $l_p = \frac{1}{7e^2}$ and $k_p = \frac{c_p}{7e^2}$. We shall show that condition (7.16) holds for all $p \in \mathbb{N}^*$. Indeed

$$\frac{k_p p^{r_1+r_2}}{(1 - l_p)\Gamma(1 + r_1)\Gamma(1 + r_2)} = \frac{c_p p^{r_1+r_2}}{(7e^2 - 1)\Gamma(1 + r_1)\Gamma(1 + r_2)} = \frac{1}{7e^2 - 1} < 1.$$

Consequently Theorem 7.8 implies that problems (7.20)–(7.21) have a unique solution defined on $[0, \infty) \times [0, \infty)$.

7.4 Functional Implicit Hyperbolic Differential Equations with Delay

7.4.1 Introduction

In the present section, we investigate the existence and uniqueness of solutions to fractional order IVP for the system

$$\overline{D}^r_\theta u(x, y) = f(x, y, u_{(x,y)}, \overline{D}^r_\theta u(x, y)); \text{ if } (x, y) \in J := [0, a] \times [0, b], \quad (7.22)$$

$$u(x, y) = \phi(x, y); \text{ if } (x, y) \in \tilde{J} := [-\alpha, a] \times [-\beta, b] \backslash (0, a] \times (0, b], \quad (7.23)$$

$$\begin{cases} u(x, 0) = \varphi(x); \ x \in [0, a], \\ u(0, y) = \psi(y); \ y \in [0, b], \end{cases} \quad (7.24)$$

where $a, b, \alpha, \beta > 0$, $f : J \times C \times \mathbb{R}^n \to \mathbb{R}^n$ is a given function, $\phi \in C(\tilde{J})$, $\varphi : [0, a] \to \mathbb{R}^n$, $\psi : [0, b] \to \mathbb{R}^n$ are given absolutely continuous functions with $\varphi(x) = \phi(x, 0)$, $\psi(y) = \phi(0, y)$ for each $x \in [0, a]$, $y \in [0, b]$, and $C := C([-\alpha, 0] \times [-\beta, 0])$ is the space of continuous functions on $[-\alpha, 0] \times [-\beta, 0]$.

Next, we consider the following fractional order IVP for the system:

$$\overline{D}^r_\theta u(x, y) = f(x, y, u_{(x,y)}, \overline{D}^r_\theta u(x, y)); \text{ if } (x, y) \in J := [0, a] \times [0, b], \quad (7.25)$$

$$u(x, y) = \phi(x, y), \text{ if } (x, y) \in \tilde{J}' := (-\infty, a] \times (-\infty, b] \backslash (0, a] \times (0, b], \quad (7.26)$$

$$\begin{cases} u(x, 0) = \varphi(x); \ x \in [0, a], \\ u(0, y) = \psi(y); \ y \in [0, b], \end{cases} \quad (7.27)$$

where φ, ψ are as in problems (7.22)–(7.24) and $\phi \in C(\tilde{J}')$, $f : J \times \mathcal{B} \times \mathbb{R}^n \to \mathbb{R}^n$ is a given continuous function, and $\mathcal{B}$ is a phase space.

The third result of this section deals with the existence and uniqueness of solutions to fractional order IVP for the system

$$\overline{D}_\theta^r u(x,y) = f(x,y,u_{(\rho_1(x,y,u_{(x,y)}),\rho_2(x,y,u_{(x,y)}))}, \overline{D}_\theta^r u(x,y)); \text{ if } (x,y) \in J, \quad (7.28)$$

$$u(x,y) = \phi(x,y); \text{ if } (x,y) \in \tilde{J}, \quad (7.29)$$

$$u(x,0) = \varphi(x),\ u(0,y) = \psi(y);\ x \in [0,a],\ y \in [0,b], \quad (7.30)$$

where $f : J \times \mathcal{C} \times \mathbb{R}^n \to \mathbb{R}^n$, $\rho_1 : J \times \mathcal{C} \to [-\alpha, a]$, $\rho_2 : J \times \mathcal{C} \to [-\beta, b]$ are given functions.

Finally we consider the following initial value problem for partial functional differential equations:

$$\overline{D}_\theta^r u(x,y) = f(x,y,u_{(\rho_1(x,y,u_{(x,y)}),\rho_2(x,y,u_{(x,y)}))}, \overline{D}_\theta^r u(x,y)); \text{ if } (x,y) \in J, \quad (7.31)$$

$$u(x,y) = \phi(x,y); \text{ if } (x,y) \in \tilde{J}', \quad (7.32)$$

$$u(x,0) = \varphi(x),\ u(0,y) = \psi(y);\ x \in [0,a],\ y \in [0,b], \quad (7.33)$$

where $f : J \times \mathcal{B} \times \mathbb{R}^n \to \mathbb{R}^n$, $\rho_1 : J \times \mathcal{B} \to (-\infty, a]$, $\rho_2 : J \times \mathcal{B} \to (-\infty, b]$ are given functions. We present two results for each of our problems, the first one is based on Banach's contraction principle and the second one on the nonlinear alternative of Leray-Schauder type.

7.4.2 *Existence Results with Finite Delay*

Let us start by defining what we mean by a solution of the problems (7.22)–(7.24).

Definition 7.9. A function $u \in C_{(\alpha,\beta)} := C([-\alpha,a] \times [-\beta,b])$ such that $u(x,y)$, $\overline{D}_{0,x}^{r_1} u(x,y)$, $\overline{D}_{0,y}^{r_2} u(x,y)$, $\overline{D}_\theta^r u(x,y)$ are continuous for $(x,y) \in J$ and $I_\theta^{1-r} u(x,y) \in AC(J)$ is said to be a solution of (7.22)–(7.24) if u satisfies (7.22) and (7.24) on J and the condition (7.23) on $\tilde{J}$.

Further, we present conditions for the existence and uniqueness of a solution of problems (7.22)–(7.24).

Theorem 7.10. *Assume*

(7.10.1) *The function* $f : J \times \mathcal{C} \times \mathbb{R}^n \to \mathbb{R}^n$ *is continuous.*

(7.10.2) *For any* $u, v \in \mathcal{C}$, $w, z \in \mathbb{R}^n$ *and* $(x,y) \in J$, *there exist constants* $\ell > 0$ *and* $0 < l < 1$ *such that*

$$\|f(x,y,u,z) - f(x,y,v,w)\| \le \ell \|u - v\|_{\mathcal{C}} + l\|z - w\|.$$

If

$$\frac{\ell a^{r_1} b^{r_2}}{(1-l)\Gamma(1+r_1)\Gamma(1+r_2)} < 1, \tag{7.34}$$

then there exists a unique solution for IVP (7.22)–(7.24) on $[-\alpha, a] \times [-\beta, b]$.

Proof. Transform the problems (7.22)–(7.23) into a fixed-point problem. Consider the operator $N : C_{(\alpha,\beta)} \to C_{(\alpha,\beta)}$ defined by

$$N(u)(x,y) = \begin{cases} \Phi(x,y); & (x,y) \in \tilde{J}, \\ \mu(x,y) + I_\theta^r g(x,y); & (x,y) \in J, \end{cases} \tag{7.35}$$

where $g \in C(J)$ such that

$$g(x,y) = f(x,y,u_{(x,y)}, g(x,y)).$$

Let $v, w \in C_{(\alpha,\beta)}$. Then, for $(x,y) \in J$, we have

$$\begin{aligned} \|N(v)(x,y) - N(w)(x,y)\| \le \frac{1}{\Gamma(r_1)\Gamma(r_2)} \int_0^x \int_0^y (x-s)^{r_1-1}(y-t)^{r_2-1} \\ \times \|g(s,t) - h(s,t)\| \mathrm{d}t\mathrm{d}s, \end{aligned} \tag{7.36}$$

where $g, h \in C(J)$ such that

$$g(x,y) = f(x,y,v_{(x,y)}, g(x,y))$$

and

$$h(x,y) = f(x,y,w_{(x,y)}, h(x,y)).$$

By (7.10.2), we get

$$\|g(x,y) - h(x,y)\| \le \ell \|v_{(x,y)} - w_{(x,y)}\|_C + l\|g(x,y) - h(x,y)\|.$$

Then

$$\begin{aligned} \|g(x,y) - h(x,y)\| &\le \frac{\ell}{1-l}\|v_{(x,y)} - w_{(x,y)}\|_C \\ &\le \frac{\ell}{1-l}\|v - w\|_\infty. \end{aligned}$$

Thus, (7.36) implies that

$$\begin{aligned} \|N(v) - N(w)\|_\infty &\le \frac{1}{\Gamma(r_1)\Gamma(r_2)} \int_0^x \int_0^y (x-s)^{r_1-1}(y-t)^{r_2-1} \frac{\ell}{1-l}\|v-w\|_\infty \mathrm{d}t\mathrm{d}s \\ &\le \frac{\ell a^{r_1} b^{r_2}}{(1-l)\Gamma(1+r_1)\Gamma(1+r_2)}\|v - w\|_\infty. \end{aligned}$$

Hence

$$\|N(v) - N(w)\|_\infty \le \frac{\ell a^{r_1} b^{r_2}}{(1-l)\Gamma(1+r_1)\Gamma(1+r_2)}\|v - w\|_\infty.$$

By (7.34), N is a contraction, and hence N has a unique fixed-point by Banach's contraction principle. □

Theorem 7.11. *Assume* (7.10.1) *and the following hypothesis hold:*

(7.11.1) There exist $p, q, d \in C(J, \mathbb{R}_+)$ *such that*

$$\|f(x, y, u, z)\| \le p(x, y) + q(x, y)\|u\|_C + d(x, y)\|z\|$$

for $(x, y) \in J$ *and each* $u \in C$, $z \in \mathbb{R}^n$.

If

$$d^* + \frac{q^* a^{r_1} b^{r_2}}{\Gamma(1+r_1)\Gamma(1+r_2)} < 1, \tag{7.37}$$

where $d^* = \sup_{(x,y)\in J} d(x, y)$ *and* $q^* = \sup_{(x,y)\in J} q(x, y)$,
then the IVP (7.22)–(7.24) have at least one solution on $[-\alpha, a] \times [-\beta, b]$.

Proof. Transform the problems (7.22)–(7.24) into a fixed-point problem. Consider the operator N defined in (7.35). We shall show that the operator N is continuous and compact .

Step 1: *N is continuous* Let $\{u_n\}_{n\in\mathbb{N}}$ be a sequence such that $u_n \to u$ in $C_{(\alpha,\beta)}$. Let $\eta > 0$ be such that $\|u_n\| \le \eta$. Then

$$\|N(u_n)(x, y) - N(u)(x, y)\| \le \frac{1}{\Gamma(r_1)\Gamma(r_2)} \int_0^x \int_0^y (x-s)^{r_1-1}(y-t)^{r_2-1} \times \|g_n(s, t) - g(s, t)\| dt ds, \tag{7.38}$$

where $g_n, g \in C(J)$ such that

$$g_n(x, y) = f(x, y, u_{n(x,y)}, g_n(x, y))$$

and

$$g(x, y) = f(x, y, u_{(x,y)}, g(x, y)).$$

Since $u_n \to u$ as $n \to \infty$ and f is a continuous function, we get

$$g_n(x, y) \to g(x, y) \text{ as } n \to \infty, \text{ for each } (x, y) \in J.$$

Hence, (7.38) gives

$$\|N(u_n) - N(u)\|_\infty \leq \frac{a^{r_1} b^{r_2}}{\Gamma(1+r_1)\Gamma(1+r_2)}\|g_n - g\|_\infty \to 0 \text{ as } n \to \infty.$$

Step 2: *N maps bounded sets into bounded sets in $C_{(\alpha,\beta)}$.* Indeed, it is enough show that for any $\eta^* > 0$, there exists a positive constant M^* such that, for each $u \in B_{\eta^*} = \{u \in C_{(\alpha,\beta)} : \|u\|_\infty \leq \eta^*\}$, we have $\|N(u)\|_\infty \leq M^*$. For $(x,y) \in J$, we have

$$\|N(u)(x,y)\| \leq \|\mu(x,y)\| + \frac{1}{\Gamma(r_1)\Gamma(r_2)} \int_0^x \int_0^y (x-s)^{r_1-1}(y-t)^{r_2-1}\|g(s,t)\|\mathrm{d}t\mathrm{d}s, \quad (7.39)$$

where $g \in C(J)$ such that

$$g(x,y) = f(x,y,u_{(x,y)}, g(x,y)).$$

By (7.11.1) we have for each $(x,y) \in J$,

$$\begin{aligned}\|g(x,y)\| &\leq p(x,y) + q(x,y)\|\mu(x,y) + I_\theta^r g(x,y)\| + d(x,y)\|g(x,y)\| \\ &\leq p^* + q^*\left(\|\mu\|_\infty + \frac{a^{r_1} b^{r_2}\|g(x,y)\|}{\Gamma(1+r_1)\Gamma(1+r_2)}\right) + d^*\|g(x,y)\|,\end{aligned}$$

where $p^* = \sup_{(x,y)\in J} p(x,y)$. Then, by (7.37) we have

$$\|g(x,y)\| \leq \frac{p^* + q^*\|\mu\|_\infty}{1 - d^* - \frac{q^* a^{r_1} b^{r_2}}{\Gamma(1+r_1)\Gamma(1+r_2)}} := M.$$

Thus, (7.39) implies that

$$\|N(u)\|_\infty \leq \|\mu\|_\infty + \frac{M a^{r_1} b^{r_2}}{\Gamma(1+r_1)\Gamma(1+r_2)} := M^*.$$

Step 3: *N maps bounded sets into equicontinuous sets in $C_{(\alpha,\beta)}$.* Let (x_1, y_1), $(x_2, y_2) \in J$, $x_1 < x_2$, $y_1 < y_2$, B_{η^*} be a bounded set of $C_{(\alpha,\beta)}$ as in Step 2, and let $u \in B_{\eta^*}$. Then

$$\|N(u)(x_2,y_2) - N(u)(x_1,y_1)\| \leq \|\mu(x_2,y_2) - \mu(x_1,y_1)\| + \frac{1}{\Gamma(r_1)\Gamma(r_2)} \int_0^{x_1} \int_0^{y_1} [(x_2-s)^{r_1-1}(y_2-t)^{r_2-1}$$

$$-(x_1-s)^{r_1-1}(y_1-t)^{r_2-1}]\|g(s,t)\|dtds$$

$$+\frac{1}{\Gamma(r_1)\Gamma(r_2)}\int_{x_1}^{x_2}\int_{y_1}^{y_2}(x_2-s)^{r_1-1}(y_2-t)^{r_2-1}\|g(s,t)\|dtds$$

$$+\frac{1}{\Gamma(r_1)\Gamma(r_2)}\int_{0}^{x_1}\int_{y_1}^{y_2}(x_2-s)^{r_1-1}(y_2-t)^{r_2-1}\|g(s,t)\|dtds$$

$$+\frac{1}{\Gamma(r_1)\Gamma(r_2)}\int_{x_1}^{x_2}\int_{0}^{y_1}(x_2-s)^{r_1-1}(y_2-t)^{r_2-1}\|g(s,t)\|dtds,$$

where $g \in C(J)$ such that

$$g(x,y) = f(x,y,u(x,y),g(x,y)).$$

But $\|g(x,y)\| \le M$. Thus

$$\|N(u)(x_2,y_2)-N(u)(x_1,y_1)\| \le \|\mu(x_2,y_2)-\mu(x_1,y_1)\|$$
$$+\frac{M}{\Gamma(1+r_1)\Gamma(1+r_2)}\left[2y_2^{r_2}(x_2-x_1)^{r_1}+2x_2^{r_1}(y_2-y_1)^{r_2}\right.$$
$$\left.+x_1^{r_1}y_1^{r_2}-x_2^{r_1}y_2^{r_2}-2(x_2-x_1)^{r_1}(y_2-y_1)^{r_2}\right].$$

As $x_1 \to x_2$, $y_1 \to y_2$ the right-hand side of the above inequality tends to zero. As a consequence of Steps 1–3, together with the Arzela-Ascoli theorem, we can conclude that N is continuous and completely continuous.

Step 4: *A priori bounds.* We now show there exists an open set $U \subseteq C_{(\alpha,\beta)}$ with $u \neq \lambda N(u)$, for $\lambda \in (0,1)$ and $u \in \partial U$. Let $u \in C_{(\alpha,\beta)}$ and $u = \lambda N(u)$ for some $0 < \lambda < 1$. Thus, for each $(x,y) \in J$, we have

$$u(x,y) = \lambda\mu(x,y) + \frac{\lambda}{\Gamma(r_1)\Gamma(r_2)}\int_0^x\int_0^y (x-s)^{r_1-1}(y-t)^{r_2-1}g(s,t)dtds.$$

This implies by (7.11.1) and as in step 2 that $\|u\| \le M^*$. Hence, for each $(x,y) \in [-\alpha,a]\times[-\beta,b]$, we have

$$\|u\|_\infty \le \max(\|\phi\|_C, M^*) := R.$$

Set

$$U = \{u \in C_{(\alpha,\beta)} : \|u\|_\infty < R+1\}.$$

By our choice of U, there is no $u \in \partial U$ such that $u = \lambda N(u)$, for $\lambda \in (0,1)$. As a consequence of Theorem 2.32, we deduce that N has a fixed-point u in $\overline{U}$ which is a solution to problem (7.22)–(7.24). □

7.4.3 Existence Results for Infinite Delay

Let the space

$$\Omega := \{u : (-\infty, a] \times (-\infty, b] \to \mathbb{R}^n : u_{(x,y)} \in \mathcal{B} \text{ for } (x,y) \in E \text{ and } u|_J \in C(J, \mathbb{R}^n)\}.$$

Definition 7.12. A function $u \in \Omega$ is said to be a solution of (7.25)–(7.27) if u satisfies (7.25) and (7.27) on J and the condition (7.26) on $\tilde{J}'$.

Our first existence result for the IVP (3.4)–(3.6) is based on the Banach contraction principle.

Theorem 7.13. *Assume that the following hypotheses hold:*

(7.13.1) There exist constants $\ell' > 0$ and $0 < l' < 1$such that

$$\|f(x,y,u,z) - f(x,y,v,w)\| \leq \ell' \|u - v\|_{\mathcal{B}} + l' \|z - w\|,$$

for any $u,\ v \in \mathcal{B},\ z, w \in \mathbb{R}^n$, and $(x,y) \in J$.

If

$$\frac{K\ell' a^{r_1} b^{r_2}}{(1 - l')\Gamma(1 + r_1)\Gamma(1 + r_2)} < 1, \tag{7.40}$$

then there exists a unique solution for IVP (7.25)–(7.27) on $(-\infty, a] \times (-\infty, b]$.

Proof. Transform the problems (7.25)–(7.27) into a fixed-point problem. Consider the operator $N : \Omega \to \Omega$ defined by

$$N(u)(x,y) = \begin{cases} \phi(x,y); & (x,y) \in \tilde{J}', \\ \mu(x,y) + I_\theta^r g(x,y); & (x,y) \in J, \end{cases} \tag{7.41}$$

where

$$g(x,y) = f(x,y,u_{(x,y)}, g(x,y)); \ (x,y) \in J.$$

Let $v(.,.) : (-\infty, a] \times (-\infty, b] \to \mathbb{R}^n$ be a function defined by

$$v(x,y) = \begin{cases} \phi(x,y), \ (x,y) \in \tilde{J}', \\ \mu(x,y), \ (x,y) \in J. \end{cases}$$

Then $v_{(x,y)} = \phi$ for all $(x, y) \in E$. For each $w \in C(J)$ with $w(x, y) = 0$ for each $(x, y) \in E$ we denote by $\overline{w}$ the function defined by

$$\overline{w}(x, y) = \begin{cases} 0, & (x, y) \in \tilde{J}', \\ w(x, y) & (x, y) \in J. \end{cases}$$

If $u(.,.)$ satisfies the integral equation

$$u(x, y) = \mu(x, y) + \frac{1}{\Gamma(r_1)\Gamma(r_2)} \int_0^x \int_0^y (x-s)^{r_1-1}(y-t)^{r_2-1} g(s,t) dt ds,$$

we can decompose $u(.,.)$ as $u(x, y) = \overline{w}(x, y) + v(x, y)$; $(x, y) \in J$, which implies $u_{(x,y)} = \overline{w}_{(x,y)} + v_{(x,y)}$, for every $(x, y) \in J$, and the function $w(.,.)$ satisfies

$$w(x, y) = \frac{1}{\Gamma(r_1)\Gamma(r_2)} \int_0^x \int_0^y (x-s)^{r_1-1}(y-t)^{r_2-1} g(s,t) dt ds,$$

where

$$g(x, y) = f(x, y, \overline{w}_{(x,y)} + v_{(x,y)}, g(x, y)); \ (x, y) \in J.$$

Set

$$C_0 = \{w \in C(J) : \ w(x, y) = 0 \ \text{for} \ (x, y) \in E\},$$

and let $\|.\|_{(a,b)}$ be the seminorm in C_0 defined by

$$\|w\|_{(a,b)} = \sup_{(x,y)\in E} \|w_{(x,y)}\|_{\mathcal{B}} + \sup_{(x,y)\in J} \|w(x, y)\| = \sup_{(x,y)\in J} \|w(x, y)\|, \ w \in C_0.$$

C_0 is a Banach space with norm $\|.\|_{(a,b)}$. Let the operator $P : C_0 \to C_0$ be defined by

$$(Pw)(x, y) = \frac{1}{\Gamma(r_1)\Gamma(r_2)} \int_0^x \int_0^y (x-s)^{r_1-1}(y-t)^{r_2-1} f(s, t, \overline{w}_{(s,t)} + v_{(s,t)}, g(s,t)) dt ds, \tag{7.42}$$

where

$$g(x, y) = f(x, y, \overline{w}_{(x,y)} + v_{(x,y)}, g(x, y)); \ (x, y) \in J.$$

The operator N has a fixed-point is equivalent to P has a fixed-point, and so we turn to proving that P has a fixed-point. We shall show that $P : C_0 \to C_0$ is a contraction map. Indeed, consider $w, w^* \in C_0$. Then we have for each $(x, y) \in J$

$$\|P(w)(x, y) - P(w^*)(x, y)\| \leq \frac{1}{\Gamma(r_1)\Gamma(r_2)} \int_0^x \int_0^y (x-s)^{r_1-1}(y-t)^{r_2-1} \times \|g(s,t) - g^*(s,t)\| dt ds,$$

where

$$g^*(x,y) = f(x,y,\overline{w^*}_{(x,y)} + v_{(x,y)}).$$

But, for each $(x,y) \in J$, we have

$$\|g(x,y) - g^*(x,y)\| \leq \frac{\ell'}{1-l'}\|\overline{w}_{(x,y)} - \overline{w^*}_{(x,y)}\|_{\mathcal{B}}.$$

Thus, we obtain that, for each $(x,y) \in J$

$$\|P(w)(x,y) - P(w^*)(x,y)\| \leq \frac{\ell'}{(1-l')\Gamma(r_1)\Gamma(r_2)} \int_0^x \int_0^y (x-s)^{r_1-1}(y-t)^{r_2-1}$$

$$\times\|\overline{w}_{(s,t)} - \overline{w^*}_{(s,t)}\|_{\mathcal{B}} dt ds$$

$$\leq \frac{\ell'}{(1-l')\Gamma(r_1)\Gamma(r_2)} \int_0^x \int_0^y (x-s)^{r_1-1}(y-t)^{r_2-1}$$

$$\times k \sup_{(s,t)\in[0,x]\times[0,y]} \|\overline{w}(s,t) - \overline{w^*}(s,t)\|_{\mathcal{B}} dt ds$$

$$\leq \frac{K\ell'}{(1-l')\Gamma(r_1)\Gamma(r_2)} \int_0^x \int_0^y (x-s)^{r_1-1}(y-t)^{r_2-1} dt ds \|\overline{w} - \overline{w^*}\|_{(a,b)}.$$

Therefore

$$\|P(w) - P(w^*)\|_{(a,b)} \leq \frac{k\ell' a^{r_1} b^{r_2}}{\Gamma(1+r_1+)\Gamma(1+r_2)}\|\overline{w} - \overline{w^*}\|_{(a,b)}.$$

and hence, by (7.40) P is a contraction. Therefore, P has a unique fixed-point by Banach's contraction principle. □

Theorem 7.14. *Assume that the following hypotheses hold:*

(7.14.1) There exist $p', q', d' \in C(J, \mathbb{R}_+)$ such that

$$\|f(x,y,u,v)\| \leq p'(x,y) + q'(x,y)\|u\|_{\mathcal{B}} + d'(x,y)\|v\|,$$

for each $(x,y) \in J$, $u \in \mathcal{B}$, and $v \in \mathbb{R}^n$.

If

$$d^{**} + \frac{q^{**} a^{r_1} b^{r_2}}{\Gamma(1+r_1)\Gamma(1+r_2)} < 1, \tag{7.43}$$

where $d^{**} = \sup_{(x,y)\in J} d'(x,y)$ and $q^{**} = \sup_{(x,y)\in J} q'(x,y)$, then the IVP (7.25)–(7.27) have at least one solution on $(-\infty, a] \times (-\infty, b]$.

Proof. Let $P : C_0 \to C_0$ be defined as in (7.42). We shall show that the operator P is continuous and completely continuous. As in Theorem 7.11, we can prove that P is continuous and completely continuous. We now show there exists an open set $U' \subseteq C_0$ with $w \neq \lambda P(w)$, for $\lambda \in (0,1)$ and $w \in \partial U'$. Let $w \in C_0$ and $w = \lambda P(w)$ for some $0 < \lambda < 1$. Thus for each $(x,y) \in J$,

$$w(x,y) = \frac{\lambda}{\Gamma(r_1)\Gamma(r_2)} \int_0^x \int_0^y (x-s)^{r_1-1}(y-t)^{r_2-1} g(s,t)\mathrm{d}t\mathrm{d}s.$$

where

$$g(x,y) = f(x,y,u_{(x,y)}, g(x,y)); \ (x,y) \in J.$$

By (7.14.1) and (7.43) we get for each $(x,y) \in J$,

$$\|g(x,y)\| \leq \frac{p^{**} + q^{**}\|\mu\|_\infty}{1 - d^{**} - \frac{q^{**}a^{r_1}b^{r_2}}{\Gamma(1+r_1)\Gamma(1+r_2)}} := M',$$

where $p^{**} = \sup\limits_{(x,y)\in J} p'(x,y)$. This implies that, for each $(x,y) \in J$, we have

$$\|w(x,y)\| \leq \frac{M' a^{r_1} b^{r_2}}{\Gamma(1+r_1)\Gamma(1+r_2)} := \widetilde{M}.$$

Hence

$$\|w\|_{(a,b)} \leq \widetilde{M}.$$

Set

$$U' = \{w \in C_0 : \|w\|_{(a,b)} < \widetilde{M} + 1\}.$$

$P : \overline{U'} \to C_0$ is continuous and completely continuous. By our choice of U', there is no $w \in \partial U'$ such that $w = \lambda P(w)$, for $\lambda \in (0,1)$. As a consequence of the nonlinear alternative of Leray-Schauder type, we deduce that N has a fixed-point which is a solution to problems (7.25)–(7.27). □

7.4.4 Existence Results with State-Dependent Delay

7.4.4.1 Finite Delay Case

Definition 7.15. A function $u \in C_{(\alpha,\beta)}$ such that $u(x,y)$, $\overline{D}_{0,x}^{r_1} u(x,y)$, $\overline{D}_{0,y}^{r_2} u(x,y)$, $\overline{D}_{\theta}^{r} u(x,y)$ are continuous for $(x,y) \in J$ and $I_\theta^{1-r} u(x,y) \in AC(J)$ is said to be a solution of (7.28)–(7.30) if u satisfies equations (7.28), (7.30) on J and the condition (7.29) on $\tilde{J}$.

Set $\mathcal{R} := \mathcal{R}_{(\rho_1^-,\rho_2^-)}$

$$= \{(\rho_1(s,t,u),\rho_2(s,t,u)) : (s,t,u) \in J \times \mathcal{C},\ \rho_i(s,t,u) \le 0;\ i = 1,2\}.$$

We always assume that $\rho_1 : J \times \mathcal{C} \to [-\alpha, a]$, $\rho_2 : J \times \mathcal{C} \to [-\beta, b]$ are continuous and the function $(s,t) \longmapsto u_{(s,t)}$ is continuous from $\mathcal{R}$ into $\mathcal{C}$.

Further, we present conditions for the existence and uniqueness of a solution of problems (7.28)–(7.30).

Theorem 7.16. *Assume that the following hypotheses hold:*

(7.16.1) The $f : J \times \mathcal{C} \times \mathbb{R}^n \to \mathbb{R}^n$ is continuous.
(7.16.2) There exist $\ell_* > 0$, $0 < l_* < 1$ such that

$$\|f(x,y,u,v) - f(x,y,\overline{u},\overline{v})\| \le \ell_*\|u-\overline{u}\|_{\mathcal{C}} + l_*\|v-\overline{v}\|;$$

for any u, $\overline{u} \in \mathcal{C}$, v, $\overline{v} \in \mathbb{R}^n$ and $(x,y) \in J$.

If

$$\frac{\ell_*\, a^{r_1} b^{r_2}}{(1-l_*)\Gamma(1+r_1)\Gamma(1+r_2)} < 1, \tag{7.44}$$

then there exists a unique solution for IVP (7.28)–(7.30) on $[-\alpha, a] \times [-\beta, b]$.

Proof. Transform the problems (7.28)–(7.30) into a fixed-point problem. Consider the operator $N : C_{(\alpha,\beta)} \to C_{(\alpha,\beta)}$ defined by

$$N(x,y) = \begin{cases} \phi(x,y); & (x,y) \in \tilde{J}, \\ \mu(x,y) + \frac{1}{\Gamma(r_1)\Gamma(r_2)} \int_0^x\int_0^y (x-s)^{r_1-1}(y-t)^{r_2-1} & \\ \times g(s,t)\mathrm{d}t\mathrm{d}s; & (x,y) \in J, \end{cases} \tag{7.45}$$

where $g \in C(J)$ such that

$$g(x,y) = f(x,y,u_{(\rho_1(x,y,u_{(x,y)}),\rho_2(x,y,u_{(x,y)})}, g(x,y)).$$

Let $u, v \in C_{(\alpha,\beta)}$. Then, for $(x,y) \in [-\alpha, a] \times [-\beta, b]$,

$$\|N(u)(x,y) - N(v)(x,y)\| \le \frac{1}{\Gamma(r_1)\Gamma(r_2)} \int_0^x\int_0^y (x-s)^{r_1-1}(y-t)^{r_2-1} \times\|g(s,t)-h(s,t)\|\mathrm{d}t\mathrm{d}s,$$

where, $g, h \in C(J)$ such that

$$g(x,y) = f(x,y,u_{(\rho_1(x,y,u_{(x,y)}),\rho_2(x,y,u_{(x,y)})}, g(x,y))$$

and

$$h(x,y) = f(x,y,v_{(\rho_1(x,y,v_{(x,y)}),\rho_2(x,y,v_{(x,y)})}, h(x,y)).$$

Since

$$\|g-h\|_\infty \leq \frac{\ell_*}{1-l_*}\|v-w\|_{C_{(a,b)}},$$

we obtain that

$$\|N(u)-N(v)\|_{C_{(a,b)}} \leq \frac{\ell_* a^{r_1} b^{r_2}}{(1-l_*)\Gamma(1+r_1)\Gamma(1+r_2)}\|v-w\|_{C_{(a,b)}}.$$

Consequently, by (7.44), N is a contraction, and hence N has a unique fixed-point by Banach's contraction principle. □

Theorem 7.17. *Assume* f (7.16.1) *and the following hypothesis holds:*

(7.17.1) There exist $p,q,d \in C(J,\mathbb{R}_+)$ *such that*

$$\|f(x,y,u,v)\| \leq p(x,y) + q(x,y)\|u\|_C + d(x,y)\|v\|;$$

for each $u \in C$, $v \in \mathbb{R}^n$ *and* $(x,y) \in J$.

If

$$d^* + \frac{q^* a^{r_1} b^{r_2}}{\Gamma(1+r_1)\Gamma(1+r_2)} < 1, \tag{7.46}$$

where $d^* = \sup_{(x,y)\in J} d(x,y)$ *and* $q^* = \sup_{(x,y)\in J} q(x,y)$. *Then the IVP (7.28)–(7.30) has at least one solution on* $[-\alpha,a]\times[-\beta,b]$.

Proof. Consider the operator N defined in (7.45). We can show that the operator N is continuous and completely continuous.

A priori bounds. We now show that there exists an open set $U \subseteq C_{(\alpha,\beta)}$ with $u \neq \lambda N(u)$, for $\lambda \in (0,1)$ and $u \in \partial U$. Let $u \in C_{(\alpha,\beta)}$ and $u = \lambda N(u)$ for some $0<\lambda<1$. Thus for each $(x,y)\in J$,

$$u(x,y) = \lambda\mu(x,y) + \frac{\lambda}{\Gamma(r_1)\Gamma(r_2)}\int_0^x\int_0^y (x-s)^{r_1-1}(y-t)^{r_2-1}g(s,t)dtds,$$

where, $g \in C(J)$ such that

$$g(x,y) = f(x,y,u_{(\rho_1(x,y,u_{(x,y)}),\rho_2(x,y,u_{(x,y)})}, g(x,y)).$$

This implies by (7.17.1) and as in step 2 (Theorem 7.11) that, for each $(x,y)\in J$, we have $\|u\|_{C_{(a,b)}} \leq M^*$.

$$U = \{u \in C_{(a,b)} : \|u\|_\infty < M^*+1\}.$$

By our choice of U, there is no $u \in \partial U$ such that $u = \lambda N(u)$, for $\lambda \in (0,1)$. As a consequence of the nonlinear alternative of Leray-Schauder type, we deduce that N has a fixed-point u in $\overline{U}$ which is a solution to problem (7.28)–(7.30). □

7.4.4.2 Infinite Delay Case

Definition 7.18. A function $u \in \Omega$ such that $u(x,y)$, $\overline{D}_{0,x}^{r_1} u(x,y), \overline{D}_{0,y}^{r_2} u(x,y)$, $\overline{D}_{\theta}^{r} u(x,y)$ are continuous for $(x,y) \in J$ and $I_{\theta}^{1-r} u(x,y) \in AC(J)$ is said to be a solution of (7.31)–(7.33) if u satisfies (7.31), (7.33) on J and the condition (7.32) on $\tilde{J}'$.

Set $\mathcal{R}' := \mathcal{R}'_{(\rho_1^-,\rho_2^-)}$

$$= \{(\rho_1(s,t,u), \rho_2(s,t,u)) : (s,t,u) \in J \times \mathcal{B}\ \rho_i(s,t,u) \leq 0;\ i = 1,2\}.$$

We always assume that $\rho_1 : J \times \mathcal{B} \to (-\infty, a]$, $\rho_2 : J \times \mathcal{B} \to (-\infty, b]$ are continuous and the function $(s,t) \longmapsto u_{(s,t)}$ is continuous from $\mathcal{R}'$ into $\mathcal{B}$.

We will need to introduce the following hypothesis:

(C_ϕ) There exists a continuous bounded function $L : \mathcal{R}'_{(\rho_1^-,\rho_2^-)} \to (0,\infty)$ such that

$$\|\phi_{(s,t)}\|_{\mathcal{B}} \leq L(s,t)\|\phi\|_{\mathcal{B}}, \text{ for any } (s,t) \in \mathcal{R}'.$$

In the sequel we will make use of the following generalization of a consequence of the phase space axioms ([147], Lemma 2.1).

Lemma 7.19. *If $u \in \Omega$, then*

$$\|u_{(s,t)}\|_{\mathcal{B}} = (M + L')\|\phi\|_{\mathcal{B}} + K \sup_{(\theta,\eta) \in [0,\max\{0,s\}] \times [0,\max\{0,t\}]} \|u(\theta,\eta)\|,$$

where

$$L' = \sup_{(s,t) \in \mathcal{R}'} L(s,t).$$

Now, we give (without proof) the existence result for the IVP (7.31)–(7.33)

Theorem 7.20. *Assume that the following hypothesis holds:*

(7.20.1) There exist $\ell_*'' > 0$, $0 < l_*'' < 1$ such that

$$\|f(x,y,u,v) - f(x,y,\overline{u},\overline{v})\| \leq \ell_*''\|u - v\|_{\mathcal{B}} + l_*''\|\overline{u} - \overline{v}\|;$$

$$\textit{for any } u,\ v \in \mathcal{B},\ \overline{u}, \overline{v} \in \mathbb{R}^n \textit{ and } (x,y) \in J.$$

If

$$\frac{\ell_*'' K a^{r_1} b^{r_2}}{(1 - l_*'')\Gamma(1 + r_1)\Gamma(1 + r_2)} < 1, \tag{7.47}$$

then there exists a unique solution for IVP (7.31)–(7.33) on $(-\infty,a] \times (-\infty,b]$.

Theorem 7.21. *Assume* (C_ϕ) *and that the following hypothesis holds:*

(7.21.1) There exist $p, q, d \in C(J, \mathbb{R}_+)$ *such that*

$$\|f(x, y, u, v)\| \le p(x, y) + q(x, y)\|u\|_{\mathcal{B}} + d(x, y)\|v\|$$

for $(x, y) \in J,\ u \in \mathcal{B}$ *and* $v \in \mathbb{R}^n$.

If

$$d^* + \frac{q^* a^{r_1} b^{r_2}}{\Gamma(1 + r_1)\Gamma(1 + r_2)} < 1, \tag{7.48}$$

where $d^* = \sup_{(x,y)\in J} d(x, y)$ *and* $q^* = \sup_{(x,y)\in J} q(x, y)$,
then the IVP (7.31)–(7.33) have at least one solution on $(-\infty, a] \times (-\infty, b]$.

7.4.5 *Examples*

7.4.5.1 Example 1

Consider the following partial hyperbolic implicit differential equations of the form:

$$\overline{D}_\theta^r u(x, y) = \frac{1}{5e^{x+y+2}(1 + |u(x-1, y-2)| + |\overline{D}_\theta^r u(x, y)|)}; \quad \text{if } (x, y) \in [0, 1] \times [0, 1], \tag{7.49}$$

$$u(x, y) = x + y^2,\ (x, y) \in [-1, 1] \times [-2, 1] \backslash (0, 1] \times (0, 1], \tag{7.50}$$

$$u(x, 0) = x,\ u(0, y) = y^2;\ x, y \in [0, 1]. \tag{7.51}$$

Set

$$f(x, y, u_{(x,y)}, \overline{D}_\theta^r u(x, y)) = \frac{1}{5e^{x+y+2}(1 + |u(x-1, y-2)| + |\overline{D}_\theta^r u(x, y)|)};$$

$(x, y) \in [0, 1] \times [0, 1]$. Clearly, the function f is continuous. For each $u, v \in C,\ \overline{u}, \overline{v} \in \mathbb{R}$ and $(x, y) \in [0, 1] \times [0, 1]$ we have

$$|f(x, y, u(x, y), v(x, y)) - f(x, y, \overline{u}(x, y), \overline{v}(x, y))| \le \frac{1}{5e^2}(\|u - \overline{u}\|_C + \|v - \overline{v}\|).$$

Hence condition (7.10.2) is satisfied with $\ell = l = \frac{1}{5e^2}$. We shall show that condition (7.34) holds with $a = b = 1$. Indeed

$$\frac{\ell a^{r_1} b^{r_2}}{(1 - l)\Gamma(1 + r_1)\Gamma(1 + r_2)} = \frac{1}{(5e^2 - 1)\Gamma(1 + r_1)\Gamma(1 + r_2)} < 1,$$

which is satisfied for each $(r_1, r_2) \in (0,1] \times (0,1]$. Consequently Theorem 7.10 implies that problems (7.49)–(7.51) have a unique solution defined on $[-1,1] \times [-2,1]$.

7.4.5.2 Example 2

Consider now the following partial hyperbolic functional implicit differential equations with infinite delay :

$$\overline{D}_\theta^r u(x,y) = \frac{e^{x+y-\gamma(x+y)}\|u_{(x,y)}\|}{2(e^{x+y}+e^{-x-y})(1+c\|u_{(x,y)}\|+|\overline{D}_\theta^r u(x,y)|)}, \quad \text{if } (x,y) \in [0,1]\times[0,1], \tag{7.52}$$

$$u(x,y) = x + y^2, \ (x,y) \in (-\infty,1]\times(-\infty,1]\backslash(0,1]\times(0,1], \tag{7.53}$$

$$u(x,0) = x, \ u(0,y) = y^2, \ x, y \in [0,1], \tag{7.54}$$

where $c = \frac{\Gamma(1+r_1)\Gamma(1+r_2)}{2}$, $r = (r_1, r_2) \in (0,1] \times (0,1]$ and γ a positive real constant. Let

$$\mathcal{B}_\gamma = \left\{ u \in C((-\infty,0]\times(-\infty,0],\mathbb{R}) : \lim_{\|(\theta,\eta)\|\to\infty} e^{\gamma(\theta+\eta)}u(\theta,\eta) \text{ exists in } \mathbb{R} \right\}.$$

The norm of $\mathcal{B}_\gamma$ is given by

$$\|u\|_\gamma = \sup_{(\theta,\eta)\in(-\infty,0]\times(-\infty,0]} e^{\gamma(\theta+\eta)}|u(\theta,\eta)|.$$

Let

$$E := [0,1]\times\{0\} \cup \{0\}\times[0,1],$$

and $u : (-\infty,1]\times(-\infty,1] \to \mathbb{R}$ such that $u_{(x,y)} \in \mathcal{B}_\gamma$ for $(x,y) \in E$. $(\mathcal{B}_\gamma, \|.\|_\gamma)$ is a Banach space and $\mathcal{B}_\gamma$ is a phase space. Set

$$f(x,y,u_{(x,y)},v) = \frac{e^{x+y-\gamma(x+y)}\|u_{(x,y)}\|}{2(e^{x+y}+e^{-x-y})(1+c\|u_{(x,y)})\|+\|v(x,y)\|}, \quad \times\,(x,y) \in [0,1]\times[0,1].$$

For each $u, \ \overline{u} \in \mathcal{B}_\gamma, \ v, \ \overline{v} \in \mathbb{R}$ and $(x,y) \in [0,1]\times[0,1]$ we have

$$\begin{aligned} &|f(x,y,u_{(x,y)},v(x,y)) - f(x,y,\overline{u}_{(x,y)},\overline{v}(x,y))| \\ &\leq \frac{e^{x+y}}{2(e^{x+y}+e^{-x-y})}(c\|u-\overline{u}\|_B + \|v-\overline{v}\|) \\ &\leq \frac{c}{2}\|u-\overline{u}\|_B + \frac{1}{2}\|v-\overline{v}\|. \end{aligned}$$

Hence condition (7.13.1) is satisfied with $\ell' = \frac{c}{2}$, $l' = \frac{1}{2}$. Since $a = b = K = 1$, we get

$$\frac{k\ell' a^{r_1} b^{r_2}}{(1-l')\Gamma(1+r_1)\Gamma(1+r_2)} = \frac{c}{\Gamma(1+r_1)\Gamma(1+r_2)} = \frac{1}{2} < 1.$$

Consequently, Theorem 7.13 implies that problems (7.52)–(7.54) have a unique solution defined on $(-\infty, 1] \times (-\infty, 1]$.

7.4.5.3 Example 3

Consider now the following fractional order hyperbolic partial functional differential equations of the form:

$$\overline{D}_\theta^r u(x,y) = \frac{|u(x-\sigma_1(u(x,y)), y-\sigma_2(u(x,y)))| + 2}{10e^{x+y+4}(1+|u(x-\sigma_1(u(x,y)), y-\sigma_2(u(x,y)))| + |\overline{D}_\theta^r u(x,y)|)},$$
$$\text{if } (x,y) \in J := [0,1] \times [0,1], \tag{7.55}$$

$$u(x,0) = x,\ u(0,y) = y^2;\ x,y \in [0,1], \tag{7.56}$$

$$u(x,y) = x + y^2,\ (x,y) \in [-1,1] \times [-2,1] \backslash [-1,0] \times [-2,0], \tag{7.57}$$

where $\sigma_1 \in C(\mathbb{R}, [0,1])$, $\sigma_2 \in C(\mathbb{R}, [0,2])$. Set

$$\rho_1(x,y,\varphi) = x - \sigma_1(\varphi(0,0)),\quad (x,y,\varphi) \in J \times C([-1,0] \times [-2,0], \mathbb{R}),$$

$$\rho_2(x,y,\varphi) = y - \sigma_2(\varphi(0,0)),\quad (x,y,\varphi) \in J \times C([-1,0] \times [-2,0], \mathbb{R}),$$

$$f(x,y,\varphi,\psi) = \frac{|\varphi| + 2}{(10e^{x+y+4})(1+|\varphi|+|\psi|)},\quad (x,y) \in [0,1] \times [0,1],$$

$\varphi \in C([-1,0]\times[-2,0], \mathbb{R})$, $\psi \in \mathbb{R}$. For each $\varphi, \overline{\varphi} \in C([-1,0]\times[-2,0], \mathbb{R})$, $\psi, \overline{\psi} \in \mathbb{R}$ and $(x,y) \in [0,1] \times [0,1]$ we have

$$|f(x,y,\varphi,\psi) - f(x,y,\overline{\varphi},\overline{\psi})| \le \frac{1}{10e^4}(\|\varphi - \overline{\varphi}\|_C + \|\psi - \overline{\psi}\|).$$

Hence the condition (7.16.2) is satisfied with $\ell_* = l_* = \dfrac{1}{10e^4}$. We shall show that condition (7.44) holds with $a = b = 1$. Indeed

$$\frac{\ell_* a^{r_1} b^{r_2}}{(1-l_*)\Gamma(1+r_1)\Gamma(1+r_2)} = \frac{1}{(10e^4-1)\Gamma(1+r_1)\Gamma(1+r_2)} < 1,$$

which is satisfied for each $(r_1, r_2) \in (0,1] \times (0,1]$. Consequently, Theorem 7.16 implies that problems (7.55)–(7.57) have a unique solution defined on $[-1,1] \times [-2,1]$.

7.4.5.4 Example 4

We consider now the following fractional order partial implicit differential equations with infinite delay of the form:

$$\overline{D}_\theta^r u(x,y) = \frac{c e^{x+y-\gamma(x+y)}}{e^{x+y}+e^{-x-y}} \times \frac{|u(x-\sigma_1(u(x,y)), y-\sigma_2(u(x,y)))|}{1+|u(x-\sigma_1(u(x,y)), y-\sigma_2(u(x,y)))|+|\overline{D}_\theta^r u(x,y)|}; \quad \text{if } (x,y)\in J, \tag{7.58}$$

$$u(x,0)=x,\ u(0,y)=y^2;\ x,y\in[0,1], \tag{7.59}$$

$$u(x,y)=x+y^2,\ (x,y)\in\tilde{J}, \tag{7.60}$$

where $J := [0,1]\times[0,1]$, $\tilde{J} := (-\infty,1]\times(-\infty,1]\setminus(0,1]\times(0,1]$, $c = 1+\frac{2}{\Gamma(1+r_1)\Gamma(1+r_2)}$, γ a positive real constant and $\sigma_1,\sigma_2\in C(\mathbb{R},[0,\infty))$.

Let the phase space

$$B_\gamma = \{u\in C((-\infty,0]\times(-\infty,0],\mathbb{R}) : \lim_{\|(\theta,\eta)\|\to\infty} e^{\gamma(\theta+\eta)}u(\theta,\eta) \text{ exists in } \mathbb{R}\},$$

defined as in Example 2. Set

$$\rho_1(x,y,\varphi) = x-\sigma_1(\varphi(0,0)),\ (x,y,\varphi)\in J\times B_\gamma,$$

$$\rho_2(x,y,\varphi) = y-\sigma_2(\varphi(0,0)),\ (x,y,\varphi)\in J\times B_\gamma,$$

$$f(x,y,\varphi,\psi) = \frac{c e^{x+y-\gamma(x+y)}|\varphi|}{(e^{x+y}+e^{-x-y})(1+|\varphi|+|\psi|)},\ (x,y)\in[0,1]\times[0,1],\ \varphi\in B_\gamma.$$

For each $\varphi,\ \overline{\varphi}\in B_\gamma,\ \psi,\ \overline{\psi}\in\mathbb{R}$ and $(x,y)\in[0,1]\times[0,1]$ we have

$$|f(x,y,\varphi,\psi)-f(x,y,\overline{\varphi},\overline{\psi})| \le \frac{1}{c}(\|\varphi-\overline{\varphi}\|_\gamma + \|\psi-\overline{\psi}\|).$$

Hence condition (7.20.1) is satisfied with $\ell''_* = l''_* = \frac{1}{c}$. Since $a=b=K=1$, we get

$$\frac{K\ell''_* a^{r_1}b^{r_2}}{(1-l''_*)\Gamma(1+r_1)\Gamma(r_2+1)} = \frac{1}{(c-1)\Gamma(1+r_1)\Gamma(1+r_2)} = \frac{1}{2} < 1,$$

for each $(r_1,r_2)\in(0,1]\times(0,1]$. Consequently Theorem 7.20 implies that problems (7.58)–(7.60) have a unique solution defined on $(-\infty,1]\times(-\infty,1]$.

7.5 Darboux Problem for Implicit Impulsive Partial Hyperbolic Differential Equations

7.5.1 Introduction

This section concerns the existence results to fractional order IVP, for the system

$$\overline{D}^{r}_{x_k}u(x,y) = f(x,y,u(x,y),\overline{D}^{r}_{x_k}u(x,y));\ \text{if } (x,y)\in J_k;\ k=0,\dots,m, \tag{7.61}$$

$$u(x_k^+,y) = u(x_k^-,y) + I_k(u(x_k^-,y));\ \text{if } y\in[0,b],\ k=1,\dots,m, \tag{7.62}$$

$$\begin{cases} u(x,0)=\varphi(x);\ x\in[0,a],\\ u(0,y)=\psi(y);\ y\in[0,b],\\ \varphi(0)=\psi(0), \end{cases} \tag{7.63}$$

where $a,b>0,\ J_0=[0,x_1]\times[0,b],\ J_k=(x_k,x_{k+1}]\times[0,b];\ k=1,\dots,m,\ 0=x_0<x_1<\cdots<x_m<x_{m+1}=a,\ f:J\times\mathbb{R}^n\times\mathbb{R}^n\to\mathbb{R}^n,\ I_k:\mathbb{R}^n\to\mathbb{R}^n;\ k=1,\dots,m,\ J:=[0,a]\times[0,b],\ \varphi\in AC([0,a])$ and $\psi\in AC([0,b])$.

We present two results for the problems (7.61)–(7.63), the first one is based on Banach's contraction principle and the second one on the nonlinear alternative of Leray-Schauder type.

7.5.2 Existence of Solutions

To define the solutions of problems (7.61)–(7.63), we shall consider the space

$$\begin{aligned} PC(J)=\{u:J\to\mathbb{R}^n : u\in C(J_k,\mathbb{R}^n);\ k=0,1,\dots,m,\ \text{and there}\\ \text{exist } u(x_k^-,y) \text{ and } u(x_k^+,y);\ k=1,\dots,m,\\ \text{with } u(x_k^-,y)=u(x_k,y) \text{ for each } y\in[0,b]\}. \end{aligned}$$

This set is a Banach space with the norm

$$\|u\|_{PC} = \sup_{(x,y)\in J}\|u(x,y)\|.$$

Definition 7.22. A function $u\in PC(J)$ such that $u(x,y),\ \overline{D}^{r_1}_{x_k,x}u(x,y),\ \overline{D}^{r_2}_{x_k,y}u(x,y),\ \overline{D}^{r}_{x_k^+}u(x,y);\ k=0,\dots,m$, are continuous for $(x,y)\in J_k;\ k=0,\dots,m$ and $I^{1-r}_{z^+}u(x,y)\in AC(J_k);\ k=0,\dots,m$ is said to be a solution of (7.61)–(7.63) if u satisfies (7.61) on $J_k;\ k=0,\dots,m$ and conditions (7.62), (7.63) are satisfied.

From Lemmas 2.15 and 7.4, we conclude the following lemma.

Lemma 7.23. *Let a function* $f : J \times \mathbb{R}^n \times \mathbb{R}^n \to \mathbb{R}^n$ *be continuous. Then problems (7.61)–(7.63) are equivalent to the problem of the solution of the equation*

$$g(x,y) = f(x,y,\xi(x,y),g(x,y)),$$

where

$$\xi(x,y) = \begin{cases} \mu(x,y) + \frac{1}{\Gamma(r_1)\Gamma(r_2)} \int_0^x \int_0^y (x-s)^{r_1-1}(y-t)^{r_2-1} g(s,t)\,dt\,ds; \\ \text{if } (x,y) \in [0,x_1] \times [0,b], \\ \mu(x,y) + \sum_{i=1}^{k} (I_i(u(x_i^-,y)) - I_i(u(x_i^-,0))) \\ + \frac{1}{\Gamma(r_1)\Gamma(r_2)} \sum_{i=1}^{k} \int_{x_{i-1}}^{x_i} \int_0^y (x_i-s)^{r_1-1}(y-t)^{r_2-1} g(s,t)\,dt\,ds \\ + \frac{1}{\Gamma(r_1)\Gamma(r_2)} \int_{x_k}^{x} \int_0^y (x-s)^{r_1-1}(y-t)^{r_2-1} g(s,t)\,dt\,ds; \\ \text{if } (x,y) \in (x_k, x_{k+1}] \times [0,b],\ k = 1,\dots,m, \end{cases}$$

$$\mu(x,y) = \varphi(x) + \psi(y) - \varphi(0).$$

And if $g \in C(J)$ *is the solution of this equation, then* $u(x,y) = \xi(x,y)$.

Further, we present conditions for the existence and uniqueness of a solution of problems (7.61)–(7.63).

Theorem 7.24. *Assume*

(7.24.1) *The function* $f : J \times \mathbb{R}^n \times \mathbb{R}^n \to \mathbb{R}^n$ *is continuous.*
(7.24.2) *For any* $u, v, w, z \in \mathbb{R}^n$ *and* $(x,y) \in J$, *there exist constants* $l > 0$ *and* $0 < l_* < 1$ *such that*

$$\|f(x,y,u,z) - f(x,y,v,w)\| \le l\|u-v\| + l_*\|z-w\|.$$

(7.24.3) *There exists a constant* $l^* > 0$ *such that*

$$\|I_k(u) - I_k(\overline{u})\| \le l^*\|u-\overline{u}\|, \text{ for each } u, \overline{u} \in \mathbb{R}^n,\ k = 1,\dots,m.$$

If

$$2ml^* + \frac{2la^{r_1}b^{r_2}}{(1-l_*)\Gamma(r_1+1)\Gamma(r_2+1)} < 1, \tag{7.64}$$

then there exists a unique solution for IVP (7.61)–(7.63) on J.

Proof. Transform the problems (7.61)–(7.63) into a fixed-point problem. Consider the operator $N : PC(J) \to PC(J)$ defined by

$$N(u)(x,y) = \mu(x,y) + \sum_{0<x_k<x} (I_k(u(x_k^-,y)) - I_k(u(x_k^-,0)))$$
$$+\frac{1}{\Gamma(r_1)\Gamma(r_2)} \sum_{0<x_k<x} \int_{x_{k-1}}^{x_k} \int_0^y (x_k-s)^{r_1-1}(y-t)^{r_2-1} g(s,t)\mathrm{d}t\mathrm{d}s$$
$$+\frac{1}{\Gamma(r_1)\Gamma(r_2)} \int_{x_k}^{x} \int_0^y (x-s)^{r_1-1}(y-t)^{r_2-1} g(s,t)\mathrm{d}t\mathrm{d}s, \qquad (7.65)$$

where $g \in C(J)$ such that

$$g(x,y) = f(x,y,u(x,y),g(x,y)),$$

By Lemma 7.23, the problem of finding the solutions of the IVP (7.61)–(7.63) is reduced to finding the solutions of the operator equation $N(u) = u$. Let $v, w \in PC(J)$. Then, for $(x,y) \in J$, we have

$$\|N(v)(x,y) - N(w)(x,y)\|$$
$$\leq \sum_{k=1}^{m} (\|I_k(v(x_k^-,y)) - I_k(w(x_k^-,y))\| + \|I_k(v(x_k^-,0)) - I_k(w(x_k^-,0))\|)$$
$$+\frac{1}{\Gamma(r_1)\Gamma(r_2)} \sum_{k=1}^{m} \int_{x_{k-1}}^{x_k} \int_0^y (x_k-s)^{r_1-1}(y-t)^{r_2-1} \|g(s,t)-h(s,t)\|\mathrm{d}t\mathrm{d}s$$
$$+\frac{1}{\Gamma(r_1)\Gamma(r_2)} \int_{x_k}^{x} \int_0^y (x-s)^{r_1-1}(y-t)^{r_2-1} \|g(s,t)-h(s,t)\|\mathrm{d}t\mathrm{d}s, \quad (7.66)$$

where $g, h \in C(J)$ such that

$$g(x,y) = f(x,y,v(x,y),g(x,y))$$

and

$$h(x,y) = f(x,y,w(x,y),h(x,y)).$$

By (7.24.2), we get

$$\|g(x,y) - h(x,y)\| \leq l\|v(x,y) - w(x,y)\| + l_*\|g(x,y) - h(x,y)\|.$$

Then

$$\begin{aligned}\|g(x,y)-h(x,y)\| &\leq \frac{l}{1-l_*}\|v(x,y)-w(x,y)\| \\ &\leq \frac{l}{1-l_*}\|v-w\|_{PC}.\end{aligned}$$

Thus, (7.24.3) and (7.66) imply that

$$\begin{aligned}&\|N(v)-N(w)\|_{PC} \\ &\leq \sum_{k=1}^{m} l^*(\|v(x_k^-,y)-w(x_k^-,y)\| + \|v(x_k^-,0)-w(x_k^-,0)\|) \\ &+ \frac{l}{(1-l_*)\Gamma(r_1)\Gamma(r_2)}\sum_{k=1}^{m}\int_{x_{k-1}}^{x_k}\int_0^y (x_k-s)^{r_1-1}(y-t)^{r_2-1}\|v-w\|_{PC}\,dt\,ds \\ &+ \frac{l}{(1-l_*)\Gamma(r_1)\Gamma(r_2)}\int_{x_k}^{x}\int_0^y (x-s)^{r_1-1}(y-t)^{r_2-1}\|v-w\|_{PC}\,dt\,ds.\end{aligned}$$

However,

$$\begin{aligned}&\frac{l}{(1-l_*)\Gamma(r_1)\Gamma(r_2)}\sum_{k=1}^{m}\int_{x_{k-1}}^{x_k}\int_0^y (x_k-s)^{r_1-1}(y-t)^{r_2-1}\|v-w\|_{PC}\,dt\,ds \\ &\leq \frac{l}{(1-l_*)\Gamma(r_1)\Gamma(r_2)}\frac{b^{r_2}}{r_2}\|v-w\|_{PC}\sum_{k=1}^{m}\int_{x_{k-1}}^{x_k}(x_k-s)^{r_1-1}ds \\ &\leq \frac{l}{(1-l_*)\Gamma(r_1)\Gamma(r_2)}\frac{b^{r_2}}{r_2}\|v-w\|_{PC}\sum_{k=1}^{m}\frac{x_{k-1}^{r_1}}{r_1} \\ &= \frac{l}{(1-l_*)\Gamma(r_1)\Gamma(r_2)}\frac{b^{r_2}}{r_2}\|v-w\|_{PC}\frac{x_{m-1}^{r_1}-x_0^{r_1}}{r_1} \\ &\leq \frac{l}{(1-l_*)\Gamma(r_1)\Gamma(r_2)}\frac{b^{r_2}}{r_2}\|v-w\|_{PC}\frac{a^{r_1}}{r_1} \\ &= \frac{la^{r_1}b^{r_2}}{(1-l_*)\Gamma(1+r_1)\Gamma(1+r_2)}\|v-w\|_{PC}.\end{aligned}$$

Then

$$\|N(v)-N(w)\|_{PC} \leq \left(2ml^* + \frac{la^{r_1}b^{r_2}}{(1-l_*)\Gamma(r_1+1)\Gamma(r_2+1)}\right.$$
$$\left.+\frac{la^{r_1}b^{r_2}}{(1-l_*)\Gamma(r_1+1)\Gamma(r_2+1)}\right)\|v-w\|_{PC}$$
$$\leq \left(2ml^* + \frac{2la^{r_1}b^{r_2}}{(1-l_*)\Gamma(r_1+1)\Gamma(r_2+1)}\right)\|v-w\|_{PC}.$$

Hence

$$\|N(v)-N(w)\|_{PC} \leq \left(2ml^* + \frac{2la^{r_1}b^{r_2}}{(1-l_*)\Gamma(r_1+1)\Gamma(r_2+1)}\right)\|v-w\|_{PC}.$$

By (7.64), N is a contraction, and hence N has a unique fixed-point by Banach's contraction principle. □

Theorem 7.25. *Assume* (7.24.1) *and the following hypotheses hold:*

(7.25.1) There exist $p,\ q,\ d \in C(J,\mathbb{R}_+)$ *such that*

$$\|f(x,y,u,z)\| \leq p(x,y) + q(x,y)\|u\| + d(x,y)\|z\|$$

for $(x,y) \in J$ *and each* $u,z \in \mathbb{R}^n$.

(7.25.2) There exists $\psi^* : [0,\infty) \to (0,\infty)$ *continuous and nondecreasing such that*

$$\|I_k(u)\| \leq \psi^*(\|u\|);\ k = 1,\ldots,m, \quad \text{for all } u \in \mathbb{R}^n.$$

(7.25.3) There exists a number $\overline{M} > 0$ *such that*

$$\frac{\overline{M}}{\|\mu\|_\infty + 2m\psi^*(\overline{M}) + \frac{2a^{r_1}b^{r_2}\left(p^*+q^*\|\mu\|_\infty+2mq^*\psi^*(\overline{M})\right)}{\left(1-d^*-\frac{2q^*a^{r_1}b^{r_2}}{\Gamma(1+r_1)\Gamma(1+r_2)}\right)\Gamma(1+r_1)\Gamma(1+r_2)}} > 1,$$

where $p^* = \sup_{(x,y)\in J} p(x,y)$, $q^* = \sup_{(x,y)\in J} q(x,y)$ *and* $d^* = \sup_{(x,y)\in J} d(x,y)$.

If

$$d^* + \frac{2q^*a^{r_1}b^{r_2}}{\Gamma(1+r_1)\Gamma(1+r_2)} < 1, \tag{7.67}$$

then the IVP (7.61)–(7.63) have at least one solution on J.

Proof. Transform the problems (7.61)–(7.63) into a fixed-point problem. Consider the operator N defined in (7.65). We shall show that the operator N is continuous and compact .

Step 1: *N is continuous.* Let $\{u_n\}_{n\in\mathbb{N}}$ be a sequence such that $u_n \to u$ in $PC(J)$. Let $\eta > 0$ be such that $\|u_n\|_{PC} \le \eta$. Then

$$\begin{aligned}
&\|N(u_n)(x,y) - N(u)(x,y)\| \\
&\le \sum_{k=1}^{m} \left(\|I_k(u_n(x_k^-,y)) - I_k(u(x_k^-,y))\| + \|I_k(u_n(x_k^-,0)) - I_k(u(x_k^-,0))\|\right) \\
&\quad + \frac{1}{\Gamma(r_1)\Gamma(r_2)} \sum_{k=1}^{m} \int_{x_{k-1}}^{x_k} \int_0^y (x_k - s)^{r_1-1}(y-t)^{r_2-1}\|g_n(s,t) - g(s,t)\| \mathrm{d}t\mathrm{d}s \\
&\quad + \frac{1}{\Gamma(r_1)\Gamma(r_2)} \int_{x_k}^{x} \int_0^y (x-s)^{r_1-1}(y-t)^{r_2-1}\|g_n(s,t) - g(s,t)\| \mathrm{d}t\mathrm{d}s, \quad (7.68)
\end{aligned}$$

where $g_n, g \in C(J)$ such that

$$g_n(x,y) = f(x,y,u_n(x,y),g_n(x,y))$$

and

$$g(x,y) = f(x,y,u(x,y),g(x,y)).$$

Since $u_n \to u$ as $n \to \infty$ and f is a continuous function, we get

$$g_n(x,y) \to g(x,y) \text{ as } n \to \infty, \text{ for each } (x,y) \in J.$$

Hence, (7.68) gives

$$\begin{aligned}
\|N(u_n) - N(u)\|_{PC} &\le 2ml^*\|u_n - u\|_{PC} \\
&\quad + \frac{2a^{r_1}b^{r_2}}{\Gamma(1+r_1)\Gamma(1+r_2)}\|g_n - g\|_\infty \to 0 \text{ as } n \to \infty.
\end{aligned}$$

Step 2: *N maps bounded sets into bounded sets in $PC(J)$.* Indeed, it is enough to show that for any $\eta^* > 0$, there exists a positive constant M^* such that for each

$$u \in B_{\eta^*} = \{u \in PC(J) : \|u\|_{PC} \le \eta^*\},$$

we have $\|N(u)\|_{PC} \le M^*$. For $(x,y) \in J$, we have

$$\begin{aligned}
\|N(u)(x,y)\| &\le \|\mu(x,y)\| + \sum_{k=1}^{m}(\|I_k(u(x_k^-,y))\| + \|I_k(u(x_k^-,0))\|) \\
&\quad + \frac{1}{\Gamma(r_1)\Gamma(r_2)} \sum_{k=1}^{m} \int_{x_{k-1}}^{x_k} \int_0^y (x_k - s)^{r_1-1}(y-t)^{r_2-1}\|g(s,t)\| \mathrm{d}t\mathrm{d}s \\
&\quad + \frac{1}{\Gamma(r_1)\Gamma(r_2)} \int_{x_k}^{x} \int_0^y (x-s)^{r_1-1}(y-t)^{r_2-1}\|g(s,t)\| \mathrm{d}t\mathrm{d}s, \quad (7.69)
\end{aligned}$$

where $g \in C(J)$ such that

$$g(x,y) = f(x,y,u(x,y),g(x,y)).$$

By (7.25.1), for each $(x,y) \in J$, we have

$$\|g(x,y)\| \le p(x,y) + q(x,y)\|\xi(x,y)\| + d(x,y)\|g(x,y)\|.$$

On the other hand, for each $(x,y) \in J$,

$$\begin{aligned}\|\xi(x,y)\| &\le \|\mu(x,y)\| + \sum_{k=1}^{m}(\|I_k(u(x_k^-,y))\| + \|I_k(u(x_k^-,0))\|)\\ &+\frac{1}{\Gamma(r_1)\Gamma(r_2)}\sum_{k=1}^{m}\int_{x_{k-1}}^{x_k}\int_0^y (x_k-s)^{r_1-1}(y-t)^{r_2-1}\|g(s,t)\|\mathrm{d}t\mathrm{d}s\\ &+\frac{1}{\Gamma(r_1)\Gamma(r_2)}\int_{x_k}^{x}\int_0^y (x-s)^{r_1-1}(y-t)^{r_2-1}\|g(s,t)\|\mathrm{d}t\mathrm{d}s\\ &\le \|\mu\|_\infty + 2m\psi^*(\eta^*) + \frac{2a^{r_1}b^{r_2}}{\Gamma(1+r_1)\Gamma(1+r_2)}\|g\|_\infty.\end{aligned}$$

Hence, for each $(x,y) \in J$, we have

$$\|g\|_\infty \le p^* + q^*\left(\|\mu\|_\infty + 2m\psi^*(\eta^*) + \frac{2a^{r_1}b^{r_2}}{\Gamma(1+r_1)\Gamma(1+r_2)}\|g\|_\infty\right) + d^*\|g\|_\infty.$$

Then, by (7.67) we have

$$\|g\|_\infty \le \frac{p^* + q^*\left(\|\mu\|_\infty + 2m\psi^*(\eta^*)\right)}{1-d^* - \frac{2q^*a^{r_1}b^{r_2}}{\Gamma(1+r_1)\Gamma(1+r_2)}} := M.$$

Thus, (7.69) implies that

$$\|N(u)\|_{PC} \le \|\mu\|_\infty + 2m\psi^*(\eta^*) + \frac{2Ma^{r_1}b^{r_2}}{\Gamma(1+r_1)\Gamma(1+r_2)} := M^*.$$

Step 3: *N maps bounded sets into equicontinuous sets in $PC(J)$.* Let (τ_1, y_1), $(\tau_2, y_2) \in J$, $\tau_1 < \tau_2$ and $y_1 < y_2$, B_{η^*} be a bounded set of $PC(J)$ as in Step 2, and let $u \in B_{\eta^*}$. Then for each $(x,y) \in J$, we have

$$
\begin{aligned}
&\|N(u)(\tau_2, y_2) - N(u)(\tau_1, y_1)\| \\
&\quad \le \|\mu(\tau_1, y_1) - \mu(\tau_2, y_2)\| + \sum_{k=1}^{m} \left(\|I_k(u(x_k^-, y_1)) - I_k(u(x_k^-, y_2))\|\right) \\
&\quad + \frac{1}{\Gamma(r_1)\Gamma(r_2)} \sum_{k=1}^{m} \int_{x_{k-1}}^{x_k} \int_0^{y_1} (x_k - s)^{r_1-1} \left[(y_2 - t)^{r_2-1} - (y_1 - t)^{r_2-1}\right] \\
&\quad \times g(s,t)\,dt\,ds \\
&\quad + \frac{1}{\Gamma(r_1)\Gamma(r_2)} \sum_{k=1}^{m} \int_{x_{k-1}}^{x_k} \int_{y_1}^{y_2} (x_k - s)^{r_1-1} (y_2 - t)^{r_2-1} \|g(s,t)\|\,dt\,ds \\
&\quad + \frac{1}{\Gamma(r_1)\Gamma(r_2)} \int_0^{\tau_1} \int_0^{y_1} \left[(\tau_2 - s)^{r_1-1}(y_2 - t)^{r_2-1} - (\tau_1 - s)^{r_1-1}(y_1 - t)^{r_2-1}\right] \\
&\quad \times g(s,t)\,dt\,ds \\
&\quad + \frac{1}{\Gamma(r_1)\Gamma(r_2)} \int_{\tau_1}^{\tau_2} \int_{y_1}^{y_2} (\tau_2 - s)^{r_1-1}(y_2 - t)^{r_2-1} \|g(s,t)\|\,dt\,ds \\
&\quad + \frac{1}{\Gamma(r_1)\Gamma(r_2)} \int_0^{\tau_1} \int_{y_1}^{y_2} (\tau_2 - s)^{r_1-1}(y_2 - t)^{r_2-1} \|g(s,t)\|\,dt\,ds \\
&\quad + \frac{1}{\Gamma(r_1)\Gamma(r_2)} \int_{\tau_1}^{\tau_2} \int_0^{y_1} (\tau_2 - s)^{r_1-1}(y_2 - t)^{r_2-1} \|g(s,t)\|\,dt\,ds,
\end{aligned}
$$

where $g \in C(J)$ such that

$$g(x, y) = f(x, y, u(x, y), g(x, y)).$$

But $\|g\|_\infty \le M$. Thus

$$
\begin{aligned}
&\|N(u)(\tau_2, y_2) - N(u)(\tau_1, y_1)\| \le \|\mu(\tau_1, y_1) - \mu(\tau_2, y_2)\| \\
&\quad + \sum_{k=1}^{m} \left(\|I_k(u(x_k^-, y_1)) - I_k(u(x_k^-, y_2))\|\right) \\
&\quad + \frac{M}{\Gamma(r_1)\Gamma(r_2)} \sum_{k=1}^{m} \int_{x_{k-1}}^{x_k} \int_0^{y_1} (x_k - s)^{r_1-1} \left[(y_2 - t)^{r_2-1} - (y_1 - t)^{r_2-1}\right] dt\,ds
\end{aligned}
$$

$$+\frac{M}{\Gamma(r_1)\Gamma(r_2)}\sum_{k=1}^{m}\int_{x_{k-1}}^{x_k}\int_{y_1}^{y_2}(x_k-s)^{r_1-1}(y_2-t)^{r_2-1}dtds$$

$$+\frac{M}{\Gamma(r_1)\Gamma(r_2)}\int_0^{\tau_1}\int_0^{y_1}\left[(\tau_2-s)^{r_1-1}(y_2-t)^{r_2-1}-(\tau_1-s)^{r_1-1}(y_1-t)^{r_2-1}\right]dtds$$

$$+\frac{M}{\Gamma(r_1)\Gamma(r_2)}\int_{\tau_1}^{\tau_2}\int_{y_1}^{y_2}(\tau_2-s)^{r_1-1}(y_2-t)^{r_2-1}dtds$$

$$+\frac{M}{\Gamma(r_1)\Gamma(r_2)}\int_0^{\tau_1}\int_{y_1}^{y_2}(\tau_2-s)^{r_1-1}(y_2-t)^{r_2-1}dtds$$

$$+\frac{M}{\Gamma(r_1)\Gamma(r_2)}\int_{\tau_1}^{\tau_2}\int_0^{y_1}(\tau_2-s)^{r_1-1}(y_2-t)^{r_2-1}dtds.$$

As $\tau_1 \longrightarrow \tau_2$ and $y_1 \longrightarrow y_2$, the right-hand side of the above inequality tends to zero. As a consequence of Steps 1 to 3, together with the Arzela-Ascoli theorem, we can conclude that N is continuous and completely continuous.

Step 4: *A priori bounds*. We now show that there exists an open set $U \subseteq PC(J)$ with $u \neq \lambda N(u)$, for $\lambda \in (0,1)$ and $u \in \partial U$. Let $u \in PC(J)$ and $u = \lambda N(u)$ for some $0 < \lambda < 1$. Thus for each $(x,y) \in J$, we have

$$\begin{aligned}\|u(x,y)\| &\leq \|\lambda\mu(x,y)\| + \sum_{k=1}^{m}\lambda(\|I_k(u(x_k^-,y))\| + \|I_k(u(x_k^-,0))\|)\\ &\quad+\frac{\lambda}{\Gamma(r_1)\Gamma(r_2)}\sum_{k=1}^{m}\int_{x_{k-1}}^{x_k}\int_0^{y}(x_k-s)^{r_1-1}(y-t)^{r_2-1}\|g(s,t)\|dtds\\ &\quad+\frac{\lambda}{\Gamma(r_1)\Gamma(r_2)}\int_{x_k}^{x}\int_0^{y}(x-s)^{r_1-1}(y-t)^{r_2-1}\|g(s,t)\|dtds\\ &\leq \|\mu\|_\infty + 2m\psi^*(\|u(x,y\|) + \frac{2a^{r_1}b^{r_2}}{\Gamma(1+r_1)\Gamma(1+r_2)}\|g\|_\infty.\end{aligned}$$

But

$$\|g\|_\infty \leq \frac{p^*+q^*\left(\|\mu\|_\infty + 2m\psi^*(\|u\|_{PC})\right)}{1-d^*-\frac{2q^*a^{r_1}b^{r_2}}{\Gamma(1+r_1)\Gamma(1+r_2)}}.$$

Thus, for each $(x, y) \in J$, we have

$$\|u\|_{PC} \leq \|\mu\|_\infty + 2m\psi^*(\|u\|_{PC}) + \frac{2a^{r_1}b^{r_2}\left(p^* + q^*\|\mu\|_\infty + 2mq^*\psi^*(\|u\|_{PC})\right)}{\left(1 - d^* - \frac{2q^*a^{r_1}b^{r_2}}{\Gamma(1+r_1)\Gamma(1+r_2)}\right)\Gamma(1+r_1)\Gamma(1+r_2)}.$$

Hence

$$\frac{\|u\|_{PC}}{\|\mu\|_\infty + 2mq^*\psi^*(\|u\|_{PC}) + \frac{2a^{r_1}b^{r_2}(p^*+q^*\|\mu\|_\infty+2m\psi^*(\|u\|_{PC}))}{\left(1-d^*-\frac{2q^*a^{r_1}b^{r_2}}{\Gamma(1+r_1)\Gamma(1+r_2)}\right)\Gamma(1+r_1)\Gamma(1+r_2)}} \leq 1.$$

By condition (7.25.3), there exists $\overline{M}$ such that $\|u\|_{PC} \neq \overline{M}$. Let

$$U = \{u \in PC(J) : \|u\|_{PC} < \overline{M} + 1\}.$$

By our choice of U, there is no $u \in \partial U$ such that $u = \lambda N(u)$, for $\lambda \in (0, 1)$. As a consequence of Theorem 2.32, we deduce that N has a fixed-point u in $\overline{U}$ which is a solution to problems (7.61)–(7.63). □

7.5.3 An Example

As an application of our results we consider the following impulsive implicit partial hyperbolic differential equations of the form:

$$\overline{D}^r_{x_k}u(x, y) = \frac{1}{10e^{x+y+2}(1 + |u(x, y)| + |\overline{D}^r_{x_k}u(x, y)|)};$$
$$\text{if } (x, y) \in J_k;\ k = 0, \ldots, m, \tag{7.70}$$

$$u(x_k^+, y) = u(x_k^-, y) + \frac{1}{6e^{x+y+4}(1 + |u(x_k^-, y)|)};\ \text{ if } y \in [0, 1],\ k = 1, \ldots, m, \tag{7.71}$$

$$u(x, 0) = x,\ u(0, y) = y^2;\ \text{if } x, y \in [0, 1]. \tag{7.72}$$

Set

$$f(x, y, u, v) = \frac{1}{10e^{x+y+2}(1 + |u| + |v|)},\ (x, y) \in [0, 1] \times [0, 1]$$

and

$$I_k(u(x_k^-, y)) = \frac{1}{6e^{x+y+4}(1 + |u(x_k^-, y)|)},\ y \in [0, 1].$$

Clearly, the function f is continuous.

For each $u, v, \overline{u}, \overline{v} \in \mathbb{R}$ and $(x, y) \in [0,1] \times [0,1]$, we have

$$|f(x, y, u, v) - f(x, y, \overline{u}, \overline{v})| \leq \frac{1}{10e^2}(|u - \overline{u}| + |v - \overline{v}|)$$

and

$$|I_k(u) - I_k(\overline{u})| \leq \frac{1}{6e^4}|u - \overline{u}|.$$

Hence conditionsd (7.24.2) and (7.24.3) are satisfied with $l = l_* = \frac{1}{10e^2}$ and $l^* = \frac{1}{6e^4}$. We shall show that condition (7.64) holds with $a = b = 1$. Indeed, if we assume, for instance, that the number of impulses $m = 3$, then we have

$$2ml^* + \frac{2la^{r_1}b^{r_2}}{(1 - l_*)\Gamma(r_1 + 1)\Gamma(r_2 + 1)} = \frac{1}{e^4} + \frac{2}{(10e^2 - 1)\Gamma(r_1 + 1)\Gamma(r_2 + 1)} < 1,$$

which is satisfied for each $(r_1, r_2) \in (0, 1] \times (0, 1]$. Consequently Theorem 7.24 implies that problems (7.70)–(7.72) have a unique solution defined on $[0, 1] \times [0, 1]$.

7.6 Implicit Impulsive Partial Hyperbolic Differential Equations with State-Dependent Delay

7.6.1 Introduction

In this section, we start by studying the existence result to fractional order IVP, for the system

$$\overline{D}^r_{x_k} u(x, y) = f(x, y, u_{(\rho_1(x,y,u_{(x,y)}),\rho_2(x,y,u_{(x,y)}))}, \overline{D}^r_{x_k} u(x, y));$$
$$\text{if } (x, y) \in J_k;\ k = 0, \ldots, m, \tag{7.73}$$

$$u(x_k^+, y) = u(x_k^-, y) + I_k(u(x_k^-, y));\ \text{if } y \in [0, b],\ k = 1, \ldots, m, \tag{7.74}$$

$$u(x, y) = \phi(x, y);\ \text{if } (x, y) \in \tilde{J} := [-\alpha, a] \times [-\beta, b] \backslash (0, a] \times (0, b], \tag{7.75}$$

$$\begin{cases} u(x, 0) = \varphi(x);\ x \in [0, a], \\ u(0, y) = \psi(y);\ y \in [0, b], \end{cases} \tag{7.76}$$

where $a, b, \alpha, \beta > 0$, $r = (r_1, r_2) \in (0, 1] \times (0, 1]$, $0 = x_0 < x_1 < \cdots < x_m < x_{m+1} = a$, $\phi \in C(\tilde{J})$, $\varphi : [0, a] \to \mathbb{R}^n$, $\psi : [0, b] \to \mathbb{R}^n$ such that $\varphi(x) = \phi(x, 0)$, $\psi(y) = \phi(0, y)$ for each $x \in [0, a]$, $y \in [0, b]$, $f : J \times C \times \mathbb{R}^n \to \mathbb{R}^n$ is a given continuous function, $\rho_1 : J \times C \to [-\alpha, a]$, $\rho_2 : J \times C \to [-\beta, b]$, $I_k : \mathbb{R}^n \to \mathbb{R}^n$, $k = 1, \ldots, m$ are given functions and C is the space defined by

$$C = C_{(\alpha,\beta)} = \{u : [-\alpha, 0] \times [-\beta, 0] \to \mathbb{R}^n : \text{ continuous and there exists}$$
$$\tau_k \in (-\alpha, 0) \text{ with } u(\tau_k^-, \tilde{y}) \text{ and } u(x_k^+, \tilde{y}),\ k = 1, \ldots, m, \text{ exists for any}$$
$$\tilde{y} \in [-\beta, 0] \text{ with } u(\tau_k^-, \tilde{y}) = u(\tau_k, \tilde{y})\}.$$

C is a Banach space with norm

$$\|u\|_C = \sup_{(x,y)\in[-\alpha,0]\times[-\beta,0]} \|u(x, y)\|.$$

Next we consider the following system of partial hyperbolic differential equations of fractional order with infinite delay:

$$\overline{D}^r_{x_k} u(x, y) = f(x, y, u_{(\rho_1(x,y,u_{(x,y)}),\rho_2(x,y,u_{(x,y)}))}, \overline{D}^r_{x_k} u(x, y));$$
$$\text{if } (x, y) \in J_k;\ k = 0, \ldots, m, \tag{7.77}$$

$$u(x_k^+, y) = u(x_k^-, y) + I_k(u(x_k^-, y)); \text{ if } y \in [0, b],\ k = 1, \ldots, m, \tag{7.78}$$

$$u(x, y) = \phi(x, y); \text{ if } (x, y) \in \tilde{J}' := (-\infty, a] \times (-\infty, b] \backslash (0, a] \times (0, b], \tag{7.79}$$

$$\begin{cases} u(x, 0) = \varphi(x); & x \in [0, a], \\ u(0, y) = \psi(y); & y \in [0, b], \end{cases} \tag{7.80}$$

where φ, ψ, I_k are as in problems (7.73)–(7.76), $f : J \times B \times \mathbb{R}^n \to \mathbb{R}^n$ is a given continuous function, $\rho_1 : J \times B \to (-\infty, a]$, $\rho_2 : J \times B \to (-\infty, b]$, $\phi \in C(\tilde{J}')$ and B is a phase space.

7.6.2 *Existence Results with Finite Delay*

Set
$\widetilde{PC} := \{u : [-\alpha, a] \times [-\beta, b] \to \mathbb{R}^n$ *continuous* : $u|_{[-\alpha,0]\times[-\beta,0]} \in C$, *and* $u|_{[0,a]\times[0,b]} \in PC\}$, which is a Banach space with the norm

$$\|u\|_{\widetilde{PC}} = \sup_{(x,y)\in[-\alpha,a]\times[-\beta,b]} \|u(x, y)\|.$$

Definition 7.26. A function $u \in \widetilde{PC}$ such that $u(x, y)$, $\overline{D}^{r_1}_{x_k,x} u(x, y)$, $\overline{D}^{r_2}_{x_k,y} u(x, y)$, $\overline{D}^r_{x_k^+} u(x, y)$; $k = 0, \ldots, m$, are continuous for $(x, y) \in J_k$; $k = 0, \ldots, m$ and $I^{1-r}_{z^+} u(x, y) \in AC(J_k)$; $k = 0, \ldots, m$ is said to be a solution of (7.73)–(7.76) if u satisfies the condition (7.75) on $\tilde{J}$, (7.73) on J_k; $k = 0, \ldots, m$ and conditions (7.74) and (7.76) are satisfied.

$$\begin{aligned}\text{Set}\mathcal{R} &:= \mathcal{R}_{(\rho_1^-,\rho_2^-)}\\ &= \{(\rho_1(s,t,u),\rho_2(s,t,u)) : (s,t,u) \in J \times C,\ \rho_i(s,t,u) \le 0;\ i = 1,2\}.\end{aligned}$$

We always assume that $\rho_1 : J \times C \to [-\alpha, a]$, $\rho_2 : J \times C \to [-\beta, b]$ are continuous and the function $(s,t) \longmapsto u_{(s,t)}$ is continuous from $\mathcal{R}$ into C.

The first result is based on the Banach fixed-point theorem.

Theorem 7.27. *Assume that*

(7.27.1) There exists constants $\ell > 0$, $0 < l < 1$ such that

$$\|f(x,y,u,v) - f(x,y,\overline{u},\overline{v})\| \le \ell\|u-\overline{u}\|_C + l\|v-\overline{v}\|;$$

for each $(x,y) \in J$, $u,\overline{u} \in C$ and $u,\overline{u} \in \mathbb{R}^n$.

(7.27.2) There exists a constant $l^* > 0$ such that

$$\|I_k(u) - I_k(\overline{u})\| \le l^*\|u-\overline{u}\|, \text{ for each } u,\overline{u} \in \mathbb{R}^n,\ k = 1,\ldots,m.$$

If

$$2ml^* + \frac{2\ell a^{r_1} b^{r_2}}{(1-l)\Gamma(r_1+1)\Gamma(r_2+1)} < 1, \tag{7.81}$$

then (7.73)–(7.76) have a unique solution on $[-\alpha,a]\times[-\beta,b]$.

Proof. We transform the problem (7.73)–(7.76) into a fixed-point problem. Consider the operator $F : \widetilde{PC} \to \widetilde{PC}$ defined by

$$F(u)(x,y) = \begin{cases} \phi(x,y) & (x,y) \in \tilde{J},\\ \mu(x,y) + \sum\limits_{0<x_k<x} (I_k(u(x_k^-,y)) - I_k(u(x_k^-,0))) & \\ +\dfrac{1}{\Gamma(r_1)\Gamma(r_2)} \sum\limits_{0<x_k<x} \int\limits_{x_{k-1}}^{x_k}\int\limits_0^y (x_k-s)^{r_1-1}(y-t)^{r_2-1} g(s,t)\mathrm{d}t\mathrm{d}s & \\ +\dfrac{1}{\Gamma(r_1)\Gamma(r_2)} \int\limits_{x_k}^{x}\int\limits_0^y (x-s)^{r_1-1}(y-t)^{r_2-1} g(s,t)\mathrm{d}t\mathrm{d}s & (x,y) \in J,\end{cases}$$

where $g \in C(J)$ such that

$$g(x,y) = f(x,y,u_{(\rho_1(x,y,u_{(x,y)}),\rho_2(x,y,u_{(x,y)}))}, g(x,y)).$$

Clearly, the fixed-points of the operator F are solutions of the problems (7.73)–(7.76). We shall use the Banach contraction principle to prove that F has a fixed-point . For this, we show that F is a contraction. Let $u, v \in \widetilde{PC}$, then for each $(x,y) \in J$, we have

$$\|F(u)(x,y) - F(v)(x,y)\|$$

$$\le \sum_{k=1}^{m} \left(\|I_k(u(x_k^-,y)) - I_k(v(x_k^-,y))\| + \|I_k(u(x_k^-,0)) - I_k(v(x_k^-,0))\|\right)$$

$$+ \frac{1}{\Gamma(r_1)\Gamma(r_2)} \sum_{k=1}^{m} \int_{x_{k-1}}^{x_k} \int_0^y (x_k - s)^{r_1-1}(y-t)^{r_2-1} \|g(s,t) - h(s,t)\| \mathrm{d}t\mathrm{d}s$$

$$+ \frac{1}{\Gamma(r_1)\Gamma(r_2)} \int_{x_k}^{x} \int_0^y (x-s)^{r_1-1}(y-t)^{r_2-1} \|g(s,t) - h(s,t)\| \mathrm{d}t\mathrm{d}s,$$

where $g, h \in C(J)$ such that

$$g(x,y) = f(x,y,u_{(\rho_1(x,y,u_{(x,y)}),\rho_2(x,y,u_{(x,y)}))}, g(x,y)),$$

$$h(x,y) = f(x,y,u_{(\rho_1(x,y,v_{(x,y)}),\rho_2(x,y,u_{(x,y)}))}, g(x,y)).$$

By (7.27.2), we get

$$\|g(x,y) - h(x,y)\| \le k\|u(x,y) - v(x,y)\| + l\|g(x,y) - h(x,y)\|.$$

Then

$$\|g(x,y) - h(x,y)\| \le \frac{\ell}{1-l}\|u(x,y) - v(x,y)\|$$
$$\le \frac{\ell}{1-l}\|u - v\|_{PC}.$$

Thus, for each $(x,y) \in J$, we have

$$\|F(u)(x,y) - F(v)(x,y)\|$$

$$\le \sum_{k=1}^{m} l^* \left(\|u(x_k^-,y) - v(x_k^-,y)\| + \|u(x_k^-,0) - v(x_k^-,0)\|\right)$$

$$+ \frac{\ell}{(1-l)\Gamma(r_1)\Gamma(r_2)} \sum_{k=1}^{m} \int_{x_{k-1}}^{x_k} \int_0^y (x_k - s)^{r_1-1}(y-t)^{r_2-1} \|u - v\|_{PC}$$

$$+ \frac{\ell}{(1-l)\Gamma(r_1)\Gamma(r_2)} \int_{x_k}^{x} \int_0^y (x-s)^{r_1-1}(y-t)^{r_2-1} \|u - v\|_{PC} \mathrm{d}t\mathrm{d}s$$

$$\le \left(2ml^* + \frac{2\ell a^{r_1} b^{r_2}}{(1-l)\Gamma(1+r_1)\Gamma(1+r_2)}\right) \|u - v\|_{PC}.$$

Hence, for each $(x, y) \in [-\alpha, a] \times [-\beta, b]$, we get

$$\|F(u)(x,y) - F(v)(x,y)\| \leq \left(2ml^* + \frac{2\ell a^{r_1} b^{r_2}}{(1-l)\Gamma(1+r_1)\Gamma(1+r_2)}\right) \|u - v\|_{\widetilde{PC}}.$$

Consequently

$$\|F(u) - F(v)\|_{\widetilde{PC}} \leq \left(2ml^* + \frac{2\ell a^{r_1} b^{r_2}}{(1-l)\Gamma(1+r_1)\Gamma(1+r_2)}\right) \|u - v\|_{\widetilde{PC}}.$$

By the condition (7.81), we conclude that F is a contraction. As a consequence of Banach's fixed-point theorem, we deduce that F has a unique fixed-point which is a solution of the problem (7.73)–(7.76). □

In the following theorem we give an existence result for the problems (7.73)–(7.76) by applying the nonlinear alternative of Leray-Schauder type.

Theorem 7.28. *Assume that the following conditions hold:*

(7.28.1) There exist continuous functions $p, q, d : J \to \mathbb{R}_+$ such that

$$\|f(x, y, u, v)\| \leq p(x, y) + q(x, y)\|u\| + d(x, y)\|v\|.$$

for each $(x, y) \in J$, $u \in C$ and $z \in \mathbb{R}^n$.

(7.28.2) There exists a continuous and nondecreasing function $\Psi : [0, \infty) \to (0, \infty)$ such that

$$\|I_k(u)\| \leq \Psi(\|u\|); \quad \textit{for all } u \in \mathbb{R}^n.$$

(7.28.3) There exists a number $\overline{M} > 0$ such that

$$\frac{\overline{M}}{\|\mu\|_\infty + 2m\Psi(\overline{M}) + \frac{2a^{r_1} b^{r_2}(p^* + q^* \overline{M})}{(1-d^*)\Gamma(1+r_1)\Gamma(1+r_2)}} > 1,$$

where $p^* = \sup_{(x,y)\in J} p(x, y)$, $q^* = \sup_{(x,y)\in J} q(x, y)$ and $d^* = \sup_{(x,y)\in J} d(x, y)$. If $d^* < 1$, then (7.73)–(7.76) have at least one solution on $[-\alpha, a] \times [-\beta, b]$.

Proof. Consider the operator F defined in Theorem 7.27. We can easily show that the operator F is continuous and completely continuous.

A priori estimate. For $\lambda \in [0, 1]$, let u be such that for each $(x, y) \in J$ we have $u(x, y) = \lambda(Fu)(x, y)$. For each $(x, y) \in J$, by (3.30.1) and (7.28.2) we have

$$\|u(x,y)\| \leq \|\mu(x,y)\| + \sum_{k=1}^{m}(\|I_k(u(x_k^-,y))\| + \|I_k(u(x_k^-,0))\|)$$
$$+\frac{1}{\Gamma(r_1)\Gamma(r_2)}\sum_{k=1}^{m}\int_{x_{k-1}}^{x_k}\int_0^y (x_k-s)^{r_1-1}(y-t)^{r_2-1}\|g(s,t)\|dtds$$
$$+\frac{1}{\Gamma(r_1)\Gamma(r_2)}\int_{x_k}^{x}\int_0^y (x-s)^{r_1-1}(y-t)^{r_2-1}\|g(s,t)\|dtds, \qquad (7.82)$$

where $g \in C(J)$ such that

$$g(x,y) = f(x,y,u_{(\rho_1(x,y,u_{(x,y)}),\rho_2(x,y,u_{(x,y)}))}, g(x,y)).$$

By (7.28.1), for each $(x,y) \in J$, we have

$$\begin{aligned}\|g(x,y)\| &\leq p(x,y) + q(x,y)\|u\|_C + d(x,y)\|g(x,y)\|\\ &\leq p^* + q^*\|u\|_C + d^*\|g(x,y)\|.\end{aligned}$$

Thus, for each $(x,y) \in J$, we have

$$\|g(x,y)\| \leq \frac{p^* + q^*\|u\|_C}{1-d^*}.$$

Then, (7.82) implies that

$$\|u\|_\infty \leq \|\mu\|_\infty + 2m\Psi(\|u\|_\infty) + \frac{2a^{r_1}b^{r_2}(p^* + q^*\|u\|_C)}{(1-d^*)\Gamma(1+r_1)\Gamma(1+r_2)}.$$

Hence

$$\frac{\|u\|_{PC}}{\|\mu\|_\infty + 2m\Psi(\|u\|_{PC}) + \frac{2a^{r_1}b^{r_2}(p^*+q^*\|u\|_{PC})}{(1-d^*)\Gamma(1+r_1)\Gamma(1+r_2)}} \leq 1.$$

By condition (7.28.3), there exists $\overline{M}$ such that $\|u\|_{PC} \neq \overline{M}$.

Let

$$U = \{u \in \widetilde{PC} : \|u\|_{\widetilde{PC}} < \overline{M}\}.$$

The operator $F : \overline{U} \to \widetilde{PC}$ is continuous and completely continuous . From the choice of U, there is no $u \in \partial U$ such that $u = \lambda F(u)$ for some $\lambda \in (0,1)$. As a consequence of the nonlinear alternative of Leray-Schauder type, we deduce that F has a fixed-point u in $\overline{U}$ which is a solution of the problem (7.73)–(7.76). □

7.6.3 Existence Results with Infinite Delay

Now we present two existence results for the problems (7.77)–(7.80). Let us start in this section by defining what we mean by a solution of the problem (7.77)–(7.80). Let the space

$$\Omega := \{u : (-\infty, a] \times (-\infty, b] \to \mathbb{R}^n : u_{(x,y)} \in B \text{ for } (x,y) \in E \text{ and } u|_J \in PC\}.$$

Definition 7.29. A function $u \in \Omega$ such that $u(x,y)$, $\overline{D}^{r_1}_{x_k,x}u(x,y)$, $\overline{D}^{r_2}_{x_k,y}u(x,y)$, $\overline{D}^{r}_{x_k}u(x,y)$ are continuous for $(x,y) \in J_k$; $k = 0,\ldots,m$ and $I^{1-r}_{x_k}u(x,y) \in AC(J_k)$; $k = 0,\ldots,m$ is said to be a solution of (7.77)–(7.80) if u satisfies the condition (7.79) on $\tilde{J}'$, (7.77) on J_k; $k = 0,\ldots,m$ and conditions (7.78) and (7.80) are satisfied.

Set $\mathcal{R}' := \mathcal{R}'_{(\rho_1^-,\rho_2^-)}$

$$= \{(\rho_1(s,t,u), \rho_2(s,t,u)) : (s,t,u) \in J \times B,\ \rho_i(s,t,u) \le 0;\ i = 1,2\}.$$

We always assume that $\rho_1 : J \times B \to (-\infty, a]$, $\rho_2 : J \times B \to (-\infty, b]$ are continuous and the function $(s,t) \longmapsto u_{(s,t)}$ is continuous from $\mathcal{R}'$ into B.

We will need to introduce the following hypothesis:

(H_ϕ) There exists a continuous bounded function $L : \mathcal{R}'_{(\rho_1^-,\rho_2^-)} \to (0,\infty)$ such that

$$\|\phi_{(s,t)}\|_B \le L(s,t)\|\phi\|_B, \text{ for any} (s,t) \in \mathcal{R}'.$$

In the sequel we will make use of the following generalization of a consequence of the phase space axioms [148].

Lemma 7.30. *If $u \in \Omega$, then*

$$\|u_{(s,t)}\|_B = (M + L')\|\phi\|_B + K \sup_{(\theta,\eta)\in[0,\max\{0,s\}]\times[0,\max\{0,t\}]} \|u(\theta,\eta)\|,$$

where

$$L' = \sup_{(s,t)\in\mathcal{R}'} L(s,t).$$

Our first existence result for the IVP (7.77)–(7.80) is based on the Banach contraction principle.

Theorem 7.31. *Assume that the following hypotheses hold:*

(7.31.1) there exist $\bar{\ell} > 0$ and $0 < \bar{l} < 1$ such that

$$\|f(x,y,u,v) - f(x,y,\overline{u},\overline{v})\| \le \bar{\ell}\|u - \overline{u}\|_B + \bar{l}\|v - \overline{v}\|,$$

for any u, $\overline{u} \in B$, $v, \overline{v} \in \mathbb{R}^n$, and $(x,y) \in J$.

(7.31.2) There exists a constant $l^* > 0$ *such that*

$$\|I_k(u) - I_k(\overline{u})\| \leq l^* \|u - \overline{u}\|, \text{ for each } u, \overline{u} \in \mathbb{R}^n, \ k = 1, \ldots, m.$$

If

$$2ml^* + \frac{2K\bar{\ell}a^{r_1}b^{r_2}}{(1-\bar{l})\Gamma(1+r_1)\Gamma(1+r_2)} < 1, \tag{7.83}$$

then there exists a unique solution for IVP (7.77)–(7.80) on $(-\infty, a] \times (-\infty, b]$.

Proof. Transform the problems (7.77)–(7.80) into a fixed-point problem. Consider the operator $N : \Omega \to \Omega$ defined by

$$N(u)(x,y) = \begin{cases} \phi(x,y), & (x,y) \in \tilde{J}', \\ \mu(x,y) + \sum\limits_{0<x_k<x} (I_k(u(x_k^-, y)) \\ \quad - I_k(u(x_k^-, 0))) \\ \quad + \frac{1}{\Gamma(r_1)\Gamma(r_2)} \sum\limits_{0<x_k<x} \int\limits_{x_{k-1}}^{x_k} \int\limits_0^y (x_k - s)^{r_1 - 1} \\ \quad \times (y-t)^{r_2-1} g(s,t) dt ds \\ \quad + \frac{1}{\Gamma(r_1)\Gamma(r_2)} \int\limits_{x_k}^{x} \int\limits_0^y (x-s)^{r_1-1} \\ \quad \times (y-t)^{r_2-1} g(s,t) dt ds, & (x,y) \in J, \end{cases} \tag{7.84}$$

where $g \in C(J)$ such that

$$g(x,y) = f(x, y, u_{(\rho_1(x,y,u_{(x,y)}), \rho_2(x,y,u_{(x,y)}))}, g(x,y)).$$

Let $v(.,.) : (-\infty, a] \times (-\infty, b] \to \mathbb{R}^n$ be a function defined by,

$$v(x,y) = \begin{cases} \phi(x,y), \ (x,y) \in \tilde{J}', \\ \mu(x,y), \ (x,y) \in J. \end{cases}$$

Then $v_{(x,y)} = \phi$ for all $(x,y) \in E$. For each $w \in C(J, \mathbb{R}^n)$ with $w(x,y) = 0$ for each $(x,y) \in E$ we denote by $\overline{w}$ the function defined by

$$\overline{w}(x,y) = \begin{cases} 0, & (x,y) \in \tilde{J}', \\ w(x,y) & (x,y) \in J. \end{cases}$$

If $u(.,.)$ satisfies the integral equation

$$u(x,y) = \mu(x,y) + \frac{1}{\Gamma(r_1)\Gamma(r_2)} \int\limits_0^x \int\limits_0^y (x-s)^{r_1-1}(y-t)^{r_2-1} g(s,t) dt ds,$$

we can decompose $u(.,.)$ as $u(x,y)=\overline{w}(x,y)+v(x,y);\ (x,y)\in J$, which implies $u_{(x,y)}=\overline{w}_{(x,y)}+v_{(x,y)}$, for every $(x,y)\in J$, and the function $w(.,.)$ satisfies

$$\begin{aligned} w(x,y) = &\sum_{0<x_k<x}(I_k(u(x_k^-,y))-I_k(u(x_k^-,0)))\\ &+\frac{1}{\Gamma(r_1)\Gamma(r_2)}\sum_{0<x_k<x}\int_{x_{k-1}}^{x_k}\int_0^y(x_k-s)^{r_1-1}(y-t)^{r_2-1}g(s,t)\mathrm{d}t\mathrm{d}s\\ &+\frac{1}{\Gamma(r_1)\Gamma(r_2)}\int_{x_k}^{x}\int_0^y(x-s)^{r_1-1}(y-t)^{r_2-1}g(s,t)\mathrm{d}t\mathrm{d}s, \end{aligned}$$

for each $(x,y)\in J$, where $g\in C(J)$ such that

$$g(x,y)=f(x,y,\overline{w}_{(\rho_1(x,y,u_{(x,y)}),\rho_2(x,y,u_{(x,y)}))}+v_{(\rho_1(x,y,u_{(x,y)}),\rho_2(x,y,u_{(x,y)}))},g(x,y)).$$

Set

$$C_0=\{w\in\Omega:\ w(x,y)=0\ \text{ for } (x,y)\in E\},$$

and let $\|.\|_{(a,b)}$ be the seminorm in C_0 defined by

$$\|w\|_{(a,b)}=\sup_{(x,y)\in E}\|w_{(x,y)}\|_B+\sup_{(x,y)\in J}\|w(x,y)\|=\sup_{(x,y)\in J}\|w(x,y)\|,\ w\in C_0.$$

C_0 is a Banach space with norm $\|.\|_{(a,b)}$. Let the operator $P: C_0\to C_0$ be defined by

$$\begin{aligned} (Pw)(x,y) = &\sum_{0<x_k<x}(I_k(u(x_k^-,y))-I_k(u(x_k^-,0)))\\ &+\frac{1}{\Gamma(r_1)\Gamma(r_2)}\sum_{0<x_k<x}\int_{x_{k-1}}^{x_k}\int_0^y(x_k-s)^{r_1-1}(y-t)^{r_2-1}g(s,t)\mathrm{d}t\mathrm{d}s\\ &+\frac{1}{\Gamma(r_1)\Gamma(r_2)}\int_{x_k}^{x}\int_0^y(x-s)^{r_1-1}(y-t)^{r_2-1}g(s,t)\mathrm{d}t\mathrm{d}s, \end{aligned}\qquad(7.85)$$

for each $(x,y)\in J$, where $g\in C(J)$ such that

$$g(x,y)=f(x,y,\overline{w}_{(\rho_1(x,y,u_{(x,y)}),\rho_2(x,y,u_{(x,y)}))}+v_{(\rho_1(x,y,u_{(x,y)}),\rho_2(x,y,u_{(x,y)}))},g(x,y))$$

The operator N has a fixed-point is equivalent to P has a fixed-point, and so we turn to proving that P has a fixed-point. We will show that $P: C_0\to C_0$ is a contraction map. Indeed, for each $u,v\in C_0$, we get

$$\|P(u)-P(v)\|_{(a,b)} \leq \left(2ml^* + \frac{2K\bar{\ell}a^{r_1}b^{r_2}}{(1-\bar{l})\Gamma(1+r_1)\Gamma(1+r_2)}\right)\|u-v\|_{(a,b)},$$

and by (7.83), P is a contraction map. Hence P has a unique fixed-point by Banach's contraction principle. □

Now we give an existence result based on the nonlinear alternative of Leray-Schauder type.

Theorem 7.32. *Assume* (H_ϕ) *and*

(7.32.1) There exist constants $\bar{p},\bar{q} > 0$ *and* $0 < \bar{d} < 1$ *such that*

$$\|f(x,y,u,v)\| \leq \bar{p}+\bar{q}\|u\|_B+\bar{d}\|v\|, \textit{ for } (x,y)\in J \textit{ and each } u\in B,\ v\in\mathbb{R}^n.$$

(7.32.2) There exist $c_k > 0;\ k=1,\ldots,m$ *such that*

$$\|I_k(u)\| \leq c_k \quad \textit{for all } u\in\mathbb{R}^n.$$

Then the IVP (7.77)–(7.80) have at least one solution on $(-\infty,a]\times(-\infty,b]$.

Proof. Let $P: C_0 \to C_0$ defined as in (7.85). As in Theorem 7.31, we can show that the operator P is continuous and completely continuous.

We now show there exists an open set $U' \subseteq C_0$ with $w \neq \lambda P(w)$, for $\lambda\in(0,1)$ and $w\in\partial U'$. By (7.32.1) for each $(x,y)\in J$, we have

$$\|g(x,y)\| \leq \frac{\bar{p}+\bar{q}\|\overline{w}+v\|_B}{1-\bar{d}}.$$

On the other hand, Lemma 7.30 implies that, for each $(s,t)\in J$ we have

$$\begin{aligned}\|\overline{w}_{(s,t)}+v_{(s,t)}\|_B &\leq \|\overline{w}_{(s,t)}\|_B + \|v_{(s,t)}\|_B\\ &\leq K\sup\{w(\tilde{s},\tilde{t}) : (\tilde{s},\tilde{t})\in[0,s]\times[0,t]\}\\ &\quad +(M+L')\|\phi\|_B + K\|\phi(0,0)\|. \end{aligned}\tag{7.86}$$

If we name $z(s,t)$ the right-hand side of (7.86), then we have

$$\|\overline{w}_{(s,t)}+v_{(s,t)}\|_B \leq z(x,y).$$

Let $w\in C_0$ and $w=\lambda P(w)$ for some $0<\lambda<1$. By (7.32.2), for each $(x,y)\in J$, we obtain

$$\begin{aligned}\|w(x,y)\| \leq \sum_{k=1}^{m} 2c_k + \frac{2}{(1-L')\Gamma(r_1)\Gamma(r_2)}\int_0^x\int_0^y (x-s)^{r_1-1}(y-t)^{r_2-1}\\ \times(\bar{p}+\bar{q}\|\overline{w}_{(s,t)}+v_{(s,t)}\|_B)dtds\end{aligned}$$

$$\leq \sum_{k=1}^{m} 2c_k + \frac{2\bar{p}a^{r_1}b^{r_2}}{(1-L')\Gamma(1+r_1)\Gamma(1+r_2)}$$
$$+\frac{2\bar{q}}{(1-L')\Gamma(r_1)\Gamma(r_2)}\int_0^x\int_0^y (x-s)^{r_1-1}(y-t)^{r_2-1}z(s,t)\mathrm{d}t\mathrm{d}s$$
$$\leq \rho + \frac{2\bar{q}}{(1-L')\Gamma(r_1)\Gamma(r_2)}\int_0^x\int_0^y (x-s)^{r_1-1}(y-t)^{r_2-1}z(s,t)\mathrm{d}t\mathrm{d}s, \tag{7.87}$$

where

$$\rho := \sum_{k=1}^{m} 2c_k + \frac{2\bar{p}a^{r_1}b^{r_2}}{(1-l')\Gamma(1+r_1)\Gamma(1+r_2)}.$$

Using (7.87) and the definition of z, for each $(x, y) \in J$ we have

$$z(x,y) \leq (M+L')\|\phi\|_B + K\|\phi(0,0)\| + \rho K$$
$$+\frac{2K\bar{q}}{(1-l')\Gamma(r_1)\Gamma(r_2)}\int_0^x\int_0^y (x-s)^{r_1-1}(y-t)^{r_2-1}z(s,t)\mathrm{d}t\mathrm{d}s.$$

$$z(x,y) \leq \rho' + \frac{2K\bar{q}}{(1-l')\Gamma(r_1)\Gamma(r_2)}\int_0^x\int_0^y (x-s)^{r_1-1}(y-t)^{r_2-1}z(s,t)\mathrm{d}t\mathrm{d}s,$$

where

$$\rho' = (M+L')\|\phi\|_B + K\|\phi(0,0)\| + \rho K.$$

Then by Lemma 2.43, there exists $\delta = \delta(r_1, r_2)$ such that we have

$$\|z(x,y)\| \leq \rho'\left(1 + \frac{2\delta K\bar{q}a^{r_1}b^{r_2}}{(1-l')\Gamma(1+r_1)\Gamma(1+r_2)}\right) := R.$$

Hence, (7.87) implies that

$$\|w\|_\infty \leq \rho + \frac{2a^{r_1}b^{r_2}\bar{q}R}{(1-l')\Gamma(1+r_1)\Gamma(1+r_2)} := R^*.$$

Set

$$U' = \left\{w \in C_0 : \|w\|_{(a,b)} < R^* + 1\right\}.$$

$P : \overline{U'} \to C_0$ is continuous and completely continuous. By our choice of U', there is no $w \in \partial U'$ such that $w = \lambda P(w)$, for $\lambda \in (0,1)$. As a consequence of the nonlinear alternative of Leray-Schauder type, we deduce that N has a fixed-point which is a solution to problems (7.77)–(7.80). □

7.6.4 Examples

7.6.4.1 Example 1

As an application of our results we consider the following impulsive partial implicit hyperbolic differential equations of the form:

$$\overline{D}^r_{x_k}u(x,y) = \frac{e^{-x-y}|u(x-\sigma_1(u(x,y)), y-\sigma_2(u(x,y)))|}{(9+e^{x+y})(1+|u(x-\sigma_1(u(x,y)), y-\sigma_2(u(x,y)))|+|\overline{D}^r_\theta u(x,y)|)};$$
$$\text{if } (x,y) \in J_k;\ k=0,1, \tag{7.88}$$

$$u\left(\left(\frac{1}{2}\right)^+, y\right) = u\left(\left(\frac{1}{2}\right)^-, y\right) + \frac{|u((\frac{1}{2})^-, y)|}{3+|u((\frac{1}{2})^-, y)|};\ y \in [0,1], \tag{7.89}$$

$$u(x,y) = x+y^2;\ (x,y) \in [-1,1]\times[-2,1]\backslash(0,1]\times(0,1], \tag{7.90}$$

$$u(x,0) = x,\ u(0,y) = y^2;\ \ x,y \in [0,1], \tag{7.91}$$

where $J_0 = [0,\frac{1}{2}]\times[0,1]$, $J_1 = (\frac{1}{2},1]\times[0,1]$, $\sigma_1 \in C(\mathbb{R},[0,1])$, $\sigma_2 \in C(\mathbb{R},[0,2])$. Set

$$\rho_1(x,y,\varphi) = x - \sigma_1(\varphi(0,0)),\ \ (x,y,\varphi) \in J \times C,$$

$$\rho_2(x,y,\varphi) = y - \sigma_2(\varphi(0,0)),\ \ (x,y,\varphi) \in J \times C,$$

where $C := C_{(1,2)}$. Set

$$f(x,y,\varphi,\psi) = \frac{e^{-x-y}|\varphi|}{(9+e^{x+y})(1+|\varphi|+|\psi|)};\ (x,y) \in [0,1]\times[0,1],\ \varphi \in C,\ \psi \in \mathbb{R},$$

and

$$I_k(u) = \frac{u}{3+u},\ \ u \in \mathbb{R}_+.$$

A simple computation shows that conditions of Theorem 7.27 are satisfied, which implies that problems (7.88)–(7.91) have a unique solution defined on $[-1,1] \times [-2,1]$.

7.6.4.2 Example 2

We consider now the following impulsive fractional order partial hyperbolic differential equations with infinite delay of the form:

$$\overline{D}_{\theta}^{r}u(x,y) = \frac{ce^{x+y-\gamma(x+y)}|u(x-\sigma_1(u(x,y)),y-\sigma_2(u(x,y)))|}{(e^{x+y}+e^{-x-y})(1+|u(x-\sigma_1(u(x,y)),y-\sigma_2(u(x,y)))|+|\overline{D}_{\theta}^{r}u(x,y)|)};$$
$$\text{if } (x,y)\in J_k;\ k=0,...m, \tag{7.92}$$

$$u\left(\left(\frac{k}{k+1}\right)^{+},y\right) = u\left(\left(\frac{k}{k+1}\right)^{-},y\right) + \frac{\left|u\left(\left(\frac{k}{k+1}\right)^{-},y\right)\right|}{3mk+\left|u\left(\left(\frac{k}{k+1}\right)^{-},y\right)\right|};$$
$$y\in[0,1],\ k=1,\dots,m, \tag{7.93}$$

$$u(x,0)=x,\ u(0,y)=y^2;\ \ x,y\in[0,1], \tag{7.94}$$

$$u(x,y)=x+y^2;\ (x,y)\in\tilde{J}:=(-\infty,1]\times(-\infty,1]\backslash(0,1]\times(0,1], \tag{7.95}$$

where $J_0 = [0,\frac{1}{2}]\times[0,1]$, $J_k = (\frac{k}{k+1},\frac{k+1}{k+2}]\times[0,1]$; $k=1,\dots,m$, $c=\frac{10}{\Gamma(1+r_1)\Gamma(1+r_2)}$, γ a positive real constant and $\sigma_1,\sigma_2\in C(\mathbb{R},[0,\infty))$. Let the phase space

$$B_\gamma = \{u\in PC((-\infty,0]\times(-\infty,0],\mathbb{R}):\ \lim_{\|(\theta,\eta)\|\to\infty} e^{\gamma(\theta+\eta)}u(\theta,\eta) \text{ exists in } \mathbb{R}\}.$$

The norm of B_γ is given by

$$\|u\|_\gamma = \sup_{(\theta,\eta)\in(-\infty,0]\times(-\infty,0]} e^{\gamma(\theta+\eta)}|u(\theta,\eta)|.$$

Set

$$\rho_1(x,y,\varphi) = x-\sigma_1(\varphi(0,0)),\ \ (x,y,\varphi)\in J\times B_\gamma,$$
$$\rho_2(x,y,\varphi) = y-\sigma_2(\varphi(0,0)),\ \ (x,y,\varphi)\in J\times B_\gamma,$$

$$f(x,y,\varphi,\psi) = \frac{ce^{x+y-\gamma(x+y)}|\varphi|}{(e^{x+y}+e^{-x-y})(1+|\varphi|+|\psi|)},$$
$$(x,y)\in[0,1]\times[0,1],\ \varphi\in B_\gamma,\ \psi\in\mathbb{R}$$

and

$$I_k(u) = \frac{u}{3mk+u};\ u\in\mathbb{R}_+,\ k=1,\dots,m.$$

We can easily show that conditions of Theorem 7.31 are satisfied, and hence problems (7.92)–(7.95) haveL a unique solution defined on $(-\infty, 1] \times (-\infty, 1]$.

7.7 Notes and Remarks

The results of Chap. 6 are taken from Abbas and Benchohra [16, 17] and Abbas et al. [31, 33, 34]. Other results may be found in [242, 244, 246, 247].

Chapter 8
Fractional Order Riemann–Liouville Integral Equations

8.1 Introduction

In this chapter, we shall present existence results for some classes of Riemann–Liouville integral equations of two variables by using some fixed-point theorems.

8.2 Uniqueness Results for Fredholm-Type Fractional Order Riemann–Liouville Integral Equations

8.2.1 Introduction

In this section we study the existence and uniqueness of solutions of the Fredholm-type Riemann–Liouville integral equation of the form

$$u(x,y) = \mu(x,y) + \frac{1}{\Gamma(r_1)\Gamma(r_2)} \int_0^a \int_0^b (a-s)^{r_1-1}(b-t)^{r_2-1}$$
$$\times f(x,y,s,t,u(s,t),({}^cD_0^r u)(s,t))\mathrm{d}t\mathrm{d}s; \text{ if } (x,y) \in J := [0,a]\times[0,b], \tag{8.1}$$

where $a, b \in (0,\infty)$, $\mu : J \to \mathbb{R}^n$, $f : J \times J \times \mathbb{R}^n \times \mathbb{R}^n \to \mathbb{R}^n$ are given continuous functions.

For $w, {}^cD_\theta^r w \in C(J)$, denote

$$\|w(x,y)\|_1 = \|w(x,y)\| + \|{}^cD_\theta^r w(x,y)\|.$$

Let E be the space of functions $w, {}^cD_\theta^r w \in C(J)$, which fulfill the following condition:

$$\exists M \geq 0 : \|w(x,y)\|_1 \leq M\mathrm{e}^{\lambda(x+y)}, \text{ for } (x,y) \in J, \tag{8.2}$$

S. Abbas et al., *Topics in Fractional Differential Equations*, Developments in Mathematics 27, DOI 10.1007/978-1-4614-4036-9_8,

where λ is a positive constant. In the space E we define the norm [210]

$$\|w\|_E = \sup_{(x,y)\in J} \left\{\|w(x,y)\|_1 e^{-\lambda(x+y)}\right\}.$$

It is easy to see that $(E, \|.\|_E)$ is a Banach space. We note that the condition (8.2) implies that

$$\|w\|_E \leq M. \tag{8.3}$$

8.2.2 *Main Results*

Let us start by defining what we mean by a solution of (8.1).

Definition 8.1. We mean by a solution of (8.1), every function $w \in C(J)$, such that the mixed derivative $D^2_{xy}(w)$ exists and is integrable on J, and w satisfies (8.1) on J.

Further, we present conditions for the uniqueness of the solution of (8.1).

Theorem 8.2. *Assume*

(8.2.1) There exist functions $\rho_1, \rho_2 : J \times J \to \mathbb{R}^+$, *such that* f *and* ${}^cD^r_\theta(f)$ *satisfy*

$$\|f(x,y,s,t,u,v) - f(x,y,s,t,\overline{u},\overline{v})\| \leq \rho_1(x,y,s,t)(\|u-\overline{u}\| + \|v-\overline{v}\|) \tag{8.4}$$

and

$$\|({}^cD^r_0 f)(x,y,s,t,u,v) - ({}^cD^r_\theta f)(x,y,s,t,\overline{u},\overline{v})\| \leq \rho_2(x,y,s,t)(\|u-\overline{u}\| + \|v-\overline{v}\|), \tag{8.5}$$

for each $(x,y),(s,t) \in J$ *and* $u, v, \overline{u}, \overline{v} \in \mathbb{R}^n$.

(8.2.2) for λ *as in (8.2), there exist nonnegative constants* $\alpha_1, \alpha_2, \alpha_3, \beta_1, \beta_2$, *and* $0 < r_3 < \min\{r_1, r_2\}$ *such that, for* $(x,y) \in J$, *we have*

$$\begin{cases} \|\mu(x,y)\|_1 \leq \alpha_1 e^{\lambda(x+y)}, \\ \int_0^a \int_0^b \|f(x,y,s,t,0,0)\|^{\frac{1}{r_3}} dt ds \leq \alpha_2^{\frac{1}{r_3}} e^{\frac{\lambda}{r_3}(x+y)}, \\ \int_0^a \int_0^b \|{}^cD^r_\theta f(x,y,s,t,0,0)\|^{\frac{1}{r_3}} dt ds \leq \alpha_3^{\frac{1}{r_3}} e^{\frac{\lambda}{r_3}(x+y)}, \end{cases} \tag{8.6}$$

and

$$\begin{cases} \int_0^a \int_0^b \rho_1^{\frac{1}{r_3}}(x,y,s,t) e^{\frac{1}{r_3}(s+t)} \mathrm{d}t\mathrm{d}s \le \beta_1^{\frac{1}{r_3}} e^{\frac{\lambda}{r_3}(x+y)}, \\ \int_0^a \int_0^b \rho_2^{\frac{1}{r_3}}(x,y,s,t) e^{\frac{\lambda}{r_3}(s+t)} \mathrm{d}t\mathrm{d}s \le \beta_2^{\frac{1}{r_3}} e^{\frac{\lambda}{r_3}(x+y)}. \end{cases} \tag{8.7}$$

If

$$\frac{(\beta_1+\beta_2) a^{(\omega_1+1)(1-r_3)} b^{(\omega_2+1)(1-r_3)}}{(\omega_1+1)^{(1-r_3)}(\omega_2+1)^{(1-r_3)} \Gamma(r_1)\Gamma(r_2)} e^{\lambda(x+y)} < 1, \tag{8.8}$$

where $\omega_1 = \frac{r_1-1}{1-r_3}$, $\omega_2 = \frac{r_2-1}{1-r_3}$, *then (8.1) has a unique solution on* J *in* E.

Proof. Let $u \in E$ and define the operator $N : E \to E$ by

$$(Nu)(x,y) = \mu(x,y) + \frac{1}{\Gamma(r_1)\Gamma(r_2)} \int_0^a \int_0^b (a-s)^{r_1-1}(b-t)^{r_2-1}$$
$$\times f(x,y,s,t,u(s,t),({}^c D_\theta^r u)(s,t)) \mathrm{d}t\mathrm{d}s. \tag{8.9}$$

Differentiating both sides of (8.9) by applying the Caputo fractional derivative, we get

$${}^c D_0^r (Nu)(x,y) = {}^c D_0^r \mu(x,y) + \frac{1}{\Gamma(r_1)\Gamma(r_2)} \int_0^a \int_0^b (a-s)^{r_1-1}(b-t)^{r_2-1}$$
$$\times {}^c D_0^r f(x,y,s,t,u(s,t),({}^c D_0^r u)(s,t)) \mathrm{d}t\mathrm{d}s. \tag{8.10}$$

Now, we show that N maps E into itself. Evidently, $N(u)$, ${}^c D_0^r (Nu)$ are continuous on J. We verify that (8.2) is fulfilled. From (8.3), (8.6), (8.7), and using the hypotheses, for each $(x,y) \in J$, we have

$$\|(Nu)(x,y)\|_1 \le \|\mu(x,y)\|_1 + \frac{1}{\Gamma(r_1)\Gamma(r_2)} \int_0^a \int_0^b (a-s)^{r_1-1}(b-t)^{r_2-1}$$
$$\times \left\| f\left(x,y,s,t,u(s,t),\left({}^c D_0^r u\right)(s,t)\right) - f(x,y,s,t,0,0) \right\| \mathrm{d}t\mathrm{d}s$$
$$+ \frac{1}{\Gamma(r_1)\Gamma(r_2)} \int_0^a \int_0^b (a-s)^{r_1-1}(b-t)^{r_2-1} \| f(x,y,s,t,0,0) \| \mathrm{d}t\mathrm{d}s$$

$$+\frac{1}{\Gamma(r_1)\Gamma(r_2)}\int_0^a\int_0^b (a-s)^{r_1-1}(b-t)^{r_2-1}$$

$$\times\|{}^cD_0^r f\left(x,y,s,t,u(s,t),\left({}^cD_0^r u\right)(s,t)\right) -{}^c D_0^r f\,(x,y,s,t,0,0)\,\|\mathrm{d}t\mathrm{d}s$$

$$+\frac{1}{\Gamma(r_1)\Gamma(r_2)}\int_0^a\int_0^b (a-s)^{r_1-1}(b-t)^{r_2-1}\|{}^cD_0^r f(x,y,s,t,0,0)\|\mathrm{d}t\mathrm{d}s$$

$$\leq \|\mu(x,y)\|_1+\frac{1}{\Gamma(r_1)\Gamma(r_2)}\left(\int_0^a\int_0^b (a-s)^{\frac{r_1-1}{1-r_3}}(b-t)^{\frac{r_2-1}{1-r_3}}\mathrm{d}t\mathrm{d}s\right)^{1-r_3}$$

$$\times\left(\int_0^a\int_0^b \|f\left(x,y,s,t,u(s,t),\left({}^cD_\theta^r u\right)(s,t)\right)-f\,(x,y,s,t,0,0)\,\|^{\frac{1}{r_3}}\mathrm{d}t\mathrm{d}s\right)^{r_3}$$

$$+\frac{1}{\Gamma(r_1)\Gamma(r_2)}\left(\int_0^a\int_0^b (a-s)^{\frac{r_1-1}{1-r_3}}(b-t)^{\frac{r_2-1}{1-r_3}}\mathrm{d}t\mathrm{d}s\right)^{1-r_3}$$

$$\times\left(\int_0^a\int_0^b \|f(x,y,s,t,0,0)\|^{\frac{1}{r_3}}\mathrm{d}t\mathrm{d}s\right)^{r_3}$$

$$+\frac{1}{\Gamma(r_1)\Gamma(r_2)}\left(\int_0^a\int_0^b (a-s)^{\frac{r_1-1}{1-r_3}}(b-t)^{\frac{r_2-1}{1-r_3}}\mathrm{d}t\mathrm{d}s\right)^{1-r_3}$$

$$\times\left(\int_0^a\int_0^b \|{}^cD_0^r f(x,y,s,t,u(s,t),({}^cD_0^r u)(s,t)) -{}^c D_0^r f(x,y,s,t,0,0)\|^{\frac{1}{r_3}}\mathrm{d}t\mathrm{d}s\right)^{r_3}$$

$$+\frac{1}{\Gamma(r_1)\Gamma(r_2)}\left(\int_0^a\int_0^b (a-s)^{\frac{r_1-1}{1-r_3}}(b-t)^{\frac{r_2-1}{1-r_3}}\mathrm{d}t\mathrm{d}s\right)^{1-r_3}$$

$$\times\left(\int_0^a\int_0^b \|{}^cD_0^r f(x,y,s,t,0,0)\|^{\frac{1}{r_3}}\mathrm{d}t\mathrm{d}s\right)^{r_3}.$$

Thus, for each $(x, y) \in J$, we obtain

$$\|(Nu)(x,y)\|_1 \leq \|\mu(x,y)\|_1 + \frac{a^{(\omega_1+1)(1-r_3)} b^{(\omega_2+1)(1-r_3)}}{(\omega_1+1)^{(1-r_3)}(\omega_2+1)^{(1-r_3)}\Gamma(r_1)\Gamma(r_2)}$$

$$\times \Bigg[\left(\int_0^a \int_0^b \|f(x,y,s,t,0,0)\|^{\frac{1}{r_3}} \mathrm{d}t\mathrm{d}s \right)^{r_3}$$

$$+ \left(\int_0^a \int_0^b \|{}^cD_0^r f(x,y,s,t,0,0)\|^{\frac{1}{r_3}} \mathrm{d}t\mathrm{d}s \right)^{r_3}$$

$$+ \left(\int_0^a \int_0^b \rho_1^{\frac{1}{r_3}}(x,y,s,t)\|u(s,t)\|_1^{\frac{1}{r_3}} \mathrm{d}t\mathrm{d}s \right)^{r_3}$$

$$+ \left(\int_0^a \int_0^b \rho_2^{\frac{1}{r_3}}(x,y,s,t)\|u(s,t)\|_1^{\frac{1}{r_3}} \mathrm{d}t\mathrm{d}s \right)^{r_3} \Bigg]$$

$$\leq \alpha_1 e^{\lambda(x+y)} + \frac{a^{(\omega_1+1)(1-r_3)} b^{(\omega_2+1)(1-r_3)}}{(\omega_1+1)^{(1-r_3)}(\omega_2+1)^{(1-r_3)}\Gamma(r_1)\Gamma(r_2)}$$

$$\times \Bigg[\alpha_2 e^{\lambda(x+y)} + \alpha_3 e^{\lambda(x+y)} + \|u\|_E \left(\int_0^a \int_0^b \rho_1^{\frac{1}{r_3}}(x,y,s,t) e^{\frac{\lambda}{r_3}(s+t)} \mathrm{d}t\mathrm{d}s \right)^{r_3}$$

$$+ \|u\|_E \left(\int_0^a \int_0^b \rho_2^{\frac{1}{r_3}}(x,y,s,t) e^{\frac{\lambda}{r_3}(s+t)} \mathrm{d}t\mathrm{d}s \right)^{r_3} \Bigg]$$

$$\leq \left[\alpha_1 + \frac{(\alpha_2 + \alpha_3 + M\beta_1 + M\beta_2) a^{(\omega_1+1)(1-r_3)} b^{(\omega_2+1)(1-r_3)}}{(\omega_1+1)^{(1-r_3)}(\omega_2+1)^{(1-r_3)}\Gamma(r_1)\Gamma(r_2)} \right] e^{\lambda(x+y)}.$$

Hence, for each $(x, y) \in J$, we get

$$\|(Nu)(x,y)\|_1 \leq \left[\alpha_1 + \frac{(\alpha_2 + \alpha_3 + M\beta_1 + M\beta_2) a^{(\omega_1+1)(1-r_3)} b^{(\omega_2+1)(1-r_3)}}{(\omega_1+1)^{(1-r_3)}(\omega_2+1)^{(1-r_3)}\Gamma(r_1)\Gamma(r_2)} \right] e^{\lambda(x+y)}. \tag{8.11}$$

From (8.11), it follows that $N(u) \in E$. This proves that the operator N maps E into itself. Next, we verify that the operator N is a contraction map. Let $u, v \in E$. From (8.9), (8.10) and using the hypotheses, for each $(x, y) \in J$, we have

$$\|(Nu)(x,y)-(Nv)(x,y)\|_1 \leq \frac{1}{\Gamma(r_1)\Gamma(r_2)}\int_0^a\int_0^b (a-s)^{r_1-1}(b-t)^{r_2-1}$$

$$\times \|f(x,y,s,t,u(s,t),({}^cD_0^r u)(s,t)) - f(x,y,s,t,v(s,t),({}^cD_0^r v)(s,t))\| \mathrm{d}t\,\mathrm{d}s$$

$$+\frac{1}{\Gamma(r_1)\Gamma(r_2)}\int_0^a\int_0^b (a-s)^{r_1-1}(b-t)^{r_2-1}$$

$$\times \|{}^cD_0^r f(x,y,s,t,u(s,t),({}^cD_0^r u)(s,t)) -{}^c D_0^r f(x,y,s,t,v(s,t),({}^cD_0^r v)(s,t))\| \mathrm{d}t\,\mathrm{d}s$$

$$\leq \frac{a^{(\omega_1+1)(1-r_3)}b^{(\omega_2+1)(1-r_3)}}{(\omega_1+1)^{(1-r_3)}(\omega_2+1)^{(1-r_3)}\Gamma(r_1)\Gamma(r_2)}$$

$$\times\left[\left(\int_0^a\int_0^b \rho_1^{\frac{1}{r_3}}(x,y,s,t)\|u(s,t)-v(s,t)\|_1^{\frac{1}{r_3}}\mathrm{d}t\,\mathrm{d}s\right)^{r_3}\right.$$

$$\left.+\left(\int_0^a\int_0^b \rho_2^{\frac{1}{r_3}}(x,y,s,t)\|u(s,t)-v(s,t)\|_1^{\frac{1}{r_3}}\mathrm{d}t\,\mathrm{d}s\right)^{r_3}\right]$$

$$\leq \frac{a^{(\omega_1+1)(1-r_3)}b^{(\omega_2+1)(1-r_3)}}{(\omega_1+1)^{(1-r_3)}(\omega_2+1)^{(1-r_3)}\Gamma(r_1)\Gamma(r_2)}\left[\left(\int_0^a\int_0^b \rho_1^{\frac{1}{r_3}}(x,y,s,t)\mathrm{e}^{\frac{\lambda}{r_3}(s+t)}\mathrm{d}t\,\mathrm{d}s\right)^{r_3}\right.$$

$$\left.+\left(\int_0^a\int_0^b \rho_2^{\frac{1}{r_3}}(x,y,s,t)\mathrm{e}^{\frac{\lambda}{r_3}(s+t)}\mathrm{d}t\,\mathrm{d}s\right)^{r_3}\right]\|u-v\|_E$$

$$\leq \frac{(\beta_1+\beta_2)a^{(\omega_1+1)(1-r_3)}b^{(\omega_2+1)(1-r_3)}}{(\omega_1+1)^{(1-r_3)}(\omega_2+1)^{(1-r_3)}\Gamma(r_1)\Gamma(r_2)}\mathrm{e}^{\lambda(x+y)}\|u-v\|_E.$$

Hence

$$\|Nu-Nv\|_E \leq \frac{(\beta_1+\beta_2)a^{(\omega_1+1)(1-r_3)}b^{(\omega_2+1)(1-r_3)}}{(\omega_1+1)^{(1-r_3)}(\omega_2+1)^{(1-r_3)}\Gamma(r_1)\Gamma(r_2)}\|u-v\|_E.$$

From (8.8), it follows that N has a unique fixed point in E by Banach contraction principle (see [95], p. 37). The fixed point of N is however a solution of (8.1). □

The next result deals with the uniqueness of a solution of (8.1) in E when the functions ρ_i; $i \in \{1,2\}$ are in the form

$$\rho_i(x,y,s,t) = h_i(s,t)c(x,y) = m_i(s,t)\mathrm{e}^{\lambda(x+y-s-t)};\ i \in \{1,2\}.$$

Theorem 8.3. *Assume*

(8.3.1) For λ as in (8.2), there exist constants $\alpha > 0,\ 0 < r_3 < \min\{r_1, r_2\}$, and strictly positive functions $m_1(x,y), m_2(x,y) \in L^{\frac{1}{r_3}}(J)$ such that for each $(x,y),(s,t) \in J,\ \mu,\ f$ and ${}^cD_0^r(f)$ satisfy

$$\|\mu(x,y)\|_1 \leq \alpha e^{\lambda(x+y)} \tag{8.12}$$

and

$$\begin{cases} \|f(x,y,s,t,0,0)\| \leq m_1(s,t)e^{\lambda(x+y)}, \\ \|{}^cD_\theta^r f(x,y,s,t,0,0)\| \leq m_2(s,t)e^{\lambda(x+y)}. \end{cases} \tag{8.13}$$

(8.3.2) There exist strictly positive functions $m_3(x,y), m_4(x,y) \in L^{\frac{1}{r_3}}(J)$ such that for each $(x,y),(s,t) \in J$ and $u,v,\overline{u},\overline{v} \in \mathbb{R}^n,\ f$ and ${}^cD_0^r(f)$ satisfy

$$\begin{aligned} \|f(x,y,s,t,u,v) - f(x,y,s,t,\overline{u},\overline{v})\| &\leq m_3(s,t) \\ \times e^{\lambda(x+y-s-t)}(\|u-\overline{u}\| + \|v-\overline{v}\|) \end{aligned} \tag{8.14}$$

and

$$\begin{aligned} \|({}^cD_0^r f)(x,y,s,t,u,v) - ({}^cD_0^r f)(x,y,s,t,\overline{u},\overline{v})\| &\leq m_4(s,t) \\ \times e^{\lambda(x+y-s-t)}(\|u-\overline{u}\| + \|v-\overline{v}\|). \end{aligned} \tag{8.15}$$

If

$$\frac{(M_3+M_4)a^{(\omega_1+1)(1-r_3)}b^{(\omega_2+1)(1-r_3)}}{(\omega_1+1)^{(1-r_3)}(\omega_2+1)^{(1-r_3)}\Gamma(r_1)\Gamma(r_2)} < 1, \tag{8.16}$$

where

$$\omega_1 = \frac{r_1-1}{1-r_3},\ \omega_2 = \frac{r_2-1}{1-r_3},\ M_i = \|m_i\|_{L^{\frac{1}{r_3}}};\ i \in \{3,4\},$$

then (8.1) has a unique solution on J in E.

Proof. Consider the operator $N : E \to E$ defined in (8.9). Now, we show that $N(u)$ maps E into itself. Evidently, $N(u)$, ${}^cD_0^r(Nu)$ are continuous on J and $N(u)$, ${}^cD_0^r(Nu) \in \mathbb{R}^n$. We verify that (8.2) is fulfilled. From (8.3), (8.14), (8.15) and using the hypotheses, for each $(x,y) \in J$, we have

$$\begin{aligned} \|(Nu)(x,y)\|_1 \leq \|\mu(x,y)\|_1 + \frac{1}{\Gamma(r_1)\Gamma(r_2)} \int_0^a \int_0^b (a-s)^{r_1-1}(b-t)^{r_2-1} \\ \times \|f(x,y,s,t,u(s,t),({}^cD_0^r u)(s,t)) - f(x,y,s,t,0,0)\| dt ds \end{aligned}$$

$$+\frac{1}{\Gamma(r_1)\Gamma(r_2)}\int_0^a\int_0^b (a-s)^{r_1-1}(b-t)^{r_2-1}\|f(x,y,s,t,0,0)\|\mathrm{d}t\mathrm{d}s$$

$$+\frac{1}{\Gamma(r_1)\Gamma(r_2)}\int_0^a\int_0^b (a-s)^{r_1-1}(b-t)^{r_2-1}$$

$$\times\|{}^cD_0^r f\left(x,y,s,t,u(s,t),\left({}^cD_0^r u\right)(s,t)\right) -{}^c D_0^r f(x,y,s,t,0,0)\|\mathrm{d}t\mathrm{d}s$$

$$+\frac{1}{\Gamma(r_1)\Gamma(r_2)}\int_0^a\int_0^b (a-s)^{r_1-1}(b-t)^{r_2-1}\|{}^cD_0^r f(x,y,s,t,0,0)\|\mathrm{d}t\mathrm{d}s$$

$$\leq \|\mu(x,y)\|_1 + \frac{1}{\Gamma(r_1)\Gamma(r_2)}\left(\int_0^a\int_0^b (a-s)^{\frac{r_1-1}{1-r_3}}(b-t)^{\frac{r_2-1}{1-r_3}}\mathrm{d}t\mathrm{d}s\right)^{1-r_3}$$

$$\times\left(\int_0^a\int_0^b \|f(x,y,s,t,u(s,t),({}^cD_\theta^r u)(s,t)) - f(x,y,s,t,0,0)\|^{\frac{1}{r_3}}\mathrm{d}t\mathrm{d}s\right)^{r_3}$$

$$+\frac{1}{\Gamma(r_1)\Gamma(r_2)}\left(\int_0^a\int_0^b (a-s)^{\frac{r_1-1}{1-r_3}}(b-t)^{\frac{r_2-1}{1-r_3}}\mathrm{d}t\mathrm{d}s\right)^{1-r_3}$$

$$\times\left(\int_0^a\int_0^b \|f(x,y,s,t,0,0)\|^{\frac{1}{r_3}}\mathrm{d}t\mathrm{d}s\right)^{r_3}$$

$$+\frac{1}{\Gamma(r_1)\Gamma(r_2)}\left(\int_0^a\int_0^b (a-s)^{\frac{r_1-1}{1-r_3}}(b-t)^{\frac{r_2-1}{1-r_3}}\mathrm{d}t\mathrm{d}s\right)^{1-r_3}$$

$$\times\left(\int_0^a\int_0^b \|{}^cD_0^r f(x,y,s,t,u(s,t),({}^cD_0^r u)(s,t)) -{}^c D_0^r f(x,y,s,t,0,0)\|^{\frac{1}{r_3}}\mathrm{d}t\mathrm{d}s\right)^{r_3}$$

$$+\frac{1}{\Gamma(r_1)\Gamma(r_2)}\left(\int_0^a\int_0^b (a-s)^{\frac{r_1-1}{1-r_3}}(b-t)^{\frac{r_2-1}{1-r_3}}\mathrm{d}t\mathrm{d}s\right)^{1-r_3}$$

$$\times\left(\int_0^a\int_0^b \|{}^cD_0^r f(x,y,s,t,0,0)\|^{\frac{1}{r_3}}\mathrm{d}t\mathrm{d}s\right)^{r_3}.$$

Thus, for each $(x, y) \in J$, we obtain

$$\|(Nu)(x,y)\|_1 \leq \|\mu(x,y)\|_1 + \frac{a^{(\omega_1+1)(1-r_3)}b^{(\omega_2+1)(1-r_3)}}{(\omega_1+1)^{(1-r_3)}(\omega_2+1)^{(1-r_3)}\Gamma(r_1)\Gamma(r_2)}$$

$$\times \left[\left(\int_0^a \int_0^b (m_1(s,t)e^{\lambda(x+y)})^{\frac{1}{r_3}} dt ds \right)^{r_3} + \left(\int_0^a \int_0^b (m_2(s,t)e^{\lambda(x+y)})^{\frac{1}{r_3}} dt ds \right)^{r_3} \right.$$

$$+ \left(\int_0^a \int_0^b (m_3(s,t)\|u(s,t)\|_1 e^{\lambda(x+y-s-t)})^{\frac{1}{r_3}} dt ds \right)^{r_3}$$

$$\left. + \left(\int_0^a \int_0^b (m_4(s,t)\|u(s,t)\|_1 e^{\lambda(x+y-s-t)})^{\frac{1}{r_3}} dt ds \right)^{r_3} \right]$$

$$\leq \alpha e^{\lambda(x+y)} + \frac{a^{(\omega_1+1)(1-r_3)}b^{(\omega_2+1)(1-r_3)}}{(\omega_1+1)^{(1-r_3)}(\omega_2+1)^{(1-r_3)}\Gamma(r_1)\Gamma(r_2)} \left[M_1 e^{\lambda(x+y)} + M_2 e^{\lambda(x+y)} \right.$$

$$+ \|u\|_E \left(\int_0^a \int_0^b (m_3(s,t)e^{\lambda(x+y)})^{\frac{1}{r_3}} dt ds \right)^{r_3}$$

$$\left. + \|u\|_E \left(\int_0^a \int_0^b (m_4(s,t)e^{\lambda(x+y)})^{\frac{1}{r_3}} dt ds \right)^{r_3} \right],$$

where

$$M_i = \|m_i\|_{L^{\frac{1}{r_3}}}; \; i \in \{1, 2\}.$$

Hence, for each $(x, y) \in J$, we get

$$\|(Nu)(x,y)\|_1 \leq \left[\alpha + \frac{\left(M_1 + M_2 + M(M_3 + M_4)\right) a^{(\omega_1+1)(1-r_3)}b^{(\omega_2+1)(1-r_3)}}{(\omega_1+1)^{(1-r_3)}(\omega_2+1)^{(1-r_3)}\Gamma(r_1)\Gamma(r_2)} \right] e^{\lambda(x+y)}. \tag{8.17}$$

From (8.17), it follows that $N(u) \in E$. This proves that the operator N maps E into itself. Next, we verify that the operator N is a contraction map. Let $u(x, y), v(x, y) \in E$. From (8.9), (8.10) and using the hypotheses, for each $(x, y) \in J$, we have

$$\|(Nu)(x,y)-(Nv)(x,y)\|_1 \le \frac{1}{\Gamma(r_1)\Gamma(r_2)}\int_0^a\int_0^b (a-s)^{r_1-1}(b-t)^{r_2-1}$$

$$\times \left\|f\left(x,y,s,t,u(s,t),\left({}^cD_0^r u\right)(s,t)\right)-f\left(x,y,s,t,v(s,t),\left({}^cD_0^r v\right)(s,t)\right)\right\|dt\,ds$$

$$+\frac{1}{\Gamma(r_1)\Gamma(r_2)}\int_0^a\int_0^b (a-s)^{r_1-1}(b-t)^{r_2-1}$$

$$\times \|{}^cD_0^r f(x,y,s,t,u(s,t),({}^cD_0^r u)(s,t)) -{}^c D_\theta^r f(x,y,s,t,v(s,t),({}^cD_0^r v)(s,t))\|dt\,ds$$

$$\le \frac{a^{(\omega_1+1)(1-r_3)}b^{(\omega_2+1)(1-r_3)}}{(\omega_1+1)^{(1-r_3)}(\omega_2+1)^{(1-r_3)}\Gamma(r_1)\Gamma(r_2)}$$

$$\times\left[\left(\int_0^a\int_0^b (m_3(s,t)\|u(s,t)-v(s,t)\|_1 e^{\lambda(x+y-s-t)})^{\frac{1}{r_3}}\,dt\,ds\right)^{r_3}\right.$$

$$\left.+\left(\int_0^a\int_0^b (m_4(s,t)\|u(s,t)-v(s,t)\|_1 e^{\lambda(x+y-s-t)})^{\frac{1}{r_3}}\,dt\,ds\right)^{r_3}\right]$$

$$\le \frac{(M_3+M_4)a^{(\omega_1+1)(1-r_3)}b^{(\omega_2+1)(1-r_3)}}{(\omega_1+1)^{(1-r_3)}(\omega_2+1)^{(1-r_3)}\Gamma(r_1)\Gamma(r_2)}e^{\lambda(w+y)}\|u-v\|_E.$$

Hence

$$\|Nu-Nv\|_E \le \frac{(M_3+M_4)a^{(\omega_1+1)(1-r_3)}b^{(\omega_2+1)(1-r_3)}}{(\omega_1+1)^{(1-r_3)}(\omega_2+1)^{(1-r_3)}\Gamma(r_1)\Gamma(r_2)}\|u-v\|_E.$$

From (8.16), it follows that N has a unique fixed point in E by Banach contraction principle (see [95], p. 37). The fixed point of N is however a solution of (8.1). □

For $w, {}^cD_{0,x}^{r_1}w, {}^cD_{0,y}^{r_2}w \in C(J)$, denote

$$\|w(x,y)\|_1 = \|w(x,y)\| + \|{}^cD_{0,x}^{r_1}w(x,y)\| + \|{}^cD_{0,y}^{r_2}w(x,y)\|.$$

Let E be the space of functions $w, {}^cD_{0,x}^{r_1}w, {}^cD_{0,y}^{r_2}w \in C(J)$, which fulfill condition (8.2).

Theorem 8.4. *Consider the following Fredholm-type Riemann–Liouville integral equation of the form:*

$$u(x,y) = \mu(x,y) + \frac{1}{\Gamma(r_1)\Gamma(r_2)}\int_0^a\int_0^b (a-s)^{r_1-1}(b-t)^{r_2-1}$$

$$\times f(x,y,s,t,u(s,t),({}^cD_{0,s}^{r_1}u)(s,t),({}^cD_{0,t}^{r_2}u)(s,t))dt\,ds;$$

$$\textit{if } (x,y)\in J := [0,a]\times[0,b]. \tag{8.18}$$

Assume

(8.4.1) *There exist functions* $\rho_1, \rho_2, \rho_3 : J \times J \to \mathbb{R}^+$, *such that* f ${}^cD_{0,x}^{r_1}(f)$, *and* ${}^cD_{0,y}^{r_2}(f)$ *satisfy*

$$\|f(x,y,s,t,u,v,w) - f(x,y,s,t,\overline{u},\overline{v},\overline{w})\| \leq \rho_1(x,y,s,t)(\|u-\overline{u}\| + \|v-\overline{v}\| + \|w-\overline{w}\|) \tag{8.19}$$

and

$$\|({}^cD_{0,x}^{r_1}f)(x,y,s,t,u,v,w) - ({}^cD_{0,x}^{r_1}f)(x,y,s,t,\overline{u},\overline{v},\overline{w})\| \leq \rho_2(x,y,s,t)(\|u-\overline{u}\| + \|v-\overline{v}\| + \|w-\overline{w}\|), \tag{8.20}$$

$$\|({}^cD_{0,y}^{r_2}f)(x,y,s,t,u,v,w) - ({}^cD_{0,y}^{r_2}f)(x,y,s,t,\overline{u},\overline{v},\overline{w})\| \leq \rho_3(x,y,s,t)(\|u-\overline{u}\| + \|v-\overline{v}\| + \|w-\overline{w}\|), \tag{8.21}$$

for each $(x,y),(s,t) \in J$ *and* $u,v,w,\overline{u},\overline{v},\overline{w} \in \mathbb{R}^n$.

(8.4.2) *for* λ *as in (8.2), there exist nonnegative constants* $\alpha_1,\alpha_2,\alpha_3,\alpha_4,\beta_1,\beta_2,\beta_3$ *and* $0 < r_3 < \min\{r_1,r_2\}$ *such that, for* $(x,y) \in J$, *we have*

$$\begin{cases} \|\mu(x,y)\|_1 \leq \alpha_1 e^{\lambda(x+y)}, \\ \int_0^a\int_0^b \|f(x,y,s,t,0,0,0)\|^{\frac{1}{r_3}} dt\,ds \leq \alpha_2^{\frac{1}{r_3}} e^{\frac{\lambda}{r_3}(x+y)}, \\ \int_0^a\int_0^b \|{}^cD_{0,x}^{r_1}f(x,y,s,t,0,0,0)\|^{\frac{1}{r_3}} dt\,ds \leq \alpha_3^{\frac{1}{r_3}} e^{\frac{\lambda}{r_3}(x+y)}, \\ \int_0^a\int_0^b \|{}^cD_{0,y}^{r_2}f(x,y,s,t,0,0,0)\|^{\frac{1}{r_3}} dt\,ds \leq \alpha_4^{\frac{1}{r_3}} e^{\frac{\lambda}{r_3}(x+y)}, \end{cases} \tag{8.22}$$

and

$$\begin{cases} \int_0^a\int_0^b \rho_1^{\frac{1}{r_3}}(x,y,s,t) e^{\frac{\lambda}{r_3}(s+t)} dt\,ds \leq \beta_1^{\frac{1}{r_3}} e^{\frac{\lambda}{r_3}(x+y)}, \\ \int_0^a\int_0^b \rho_2^{\frac{1}{r_3}}(x,y,s,t) e^{\frac{\lambda}{r_3}(s+t)} dt\,ds \leq \beta_2^{\frac{1}{r_3}} e^{\frac{\lambda}{r_3}(x+y)}, \\ \int_0^a\int_0^b \rho_3^{\frac{1}{r_3}}(x,y,s,t) e^{\frac{\lambda}{r_3}(s+t)} dt\,ds \leq \beta_3^2 e^{2\lambda(x+y)}. \end{cases} \tag{8.23}$$

If

$$\frac{(\beta_1+\beta_2+\beta_3)a^{(\omega_1+1)(1-r_3)}b^{(\omega_2+1)(1-r_3)}}{(\omega_1+1)^{(1-r_3)}(\omega_2+1)^{(1-r_3)}\Gamma(r_1)\Gamma(r_2)}e^{\lambda(x+y)} < 1, \tag{8.24}$$

where $\omega_1 = \frac{r_1-1}{1-r_3}$, $\omega_2 = \frac{r_2-1}{1-r_3}$, *then (8.18) has a unique solution on* J *in* E.

Theorem 8.5. *Assume*

(8.5.1) *For* λ *as in (8.2), there exist constants* $\alpha > 0$, $0 < r_3 < \min\{r_1, r_2\}$, *and strictly positive functions* $m_1(x,y), m_2(x,y) \in L^{\frac{1}{r_3}}(J)$ *such that for each* $(x,y),(s,t) \in J$, μ, f *and* ${}^cD^r_\theta(f)$ *satisfy*

$$\|\mu(x,y)\|_1 \le \alpha e^{\lambda(x+y)} \tag{8.25}$$

and

$$\begin{cases} \|f(x,y,s,t,0,0,0)\| \le m_1(s,t)e^{\lambda(x+y)}, \\ \|{}^cD^r_\theta f(x,y,s,t,0,0,0)\| \le m_2(s,t)e^{\lambda(x+y)}. \end{cases} \tag{8.26}$$

(8.5.2) *There exist strictly positive functions* $m_3(x,y), m_4(x,y), m_5(x,y) \in L^{\frac{1}{r_3}}(J)$ *such that for each* $(x,y),(s,t) \in J$ *and* $u, v, \overline{u}, \overline{v} \in \mathbb{R}^n$, f *and* ${}^cD^r_\theta(f)$ *satisfy*

$$\begin{aligned} &\|f(x,y,s,t,u,v,w) - f(x,y,s,t,\overline{u},\overline{v}),\overline{w}\| \le m_3(s,t)e^{\lambda(x+y-s-t)} \\ &\times(\|u-\overline{u}\| + \|v-\overline{v}\| + \|w-\overline{w}\|), \end{aligned} \tag{8.27}$$

$$\begin{aligned} &\|({}^cD^{r_1}_{0,x}f)(x,y,s,t,u,v,w) - ({}^cD^{r_1}_{0,x}f)(x,y,s,t,\overline{u},\overline{v},\overline{w})\| \le m_4(s,t) \\ &\times e^{\lambda(x+y-s-t)}(\|u-\overline{u}\| + \|v-\overline{v}\| + \|w-\overline{w}\|) \end{aligned} \tag{8.28}$$

and

$$\begin{aligned} &\|({}^cD^{r_2}_{0,y}f)(x,y,s,t,u,v,w) - ({}^cD^{r_2}_{0,y}f)(x,y,s,t,\overline{u},\overline{v},\overline{w})\| \le m_5(s,t) \\ &\times e^{\lambda(x+y-s-t)}(\|u-\overline{u}\| + \|v-\overline{v}\| + \|w-\overline{w}\|). \end{aligned} \tag{8.29}$$

If

$$\frac{(M_3+M_4+M_5)a^{(\omega_1+1)(1-r_3)}b^{(\omega_2+1)(1-r_3)}}{(\omega_1+1)^{(1-r_3)}(\omega_2+1)^{(1-r_3)}\Gamma(r_1)\Gamma(r_2)} < 1, \tag{8.30}$$

where

$$\omega_1 = \frac{r_1-1}{1-r_3}, \quad \omega_2 = \frac{r_2-1}{1-r_3}, \quad M_i = \|m_i\|_{L^{\frac{1}{r_3}}};\ i \in \{3,4,5\},$$

then (8.18) has a unique solution on J *in* E.

8.3 Fractional Order Riemann–Liouville Integral Equations with Multiple Time Delay

8.3.1 Introduction

In this section we investigate the existence and uniqueness of solutions for the following fractional order integral equations for the system:

$$u(x,y)=\sum_{i=1}^{m} g_i(x,y)u(x-\xi_i,y-\mu_i)+I_0^r f(x,y,u(x,y));$$
$$\text{if } (x,y)\in J:=[0,a]\times[0,b], \tag{8.31}$$
$$u(x,y)=\Phi(x,y);\ \text{if } (x,y)\in\tilde{J}:=[-\xi,a]\times[-\mu,b]\backslash(0,a]\times(0,b], \tag{8.32}$$

where $a,b>0$, $\xi_i,\mu_i\geq 0;\ i=1\ldots,m,\ \xi=\max_{i=1\ldots,m}\{\xi_i\},\ \mu=\max_{i=1\ldots,m}\{\mu_i\}$, I_0^r is the left-sided mixed Riemann–Liouville integral of order $r=(r_1,r_2)\in(0,\infty)\times(0,\infty)$, $f:J\times\mathbb{R}^n\to\mathbb{R}^n$, $g_i:J\to\mathbb{R}^n;\ i=1\ldots m$ are given continuous functions, and $\Phi:\tilde{J}\to\mathbb{R}^n$ is a given continuous function such that

$$\Phi(x,0)=\sum_{i=1}^{m} g_i(x,0)\Phi(x-\xi_i,-\mu_i);\ x\in[0,a],$$

and

$$\Phi(0,y)=\sum_{i=1}^{m} g_i(0,y)\Phi(-\xi_i,y-\mu_i);\ y\in[0,b].$$

We present three results for the problems (8.31)–(8.32), the first one is based on Schauder's fixed-point theorem, the second one is a uniqueness of the solution by using the Banach fixed-point theorem, and the last one on the nonlinear alternative of Leray–Schauder type.

8.3.2 Existence of Solutions

Set $C:=C([-\xi,a]\times[-\mu,b])$. C is a Banach space endowed with the norm

$$\|w\|_C=\sup_{(x,y)\in[-\xi,a]\times[-\mu,b]}\|w(x,y)\|.$$

Definition 8.6. A function $u\in C$ is said to be a solution of (8.31)–(8.32) if u satisfies (8.31) on J and condition (8.32) on $\tilde{J}$.

Set

$$B = \max_{i=1\ldots m}\{\sup_{(x,y)\in J}\|g_i(x,y)\|\}.$$

Theorem 8.7. *Assume*

(8.7.1) There exists a positive function $h \in C(J)$ such that

$$\|f(x,y,u)\| \le h(x,y), \text{ for all } (x,y) \in J \text{ and } u \in \mathbb{R}^n.$$

If $mB < 1$, then problems (8.31)–(8.32) have at least one solution u on $[-\xi,a] \times [-\mu,b]$.

Proof. Transform problems (8.31)–(8.32) into a fixed-point problem. Consider the operator $N : C \to C$ defined by,

$$N(u)(x,y) = \begin{cases} \Phi(x,y); & (x,y) \in \tilde{J}, \\ \sum_{i=1}^{m} g_i(x,y)u(x-\xi_i, y-\mu_i) + I_\theta^r f(x,y,u(x,y)); & (x,y) \in J. \end{cases} \tag{8.33}$$

The problem of finding the solutions of problems (8.31)–(8.32) is reduced to finding the solutions of the operator equation $N(u) = u$. Let $R \ge \text{Max}||\Phi||, \frac{R^*}{1-mB}$ where

$$R^* = \frac{a^{r_1}b^{r_2}h^*}{\Gamma(1+r_1)\Gamma(1+r_2)},$$

and $h^* = \|h\|_\infty$, and consider the set

$$B_R = \{u \in C : \|u\|_C \le R\}.$$

It is clear that B_R is a closed bounded and convex subset of C. For every $u \in B_R$ and $(x,y) \in J$ we obtain by (8.7.1) that

$$\begin{aligned}
\|N(u)(x,y)\| &\le \sum_{i=1}^{m}\|g_i(x,y)\|\|u(x-\xi_i, y-\mu_i)\| \\
&\quad + \frac{1}{\Gamma(r_1)\Gamma(r_2)}\int_0^x\int_0^y (x-s)^{r_1-1}(y-t)^{r_2-1}\|f(s,t,u(s,t))\|dtds \\
&\le mB\|u\|_C + \frac{1}{\Gamma(r_1)\Gamma(r_2)}\int_0^x\int_0^y (x-s)^{r_1-1}(y-t)^{r_2-1}h(s,t)dtds \\
&\le mB\|u\|_C + h^*\frac{a^{r_1}b^{r_2}}{\Gamma(1+r_1)\Gamma(1+r_2)} \\
&\le mBR + (1-mB)R = R.
\end{aligned}$$

On the other hand, for every $u \in B_R$ and $(x, y) \in \tilde{J}$, we obtain

$$\|N(u)(x,y)\| = \|\Phi(x,y)\| \leq R,$$

So we obtain that

$$\|N(u)\|_C \leq R.$$

Hence, $N(B_R) \subseteq B_R$. Since f is bounded on B_R, thus $N(B_R)$ is equicontinuous. Schauder fixed-point theorem shows that N has at least one fixed point $u^* \in B_R$ which corresponds to the solution of (8.31)–(8.32). □

For the uniqueness we prove the following theorem.

Theorem 8.8. *Assume that following hypothesis hold:*

(8.8.1) There exists a positive function $l \in C(J)$ such that

$$\|f(x,y,u) - f(x,y,v)\| \leq l(x,y)\|u - v\|,$$

for each $(x, y) \in J$ and $u, v \in \mathbb{R}^n$.

If

$$\frac{mB\Gamma(1+r_1)\Gamma(1+r_2) + a^{r_1}b^{r_2}l^*}{\Gamma(1+r_1)\Gamma(1+r_2)} < 1, \tag{8.34}$$

where $l^* = \|l\|_\infty$, then problems (8.31)–(8.32) have a unique solution on $[-\xi, a] \times [-\mu, b]$.

Proof. Consider the operator N defined in (8.33). Then by (8.8.1), for every $u, v \in C$ and $(x, y) \in J$ we have

$$\begin{aligned}
\|N(u)(x,y) - N(v)(x,y)\| &\leq \sum_{i=1}^{m} \|g_i(x,y)\| \|u(x-\xi_i, y-\mu_i) - v(x-\xi_i, y)\| \\
&\quad + \frac{1}{\Gamma(r_1)\Gamma(r_2)} \int_0^x \int_0^y (x-s)^{r_1-1}(y-t)^{r_2-1} \\
&\quad \times \|f(s,t,u(s,t)) - f(s,t,v(s,t))\| \mathrm{d}t\,\mathrm{d}s \\
&\leq mB\|u - v\|_\infty \\
&\quad + \frac{1}{\Gamma(r_1)\Gamma(r_2)} \int_0^x \int_0^y (x-s)^{r_1-1}(y-t)^{r_2-1} \\
&\quad \times l(s,t)\|u - v\|_C \mathrm{d}t\,\mathrm{d}s
\end{aligned}$$

$$\leq mB\|u-v\|_\infty + l^*\frac{a^{r_1}b^{r_2}}{\Gamma(1+r_1)\Gamma(1+r_2)}\|u-v\|_C$$

$$= \left(mB + \frac{l^*a^{r_1}b^{r_2}}{\Gamma(1+r_1)\Gamma(1+r_2)}\right)\|u-v\|_C.$$

Thus

$$\|N(u)-N(v)\|_C \leq \frac{mB\Gamma(1+r_1)\Gamma(1+r_2)+a^{r_1}b^{r_2}l^*}{\Gamma(1+r_1)\Gamma(1+r_2)}\|u-v\|_C$$

Hence by (8.34), we have that N is a contraction mapping. Then in view of Banach fixed-point theorem, N has a unique fixed point which is corresponding to the solution of problems (8.31)–(8.32). □

Now, we present an existence result for the problems (8.31)–(8.32) based on the nonlinear alternative of Leray–Schauder type.

Theorem 8.9. *Assume*

(8.9.1) There exist positive functions $p, q \in C(J)$ such that

$$\|f(x,y,u)\| \leq p(x,y) + q(x,y)\|u\|, \text{ for all } (x,y) \in J \text{ and } u \in \mathbb{R}^n.$$

If $mB < 1$, then problems (8.31)–(8.32) have at least one solution on $[-\xi, a] \times [-\mu, b]$.

Proof. Consider the operator N defined in (8.33). We shall show that the operator N is completely continuous. By the continuity of f and the Arzela-Ascoli Theorem, we can easily obtain that N is completely continuous.

A priori bounds. We shall show there exists an open set $U \subseteq C$ with $u \neq \lambda N(u)$, for $\lambda \in (0,1)$ and $u \in \partial U$. Let $u \in C$ and $u = \lambda N(u)$ for some $0 < \lambda < 1$. Thus for each $(x,y) \in J$, we have

$$u(x,y) = \lambda \sum_{i=1}^{m} g_i(x,y)u(x-\xi_i, y-\mu_i) + \lambda I_\theta^r f(x,y,u(x,y)).$$

This implies by (8.9.1) that, for each $(x,y) \in J$, we have

$$\|u(x,y)\| \leq mB\|u(x,y)\| + \frac{p^*a^{r_1}b^{r_2}}{\Gamma(1+r_1)\Gamma(1+r_2)}$$

$$+\frac{q^*}{\Gamma(r_1)\Gamma(r_2)}\int_0^x\int_0^y (x-s)^{r_1-1}(y-t)^{r_2-1}u(s,t)dtds,$$

where $p^* = \|p\|_\infty$ and $q^* = \|q\|_\infty$. Thus, for each $(x, y) \in J$, we get

$$\begin{aligned}\|u(x,y)\| &\leq \frac{p^* a^{r_1} b^{r_2}}{(1-mB)\Gamma(1+r_1)\Gamma(1+r_2)} \\ &\quad + \frac{q^*}{(1-mB)\Gamma(r_1)\Gamma(r_2)} \int_0^x \int_0^y (x-s)^{r_1-1}(y-t)^{r_2-1} u(s,t)\mathrm{d}t\mathrm{d}s \\ &\leq w + c\int_0^x \int_0^y (x-s)^{r_1-1}(y-t)^{r_2-1} u(s,t)\mathrm{d}t\mathrm{d}s,\end{aligned}$$

where

$$w := \frac{p^* a^{r_1} b^{r_2}}{(1-mB)\Gamma(1+r_1)\Gamma(1+r_2)}$$

and

$$c := \frac{q^*}{(1-mB)\Gamma(r_1)\Gamma(r_2)}.$$

From Lemma 2.43, there exists $\delta := \delta(r_1, r_2) > 0$ such that, for each $(x, y) \in J$, we get

$$\begin{aligned}\|u\|_\infty &\leq w\left(1 + c\delta \int_0^x \int_0^y (x-s)^{r_1-1}(y-t)^{r_2-1}\mathrm{d}t\mathrm{d}s\right) \\ &\leq w\left(1 + \frac{c\delta a^{r_1} b^{r_2}}{r_1 r_2}\right) := \widetilde{M}.\end{aligned}$$

Set $M^* := \max\{\|\Phi\|, \widetilde{M}\}$ and

$$U = \{u \in C : \|u\|_C < M^* + 1\}.$$

By our choice of U, there is no $u \in \partial U$ such that $u = \lambda N(u)$, for $\lambda \in (0, 1)$. As a consequence of Theorem 2.32, we deduce that N has a fixed point u in $\overline{U}$ which is a solution to problem (8.31)–(8.32). □

8.3.3 Examples

8.3.3.1 Example 1

As an application of our results we consider the following system of fractional integral equations of the form:

$$u(x,y) = \frac{x^3y}{8}u\left(x-\frac{3}{4}, y-3\right) + \frac{x^4y^2}{12}u\left(x-2, y-\frac{1}{2}\right) + \frac{1}{4}u\left(x-1, y-\frac{3}{2}\right)$$
$$+I_0^r f(x,y,u);\ \text{if } (x,y) \in J := [0,1]\times[0,1], \tag{8.35}$$

$$u(x,y) = 0;\ \text{if } (x,y) \in \tilde{J} := [-2,1]\times[-3,1]\backslash(0,1]\times(0,1], \tag{8.36}$$

where $m = 3,\ r = (\frac{1}{2}, \frac{1}{5})$ and

$$f(x,y,u) = \mathrm{e}^{x+y}\frac{1}{1+|u|}.$$

Set

$$g_1(x,y) = \frac{x^3y}{8},\ g_2(x,y) = \frac{x^4y^2}{12},\ g_3(x,y) = \frac{1}{4}.$$

We have $B = \frac{1}{4}$ and

$$|f(x,y,u)| \le \mathrm{e}^{x+y};\ \text{for all } (x,y)\in J \text{ and } u \in \mathbb{R}.$$

Then condition (8.7.1) is satisfied and $mB = \frac{3}{4} < 1$. In view of Theorem 8.7, problems (8.35)–(8.35) have a solution defined on $[-2,1]\times[-3,1]$.

8.3.3.2 Example 2

Consider the fractional integral equation

$$u(x,y) = \frac{x^3y}{8}u\left(x-1, y-\frac{1}{2}\right) + \frac{x^4y^2}{12}u\left(x-\frac{2}{5}, y-\frac{3}{4}\right) + \frac{1}{8}u(x-3, y-2)$$
$$+I_0^r f(x,y,u);\ \text{if } (x,y) \in J := [0,1]\times[0,1], \tag{8.37}$$

$$u(x,y) = \Phi(x,y);\ \text{if } (x,y) \in \tilde{J} := [-3,1]\times[-2,1]\backslash(0,1]\times(0,1], \tag{8.38}$$

where $m = 3,\ r = (\frac{1}{2}, \frac{1}{5}),\ f(x,y,u) = \frac{x+y}{20}\frac{|u|}{1+|u|}$ and $\Phi : \tilde{J} \to \mathbb{R}$ is continuous with

$$\Phi(x,0) = \frac{1}{8}\Phi(x-3,-2),\ \Phi(0,y) = \frac{1}{8}\Phi(-3,y-2);\ x,y\in[0,1]. \tag{8.39}$$

Notice that condition (8.39) is satisfied by $\Phi \equiv 0$.

Set

$$g_1(x,y) = \frac{x^3 y}{8}, \; g_2(x,y) = \frac{x^4 y^2}{12}, \; g_3(x,y) = \frac{1}{8}.$$

We have $B = \frac{1}{8}$. It is clear that f satisfies (8.8.1) with $l^* = \frac{1}{10}$. A simple computation shows that condition (8.34) is satisfied. Hence by Theorem 8.8, problems (8.37)–(8.38) have a unique solution defined on $[-3,1] \times [-2,1]$.

8.4 Nonlinear Quadratic Volterra Riemann–Liouville Integral Equations of Fractional Order

8.4.1 *Introduction*

This section deals with the existence of solutions to the following nonlinear quadratic Volterra integral equation of Riemann–Liouville fractional order:

$$u(x,y) = f(x,y,u(x,y)) \Bigg[\mu(x,y) + \frac{1}{\Gamma(r_1)\Gamma(r_2)} \times \int_0^x \int_0^y (x-s)^{r_1-1}(y-t)^{r_2-1} g(x,y,s,t,u(s,t))\, dt ds \Bigg],$$

$$\text{if } (x,y) \in J := [0,a] \times [0,b], \tag{8.40}$$

where $a, b > 0$, $r_1, r_2 \in (0,\infty)$, $f : J \times \mathbb{R} \to \mathbb{R}$, $g : \mathcal{D} \times \mathbb{R} \to \mathbb{R}$ and $\mu : J \to \mathbb{R}$ are given continuous functions, where

$$\mathcal{D} = \{((x,y),(s,t)) \in J \times J : s \le x \text{ and } t \le y\}.$$

8.4.2 *Existence of Solutions*

For proving our existence result, we employ the following hybrid fixed-point theorem of Dhage [109].

Theorem 8.10 (Dhage [109]). *Let D be a closed-convex and bounded subset of the Banach algebra X and let $F, G : D \to X$ be two operators satisfying:*

(a) A is Lipschitz with the Lipschitz constant λ.
(b) B is completely continuous.
(c) $Au\,Bz \in D$ for all $u, z \in D$.
(d) $\lambda M < 1$ where $M = \|B(D)\| = \sup_{u \in D} \|B(u)\|$.

Then the operator equation $AuBu = u$ has a solution and the set of all solutions is compact in D.

Now, we are concerned with the existence of solutions for (8.40). The following hypotheses will be used in the sequel.

(8.11.1) There exists a positive continuous function $\alpha : J \to \mathbb{R}$ such that

$$|f(x,y,u) - f(x,y,v)| \leq \alpha(x,y)|u-v|, \quad \text{for all } (x,y) \in J, \text{ and } u, v \in \mathbb{R}.$$

(8.11.2) There exists a positive continuous function $\beta : J \to \mathbb{R}$ and a positive bounded continuous function $h : [0,\infty) \to \mathbb{R}$ with $h(0) = 0$ such that for all $((x,y),(s,t)) \in \mathcal{D}$, and $u, v \in \mathbb{R}$,

$$|g(x,y,s,t,u) - g(x,y,s,t,v)| \leq \beta(x,y)h(|u-v|).$$

Set

$$K = \sup_{\eta > 0} \frac{a^{r_1} b^{r_2}[g^* + \|\beta\|_\infty h(\eta)]}{\Gamma(1+r_1)\Gamma(1+r_2)},$$

where $g^* = \sup_{((x,y),(s,t)) \in \mathcal{D}} g(x,y,s,t,0)$.

Theorem 8.11. *Assume that hypotheses (8.11.1) and (8.11.2) hold. If*

$$\|\alpha\|_\infty (K + \|\mu\|_\infty) < 1, \tag{8.41}$$

then (8.40) has at least one solution on J.

Proof. Consider the closed ball $D := \{u \in C(J) : \|u\|_\infty \leq \rho\}$, where

$$\rho = \frac{f^*(K + \|\mu\|_\infty)}{1 - \|\alpha\|_\infty (K + \|\mu\|_\infty)} > 0, \text{ and } f^* = \sup_{(x,y) \in J} \|f(x,y,0)\|.$$

Let us define two operators A and B on D by

$$Au(x,y) = f(x,y,u(x,y)); \quad (x,y) \in J, \tag{8.42}$$

$$Bu(x,y) = \mu(x,y) + \frac{1}{\Gamma(r_1)\Gamma(r_2)} \times \int_0^x \int_0^y (x-s)^{r_1-1}(y-t)^{r_2-1} g(x,y,s,t,u(s,t))\,dt\,ds; \ (x,y) \in J. \tag{8.43}$$

Clearly A and B define the operators $A,\ B : D \to C(J)$. Now solving (8.40) is equivalent to solving the operator equation

$$Au(x,y)\,Bu(x,y) = u(x,y); \quad (x,y) \in J. \tag{8.44}$$

We show that operators A and B satisfy all the assumptions of Theorem 8.10.

First we shall show that A is a Lipschitz. Let $u_1, u_2 \in D$. Then by (8.11.1), for all $(x,y) \in J$, we get

$$\begin{aligned} |Au_1(x,y) - Au_2(x,y)| &= |f(x,y,u_1(x,y)) - f(x,y,u_2(x,y))| \\ &\leq \alpha(x,y)|u_1(x,y) - u_2(x,y)|. \end{aligned}$$

Taking the maximum over (x,y) in the above inequality yields

$$\|Au_1 - Au_2\|_\infty \leq \|\alpha\|_\infty \|u_1 - u_2\|_\infty,$$

and so A is a Lipschitz with a Lipschitz constant $\|\alpha\|_\infty$.

Next, we show that B is a continuous and compact operator on D. The proof will be given in several steps.

Step 1: *B is continuous.* Let $\{u_n\}$ be a sequence such that $u_n \to u$ in D. Then, for each $(x,y) \in J$, we have

$$\begin{aligned} &|B(u_n)(x,y) - B(u)(x,y)| \\ &\leq \frac{1}{\Gamma(r_1)\Gamma(r_2)} \int_0^x \int_0^y |x-s|^{r_1-1}|y-t|^{r_2-1}|g(x,y,s,t,u_n(s,t)) \\ &\quad - g(x,y,s,t,u(s,t))|\mathrm{d}t\,\mathrm{d}s \\ &\leq \frac{1}{\Gamma(r_1)\Gamma(r_2)} \int_0^a \int_0^b |x-s|^{r_1-1}|y-t|^{r_2-1} \sup_{(x,y)\in J} \beta(x,y)h(\|u_n - u\|_\infty)\mathrm{d}t\,\mathrm{d}s \\ &\leq \frac{\|\beta\|_\infty h(\|u_n - u\|_\infty)}{\Gamma(r_1)\Gamma(r_2)} \int_0^a \int_0^b |x-s|^{r_1-1}|y-t|^{r_2-1}\mathrm{d}t\,\mathrm{d}s \\ &\leq \frac{a^{r_1}b^{r_2}\|\,\|\beta\|_\infty h(\|u_n - u\|_\infty)}{\Gamma(1+r_1)\Gamma(1+r_2)}. \end{aligned}$$

Since h is a continuous function, we have

$$\|B(u_n) - B(u)\|_\infty \leq \frac{a^{r_1}b^{r_2}\|\beta\|_\infty h(\|u_n - u\|_\infty)}{\Gamma(1+r_1)\Gamma(1+r_2)} \to 0 \text{ as } n \to \infty.$$

Step 2: *$B(D)$ is bounded.* Indeed, it is enough show that there exists a positive constant M^* such that, for each $u \in D$, we have $\|B(u)\| \leq M^*$. Let $u \in D$ be arbitrary; then for each $(x, y) \in J$, we have

$$\begin{aligned}
|Bu(x,y)| &\leq |\mu(x,y)| + \frac{1}{\Gamma(r_1)\Gamma(r_2)} \\
&\quad \times \int_0^x \int_0^y (x-s)^{r_1-1}(y-t)^{r_2-1}|g(x,y,s,t,u(s,t))|\mathrm{d}t\mathrm{d}s \\
&\leq \|\mu\|_\infty + \frac{a^{r_1}b^{r_2}[g^* + \|\beta\|_\infty h(\rho)]}{\Gamma(1+r_1)\Gamma(1+r_2)} \\
&\leq K + \|\mu\|_\infty := M^*.
\end{aligned}$$

Step 3: *$B(D)$ is equicontinuous.* Let $(x_1, y_1), (x_2, y_2) \in J$, $x_1 < x_2$, $y_1 < y_2$ and $u \in D$. Then

$$\begin{aligned}
&|B(u)(x_2,y_2) - B(u)(x_1,y_1)| = |\mu(x_1,y_1) - \mu(x_2,y_2)| \\
&\quad + \Bigg| \frac{1}{\Gamma(r_1)\Gamma(r_2)} \int_0^{x_1}\int_0^{y_1} \left[(x_2-s)^{r_1-1}(y_2-t)^{r_2-1} - (x_1-s)^{r_1-1}(y_1-t)^{r_2-1}\right] \\
&\quad \times g(x,y,s,t,u(s,t))\mathrm{d}t\mathrm{d}s \\
&\quad + \frac{1}{\Gamma(r_1)\Gamma(r_2)} \int_{x_1}^{x_2}\int_{y_1}^{y_2} (x_2-s)^{r_1-1}(y_2-t)^{r_2-1} g(x,y,s,t,u(s,t))\mathrm{d}t\mathrm{d}s \\
&\quad + \frac{1}{\Gamma(r_1)\Gamma(r_2)} \int_0^{x_1}\int_{y_1}^{y_2} (x_2-s)^{r_1-1}(y_2-t)^{r_2-1} g(x,y,s,t,u(s,t))\mathrm{d}t\mathrm{d}s \\
&\quad + \frac{1}{\Gamma(r_1)\Gamma(r_2)} \int_{x_1}^{x_2}\int_0^{y_1} (x_2-s)^{r_1-1}(y_2-t)^{r_2-1} g(x,y,s,t,u(s,t))\mathrm{d}t\mathrm{d}s \Bigg| \\
&\leq |\mu(x_1,y_1) - \mu(x_2,y_2)| + \frac{g^* + \|\beta\|_\infty h(\rho)}{\Gamma(r_1)\Gamma(r_2)} \\
&\quad \times \int_0^x\int_0^y \left[(x_1-s)^{r_1-1}(y_1-t)^{r_2-1} - (x_2-s)^{r_1-1}(y_2-t)^{r_2-1}\right]\mathrm{d}t\mathrm{d}s \\
&\quad + \frac{g^* + \|\beta\|_\infty h(\rho)}{\Gamma(r_1)\Gamma(r_2)} \int_{x_1}^{x_2}\int_{y_1}^{y_2} (x_2-s)^{r_1-1}(y_2-t)^{r_2-1}\mathrm{d}t\mathrm{d}s
\end{aligned}$$

$$+\frac{g^* + \|\beta\|_\infty h(\rho)}{\Gamma(r_1)\Gamma(r_2)} \int_0^{x_1}\int_{y_1}^{y_2} (x_2-s)^{r_1-1}(y_2-t)^{r_2-1}\mathrm{d}t\mathrm{d}s$$

$$+\frac{g^* + \|\beta\|_\infty h(\rho)}{\Gamma(r_1)\Gamma(r_2)} \int_{x_1}^{x_2}\int_0^{y_1} (x_2-s)^{r_1-1}(y_2-t)^{r_2-1}\mathrm{d}t\mathrm{d}s$$

$$\leq |\mu(x_1,y_1)-\mu(x_2,y_2)|$$

$$+\frac{[g^* + \|\beta\|_\infty h(\rho)]a^{r_1}b^{r_2}}{\Gamma(1+r_1)\Gamma(1+r_2)}\left[2y_2^{r_2}(x_2-x_1)^{r_1}+2x_2^{r_1}(y_2-y_1)^{r_2}\right.$$

$$\left.+x_1^{r_1}y_1^{r_2}-x_2^{r_1}y_2^{r_2}-2(x_2-x_1)^{r_1}(y_2-y_1)^{r_2}\right].$$

As $x_1 \to x_2$, $y_1 \to y_2$ the right-hand side of the above inequality tends to zero. As a consequence of steps 1–3 together with the Arzelá-Ascoli theorem, we can conclude that B is continuous and compact.

Next, we show that $Au\ Bu \in D$ for all $u \in D$. Let $u \in D$ be arbitrary, then for each $(x,y) \in J$,

$$|Au(x,y)Bu(x,y)| \leq (f^* + \rho\|\alpha\|_\infty)\left[|\mu(x,y)| + \frac{1}{\Gamma(r_1)\Gamma(r_2)}\right.$$

$$\left.\times \int_0^x\int_0^y (x-s)^{r_1-1}(y-t)^{r_2-1}|g(x,y,s,t,u(s,t))|\mathrm{d}t\mathrm{d}s\right]$$

$$\leq (f^* + \rho\|\alpha\|_\infty)\left[\|\mu\|_\infty + \frac{a^{r_1}b^{r_2}[g^* + \|\beta\|_\infty h(\rho)]}{\Gamma(1+r_1)\Gamma(1+r_2)}\right]$$

$$\leq (f^* + \rho\|\alpha\|_\infty)(\|\mu\|_\infty + K)$$

$$= \frac{f^*(k + \|\mu\|_\infty)}{1-\|\alpha\|_\infty(K+\|\mu\|_\infty)}$$

$$= \rho.$$

Also, we have

$$M = \|B(D)\| \leq K + \|\mu\|_\infty,$$

and therefore, by (8.41), we get

$$M\|\alpha\|_\infty \leq \|\alpha\|_\infty(K+\|\mu\|_\infty) < 1.$$

Now we apply Theorem 8.10 to conclude that (8.40) has a solution on J and the set of all solutions is compact in D. □

8.4.3 An Example

Consider the following quadratic Volterra integral equation of fractional order:

$$u(x,y) = \left[xy^2 + xyu(x,y)\right]\left[xye^{-(x^2+y^2)} + \frac{1}{\Gamma(\frac{2}{3})\Gamma(\frac{3}{4})} \times \int_0^x\int_0^y (x-s)^{-\frac{1}{3}}(y-t)^{-\frac{1}{4}}\left(xt + \frac{e^{-(x+y+s+t)}}{1+u^{\frac{2}{3}}(s,t)}\right)dt\,ds\right];\ x,y \in [0,1], \tag{8.45}$$

where $r = (\frac{2}{3}, \frac{3}{4})$, $\mu : [0,1]\times[0,1] \to \mathbb{R}$, $f : [0,1]\times[0,1]\times\mathbb{R} \to \mathbb{R}$ and $g : \mathcal{D}\times\mathbb{R} \to \mathbb{R}$ defined by

$$\mu(x,y) = xye^{-(x^2+y^2)},\quad f(x,y,u) = xy^2 + xyu,$$

and

$$g(x,y,s,t,u) = xt + \frac{1}{1+u^{\frac{2}{3}}(s,t)}e^{-(x+y+s+t)}.$$

Here the set $\mathcal{D}$ is defined by

$$\mathcal{D} = \{((x,y),(s,t)) \in [0,1]\times[0,1]\times[0,1]\times[0,1] : s \le x \text{ and } t \le y\}.$$

Then we can easily check that the assumptions of Theorem 8.11 are satisfied. In fact, we have that the function f is continuous and satisfies assumption (8.11.1), where $\alpha(x,y) = xy$, then $\|\alpha\|_\infty = 1$ and $f^* = 1$. Next, let us notice that the function g satisfies assumption (A2), where $\beta(x,y) = e^{-(x+y)}$, $h(\rho) = \frac{1}{1+\rho^{\frac{2}{3}}}$, and $g^* = 1$. Also, condition (8.41) is satisfied. Indeed,

$$K = \sup_{\eta>0}\frac{a^{r_1}b^{r_2}[g^* + \|\beta\|_\infty h(\eta)]}{\Gamma(1+r_1)\Gamma(1+r_2)} = \frac{1}{\Gamma(\frac{5}{3})\Gamma(\frac{7}{4})},\quad \|\mu\|_\infty \le e^{-2},$$

and

$$\|\alpha\|_\infty(K + \|\mu\|_\infty) \le \frac{1}{\Gamma(\frac{5}{3})\Gamma(\frac{7}{4})} + e^{-2} < 0.64 + 0.13 = 0.77 < 1.$$

Hence by Theorem 8.11, (8.45) has a solution defined on $[0,1]\times[0,1]$.

8.5 Asymptotic Stability of Solutions of Nonlinear Quadratic Volterra Integral Equations of Fractional Order

8.5.1 *Introduction*

This paper deals with the existence of solutions to the following nonlinear quadratic Volterra integral equation of Riemann–Liouville fractional order:

$$u(t,x) = f(t,x,u(t,x),u(\alpha(t),x)) + \frac{1}{\Gamma(r)}\int_0^{\beta(t)} (\beta(t)-s)^{r-1}$$

$$\times g(t,x,s,u(s,x),u(\gamma(s),x))\mathrm{d}s, \quad (t,x) \in \mathbb{R}_+ \times [0,b], \tag{8.46}$$

where $b > 0$, $r \in (0,\infty)$, α, β, $\gamma : \mathbb{R}_+ \to \mathbb{R}_+$, $f : \mathbb{R}_+ \times [0,b] \times \mathbb{R} \times \mathbb{R} \to \mathbb{R}$ and $g : \mathbb{R}_+ \times [0,b] \times \mathbb{R}_+ \times \mathbb{R} \times \mathbb{R} \to \mathbb{R}$ are given continuous functions. Our investigations are conducted in a Banach space with an application of Schauder's fixed-point theorem for the existence of solutions of (8.46). Also, we obtain some results about the asymptotic stability of solutions of the equation in question. Finally, we present an example illustrating the applicability of the imposed conditions.

By $BC := BC(\mathbb{R}_+ \times [0,b])$ we denote the Banach space of all bounded and continuous functions from $\mathbb{R}_+ \times [0,b]$ into $\mathbb{R}$ equipped with the standard norm

$$\|u\|_{BC} = \sup_{(t,x)\in\mathbb{R}_+\times[0,b]} |u(t,x)|.$$

For $u_0 \in BC$ and $\eta \in (0,\infty)$, we denote by $B(u_0,\eta)$, the closed ball in BC centered at u_0 with radius η.

Let $\emptyset \neq \Omega \subset BC$, and let $G : \Omega \to \Omega$, and consider the solutions of equation

$$(Gu)(t,x)) = u(t,x). \tag{8.47}$$

Now we review the concept of attractivity of solutions for (8.46).

Definition 8.12 ([63]). Solutions of (8.47) are locally attractive if there exists a ball $B(u_0,\eta)$ in the space BC such that, for arbitrary solutions $v = v(t,x)$ and $w = w(t,x)$ of (8.47) belonging to $B(u_0,\eta) \bigcap \Omega$, we have that, for each $x \in [0,b]$,

$$\lim_{t\to\infty}\Big(v(t,x) - w(t,x)\Big) = 0. \tag{8.48}$$

When the limit (8.48) is uniform with respect to $B(u_0,\eta) \cap \Omega$, solutions of (8.47) are said to be uniformly locally attractive (or equivalently that solutions of (8.47) are locally asymptotically stable).

Definition 8.13 ([63]). The solution $v = v(t, x)$ of (8.47) is said to be globally attractive if (8.48) holds for each solution $w = w(t, x)$ of (8.47). If condition (8.48) is satisfied uniformly with respect to the set Ω, solutions of (8.47) are said to be globally asymptotically stable (or uniformly globally attractive).

Lemma 8.14 [Co]. *Let $D \subset BC$. Then D is relatively compact in BC if the following conditions hold:*

(a) D is uniformly bounded in BC.
(b) The functions belonging to D are almost equicontinuous on $\mathbb{R}_+ \times [0, b]$, i.e., equicontinuous on every compact of $\mathbb{R}_+ \times [0, b]$.
(c) The functions from D are equiconvergent, i.e., given $\epsilon > 0$, $x \in [0, b]$ there corresponds $T(\epsilon, x) > 0$ such that $|u(t, x) - \lim_{t\to\infty} u(t, x)| < \epsilon$ for any $t \geq T(\epsilon, x)$ and $u \in D$.

8.5.2 *Main Results*

In this section, we are concerned with the existence and global asymptotic stability of solutions for (8.46). The following hypotheses will be used in the sequel.

(8.15.1) The functions $\alpha, \beta, \gamma : \mathbb{R}_+ \to \mathbb{R}_+$ are continuous and $\lim_{t\to\infty} \alpha(t) = \infty$.
(8.15.2) The function f is continuous and there exist positive constants M and L such that $M < \frac{L}{2}$ and

$$|f(t, x, u_1, u_2) - f(t, x, v_1, v_2)| \leq \frac{M(|u_1 - v_1| + |u_2 - v_2|)}{(1 + \alpha(t))(L + |u_1 - v_1| + |u_2 - v_2|)},$$

for $(t, x) \in \mathbb{R}_+ \times [0, b]$ and for $u_1, u_2, v_1, v_2 \in \mathbb{R}$.
(8.15.3) The function $t \to f(t, x, 0, 0)$ is bounded on $\mathbb{R}_+ \times [0, b]$ with

$$f^* = \sup_{(t,x)\in\mathbb{R}_+\times[0,b]} f(t, x, 0, 0)$$

and

$$\lim_{t\to\infty} |f(t, x, 0, 0)| = 0, \ x \in [0, b].$$

(8.15.4) The function g is continuous and there exist functions $p, q : \mathbb{R}_+ \times [0, b] \to \mathbb{R}_+$ such that

$$|g(t, x, s, u, v)| \leq \frac{p(t, x)q(s, x)}{1 + \alpha(t) + |u| + |v|},$$

for $(t, x) \in \mathbb{R}_+ \times [0, b]$, $s \in \mathbb{R}_+$ and for $u \in \mathbb{R}$. Moreover, assume that

$$\lim_{t\to\infty} p(t, x) \int_0^{\beta(t)} (\beta(t) - s)^{r-1} q(s, x) ds = 0, \ \ x \in [0, b].$$

Theorem 8.15. *Assume that hypotheses* (8.15.1) – (8.15.4) *hold. Then (8.46) has at least one solution in the space BC. Moreover, solutions of (8.46) are locally asymptotically stable.*

Proof. Set $d^* := \sup_{(t,x)\in\mathbb{R}_+\times[0,b]} d(t,x)$ where

$$d(t,x) = \frac{p(t,x)}{\Gamma(r)} \int_0^{\beta(t)} (\beta(t)-s)^{r-1} q(s,x)\mathrm{d}s.$$

From hypothesis (8.15.4), we infer that d^* is finite. Let us define the operator N such that, for any $u \in BC$,

$$(Nu)(t,x) = f(t,x,u(t,x),u(\alpha(t),x)) + \frac{1}{\Gamma(r)} \int_0^{\beta(t)} (\beta(t)-s)^{r-1}$$
$$\times g(t,x,s,u(s,x),u(\gamma(s),x))\mathrm{d}s, \quad (t,x) \in \mathbb{R}_+ \times [0,b]. \quad (8.49)$$

By considering the assumptions of this theorem, we infer that $N(u)$ is continuous on $\mathbb{R}_+ \times [0,b]$. Now we prove that $N(u) \in BC$ for any $u \in BC$. For arbitrarily fixed $(t,x) \in \mathbb{R}_+ \times [0,b]$ we have

$$\begin{aligned}
|(Nu)(t,x)| &= \Bigg| f(t,x,u(t,x),u(\alpha(t),x)) \\
&\quad + \frac{1}{\Gamma(r)} \int_0^{\beta(t)} (\beta(t)-s)^{r-1} g(t,x,s,u(s,x),u(\gamma(s),x))\mathrm{d}s \Bigg| \\
&\le |f(t,x,u(t,x),u(\alpha(t),x)) - f(t,x,0,0) + f(t,x,0,0)| \\
&\quad + \left| \frac{1}{\Gamma(r)} \int_0^{\beta(t)} (\beta(t)-s)^{r-1} g(t,x,s,u(s,x),u(\gamma(s),x))\mathrm{d}s \right| \\
&\le \frac{M(|u(t,x)| + |u(\alpha(t),x)|)}{(1+\alpha(t))(L + |u(t,x)| + |u(\alpha(t),x)|)} + \left| f(t,x,0,0) \right| \\
&\quad + \frac{p(t,x)}{\Gamma(r)} \int_0^{\beta(t)} \frac{(\beta(t)-s)^{r-1} q(s,x)}{1+\alpha(t)+|u(s,x)|+|u(\gamma(s),x)|}\mathrm{d}s \\
&\le \frac{2M\|u\|}{L+2\|u\|} + f^* + d^*.
\end{aligned}$$

Thus

$$\|N(u)\| \leq M + f^* + d^*. \tag{8.50}$$

Hence $N(u) \in BC$. Equation (8.50) yields that N transforms the ball $B_\eta := B(0,\eta)$ into itself where $\eta = M + f^* + d^*$. We shall show that $N : B_\eta \to B_\eta$ satisfies the assumptions of Schauder's fixed-point theorem [136]. The proof will be given in several steps.

Step 1: *N is continuous.* Let $\{u_n\}_{n\in\mathbb{N}}$ be a sequence such that $u_n \to u$ in B_η. Then, for each $(t,x) \in \mathbb{R}_+ \times [0,b]$, we have

$$\begin{aligned}
&|(Nu_n)(t,x) - (Nu)(t,x)| \\
&\quad \leq |f(t,x,u_n(t,x),u_n(\alpha(t),x)) - f(t,x,u(t,x),u(\alpha(t),x))| \\
&\qquad + \frac{1}{\Gamma(r)} \int_0^{\beta(t)} (\beta(t)-s)^{r-1} |g(t,x,s,u_n(s,x),u_n(\gamma(s),x)) \\
&\qquad - g(t,x,s,u(s,x),u(\gamma(s),x))| ds \\
&\quad \leq \frac{2M\|u_n - u\|}{(1+\alpha(t))(L + 2\|u_n - u\|)} \\
&\qquad + \frac{1}{\Gamma(r)} \int_0^{\beta(t)} (\beta(t)-s)^{r-1} |g(t,x,s,u_n(s,x),u_n(\gamma(s),x)) \\
&\qquad - g(t,x,s,u(s,x),u(\gamma(s),x))| ds.
\end{aligned} \tag{8.51}$$

Case 1. If $(t,x) \in [0,T] \times [0,b]$; $T > 0$, then, since $u_n \to u$ as $n \to \infty$ and g is continuous, (8.51) gives

$$\|N(u_n) - N(u)\|_{BC} \to 0 \quad \text{as } n \to \infty.$$

Case 2. If $(t,x) \in (T,\infty) \times [0,b]$; $T > 0$, then from (8.15.4) and (8.51), for each $(t,x) \in \mathbb{R}_+ \times [0,b]$, we get

$$\begin{aligned}
&|(Nu_n)(t,x) - (Nu)(t,x)| \\
&\leq \frac{2M\|u_n - u\|}{L + 2\|u_n - u\|} + \frac{p(t,x)}{\Gamma(r)} \\
&\times \int_0^{\beta(t)} \frac{(\beta(t)-s)^{r-1} q(s,x)}{(1+\alpha(t)+|u_n(s)|+|u_n(\gamma(s))|)(1+\alpha(t)+|u(s)|+|u(\gamma(s))|)} ds
\end{aligned}$$

$$\leq \frac{M\|u_n-u\|}{L+\|u_n-u\|}+\frac{p(t,x)}{\Gamma(r)}\int_0^{\beta(t)}(\beta(t)-s)^{r-1}q(s,x)\mathrm{d}s$$
$$\leq \frac{2M\|u_n-u\|}{L+2\|u_n-u\|}+d(t,x). \tag{8.52}$$

Since $u_n \to u$ as $n \to \infty$ and $t \to \infty$, then (8.52) gives

$$\|N(u_n)-N(u)\|_{BC} \to 0 \quad \text{as } n \to \infty.$$

Step 2: *$N(B_\eta)$ is uniformly bounded.* This is clear since $N(B_\eta) \subset B_\eta$ and B_η is bounded.

Step 3: *$N(B_\eta)$ is equicontinuous on every compact subset* $[0,a]\times[0,b]$ *of* $\mathbb{R}_+ \times [0,b]$, $a>0$. Let $(t_1,x_1),(t_2,x_2)\in[0,a]\times[0,b]$, $t_1<t_2$, $x_1<x_2$ and let $u\in B_\eta$. Also without loss of generality suppose that $\beta(t_1)\leq\beta(t_2)$. Thus we have

$$\begin{aligned}
&|(Nu)(t_2,x_2)-(Nu)(t_1,x_1)|\\
&\quad\leq |f(t_2,x_2,u(t_2,x_2),u(\alpha(t_2),x_2))-f(t_2,x_2,u(t_1,x_1),u(\alpha(t_1),x_1))|\\
&\qquad+|f(t_2,x_2,u(t_1,x_1),u(\alpha(t_1),x_1))-f(t_1,x_1,u(t_1,x_1),u(\alpha(t_1),x_1))|\\
&\qquad+\frac{1}{\Gamma(r)}\left|\int_0^{\beta(t_2)}(\beta(t_2)-s)^{r-1}\times[g(t_2,x_2,s,u(t_2,x_2),u(\gamma(s),x_2))\right.\\
&\qquad\qquad\left.-g(t_1,x_1,s,u(t_1,x_1),u(\gamma(s),x_1))]\,\mathrm{d}s\right|\\
&\qquad+\frac{1}{\Gamma(r)}\left|\int_0^{\beta(t_2)}(\beta(t_2)-s)^{r-1}g(t_1,x_1,s,u(t_1,x_1),u(\gamma(s),x_1))\mathrm{d}s\right.\\
&\qquad\left.-\int_0^{\beta(t_1)}(\beta(t_2)-s)^{r-1}g(t_1,x_1,s,u(t_1,x_1),u(\gamma(s),x_1))\mathrm{d}s\right|\\
&\qquad+\frac{1}{\Gamma(r)}\left|\int_0^{\beta(t_1)}(\beta(t_2)-s)^{r-1}g(t_1,x_1,s,u(t_1,x_1),u(\gamma(s),x_1))\mathrm{d}s\right.\\
&\qquad\left.-\int_0^{\beta(t_1)}(\beta(t_1)-s)^{r-1}g(t_1,x_1,s,u(t_1,x_1),u(\gamma(s),x_1))\mathrm{d}s\right|
\end{aligned}$$

$$
\begin{aligned}
&\le \frac{M\,|u(\alpha(t_2),x_2)-u(\alpha(t_1),x_1)|}{(1+\alpha(t_2))(L+|u(\alpha(t_2),x_2)-u(\alpha(t_1),x_1)|)}\\
&\quad+|f(t_2,x_2,u(t_1,x_1),u(\alpha(t_1),x_1))-f(t_1,x_1,u(t_1,x_1),u(\alpha(t_1),x_1))|\\
&\quad+\frac{1}{\Gamma(r)}\int_0^{\beta(t_2)}(\beta(t_2)-s)^{r-1}\\
&\quad\times|g(t_2,x_2,s,u(t_2,x_2),u(\gamma(s),x_2))-g(t_1,x_1,s,u(t_1,x_1),u(\gamma(s),x_1))|\,\mathrm{d}s\\
&\quad+\frac{1}{\Gamma(r)}\int_{\beta(t_1)}^{\beta(t_2)}(\beta(t_2)-s)^{r-1}\,|g(t_1,x_1,s,u(t_1,x_1),u(\gamma(s),x_1))|\,\mathrm{d}s\\
&\quad+\frac{1}{\Gamma(r)}\int_0^{\beta(t_1)}\left|(\beta(t_2)-s)^{r-1}-(\beta(t_1)-s)^{r-1}\right|\\
&\quad\times|g(t_1,x_1,s,u(t_1,x_1),u(\gamma(s),x_1))|\,\mathrm{d}s\\
&\le \frac{M(|u(t_2,x_2)-u(t_1,x_1)|+|u(\alpha(t_2),x_2)-u(\alpha(t_1),x_1)|)}{L+|u(t_2,x_2)-u(t_1,x_1)|+|u(\alpha(t_2),x_2)-u(\alpha(t_1),x_1)||}\\
&\quad+|f(t_2,x_2,u(\alpha(t_1),x_1))-f(t_1,x_1,u(\alpha(t_1),x_1))|\\
&\quad+\frac{1}{\Gamma(r)}\int_0^{\beta(t_2)}(\beta(t_2)-s)^{r-1}\\
&\quad\times\Big|g(t_2,x_2,s,u(s,x_2),u(\gamma(s),x_2))-g(t_1,x_1,s,u(s,x_1),u(\gamma(s),x_1))\Big|\mathrm{d}s\\
&\quad+\frac{p(t_1,x_1)}{\Gamma(r)}\int_{\beta(t_1)}^{\beta(t_2)}(\beta(t_2)-s)^{r-1}q(s,x_1)\mathrm{d}s\\
&\quad+\frac{p(t_1,x_1)}{\Gamma(r)}\int_0^{\beta(t_1)}\left|(\beta(t_2)-s)^{r-1}-(\beta(t_1)-s)^{r-1}\right|q(s,x_1)\mathrm{d}s.
\end{aligned}
$$

From continuity of α,β,f,g and as $t_1\longrightarrow t_2$ and $x_1\longrightarrow x_2$, the right-hand side of the above inequality tends to zero.

Step 4: *$N(B_\eta)$ is equiconvergent.* Let $(t,x)\in\mathbb{R}_+\times[0,b]$ and $u\in B_\eta$, then we have

$$
|(Nu)(t,x)|\le|f(t,x,u(t,x),u(\alpha(t),x))-f(t,x,0,0)+f(t,x,0,0)|
$$

$$+\left|\frac{1}{\Gamma(r)}\int_0^{\beta(t)}(\beta(t)-s)^{r-1}g(t,x,s,u(s,x),u(\gamma(s),x))\mathrm{d}s\right|$$

$$\leq \frac{M(|u(t,x)|+|u(\alpha(t),x)|)}{(1+\alpha(t))(L+|u(t,x)|+|u(\alpha(t),x)|)}+|f(t,x,0,0)|$$

$$+\frac{p(t,x)}{\Gamma(r)}\int_0^{\beta(t)}\frac{(\beta(t)-s)^{r-1}q(s,x)}{1+\alpha(t)+|u(s,x)|+|u(\gamma(s),x)|}\mathrm{d}s$$

$$\leq \frac{M}{1+\alpha(t)}+|f(t,x,0,0)|$$

$$+\frac{1}{1+\alpha(t)}\left(\frac{p(t,x)}{\Gamma(r)}\int_0^{\beta(t)}(\beta(t)-s)^{r-1}q(s,x)\mathrm{d}s\right)$$

$$\leq \frac{M}{1+\alpha(t)}+|f(t,x,0,0)|+\frac{d^*}{1+\alpha(t)}.$$

Thus, for each $x\in[0,b]$, we get

$$|(Nu)(t,x)|\longrightarrow 0,\ as\ t\longrightarrow+\infty.$$

Hence,

$$|(Nu)(t,x)-u(+\infty,x)|\longrightarrow 0,\ as\ t\longrightarrow+\infty.$$

As a consequence of Steps 1–4 together with the Lemma 8.14, we can conclude that $N : B_\eta \to B_\eta$ is continuous and compact. From an application of Schauder's theorem [136], we deduce that N has a fixed point u which is a solution of (8.46).

Now we investigate the uniform local attractivity for solutions of (8.46). Let us assume that u_0 is a solution of (8.46) with the conditions of this theorem. Consider the ball $B(u_0,\eta)$ with $\eta^*=\frac{LM^*}{L-2M}$, where

$$M^*:=\frac{1}{\Gamma(r)}\sup_{(t,x)\in\mathbb{R}_+\times[0,b]}\left\{\int_0^{\beta(t)}(\beta(t)-s)^{r-1}|g(t,x,s,u(s,x),u(\gamma(s),x))\right.$$

$$\left.-g(t,x,s,u_0(s,x),u_0(\gamma(s),x))|\mathrm{d}s;\ u\in BC\right\}.$$

Taking $u \in B(u_0, \eta^*)$, we have

$$\begin{aligned}|(Nu)(t,x) - u_0(t,x)| &= |(Nu)(t,x) - (Nu_0)(t,x)| \\ &\leq |f(t,x,u(t,x),u(\alpha(t),x)) - f(t,x,u_0(t,x),u_0(\alpha(t),x))| \\ &\quad + \frac{1}{\Gamma(r)} \int_0^{\beta(t)} (\beta(t)-s)^{r-1} |g(t,x,s,u(s,x),u(\gamma(s),x)) \\ &\quad - g(t,x,s,u_0(s,x),u_0(\gamma(s),x))| \mathrm{d}s \\ &\leq \frac{2M\|u-u_0\|}{L+2\|u-u_0\|} \\ &\quad + \frac{1}{\Gamma(r)} \int_0^{\beta(t)} (\beta(t)-s)^{r-1} |g\,(t,x,s,u(s,x),u(\gamma(s),x)) \\ &\quad - g(t,x,s,u_0(s,x),u_0(\gamma(s,x)))| \mathrm{d}s \\ &\leq \frac{2M}{L}\eta^* + M^* = \eta^*.\end{aligned}$$

Thus we observe that N is a continuous function such that $N(B(u_0, \eta^*)) \subset B(u_0, \eta^*)$. Moreover, if u is a solution of (8.46), then

$$\begin{aligned}|u(t,x) - u_0(t,x)| &= |(Nu)(t,x) - (Nu_0)(t,x)| \\ &\leq |f(t,x,u(t,x),u(\alpha(t),x)) - f(t,x,u_0(t,x),u_0(\alpha(t),x))| \\ &\quad + \frac{1}{\Gamma(r)} \int_0^{\beta(t)} (\beta(t)-s)^{r-1} |g(t,x,s,u(s,x),u(\gamma(s),x)) \\ &\quad - g(t,x,s,u_0(s,x),u_0(\gamma(s),x))| \mathrm{d}s \\ &\leq \frac{M(|u(t,x)-u_0(t,x)| + |u(\alpha(t),x)-u_0(\alpha(t),x)|)}{L + |u(t,x)-u_0(t,x)| + |u(\alpha(t),x)-u_0(\alpha(t),x)|} \\ &\quad + \frac{p(t,x)}{\Gamma(r)} \int_0^{\beta(t)} \frac{(\beta(t)-s)^{r-1} q(s,x)}{(1+|u(s,x)|+|u(\gamma(s),x)|)(1+|u_0(s,x)|+|u_0(\gamma(s),x)|)} \mathrm{d}s \\ &\leq \frac{M}{L}(|u(t,x)-u_0(t,x)| + |u(\alpha(t),x)-u_0(\alpha(t),x)|) \\ &\quad + \frac{p(t,x)}{\Gamma(r)} \int_0^{\beta(t)} (\beta(t)-s)^{r-1} q(s,x) \mathrm{d}s. \qquad (8.53)\end{aligned}$$

By using (8.53) and the fact that $\alpha(t) \longrightarrow \infty$ as $t \longrightarrow \infty$, we deduce that

$$\lim_{t\to\infty} |u(t,x) - u_0(t,x)| \le \lim_{t\to\infty} \frac{Lp(t,x)}{(L-2M)\Gamma(r)} \int_0^{\beta(t)} (\beta(t)-s)^{r-1} q(s,x) ds = 0.$$

Consequently, all solutions of (8.46) are locally asymptotically stable. □

8.5.3 An Example

As an application of our results we consider the following integral equation of fractional order:

$$u(t,x) = \frac{tx}{4(1+t+t^2+t^3)} \sin(u(t,x)) + \frac{1}{\Gamma(\frac{1}{3})} \times \int_0^t (t-s)^{\frac{-2}{3}} \frac{\ln(1+2sx|u(s,x)|)}{(1+t+2|u(s,x)|)^2(1+x^2+t^4)} ds, \ (t,x) \in \mathbb{R}_+ \times [0,1], \tag{8.54}$$

where $r = \frac{1}{3}$, $\alpha(t) = \beta(t) = \delta(t) = t$ and

$$f(t,x,u,v) = \frac{tx(\sin(u) + \sin(v))}{8(1+t)(1+t^2)},$$

$$g(t,x,s,u,v) = \frac{\ln(1+sx(|u|+|v|))}{(1+t+|u|+|v|)^2(1+x^2+t^4)};$$

for $(t,x) \in \mathbb{R}_+ \times [0,1]$, and $u \in \mathbb{R}$.

Then we can easily check that the assumptions of Theorem 8.15 are satisfied. In fact, we have that the function f is continuous and satisfies assumption (8.15.2), where $M = \frac{1}{8}$, $L = 1$, and also f satisfies assumption (8.15.3), with $f^* = 0$. Next, let us notice that the function g satisfies assumption (8.15.4), where $p(t,x) = \frac{1}{1+x^2+t^4}$ and $q(s,x) = sx$. Also,

$$\lim_{t\to\infty} p(t,x) \int_0^{\beta(t)} (\beta(t)-s)^{r-1} q(s,x) ds = \lim_{t\to\infty} \frac{x}{1+x^2+t^4} \int_0^t s(t-s)^{\frac{-2}{3}} ds = 0.$$

Hence by Theorem 8.15, (8.54) has a solution defined on $\mathbb{R}_+ \times [0,1]$ and solutions of this equation are locally asymptotically stable.

8.6 Attractivity Results for Nonlinear Fractional Order Riemann–Liouville Integral Equations in Banach Algebras

8.6.1 Introduction

In this section, we present some results concerning the existence and local asymptotic attractivity of solutions to the following nonlinear quadratic Volterra integral equation of Riemann–Liouville fractional order:

$$u(t,x) = f(t,x,u(t,x),u(\alpha(t),x))\left[\mu(t,x) + \frac{1}{\Gamma(r)}\right.$$
$$\left.\times \int_0^{\beta(t)} (\beta(t)-s)^{r-1} g(t,x,s,u(s,x),u(\gamma(s),x))\mathrm{d}s\right];$$
$$\text{if } (t,x) \in \mathbb{R}_+ \times [0,b], \tag{8.55}$$

where $b > 0$, $r \in (0,\infty)$, $\mu : \mathbb{R}_+ \times [0,b] \to \mathbb{R}$, α, β, $\gamma : \mathbb{R}_+ \to \mathbb{R}_+$, $f : \mathbb{R}_+ \times [0,b] \times \mathbb{R} \times \mathbb{R} \to \mathbb{R}$, $g : \mathbb{R}_+ \times [0,b] \times \mathbb{R}_+ \times \mathbb{R} \times \mathbb{R} \to \mathbb{R}$ are given functions. Our investigations of the existence and the local asymptotic attractivity of solutions of (8.55) are placed in Banach algebras, with the application a fixed-point theorem of Dhage [109]. Also, we present an example illustrating the applicability of the imposed conditions.

By $BC := BC(\mathbb{R}_+ \times [0,b])$ we denote the Banach space of all bounded and continuous functions from $\mathbb{R}_+ \times [0,b]$ into $\mathbb{R}$ equipped with the standard norm

$$\|u\|_{BC} = \sup_{(t,x)\in\mathbb{R}_+\times[0,b]} |u(t,x)|.$$

Define a multiplication " $\cdot$ " by

$$(u \cdot v)(t,x) = u(t,x)v(t,x); \quad \text{for } (t,x) \in \mathbb{R}_+ \times [0,b].$$

Then BC is a Banach algebra with the above norm and multiplication.

For $u_0 \in BC$ and $\eta \in (0,\infty)$, we denote by $B(u_0,\eta)$, the closed ball in BC centered at u_0 with radius η.

We use the following fixed-point theorem by Dhage for proving the existence of solutions for our problem.

Theorem 8.16 (Dhage [109]). *Let S be a closed, convex and bounded subset of a Banach algebra X and let $A, B : S \to X$ be two operators such that*

(a) *A is Lipschitz with the Lipschitz constant k*
(b) *B is completely continuous*
(c) $Mk < 1$, *where* $M = \|B(S)\| := \sup\{\|Bu\| : u \in S\}$

Then the operator equation $AuBu = u$ *has a solution and the set of all solutions is compact in* S.

8.6.2 Main Results

In this section, we are concerned with the existence and global asymptotic stability of solutions for (8.55). The following hypotheses will be used in the sequel:

(8.18.1) The function μ is continuous and bounded with $\mu^* = \sup_{(t,x)\in\mathbb{R}_+\times[0,b]} |\mu(t,x)|$, and $\lim_{t\to\infty}\mu(t,x) = 0$; *for* $x \in [0,b]$.
(8.18.2) The functions $\alpha, \beta : \mathbb{R}_+ \to \mathbb{R}_+$ are continuous and the function $\gamma : \mathbb{R}_+ \to \mathbb{R}_+$ is measurable.
(8.18.3) The function f is continuous and there exists a positive constant L such that

$$|f(t,x,u_1,u_2) - f(t,x,v_1,v_2)| \le L(|u_1 - v_1| + |u_2 - v_2|),$$

for $(t,x) \in \mathbb{R}_+ \times [0,b]$ and for $u_1, u_2, v_1, v_2 \in \mathbb{R}$.
(8.18.4) The function $t \to f(t,x,0,0)$ is bounded on $\mathbb{R}_+ \times [0,b]$ with

$$f^* = \sup_{(t,x)\in\mathbb{R}_+\times[0,b]} f(t,x,0,0).$$

(8.18.5) The function g is continuous and there exist functions $p, q : \mathbb{R}_+ \times [0,b] \to \mathbb{R}_+$ such that

$$|g(t,x,s,u,v)| \le \frac{p(t,x)q(s,x)}{1 + |u| + |v|},$$

for $(t,x) \in \mathbb{R}_+ \times [0,b]$, $s \in \mathbb{R}_+$ and for $u \in \mathbb{R}$. Moreover, assume that

$$\lim_{t\to\infty} p(t,x) \int_0^{\beta(t)} (\beta(t) - s)^{r-1} q(s,x)ds = 0; \ \ for\ x \in [0,b].$$

Remark 8.17. Set $d^* := \sup_{(t,x)\in\mathbb{R}_+\times[0,b]} d(t,x)$ where

$$d(t,x) = \frac{p(t,x)}{\Gamma(r)} \int_0^{\beta(t)} (\beta(t) - s)^{r-1} q(s,x)ds.$$

From hypothesis (8.18.5), we infer that d^* is finite.

Set $B_\eta := B(0,\eta) \subset BC$, where $\eta = \frac{f^*(\mu^*+d^*)}{1-2L(\mu^*+d^*)}$.

Theorem 8.18. *Assume that hypotheses (8.18.1)–(8.18.5) hold. Furthermore, if*

$$L(\mu^* + d^*) < \frac{1}{2},$$

where the numbers d^* is defined in Remark 8.17, then (8.55) has at least one solution in the space BC and the set of all solutions is compact in B_η. Moreover, solutions of (8.55) are uniformly locally asymptotically attractive on $\mathbb{R}_+ \times [0,b]$.

Proof. Define two mappings A on BC and B on B_η by

$$(Au)(t,x) = f(t,x,u(t,x),u(\alpha(t),x)) \tag{8.56}$$

and

$$(Bu)(t,x) = \mu(t,x) + \frac{1}{\Gamma(r)} \int_0^{\beta(t)} (\beta(t)-s)^{r-1} g(t,x,s,u(s,x),u(\gamma(s),x))ds, \tag{8.57}$$

for all $(t,x) \in \mathbb{R}_+ \times [0,b]$. Then (8.55) is equivalent to the operator equation

$$(Au)(t,x)(Bu)(t,x) = u(t,x);\ (t,x) \in \mathbb{R}_+ \times [0,b].$$

Since the hypothesis (8.18.3) holds, the mapping A is well defined and the function $A(u)$ is continuous and bounded on $\mathbb{R}_+ \times [0,b]$. Again, since the functions μ and β are continuous, the function $B(u)$ is also continuous and bounded in view of hypothesis (8.18.5). Therefore A and B define the operators $A : B_\eta \to BC$. We shall show that A and B satisfy all the requirements of Theorem 8.16 on B_η.

Step 1: *A is a Lipschitz operator on B_η.* Let $u, v \in BC$. Then by hypothesis (8.18.3), for each $(t,x) \in \mathbb{R}_+ \times [0,b]$, we have

$$\begin{aligned} |(Au)(t,x) - (Av)(t,x)| &\le |f(t,x,u(t,x),u(\alpha(t),x)) \\ &\quad - f(t,x,v(t,x),v(\alpha(t),x))| \\ &\le 2L\|u-v\|_{BC} \end{aligned}$$

Thus, for all $u, v \in BC$, we get

$$\|A(u) - A(v)\|_{BC} \le 2L\|u-v\|_{BC}.$$

This shows that A is a Lipschitz on B_η with the Lipschitz constant $2L$.

Step 2: *B is a continuous and compact operator on B_η.* The proof will be given in several claims.

Claim 1. *B is continuous on B_η.* Let $\{u_n\}_{n\in\mathbf{N}}$ be a sequence such that $u_n \to u$ in B_η. Then, for each $(t,x) \in \mathbb{R}_+ \times [0,b]$, we have

$$\begin{aligned}
&|(Bu_n)(t,x) - (Bu)(t,x)| \\
&\le \frac{1}{\Gamma(r)} \int_0^{\beta(t)} (\beta(t)-s)^{r-1} |g(t,x,s,u_n(s,x),u_n(\gamma(s),x)) \\
&\quad - g(t,x,s,u(s,x),u(\gamma(s),x))| \mathrm{d}s \\
&\le \frac{1}{\Gamma(r)} \int_0^{\beta(t)} (\beta(t)-s)^{r-1} \|g(t,x,s,u_n(s,x),u_n(\gamma(s),x)) \\
&\quad - g(t,x,s,u(s,x),u(\gamma(s),x))\| \mathrm{d}s.
\end{aligned} \tag{8.58}$$

Case 3. If $(t,x) \in [0,T] \times [0,b]$; $T > 0$, then, since $u_n \to u$ as $n \to \infty$ and g is continuous, (8.56) gives

$$\|N(u_n) - N(u)\|_{BC} \to 0 \quad \text{as } n \to \infty.$$

Case 4. If $(t,x) \in (T,\infty) \times [0,b]$; $T > 0$, then from (8.18.5) and (8.58), for each $(t,x) \in \mathbb{R}_+ \times [0,b]$ we get

$$\begin{aligned}
&|(Bu_n)(t,x) - (Bu)(t,x)| \\
&\le \frac{p(t,x)}{\Gamma(r)} \int_0^{\beta(t)} \frac{(\beta(t)-s)^{r-1} q(s,x)}{(1+|u_n(s)|+|u_n(\gamma(s))|)(1+|u(s)|+|u(\gamma(s))|)} \mathrm{d}s \\
&\le \frac{p(t,x)}{\Gamma(r)} \int_0^{\beta(t)} (\beta(t)-s)^{r-1} q(s,x) \mathrm{d}s = d(t,x).
\end{aligned} \tag{8.59}$$

Since $u_n \to u$ as $n \to \infty$ and $t \to \infty$, then (8.59) gives

$$\|B(u_n) - B(u)\|_{BC} \to 0 \quad \text{as } n \to \infty.$$

Claim 2. *$B(B_\eta)$ is uniformly bounded.* This is clear since $B(B_\eta) \subset B_\eta$ and B_η is bounded.

Claim 3. *$B(B_\eta)$ is equicontinuous on every compact $[0,a] \times [0,b]$; $a > 0$ of $\mathbb{R}_+ \times [0,b]$.* Let $(t_1,x_1), (t_2,x_2) \in [0,a] \times [0,b]$, $t_1 < t_2$, $x_1 < x_2$ and let $u \in B_\eta$. Also without lose of generality suppose that $\beta(t_1) \le \beta(t_2)$, thus we have

$$
\begin{aligned}
|(Bu)(t_2,x_2)-(Bu)(t_1,x_1)| &\le |\mu(t_2,x_2)-\mu(t_1,x_1)| \\
&+\frac{1}{\Gamma(r)}\left|\int_0^{\beta(t_2)}(\beta(t_2)-s)^{r-1}\right. \\
&\left.\times[g(t_2,x_2,s,u(t_2,x_2),u(\gamma(s),x_2))-g(t_1,x_1,s,u(t_1,x_1),u(\gamma(s),x_1))]\,\mathrm{d}s\right| \\
&+\frac{1}{\Gamma(r)}\left|\int_0^{\beta(t_2)}(\beta(t_2)-s)^{r-1}g(t_1,x_1,s,u(t_1,x_1),u(\gamma(s),x_1))\mathrm{d}s\right. \\
&\left.-\int_0^{\beta(t_1)}(\beta(t_2)-s)^{r-1}g(t_1,x_1,s,u(t_1,x_1),u(\gamma(s),x_1))\mathrm{d}s\right| \\
&+\frac{1}{\Gamma(r)}\left|\int_0^{\beta(t_1)}(\beta(t_2)-s)^{r-1}g(t_1,x_1,s,u(t_1,x_1),u(\gamma(s),x_1))\mathrm{d}s\right. \\
&\left.-\int_0^{\beta(t_1)}(\beta(t_1)-s)^{r-1}g(t_1,x_1,s,u(t_1,x_1),u(\gamma(s),x_1))\mathrm{d}s\right| \\
&\le |\mu(t_2,x_2)-\mu(t_1,x_1)|+\frac{1}{\Gamma(r)}\int_0^{\beta(t_2)}(\beta(t_2)-s)^{r-1} \\
&\times|g(t_2,x_2,s,u(t_2,x_2),u(\gamma(s),x_2))-g(t_1,x_1,s,u(t_1,x_1),u(\gamma(s),x_1))|\,\mathrm{d}s \\
&+\frac{1}{\Gamma(r)}\int_{\beta(t_1)}^{\beta(t_2)}(\beta(t_2)-s)^{r-1}\,|g(t_1,x_1,s,u(t_1,x_1),u(\gamma(s),x_1))|\,\mathrm{d}s \\
&+\frac{1}{\Gamma(r)}\int_0^{\beta(t_1)}\left|(\beta(t_2)-s)^{r-1}-(\beta(t_1)-s)^{r-1}\right| \\
&\times|g(t_1,x_1,s,u(t_1,x_1),u(\gamma(s),x_1))|\,\mathrm{d}s \\
&\le |\mu(t_2,x_2)-\mu(t_1,x_1)|+\frac{1}{\Gamma(r)}\int_0^{\beta(t_2)}(\beta(t_2)-s)^{r-1} \\
&\times|g(t_2,x_2,s,u(s,x_2),u(\gamma(s),x_2))-g(t_1,x_1,s,u(s,x_1),u(\gamma(s),x_1))|\,\mathrm{d}s
\end{aligned}
$$

$$+\frac{p(t_1,x_1)}{\Gamma(r)}\int\limits_{\beta(t_1)}^{\beta(t_2)}(\beta(t_2)-s)^{r-1}q(s,x_1)\mathrm{d}s$$

$$+\frac{p(t_1,x_1)}{\Gamma(r)}\int\limits_{0}^{\beta(t_1)}\left|(\beta(t_2)-s)^{r-1}-(\beta(t_1)-s)^{r-1}\right|q(s,x_1)\mathrm{d}s.$$

From continuity of α,β,g and as $t_1\longrightarrow t_2,\ x_1\longrightarrow x_2$, the right-hand side of the above inequality tends to zero.

Claim 4. *$B(B_\eta)$ is equiconvergent.* Let $(t,x)\in\mathbb{R}_+\times[0,b]$ and $u\in B_\eta$, then we have

$$\begin{aligned}|(Bu)(t,x)| &\le |\mu(t,x)|+\Bigg|\frac{1}{\Gamma(r)}\\ &\qquad\times\int\limits_0^{\beta(t)}(\beta(t)-s)^{r-1}g(t,x,s,u(s,x),u(\gamma(s),x))\mathrm{d}s\Bigg|\\ &\le|\mu(t,x)|+\frac{p(t,x)}{\Gamma(r)}\int\limits_0^{\beta(t)}\frac{(\beta(t)-s)^{r-1}q(s,x)}{1+|u(s,x)|+|u(\gamma(s),x)|}\mathrm{d}s\\ &\le|\mu(t,x)|+\frac{p(t,x)}{\Gamma(r)}\int\limits_0^{\beta(t)}(\beta(t)-s)^{r-1}q(s,x)\mathrm{d}s\\ &\le|\mu(t,x)|+d(t,x).\end{aligned}$$

Thus, for each $x\in[0,b]$, we get

$$|(Bu)(t,x)|\longrightarrow 0,\ as\ t\longrightarrow+\infty.$$

As a consequence of Claims 1 to 4 together with Lemma 8.14, we can conclude that $B:B_\eta\to BC$ is continuous and compact.

Step 3: *$AuBu\in B_\eta$ for all $u\in B_\eta$.* Let $u\in B_\eta$ be arbitrary, then for all $(t,x)\in\mathbb{R}_+\times[0,b]$ we have

$$\begin{aligned}|(Au)(t,x)(Bu)(t,x)| &\le |(Au)(t,x)|+|(Bv)(t,x)|\\ &\le(|f(t,x,u(t,x),u(\alpha(t),x))-f(t,x,0,0)|+|f(t,x,0,0)|)\\ &\quad\times\left[|\mu(t,x)|+\frac{1}{\Gamma(r)}\int\limits_0^{\beta(t)}(\beta(t)-s)^{r-1}|g(t,x,s,v(s,x),v(\gamma(s),x))|\mathrm{d}s\right]\end{aligned}$$

$$\leq (2L\|u\| + f^*)\left(\mu^* + \frac{p(t,x)}{\Gamma(r)}\int_0^{\beta(t)}(\beta(t)-s)^{r-1}q(s,x)\mathrm{d}s\right)$$
$$\leq (2L\eta + f^*)(\mu^* + d^*) = \eta.$$

Hence, we obtain that $AuBu \in B_\eta$ for all $u \in B_\eta$. Also, one has

$$M = \|B(B_\eta(u)\| = \sup_{u\in B_\eta}\|B(u)\| \leq \mu^* + d^*,$$

and therefore, $Mk = 2L(\mu^* + d^*) < 1$. As a consequence of Steps 1–3 together with Theorem 8.16, we deduce that AB has a fixed point u in B_η which is a solution of (8.55) and the set of all solutions is compact in B_η.

Finally, we show the uniform locally asymptotic attractivity of the solutions of (8.55) on $\mathbb{R}_+ \times [0,b]$. Let u and v be any two solutions of (8.55) in B_η, then for each $(t,x) \in \mathbb{R}_+ \times [0,b]$ we have

$$|u(t,x) - v(t,x)| \leq |f(t,x,u(t,x),u(\alpha(t),x))| \Bigg| \mu(t,x)$$
$$+\frac{1}{\Gamma(r)}\int_0^{\beta(t)}(\beta(t)-s)^{r-1}|g(t,x,s,u(s,x),u(\gamma(s),x))|\mathrm{d}s\Bigg|$$
$$+|f(t,x,v(t,x),v(\alpha(t),x))| \Bigg| \mu(t,x)$$
$$+\frac{1}{\Gamma(r)}\int_0^{\beta(t)}(\beta(t)-s)^{r-1}|g(t,x,s,v(s,x),v(\gamma(s),x))|\mathrm{d}s\Bigg|$$
$$\leq 2(2L\eta + f^*)\left(\mu(t,x) + \frac{p(t,x)}{\Gamma(r)}\int_0^{\beta(t)}(\beta(t)-s)^{r-1}q(s,x)\mathrm{d}s\right). \quad (8.60)$$

By using (8.60), we deduce that

$$\lim_{t\to\infty}(u(t,x) - v(t,x)) = 0.$$

Consequently, (8.55) has a solution and all solutions are uniformly locally asymptotically attractive $\mathbb{R}_+ \times [0,b]$. □

8.6.3 An Example

As an application of our results we consider the following integral equation of fractional order:

$$u(t,x) = \left(1 + \left|\sin\left(\frac{u(t,x)}{8}\right)\right| + \left|\sin\left(\frac{u(2t,x)}{8}\right)\right|\right)\left[\frac{1}{1+t+x^2}\right.$$
$$\left. + \frac{1}{\Gamma(\frac{1}{3})}\int_0^t \frac{sxe^{-t}(t-s)^{\frac{-2}{3}}}{2+|u(s,x)|+|u(s^2,x)|}ds\right];\ (t,x)\in\mathbb{R}_+\times[0,1]. \quad (8.61)$$

Let

$$r = \frac{1}{3},\ \alpha(t) = 2t,\ \beta(t) = t,\ \gamma(t) = t^2,$$

$$\mu(t,x) = \frac{1}{1+t+x^2},$$

$$f(t,x,u,v) = 1 + \left|\sin\left(\frac{|u|}{8}\right)\right| + \left|\sin\left(\frac{|v|}{8}\right)\right|$$

and

$$g(t,x,s,u,v) = \frac{sxe^{-t}}{2+|u|+|v|}\ \text{for } (t,x)\in\mathbb{R}_+\times[0,1],\ \text{and } u,v\in\mathbb{R}.$$

Then we can easily check that the assumptions of Theorem 8.18 are satisfied. In fact, the function μ satisfies assumption (8.18.1) with $\mu^* = 1$. The function f is continuous and satisfies assumption (8.18.3), where $L = \frac{1}{8}$ and also f satisfies assumption (8.18.4) with $f^* = 1$. Next, let us notice that the function g satisfies assumption (8.18.5), where $p(t,x) = e^{-t}$ and $q(s,x) = \frac{sx}{2}$. Also,

$$\lim_{t\to\infty} p(t,x)\int_0^{\beta(t)} (\beta(t)-s)^{r-1}q(s,x)ds = \lim_{t\to\infty} xe^{-t}\int_0^t s(t-s)^{\frac{-2}{3}}ds = 0.$$

Here

$$d^* = \frac{1}{\Gamma(\frac{1}{3})}\sup_{(t,x)\in\mathbb{R}_+\times[0,1]} xe^{-t}\int_0^t s(t-s)^{\frac{-2}{3}}ds$$
$$\leq \frac{1}{\Gamma(\frac{1}{3})}\sup_{t\in\mathbb{R}_+}\frac{e^{-t}}{2}t^{\frac{4}{3}} = \frac{e^{-\frac{4}{3}}}{2\Gamma(\frac{1}{3})}\left(\frac{4}{3}\right)^{\frac{4}{3}} < 2.$$

A simple computation gives

$$L(\mu^* + d^*) < \frac{3}{8} < \frac{1}{2}.$$

Hence by Theorem 8.18, (8.61) has a solution defined on $\mathbb{R}_+ \times [0, 1]$ and all these solutions are uniformly locally asymptotically attractive on $\mathbb{R}_+ \times [0, 1]$.

8.7 Notes and Remarks

The results of Chap. 6 are taken from Abbas et al. [20–23, 28, 29]. Other results may be found in [1, 63, 79, 111, 155, 210–212].

References

1. M.I. Abbas, Onthe existence of locally attractive solutions of a nonlinear quadratic volterra integral equation of fractional order. Adv. Diff. Equ. **2010**, 1–11 (2010)
2. S. Abbas, R.P. Agarwal, M. Benchohra, Darboux problem for impulsive partial hyperbolic differential equations of fractional order with variable times and infinite delay. Nonlinear Anal. Hybrid Syst. **4**, 818–829 (2010)
3. S. Abbas, R.P. Agarwal, M. Benchohra, Impulsive discontinuous partialhyperbolic differential equations of fractional order on Banach Algebras. Electron. J. Differ. Equat. **2010**(91), 1–17 (2010)
4. S. Abbas, R.P. Agarwal, M. Benchohra, Existence theory for partial hyperbolic differential inclusions with finite delay involving the Caputo fractional derivative, (submitted)
5. S. Abbas, M. Benchohra, Partial hyperbolic differential equations with finite delay involving the Caputo fractional derivative. Commun. Math. Anal. **7**, 62–72 (2009)
6. S. Abbas, M. Benchohra, Darboux problem for perturbed partial differential equations of fractional order with finite delay. Nonlinear Anal. Hybrid Syst. **3**, 597–604 (2009)
7. S. Abbas, M. Benchohra, Upper and lower solutions method for impulsive partial hyperbolic differential equations with fractional order. Nonlinear Anal. Hybrid Syst. **4**, 406–413 (2010)
8. S. Abbas, M. Benchohra, The method of upper and lower solutions for partial hyperbolic fractional order differential inclusions with impulses. Discuss. Math. Differ. Incl. Control Optim. **30**(1), 141–161 (2010)
9. S. Abbas, M. Benchohra, Impulsive partial hyperbolic differential inclusions of fractional order. Demonstratio Math. **XLIII**(4), 775–797 (2010)
10. S. Abbas, M. Benchohra, Darboux problem for partial functional differential equations with infinite delay and Caputo's fractional derivative, Adv. Dynamical Syst. Appl. **5**(1), 1–19 (2010)
11. S. Abbas, M. Benchohra, Impulsive partial hyperbolic functional differential equations of fractional order with state-dependent delay. Frac. Calc. Appl. Anal. **13**(3), 225–244 (2010)
12. S. Abbas, M. Benchohra, Upper and lower solutions method for the darboux problem for fractional order partial differential inclusions. Int. J. Modern Math. **5**(3), 327–338 (2010)
13. S. Abbas, M. Benchohra, Existence theory for impulsive partial hyperbolic differential equations of fractional order at variable times. Fixed Point Theory. **12**(1), 3–16 (2011)
14. S. Abbas, M. Benchohra, Upper and lower solutions method for partial hyperbolic functional differential equations with Caputo's fractional derivative. Libertas Math. **31**, 103–110 (2011)
15. S. Abbas, M. Benchohra, Existence results for fractional order partial hyperbolic functional differential inclusions, (submitted)
16. S. Abbas, M. Benchohra, A global uniqueness result for fractional order implicit differential equations. Math. Univ. Comen (submitted)

S. Abbas et al., *Topics in Fractional Differential Equations*, Developments in Mathematics 27, DOI 10.1007/978-1-4614-4036-9,

17. S. Abbas, M. Benchohra, Darboux problem for implicit impulsive partial hyperbolic differential equations. Electron. J. Differ. Equat. **2011**, 15 (2011)
18. S. Abbas, M. Benchohra, On the set of solutions of fractional order Riemann-Liouville integral inclusions. Demonstratio Math. (to appear)
19. S. Abbas, M. Benchohra, On the set of solutions for the Darboux problem for fractional order partial hyperbolic functional differential inclusions. Fixed Point Theory (to appear)
20. S. Abbas, M. Benchohra, Uniqueness results for Fredholm type fractional order Riemann-Liouville integral equations (submitted)
21. S. Abbas, M. Benchohra, Fractional order Riemann-Liouville integral equations with multiple time delay. Appl. Math. E-Notes (to appear)
22. S. Abbas, M. Benchohra, Nonlinear quadratic Volterra Riemann-Liouville integral equations of fractional order. Nonlinear Anal. Forum **17**, 1–9 (2012)
23. S. Abbas, M. Benchohra, On the set of solutions of nonlinear fractional order Riemann-Liouville functional integral equations in Banach algebras (submitted)
24. S. Abbas, M. Benchohra, Fractional order Riemann-Liouville integral inclusions with two independent variables and multiple time delay. Opuscula Math. (to appear)
25. S. Abbas, M. Benchohra, L. Gorniewicz, Existence theory for impulsive partial hyperbolic functional differential equations involving the Caputo fractional derivative. Sci. Math. Jpn.. online e- 2010, 271–282
26. S. Abbas, M. Benchohra, L. Gorniewicz, Fractional order impulsive partial hyperbolic differential inclusions with variable times. Discussions Mathe. Differ. Inclu. Contr. Optimiz. **31**(1), 91–114 (2011)
27. S. Abbas, M. Benchohra, L. Gorniewicz, Fractional order impulsive partial hyperbolic functional differential equations with variable times and state-dependent delay. Math. Bulletin **7**, 317–350 (2010)
28. S. Abbas, M. Benchohra, J. Henderson, Global asymptotic stability of solutions of nonlinear quadratic Volterra integral equations of fractional order. Comm. Appl. Nonlinear Anal. **19**, 79–89 (2012)
29. S. Abbas, M. Benchohra, J. Henderson, Attractivity results for nonlinear fractional order Riemann-Liouville integral equations in Banach algebras, (submitted)
30. S. Abbas, M. Benchohra, J.J. Nieto, Global uniqueness results for fractional order partial hyperbolic functional differential equations. Adv. in Difference Equ. **2011**, Art. ID 379876, 25 pp
31. S. Abbas, M. Benchohra, J.J. Nieto, Functional implicit hyperbolic fractional order differential equations with delay, (submitted)
32. S. Abbas, M. Benchohra, G.M. N'Guérékata, B.A. Slimani, Darboux problem for fractional order discontinuous hyperbolic partial differential equations in Banach algebras. Complex Variables and Elliptic Equations **57**(2–4), 337–350 (2012)
33. S. Abbas, M. Benchohra, J.J. Trujillo, Fractional order impulsive hyperbolic implicit differential equations with state-dependent delay (submitted)
34. S. Abbas, M. Benchohra, A.N. Vityuk, On fractional order derivatives and Darboux problem for implicit differential equations. Frac. Calc. Appl. Anal. **15**(2), 168–182 (2012)
35. S. Abbas, M. Benchohra, Y. Zhou, Darboux problem for fractional order neutral functional partial hyperbolic differential equations, Int. J. Dynamical Systems Differential Equations. **2**(3&4), 301–312 (2009)
36. S. Abbas, M. Benchohra, Y. Zhou, Fractional order partial functional differential inclusions with infinite delay. Proc. A. Razmadze Math. Inst. **154**, 1–19 (2010)
37. S. Abbas, M. Benchohra, Y. Zhou, Fractional order partial hyperbolic functional differential equations with state-dependent delay. Int. J. Dyn. Syst. Differ. Equat. **3**(4), 459–490 (2011)
38. N.H. Abel, Solutions de quelques problèmes à l'aide d'intégrales définies (1823). *Œuvres complètes de Niels Henrik Abel*, **1**, Grondahl, Christiania, 1881, 11–18
39. R.P Agarwal, M. Benchohra, S. Hamani, Boundary value problems for fractional differential equations. Georgian. Math. J. **16**, 401–411 (2009)

40. R.P Agarwal, M. Benchohra, S. Hamani, A survey on existence result for boundary value problems of nonlinear fractional differential equations and inclusions. Acta. Appl. Math. **109**, 973–1033 (2010)
41. R.P. Agarwal, M. Benchohra, B.A. Slimani, Existence results for differential equations with fractional order and impulses. Mem. Differ. Equat. Math. Phys. **44**, 1–21 (2008)
42. R.P. Agarwal, M. Meehan, D. O'Regan, Fixed Point Theory and Applications, in *Cambridge Tracts in Mathematics*, vol. 141 (Cambridge University Press, Cambridge, 2001)
43. R.P. Agarwal, D. ORegan, S. Stanek, Positive solutions for Dirichlet problem of singular nonlinear fractional differential equations. J. Math. Anal. Appl. **371**, 57–68 (2010)
44. R.P Agarwal, Y. Zhou, Y. He, Existence of fractional neutral functional differential equations. Comput. Math. Appl. **59**(3), 1095–1100 (2010)
45. R.P. Agarwal, Y. Zhou, J. Wang, X. Luo, Fractional functional differential equations with causal operators in Banach spaces. Math. Comput. Model. **54**(5–6), 1440–1452 (2011)
46. O.P. Agrawal, O. Defterli, D. Baleanu, Fractional optimal control problems with several state and control variables. J. Vib. Contr. **16**(13), 1967–1976 (2010)
47. B. Ahmad, J.J. Nieto, Existence of solutions for impulsive anti-periodic boundary value problems of fractional order. Taiwanese J. Math. **15**(3), 981–993 (2011)
48. B. Ahmad, S. Sivasundaram, Existence results for nonlinear impulsive hybrid boundary value problems involving fractional differential equations. Nonlinear Anal. Hybrid Syst. **3**, 251–258 (2009)
49. E. Ait Dads, M. Benchohra, S. Hamani, Impulsive fractional differential inclusions involving the Caputo fractional derivative. Fract. Calc. Appl. Anal. **12**(1), 15–38 (2009)
50. R. Almeida, D.F.M. Torres, Fractional variational calculus for nondifferentiable functions. Comput. Math. Appl. **61**, 3097–3104 (2011)
51. R. Almeida, D.F.M. Torres, Necessary and sufficient conditions for the fractional calculus of variations with Caputo derivatives. Commun. Nonlinear Sci. Numer. Simul. **16**, 1490–1500 (2011)
52. G.A. Anastassiou, in *Advances on Fractional Inequalities* (Springer, New York, 2011)
53. D. Araya, C. Lizama, Almost automorphic mild solutions to fractional differential equations. Nonlinear Anal. **69**, 3692–3705 (2008)
54. S. Arshad, V. Lupulescu, On the fractional differential equations with uncertainty. Nonlinear Anal. **74**, 3685–3693 (2011)
55. J.P. Aubin, Impulse differential inclusions and hybrid systems: a viability ap- proach, Lecture Notes, Universit Paris-Dauphine (2002)
56. J.P. Aubin, A. Cellina, in *Differential Inclusions* (Springer, Berlin, 1984)
57. J.P. Aubin, H. Frankowska, in *Set-Valued Analysis* (Birkhauser, Boston, 1990)
58. I. Bajo, E. Liz, Periodic boundary value problem for first order differential equations with impulses at variable times. J. Math. Anal. Appl. **204**, 65–73 (1996)
59. K. Balachandran, S. Kiruthika, J.J. Trujillo, Existence results for fractional impulsive integrodifferetial equations in Banach spaces. Comm. Nonlinear Sci. Numer. Simul. **16**, 1970–1977 (2011)
60. K. Balachandran, J.J. Trujillo, The nonlocal Cauchy problem for nonlinear fractional integrodifferential equations in Banach spaces. Nonlinear Anal. **72**, 4587-4593 (2010)
61. D. Baleanu, K. Diethelm, E. Scalas, J.J. Trujillo, in *Fractional Calculus Models and Numerical Methods* (World Scientific Publishing, New York, 2012)
62. D. Baleanu, S.I. Vacaru, Fractional curve flows and solitonic hierarchies in gravity and geometric mechanics. J. Math. Phys. **52**(5), 053514, 15 (2011)
63. J. Banaś, B.C. Dhage, Global asymptotic stability of solutions of a functional integral equation. Nonlinear Anal. **69**(7), 1945–1952 (2008)
64. E. Bazhlekova, in *Fractional Evolution Equations in Banach Spaces* (University Press Facilities, Eindhoven University of Technology, 2001)
65. A. Belarbi, M. Benchohra, Existence theory for perturbed impulsive hyperbolic differential inclusions with variable times. J. Math. Anal. Appl. **327**, 1116–1129 (2007)

66. A. Belarbi, M. Benchohra, A. Ouahab, Uniqueness results for fractional functional differential equations with infinite delay in Fréchet spaces. Appl. Anal. **85**, 1459–1470 (2006)
67. M. Benchohra, J.R. Graef, S. Hamani, Existence results for boundary value problems of nonlinear fractional differential equations with integral conditions. Appl. Anal. **87**(7), 851–863 (2008)
68. M. Benchohra, J.R. Graef, F-Z. Mostefai, Weak solutions for nonlinear fractional differential equations on reflexive Banach spaces. Electron. J. Qual. Theory Differ. Equat. **2010**(54), 10 pp
69. M. Benchohra, S. Hamani, S.K. Ntouyas, boundary value problems for differential equations with fractional order. Surv. Math. Appl. **3**, 1–12 (2008)
70. M. Benchohra, J. Henderson, S.K. Ntouyas, in *Impulsive Differential Equations and Inclusions*, vol. 2 (Hindawi Publishing Corporation, New York, 2006)
71. M. Benchohra, J. Henderson, S.K. Ntouyas, A. Ouahab, Existence results for functional differential equations of fractional order. J. Math. Anal. Appl. **338**, 1340–1350 (2008)
72. M. Benchohra, J. Henderson, S.K. Ntouyas, A. Ouahab, On first order impulsive dynamic equations on time scales. J. Difference Equ. Appl. **10**, 541–548 (2004)
73. M. Benchohra, J.J. Nieto, D. Seba, Measure of noncompactness and hyperbolic partial fractional differential equations in Banach spaces. Panamer. Math. J. **20**(3), 27–37 (2010)
74. M. Benchohra, S.K. Ntouyas, An existence theorem for an hyperbolic differential inclusion in Banach spaces. Discuss. Math. Differ. Incl. Contr. Optim. **22**, 5–16 (2002)
75. M. Benchohra, S.K. Ntouyas, On an hyperbolic functional differential inclusion in Banach spaces. Fasc. Math. **33**, 27–35 (2002)
76. M. Benchohra, S.K. Ntouyas, An existence result for hyperbolic functional differential inclusions. Comment. Math. Prace Mat. **42**, 1–16 (2002)
77. M. Benchohra, B.A. Slimani, Existence and uniqueness of solutions to impulsive fractional differential equations. Electron. J. Differ. Equat. **2009**(10), 11 (2009)
78. F. Berhoun, A contribution of some classes of impulsive differential equations with integer and non integer order, Doctorate thesis, University of Sidi Bel Abbes, 2010
79. A. Bica, V.A. Caus, S. Muresan, Application of a trapezoid inequality to neutral Fredholm integro-differential equations in Banach spaces. J. Inequal. Pure Appl. Math. **7**, 5 (2006), Art. 173
80. F.S. De Blasi, G. Pianigiani, V. Staicu: On the solution sets of some nonconvex hyperbolic differential inclusions. Czechoslovak Math. J. **45**, 107–116 (1995)
81. H.F. Bohnenblust, S. Karlin, On a theorem of ville. Contribution to the theory of games, in *Annals of Mathematics Studies*, vol. 24 (Priceton University Press, Princeton. N. G., 1950), pp. 155–160
82. A. Bressan, G. Colombo, Extensions and selections of maps with decomposable values. Studia Math. **90**, 69–86 (1988)
83. T.A. Burton, Fractional differential equations and Lyapunov functionals. Nonlinear Anal. **74**, 5648–5662 (2011)
84. T.A. Burton, C. Kirk, A fixed point theorem of Krasnoselskii-Schaefer type. Math. Nachr. **189**, 23–31 (1998)
85. L. Byszewski, Existence and uniqueness of solutions of nonlocal problems for hyperbolic equation $u_{xt} = F(x, t, u, u_x)$. J. Appl. Math. Stochastic Anal. **3**, 163–168 (1990)
86. L. Byszewski, Theorem about existence and uniqueness of continuous solutions of nonlocal problem for nonlinear hyperbolic equation. Appl. Anal. **40**, 173–180 (1991)
87. L. Byszewski, Existence and uniqueness of mild and classical solutions of semilinear functional differential evolution nonlocal Cauchy problem, Selected Problems in Mathematics, Cracow Univ. of Tech. Monographs, Anniversary Issue **6**, 25–33 (1995)
88. L. Byszewski, V. Lakshmikantam, Monotone iterative technique for non-local hyperbolic differential problem. J. Math. Phys. Sci **26**, 345–359 (1992)
89. L. Byszewski, S.N. Papageorgiou, An application of a noncompactness technique to an investigation of the existence of solutions to a nonlocal multivalued Darboux problem. J. Appl. Math. Stoch. Anal. **12**, 179–180 (1999)

90. M. Caputo, Linear models of dissipation whose Q is almost frequency independent II. Geophys. J. Roy. Astron. Soc. **13**, 529–539 (1967), reprinted in Fract. Calc. Appl. Anal. **11**, 4–14 (2008)
91. M. Caputo, Linear models of dissipation whose $\mathbb{Q}$ is almost frequency independent-II. Geophys. J. R. Astr. Soc. **13**, 529–539 (1967)
92. M. Caputo, in *Elasticità e Dissipazione* (Zanichelli, Bologna, 1969)
93. C. Castaing, M. Valadier, in *Convex Analysis and Measurable Multifunctions*. Lecture Notes in Mathematics, vol. 580 (Springer, Berlin, 1977)
94. Y.-K. Chang, J.J. Nieto, Some new existence results for fractional differential inclusions with boundary conditions. Math. Comput. Model. **49**, 605–609 (2009)
95. C. Corduneanu, in *Integral Equations and Applications* (Cambridge University Press, Cambridge, 1991)
96. H. Covitz, S.B. Nadler Jr., Multivalued contraction mappings in generalized metric spaces. Israel J. Math. **8**, 5–11 (1970)
97. T. Czlapinski, On the Darboux problem for partial differential-functional equations with infinite delay at derivatives. Nonlinear Anal. **44**, 389–398 (2001)
98. T. Czlapinski, Existence of solutions of the Darboux problem for partial differential-functional equations with infinite delay in a Banach space. Comment. Math. Prace Mat. **35**, 111–122 (1995)
99. M.F. Danca, K. Diethelm, Kai. Fractional-order attractors synthesis via parameter switchings. Commun. Nonlinear Sci. Numer. Simul. **15**(12), 3745–3753 (2010)
100. M.A. Darwish, J. Henderson, D. O'Regan, Existence and asymptotic stability of solutions of a perturbed fractional functional-integral equation with linear modification of the argument. Bull. Korean Math. Soc. **48**, 539–553 (2011)
101. M. Dawidowski, I. Kubiaczyk, An existence theorem for the generalized hyperbolic equation $z''_{xy} \in F(x, y, z)$ in Banach space. Ann. Soc. Math. Pol. Ser. I Comment. Math. **30**(1), 41–49 (1990)
102. A. Debbouche, Fractional evolution integro-differential systems with nonlocal conditions. Adv. Dyn. Syst. Appl. **5**(1), 49–60 (2010)
103. A. Debbouche, D. Baleanu, Controllability of fractional evolution nonlocal impulsive quasilinear delay integro-differential systems. Comput. Math. Appl. **62**, 1442–1450 (2011)
104. K. Deimling, in *Multivalued Differential Equations* (Walter De Gruyter, Berlin, 1992)
105. D. Delbosco, L. Rodino, Existence and uniqueness for a nonlinear fractional differential equation. J. Math. Anal. Appl. **204**, 609–625 (1996)
106. Z. Denton, A.S. Vatsala, Monotone iterative technique for finite systems of nonlinear Riemann-Liouville fractional differential equations. Opuscula Math. **31**(3), 327–339 (2011)
107. B.C. Dhage, A nonlinear alternative in Banach algebras with applications to functional differential equations. Nonlinear Funct. Anal. Appl. **8**, 563–575 (2004)
108. B.C. Dhage, Some algebraic fixed point theorems for multi-valued mappings with applications. Diss. Math. Differ. Inclusions Contr. Optim. **26**, 5–55 (2006)
109. B.C. Dhage, Nonlinear functional boundary value problems in Banach algebras involving Carathéodories. Kyungpook Math. J. **46**(4), 527–541 (2006)
110. B.C. Dhage, Existence theorems for hyperbolic differential inclusions in Banach algebras. J. Math. Anal. Appl. **335**, 225–242 (2007)
111. B.C. Dhage, Attractivity and positivity results for nonlinear functional integral equations via measure of noncompactness. Diff. Equ. Appl. **2**(3), 299–318 (2010)
112. T. Diagana, G.M. Mophou, G.M. N'Guérékata, On the existence of mild solutions to some semilinear fractional integro-differential equations. Electron. J. Qual. Theory Differ. Equ. **2010** (58), 17
113. K. Diethelm, in *The Analysis of Fractional Differential Equations*. Lecture Notes in Mathematics (Springer, Berlin, 2010)
114. K. Diethelm, N.J. Ford, Analysis of fractional differential equations. J. Math. Anal. Appl. **265**, 229–248 (2002)

115. K. Diethelm, A.D. Freed, On the solution of nonlinear fractional order differential equations used in the modeling of viscoplasticity, in *Scientifice Computing in Chemical Engineering II-Computational Fluid Dynamics, Reaction Engineering and Molecular Properties*, ed. by F. Keil, W. Mackens, H. Voss, J. Werther (Springer, Heidelberg, 1999), pp. 217–224
116. X. Dong, J. Wang, Y. Zhou, Yong. On nonlocal problems for fractional differential equations in Banach spaces. Opuscula Math. **31**(3), 341–357 (2011)
117. S. Dugowson, L'élaboration par Riemann d'une définition de la dérivation d'ordre non entier. revue d'histoire des Mathématiques **3**, 49–97 (1997)
118. M.M. El-Borai, On some fractional evolution equations with nonlocal conditions. Int. J. Pure Appl. Math. **24**, 405–413 (2005)
119. M.M. El-Borai, The fundamental solutions for fractional evolution equations of parabolic type. J. Appl. Math. Stoch. Anal. **2004**(3), 197–211
120. M.M. El-Borai, K. El-Said El-Nadi, E.G. El-Akabawy On some fractional evolution equations. Comput. Math. Appl. **59**(3), 1352–1355 (2010)
121. M.M. El-Borai, K. El-Nadi, H.A. Fouad, On some fractional stochastic delay differential equations. Comput. Math. Appl. **59**(3), 1165–1170 (2010)
122. A.M.A. El-Sayed, Fractional order evolution equations. J. Fract. Calc. **7**, 89–100 (1995)
123. A.M.A. El-Sayed, Fractional order diffusion-wave equations. Int. J. Theo. Phys. **35**, 311–322 (1996)
124. A.M.A. El-Sayed, Nonlinear functional differential equations of arbitrary orders. Nonlinear Anal. **33**, 181–186 (1998)
125. J.B.J. Fourier, Théorie Analytique de la Chaleur, Didot, Paris, 499–508 (1822)
126. M. Frigon, Théorèmes d'existence de solutions d'inclusions différentielles, Topological Methods in *Differential Equations and Inclusions*, NATO ASI Series C, vol. 472, ed. by A. Granas, M. Frigon (Kluwer Academic Publishers, Dordrecht, 1995), pp. 51–87
127. M. Frigon, A. Granas, Théorèmes d'existence pour des inclusions différentielles sans convexité. C. R. Acad. Sci. Paris, Ser. I **310**, 819–822 (1990)
128. M. Frigon, D. O'Regan, Impulsive differential equations with variable times. Nonlinear Anal. **26**, 1913–1922 (1996)
129. M. Frigon, D. O'Regan, First order impulsive initial and periodic problems with variable moments. J. Math. Anal. Appl. **233**, 730–739 (1999)
130. M. Frigon, D. O'Regan, Second order Sturm-Liouville BVP's with impulses at variable moments. Dynam. Contin. Discrete Impuls. Syst. **8** (2), 149–159 (2001)
131. K.M. Furati, N.-eddine Tatar, Behavior of solutions for a weighted Cauchy-type fractional differential problem. J. Frac. Calc. **28**, 23–42 (2005)
132. K.M. Furati, N.-eddine Tatar, Power type estimates for a nonlinear fractional differential equation. Nonlinear Anal. **62**, 1025–1036 (2005)
133. L. Gaul, P. Klein, S. Kempfle, Damping description involving fractional operators. Mech. Syst. Signal Process. **5**, 81–88 (1991)
134. W.G. Glockle, T.F. Nonnenmacher, A fractional calculus approach of self-similar protein dynamics. Biophys. J. **68**, 46–53 (1995)
135. L. Gorniewicz, in *Topological Fixed Point Theory of Multivalued Mappings, Mathematics and its Applications*, vol. 495 (Kluwer Academic Publishers, Dordrecht, 1999)
136. A. Granas, J. Dugundji, in *Fixed Point Theory* (Springer, New York, 2003)
137. A.K. Grunwald, Dérivationen und deren Anwendung. Zeitschrift für Mathematik und Phisik, **12**, 441–480 (1867)
138. J. Hale, J. Kato, Phase space for retarded equationswith infinite delay. Funkcial. Ekvac. **21**, 11–41 (1978)
139. J.K. Hale, S. Verduyn Lunel, in *Introduction to Functional -Differential Equations*. Applied Mathematical Sciences, vol. 99 (Springer, New York, 1993)
140. F. Hartung, Differentiability of solutions with respect to parameters in neutral differential equations with state-dependent delays. J. Math. Anal. Appl. **324**(1), 504–524 (2006)
141. F. Hartung, Linearized stability in periodic functional differential equations with state-dependent delays. J. Comput. Appl. Math. **174**(2), 201–211 (2005)

142. D. Henry, in *Geometric Theory of Semilinear Parabolic Partial Differential Equations* (Springer, Berlin, 1989)
143. S. Heikkila, V. Lakshmikantham, in *Monotone Iterative Technique for Nonlinear Discontinuous Differential Equations* (Marcel Dekker Inc., New York, 1994)
144. J. Henderson, A. Ouahab, Fractional functional differential inclusions with finite delay. Nonlinear Anal. **70** (2009) 2091–2105
145. J. Henderson, A. Ouahab, Impulsive differential inclusions with fractional order. Comput. Math. Appl. **59**, 1191–1226 (2010)
146. J. Henderson, C. Tisdell, Topological transversality and boundary value problems on time scales. J. Math. Anal. Appl. **289**, 110–125 (2004)
147. E. Hernández, A. Prokopczyk, L. Ladeira, A note on partial functional differential equations with state-dependent delay. Nonlinear Anal. Real World Applications **7**, 510–519 (2006)
148. E. Hernandez M., R. Sakthivel, S. Tanaka Aki, Existence results for impulsive evolution differential equations with state-dependent delay. Electron. J. Differ. Equat. **2008** (28), 1–11 (2008)
149. M.A.E. Herzallah, D. Baleanu, Fractional-order variational calculus with generalized boundary conditions. Adv. Difference Equ. Article ID 357580, 9 p **2011**
150. M.A.E. Herzallah, A.M.A. El-Sayed, D. Baleanu, Perturbation for fractional-order evolution equation. Nonlinear Dynam. **62**(3), 593–600 (2010)
151. R. Hilfer, in *Applications of Fractional Calculus in Physics* (World Scientific, Singapore, 2000)
152. Y. Hino, S. Murakami, T. Naito, in *Functional Differential Equations with Infinite Delay*. Lecture Notes in Mathematics, vol. 1473 (Springer, Berlin, 1991)
153. Sh. Hu, N. Papageorgiou, in *Handbook of Multivalued Analysis, Theory* **I** (Kluwer, Dordrecht, 1997)
154. R.W. Ibrahim, Existence and uniqueness of holomorphic solutions for fractional Cauchy problem. J. Math. Anal. Appl. **380**, 232–240 (2011)
155. R.W. Ibrahim, H.A. Jalab, Existence of the solution of fractiona integral inclusion with time delay. Misk. Math. Notes **11**(2), 139–150 (2010)
156. T. Kaczorek, in *Selected Problems of Fractional Systems Theory* (Springer, London, 2011)
157. A. Kadem, D. Baleanu, Homotopy perturbation method for the coupled fractional Lotka-Volterra equations. Romanian J. Phys. **56**(3–4), 332–338 (2011)
158. Z. Kamont, in *Hyperbolic Functional Differential Inequalities and Applications* (Kluwer Academic Publishers, Dordrecht, 1999)
159. Z. Kamont, K. Kropielnicka, Differential difference inequalities related to hyperbolic functional differential systems and applications. Math. Inequal. Appl. **8**(4), 655–674 (2005)
160. S.K. Kaul, V. Lakshmikantham, S. Leela, Extremal solutions, comparison principle and stability criteria for impulsive differential equations with variable times. Nonlinear Anal. **22**, 1263–1270 (1994)
161. S.K. Kaul, X.Z. Liu, Vector Lyapunov functions for impulsive differential systems with variable times. Dynam. Contin. Discrete Impuls. Syst. **6**, 25–38 (1999)
162. S.K. Kaul, X.Z. Liu, Impulsive integro-differential equations with variable times. Nonlinear Stud. **8**, 21–32 (2001)
163. E.R. Kaufmann, E. Mboumi, Positive solutions of a boundary value problem for a nonlinear fractional differential equation. Electron. J. Qual. Theory Differ. Equat. (3), 11 (2007)
164. A.A. Kilbas, B. Bonilla, J. Trujillo, Nonlinear differential equations of fractional order in a space of integrable functions. Dokl. Ross. Akad. Nauk **374**(4), 445–449 (2000)
165. A.A. Kilbas, S.A. Marzan, Nonlinear differential equations with the Caputo fractional derivative in the space of continuously differentiable functions. Differ. Equat. **41**, 84–89 (2005)
166. A.A. Kilbas, Hari M. Srivastava, and Juan J. Trujillo, in *Theory and Applications of Fractional Differential Equations*. North-Holland Mathematics Studies, vol. 204 (Elsevier Science B.V., Amsterdam, 2006)
167. M. Kirane, M. Medved, N. Tatar, Semilinear Volterra integrodifferential problems with fractional derivatives in the nonlinearities. Abstr. Appl. Anal. **2011**, Art. ID 510314, 11 pp

168. V.S. Kiryakova, Y.F. Luchko, The multi-index Mittag-Leffler functions and their appplications for solving fractional order problems in applied analysis. Application of mathematics in technical and natural sciences, 597–613, AIP Conf. Proc., 1301, Amer. Inst. Phys., Melville, NY, 2010
169. M. Kisielewicz, in *Differential Inclusions and Optimal Control* (Kluwer, Dordrecht, The Netherlands, 1991)
170. S. Labidi, N. Tatar, Blow-up of solutions for a nonlinear beam equation with fractional feedback. Nonlinear Anal. **74**(4), 1402–1409 (2011)
171. S.F. Lacroix, Traité du Calcul Différentiel et du Calcul Intégral, Courcier, Paris, t.3 (1819), 409–410
172. G.S. Ladde, V. Lakshmikanthan, A.S. Vatsala, in *Monotone Iterative Techniques for Nonliner Differential Equations* (Pitman Advanced Publishing Program, London, 1985)
173. V. Lakshmikantham, Theory of fractional differential equations. Nonlinear Anal. **60**, 3337–3343 (2008)
174. V. Lakshmikantham, D.D. Bainov, P.S. Simeonov, in *Theory of Impulsive Differential Equations* (World Scientific, Singapore, 1989)
175. V. Lakshmikantham, S. Leela, J. Vasundhara, in *Theory of Fractional Dynamic Systems* (Cambridge Academic Publishers, Cambridge, 2009)
176. V. Lakshmikantham, S.G. Pandit, The method of upper, lower solutions and hyperbolic partial differential equations. J. Math. Anal. Appl. **105**, 466–477 (1985)
177. V. Lakshmikantham, N.S. Papageorgiou, J. Vasundhara, The method of upper and lower solutions and monotone technique for impulsive differential equations with variable moments. Appl. Anal. **15**, 41–58 (1993)
178. V. Lakshmikantham, A.S. Vatsala, Basic theory of fractional differential equations. Nonlinear Anal. **69**, 2677–2682 (2008)
179. V. Lakshmikantham, L. Wen, B. Zhang, in *Theory of Differential Equations with Unbounded Delay*. Mathematics and its Applications (Kluwer Academic Publishers, Dordrecht, 1994)
180. A. Lasota, Z. Opial, An application of the Kakutani-Ky Fan theorem in the theory of ordinary differential equations. Bull. Acad. Pol. Sci. Ser. Sci. Math. Astronom. Phys. **13**, 781–786 (1965)
181. G.W. Leibniz, Letter from Hanover, Germany, Deptember 30, 1695 to G.A. L'Hospital, in *JLeibnizen Mathematische Schriften*, vol. 2 (Olms Verlag, Hildesheim, Germany, 1962), pp. 301–302. First published in 1849
182. F. Li, G.M. N'Guérékata, An existence result for neutral delay integrodifferential equations with fractional order and nonlocal conditions Abst. Appl. Anal. (2011), Article ID 952782, 20 pages
183. Y. Li, Y. Chen, I. Podlubny, Stability of fractional-order nonlinear dynamic systems: Lyapunov direct method and generalized Mittag-Leffler stability. Comput. Math. Appl. **59**(5), 1810–1821 (2010)
184. T.C. Lim, On fixed point stability for set-valued contractive mappings with applications to generalized differential equations. J. Math. Anal. Appl. **110**, 436–441 (1985)
185. J. Liouville, Mémoire sur le calcul des différentielles à indices quelconques. J. l'Ecole Roy. Polytéchn. **13**, 529–539 (1832)
186. Y. Luchko, Initial-boundary-value problems for the generalized multi-term time-fractional diffusion equation. J. Math. Anal. Appl. **374**, 538–548 (2011)
187. R. Magin, in *Fractional Calculus in Bioengineering* (Begell House Publishers, Redding, 2006)
188. R. Magin, M.D. Ortigueira, I. Podlubny, J.J. Trujillo, On the fractional signals and systems. Signal Process. **91**, 350–371 (2011)
189. F. Mainardi, Fractional calculus: Some basic problems in continuum and statistical mechanics, in *Fractals and Fractional Calculus in Continuum Mechanics* ed. by A. Carpinteri, F. Mainardi (Springer-Verlag, Wien, 1997), pp. 291–348
190. S. Marano, V. Staicu, On the set of solutions to a class of nonconvex nonclosed differential inclusions. Acta Math. Hungar. **76**, 287–301 (1997)

191. F. Metzler, W. Schick, H.G. Kilian, T.F. Nonnenmacher, Relaxation in filled polymers: A fractional calculus approach. J. Chem. Phys. **103**, 7180–7186 (1995)
192. K.S. Miller, B. Ross, in *An Introduction to the Fractional Calculus and Differential Equations* (Wiley, New York, 1993)
193. V.D. Milman, A.A. Myshkis, On the stability of motion in the presence of impulses. Sib. Math. J. **1**, 233–237 (1960), [in Russian]
194. V.D. Milman, A.A. Myshkis, Random impulses in linear dynamical systems, in *Approximante Methods for Solving Differential Equations* (Publishing House of the Academy of Sciences of Ukainian SSR, Kiev, 1963), pp. 64–81, [in Russian]
195. G.M. Mittag-Leffler, Sur la nouvelle function E_α. C. R. Acad. Sci. Paris **137**, 554–558 (1903)
196. G.M. Mittag-Leffler, Sopra la funzione $E_\alpha(x)$. Rend. Accad. Lincei, ser. 5 **13**, 3–5 (1904)
197. K. Moaddy, S. Momani, I. Hashim, The non-standard finite difference scheme for linear fractional PDEs in fluid mechanics. Comput. Math. Appl. **61**(4), 1209–1216 (2011)
198. G.M. Mophou, Optimal control of fractional diffusion equation. Comput. Math. Appl. **61**, 68–78 (2011)
199. M. Mophou, O. Nakoulima, G.M. N'Guérékata, Existence results for some fractional differential equations with nonlocal conditions. Nonlinear Stud. **17**, 15–22 (2010)
200. G.M. Mophou, G.M. N'Guérékata, Existence of the mild solution for some fractional differential equations with nonlocal conditions. Semigroup Forum **79**, 315–322 (2009)
201. G.M. Mophou, G.M. N'Guérékata, On some classes of almost automorphic functions and applications to fractional differential equations. Comput. Math. Appl. **59**, 1310–1317 (2010)
202. G.M. Mophou, G.M. N'Guérékata, On integral solutions of some nonlocal fractional differential equations with nondense domain. Nonlinear Anal. **71**, 4668–4675 (2009)
203. G.M. Mophou, G.N. N'Guérékata, Controllability of semilinear neutral fractional functional evolution equations with infinite delay. Nonlinear Stud. **18**, 195–209 (2011)
204. G.M. Mophou, G.M. N'Guérékata, V. Valmorin, Pseudo almost automorphic solutions of a neutral functional fractional differential equations. Intern. J. Evol. Equ. **4**, 129–139 (2009)
205. S. Muslih, O.P. Agrawal, Riesz fractional derivatives and fractional dimensional space. Int. J. Theor. Phys. **49**(2), 270–275 (2010)
206. S. Muslih, O.P. Agrawal, D. Baleanu, A fractional Schrdinger equation and its solution. Int. J. Theor. Phys. **49**(8), 1746–1752 (2010)
207. J.J. Nieto, Maximum principles for fractional differential equations derived from Mittag-Leffler functions. Appl. Math. Lett. **23**, 1248–1251 (2010)
208. K.B. Oldham, J. Spanier, in *The Fractional Calculus* (Academic Press, New York, 1974)
209. M.D. Ortigueira, in *Fractional Calculus for Scientists and Engineers* (Springer, Dordrecht, 2011)
210. B.G. Pachpatte, On Volterra-Fredholm integral equation in two variables. Demonstratio Math. **XL**(4), 839–852 (2007)
211. B.G. Pachpatte, On Fredholm type integrodifferential equation. Tamkang J. Math. **39**(1), 85–94 (2008)
212. B.G. Pachpatte, On Fredholm type integral equation in two variables. Diff. Equ. Appl. **1**, 27–39 (2009)
213. S.G. Pandit, Monotone methods for systems of nonlinear hyperbolic problems in two independent variables. Nonlinear Anal. **30**, 235–272 (1997)
214. I. Podlubny, in *Fractional Differential Equations*. Mathematics in Science and Engineering, vol. 198 (Academic Press, San Diego, 1999)
215. I. Podlubny, Geometric and physical interpretation of fractional integration and fractional differentiation. Fract. Calculus Appl. Anal. **5**, 367–386 (2002)
216. I. Podlubny, I. Petraš, B.M. Vinagre, P. O'Leary, L. Dorčak, Analogue realizations of fractional-order controllers. Fractional order calculus and its applications. Nonlinear Dynam. **29**, 281–296 (2002)
217. J.D. Ramrez, A.S. Vatsala, Monotone method for nonlinear Caputo fractional boundary value problems. Dynam. Systems Appl. **20**(1), 73–88 (2011)

218. A. Razminia, V.J. Majd, D. Baleanu, Chaotic incommensurate fractional order Rssler system: Active control and synchronization. Adv. Difference Equat. **2011**(15), 12 (2011)
219. M. Rivero, J.J. Trujillo, L. Vzquez, M.P. Velasco, Fractional dynamics of populations. Appl. Math. Comput. **218**, 1089–1095 (2011)
220. J. Sabatier, O. Agrawal, J. Machado (eds.), in *Advances in Fractional Calculus. Theoretical Developments and Applications in Physics and Engineering* (Springer, Dordrecht, 2007)
221. J. Sabatier, M. Merveillaut, R. Malti, A. Oustaloup, How to impose physically coherent initial conditions to a fractional system? Commun. Nonlinear Sci. Numer. Simul. **15**(5), 1318–1326 (2010)
222. H.A.H. Salem, On the fractional order m-point boundary value problem in reflexive Banach spaces and weak topologies. Comput. Math. Appl. **224**, 565–572 (2009)
223. H.A.H. Salem, On the fractional calculus in abstract spaces and their applications to the Dirichlet-type problem of fractional order. Comput. Math. Appl. **59**(3), 1278–1293 (2010)
224. H.A.H. Salem Global monotonic solutions of multi term fractional differential equations. Appl. Math. Comput. **217**(14), 6597–6603 (2011)
225. S.G. Samko, A.A. Kilbas, O.I. Marichev, in *Fractional Integrals and Derivatives. Theory and Applications* (Gordon and Breach, Yverdon, 1993)
226. N. Samko, S. Samko, B. Vakulov, Fractional integrals and hypersingular integrals in variable order Hlder spaces on homogeneous spaces. J. Funct. Spaces Appl. **8**(3), 215–244 (2010)
227. A.M. Samoilenko, N.A. Perestyuk, in *Impulsive Differential Equations* (World Scientific, Singapore, 1995)
228. N.P. Semenchuk, On one class of differential equations of noninteger order. Differents. Uravn. **10**, 1831–1833 (1982)
229. H. Sheng, Y. Chen, T. Qiu, in *Fractional Processes and Fractional-order Signal Processing; Techniques and Applications* (Springer-Verlag, London, 2011)
230. B.A. Slimani, A contribution to fractional order differential equations and inclusions with impulses, Doctorate thesis, University of Sidi Bel Abbes, 2009
231. V.E. Tarasov, in *Fractional dynamics: Application of Fractional Calculus to Dynamics of Particles, Fields and Media* (Springer, Heidelberg, 2010)
232. V.E. Tarasov, Fractional dynamics of relativistic particle. Int. J. Theor. Phys. **49**(2), 293–303 (2010)
233. V.E. Tarasov, M. Edelman, Fractional dissipative standard map. Chaos **20**(2), 023127, 7 (2010)
234. J.A. Tenreiro Machado, Time-delay and fractional derivatives. Adv. Difference Equ. **2011**, Art. ID 934094, 12 pp
235. J.A. Tenreiro Machado. Entropy analysis of integer and fractional dynamical systems. Nonlinear Dynam. **62**(1–2), 371–378 (2010)
236. J.A. Tenreiro Machado. Time-delay and fractional derivatives. Adv. Difference Equ. (2011), Art. ID 934094, 12 pp
237. J.A. Tenreiro Machado, V. Kiryakova, F. Mainardi, A poster about the old history of fractional calculus. Fract. Calc. Appl. Anal. **13**(4), 447–454 (2010)
238. J.A. Tenreiro Machado, V. Kiryakova, F. Mainardi, Recent history of fractional calculus. Commun. Nonlinear Sci. Numer. Simul. **16**(3), 1140–1153 (2011)
239. J.A. Tenreiro Machado, V. Kiryakova, F. Mainardi, A poster about the old history of fractional calculus. Fract. Calc. Appl. Anal. **13**(4), 447–454 (2010)
240. J.C. Trigeassou, N. Maamri, J. Sabatier, A.A. Oustaloup, Lyapunov approach to the stability of fractional differential equations. Signal Process. **91**, 437–445 (2011)
241. L. Vzquez. From Newton's equation to fractional diffusion and wave equations. Adv. Difference Equ. **2011**, Art. ID 169421, 13 pp
242. A.N. Vityuk, On solutions of hyperbolic differential inclusions with a nonconvex right-hand side (Russian) Ukran. Mat. Zh. **47**(4), 531–534 (1995); translation in Ukrainian Math. J. 47 (1995), no. 4, 617–621 (1996)
243. A.N. Vityuk, Existence of Solutions of partial differential inclusions of fractional order. Izv. Vyssh. Uchebn. Ser. Mat. **8**, 13–19 (1997)

244. A.N. Vityuk, A.V. Golushkov, Existence of solutions of systems of partial differential equations of fractional order. Nonlinear Oscil. **7**(3), 318–325 (2004)
245. A.N. Vityuk, A.V. Golushkov, The Darboux problem for a differential equation containing a fractional derivative. Nonlinear Oscil. **8**, 450–462 (2005)
246. A.N. Vityuk, A.V. Mykhailenko, On one class of differential quations of fractional order. Nonlinear Oscil. **11**(3) (2008), 307–319
247. A.N. Vityuk, A.V. Mykhailenko, The Darboux problem for an implicit fractional-order differential equation. J. Math. Sci. **175**(4), 391–401 (2011)
248. J. Wang, Y. Zhou, W. Wei, A class of fractional delay nonlinear integrodifferential controlled systems in Banach spaces. Commun. Nonlinear Sci. Numer. Simul. **16**(10), 4049–4059 (2011)
249. C. Yu, G. Gao, Existence of fractional differential equations. J. Math. Anal. Appl. **310**, 26–29 (2005)
250. G. Zaslavsky, in *Hamiltonian Chaos and Fractional Dynamics* (Oxford University Press, New York, 2005)
251. S. Zhang, Positive solutions for boundary-value problems of nonlinear fractional diffrential equations. Electron. J. Differ. Equat. (36), 1–12 (2006)
252. S. Zhang, Existence of positive solutions of a singular partial differential equation. Math. Bohemica **133**(1), 29–40 (2008)
253. Y. Zhou, Existence and uniqueness of fractional functional differential equations with unbounded delay. Int. J. Dyn. Syst. Differ. Equat. **1**(4), 239–244 (2008)
254. Y. Zhou, F. Jiao, J. Li, Existence and uniqueness for p-type fractional neutral differential equations. Nonlinear Anal. **71**, 2724–2733 (2009)
255. Y. Zhou, F. Jiao, J. Li, Existence and uniqueness for fractional neutral differential equations with infinite delay. Nonlinear Anal. **71**, 3249–3256 (2009)

Index

S. Abbas et al., *Topics in Fractional Differential Equations*, Developments in Mathematics 27, DOI 10.1007/978-1-4614-4036-9,